Few-Body Systems

Editor in Chief: W. Plessas, Graz

Supplement 14

Few-Body Problems in Physics '02

Proceedings of the XVIIIth European Conference on Few-Body Problems in Physics, Bled, Slovenia, September 8–14, 2002

Edited by
R. Krivec, B. Golli, M. Rosina, and *S. Širca*

SpringerWienNewYork

Dr. Rajmund Krivec
Dr. Mitja Rosina
Dr. Bojan Golli
Dr. Simon Širca
J. Stefan Institute
University of Ljubljana
Ljubljana, Slovenia

Printed in Austria

Typesetting: Camera-ready by authors
Printing: Druckerei Novographic, A-1230 Wien
Binding: Papyrus, A-1100 Wien

Printed on acid-free and chlorine-free bleached paper
SPIN: 10892019

With 156 Figures

CIP data applied for

ISSN 0177-8811
ISBN 3-211-83900-3 Springer-Verlag Wien New York

INTERNATIONAL ADVISORY COMMITTEE

J. Adam, Prague
H. Arenhövel, Mainz
B.L.G. Bakker, Amsterdam
M. Beyer, Rostock
R.F. Bishop, Manchester
B. Desplanques, Grenoble
A. Fonseca, Lisbon
J. Friar, Los Alamos
A. Gal, Jerusalem
W. Glöckle, Bochum
R.S. Hayano, Tokyo
K. de Jager, JLab
B. Jonson, Göteborg
N. Kalantar, Groningen
K. Langanke, Aarhus
R.G. Lovas, Debrecen
V.B. Mandelzweig, Jerusalem
W. Plessas, Graz
L.I. Ponomarev, Moscow
D.O. Riska, Helsinki
S. Rosati, Pisa
J.M. Rost, Dresden
P.U. Sauer, Hannover
I. Sick, Basel
A. Stadler, Évora
F. Stancu, Liège
R. Timmermans, Groningen
J. Tjon, Utrecht
J.W. van Orden, JLab
V. Vento, Valencia

LOCAL ORGANISING COMMITTEE

R. Krivec (chairman), B. Golli, M. Rosina, S. Širca

Organisers:

Jožef Stefan Institute
University of Ljubljana

Sponsors:

European Physical Society
Adria Airways
Ministry of Science and Education of Slovenia
Slovenian Institute of Quality
Marand d.o.o.
Čebelarstvo Božnar d.o.o.

Preface

In this Supplement we have collected the invited and contributed talks presented at the XVIII European Conference on Few-Body Problems in Physics, organised by the Jožef Stefan Institute and the University of Ljubljana, Slovenia. The Conference, sponsored by the European Physical Society, took place at the lakeside resort of Bled from 8 to 14 September, 2002. This meeting was a part of the series of European Few-Body Conferences, previously held in Évora/Portugal (2000), Autrans/France (1998), Peniscola/Spain (1995), ...

Our aim was to emphasise, to a larger extent than at previous Conferences, the interdisciplinarity of research fields of the Few-Body community. To promote a richer exchange of ideas, we therefore strived to avoid parallel sessions as much as possible. On the other hand, to promote the participation of young scientists who we feel will eventually shape the future of Few-Body Physics, we wished to give almost all attendees the opportunity to speak.

We hoped to resolve this timing conflict by placing the invited 30-minute talks as well as the short 10-minute talks into plenary sessions, while allocating the more detailed and technical 20-minute talks into a few parallel sessions. Although parallel processing could not be avoided altogether, both the review talks and the short "punch-line" contributions were presented to the complete audience. In these Proceedings, contributions to a given Topic appear in the 6–4–2-page (30–20–10-minute) order. Within these sections, the papers are sorted alphabetically according to the Speaker. In the Author index, the page-references of the Speakers are denoted in bold face.

We wish to thank all those who helped to make the Conference a success. Special thanks are due to the members of the International Advisory Committee for their advice and encouragement, and to the Chairpersons for moderating the sessions in a perpetual trade-off between allocated time and curiosity. We would like to acknowledge the contribution of the sponsors: European Physical Society, Adria Airways, Ministry of Science and Education of Slovenia, Slovenian Institute of Quality, Jožef Stefan Institute, Marand d.o.o., Čebelarstvo Božnar d.o.o., and University of Ljubljana. It was our pleasure to have the Albatros Travel Agency working with us. They relieved us from mundane worries and solved many problems in real time, allowing the members of the Organising Committee to even attend talks.

Dear Participants, we were happy to see so many of you come to Bled, to deliver such excellent talks, and to submit written contributions. We have tried to edit this Volume meticulously, and are honoured to be now able to place it into your hands.

The Organising Committee

Rajmund Krivec
Mitja Rosina
Bojan Golli
Simon Širca

Contents

Two-Nucleon Systems and NN-Interaction

Three- and Four-Nucleon Systems

Atomic and Molecular Processes

New Developments in Computational Methods

Interdisciplinary

Hyperon-Nucleon Interaction and Hypernuclei

Baryonic Excitation and Meson Production

Physics with Electro-Weak Probes

Spin Structure of Few-Body Systems

Few-Body Correlations in Medium

Relativity in Few-Body Physics

Few-Body Systems Suppl. 14, 1–6 (2003)

Few-Body Systems

Baryons as Solitons in Chiral Quark Models

B. Golli[1,*], W. Broniowski[2], G. Ripka[3]

[1] Faculty of Education, University of Ljubljana and J. Stefan Institute, Jamova 39, P.O. Box 3000, 1001 Ljubljana, Slovenia

[2] H. Niewodniczański Institute of Nuclear Physics, PL-31342 Kraków, Poland

[3] Service de Physique Théorique, Centre d'Etudes de Saclay, F-91191 Gif-sur-Yvette Cedex, France, and ECT*, Villa Tambosi, I-38050 Villazano (Trento), Italy

Abstract. We describe the formation of solitons in NJL-type models and discuss the influence of the regularization scheme on the stability of the solution. We concentrate on models with non-local regulators in which stable solutions exist without introducing additional constraints.

1 Introduction

The quark models of baryons can be divided into two main classes: (i) *constituent quark models* describing baryons as bound states of three massive quarks interacting via a phenomenological potential and (ii) *chiral models* with "bare" quarks surrounded by a cloud of either effective meson fields or $q\bar{q}$ pairs. The chiral model distinguishes itself from the constituent quark models in that the baryon masses, as well as the constituent quark masses are generated dynamically. The two classes of models describe hadrons at two different levels: at a higher level, the constituent quark model successfully predicts the spectrum of baryons and their excited states; at a lower level, the chiral models are able to describe the vacuum, meson and baryon sectors with the same effective Lagrangian and they model the origin of the masses of constituent quarks and in particular the role the chiral mesons play in the constituent quark model.

We present here the calculation of solitons in NJL-type models and discuss how different regularization schemes influence the stability of solutions. We focus on the version with non-local quark interactions as suggested from the instanton-liquid model which supports stable solutions without introducing artificial constraints as in previous calculations [1, 2].

* *E-mail address:* bojan.golli@ijs.si

2 Basic Properties of the NJL

The NJL model assumes an attractive interaction between quarks constructed in terms of quark bilinears in a form that obeys the chiral symmetry. To describe low energy phenomena, it is enough to consider the non-strange quarks and keep only the scalar and pseudoscalar terms. (For a review of the model and its applications see ref. [3].) In the chiral limit, the Lagrangian becomes:

$$\mathcal{L}_{NJL} = \bar{q}\,(\mathrm{i}\partial_\mu \gamma^\mu)\, q + \frac{G}{2}\left[(\bar{q}q)^2 + (\bar{q}\,\mathrm{i}\gamma_5\tau_a q)^2\right] , \tag{1}$$

where $a = 1, \ldots 3$ is the isospin index. The interaction is point-like and the UV divergences have to be removed by a suitable regularization prescription.

To solve the model in the mean-field (Hartree) approximation it is practical to introduce auxiliary chiral fields

$$S = -G\,\bar{q}q \qquad \text{and} \qquad P_a = -G\,\bar{q}\,\mathrm{i}\gamma_5\tau_a q , \tag{2}$$

such that the Lagrangian takes the *semi-bosonized* form:

$$\mathcal{L}_{NJL} = \bar{q}\beta\,(\mathrm{i}\partial_t - h)\, q - \frac{1}{2G}\left(S^2 + P_a^2\right) , \tag{3}$$

$$h = -\mathrm{i}\boldsymbol{\alpha}\cdot\boldsymbol{\nabla} + \beta(S + \mathrm{i}\gamma_5\tau_a P_a) . \tag{4}$$

The vacuum in the Hartree approximation is calculated by assuming a constant value, M, for $\langle S\rangle$ and evaluating $\langle\bar{q}q\rangle$ in (2) by summing up the bubble graphs or, equivalently, by performing functional integration over the quark fields. This yields the so called "gap equation":

$$\frac{1}{G} = 24\sum_{\Lambda}\frac{1}{k^2 + M^2} , \tag{5}$$

where $\sum_\Lambda$ denotes a regularized sum over quark states with a cut-off parameter Λ. Starting from a zero mass we have generated a finite quark mass. The model thus exhibits the *spontaneous breaking of the chiral symmetry*, yielding a nonzero value for the vacuum quark condensate $\langle\bar{q}q\rangle = -M/G$, and three Goldstone bosons (pions) corresponding to small oscillation of the pseudoscalar field around its vacuum value $\langle P_a\rangle = 0$. The model predicts the pion decay constant, f_π, and for a finite current quark mass, $m \neq 0$, also a finite pion mass satisfying the Gell-Mann–Oakes–Renner relation. Two of the three free parameters, G, Λ, or m can be eliminated by fitting f_π and m_π. The remaining free parameter is usually expressed in term of M, the so-called *constituent mass*.

In calculations of mesons and baryons several regularization schemes have been used including a *sharp cut-off* in 3- or 4-dimensions, the *proper-time* and *Pauli-Villars* regularizations and *non-local regulators* as suggested from the instanton-liquid model. All of them require a rather low cut-off parameter, $\Lambda \sim 600$ MeV, which indicates that it may be derivable from the underlying theory.

Indeed, the instanton-liquid model [4] predicts a non-local interaction between quarks with Λ related to the inter-instanton spacing $\rho^{-1} \sim 600$ MeV. The calculated vacuum properties (quark condensate, gluon condensate) suggest the value M should lie in the range between 300 and 400 MeV.

The soliton, corresponding to a baryon, is constructed around three valence quarks that polarize the vacuum generating a spatially non-trivial configuration. The resulting mean-field potential may make the system stable against decaying into free quarks. This is not a mechanism of confinement but simply a mechanism to bind quarks. If this scenario is realized it means that the interactions of the quarks with the chiral field may play an equally important role in the formation of the baryon as the color confining forces.

The existence of a stable solution is closely related to the choice of regularization scheme. A simple sharp cut-off does not yield stable solutions. Using the proper-time or Pauli-Villars regularization a stable soliton can be formed only by restricting the chiral fields to lie on the chiral circle, i.e. $S(\boldsymbol{r})^2 + P_a(\boldsymbol{r})^2 = M^2$, otherwise the scalar field acquires an arbitrarily low value at the origin producing a strongly localized state with a zero energy. Such a constraint does not have any justification in the model. It is only in the version with non-local regulators that solitons exist without introducing additional constraints.

3 Solitons in the NJL Model with Non-Local Regulators

In a non-local version of the NJL, the $q\bar{q}$ interaction is smeared by replacing the quark fields in the interaction part of (1) by *delocalized quark fields*:

$$\psi(x) = \langle x|\psi\rangle = \langle x|r|q\rangle = \sum_k r_k e^{\mathrm{i}k.x}\langle k|q\rangle \ . \tag{6}$$

Here r_k is a regulator, diagonal in k-space, and k is the *Euclidean* 4-momentum, $k^2 = \omega^2 + \boldsymbol{k}^2$. The instanton-liquid model predicts a form of the regulator. For model calculations it is simpler to take a Gaussian form, $r_k = e^{-k^2/2\Lambda^2}$. Using the Euclidean formulation the Dirac Hamiltonian takes the form

$$h = -\mathrm{i}\boldsymbol{\alpha}\cdot\boldsymbol{\nabla} + r\beta S r + \mathrm{i}\gamma_5\beta\tau_a r P_a r \ , \qquad r = r\left(-\partial_t^2 - \boldsymbol{\nabla}^2, \Lambda\right) \ . \tag{7}$$

The key point is the construction of the valence orbit. Let us first study the free quark propagator. In the chiral limit ($m = 0$) the inverse propagator takes the form $k_\mu\gamma_\mu + r(k^2)^2 M$. The pole (setting $\boldsymbol{k}^2 = 0$, $k_0^2 = -M_q^2$) occurs at the solution of

$$M_q{}^2 = M^2 \mathrm{e}^{2M_q{}^2/\Lambda^2} \ . \tag{8}$$

The most striking feature of (8) is that the solution exists only below a critical value of M. Above this value, free on-shell quarks do not exist in the vacuum.

The treatment of the quark propagator in a spatially non-homogeneous field configuration of the soliton is a non-trivial problem because of the presence of the time-dependent regulator in the Dirac Hamiltonian. For stationary chiral fields S and P_a the Dirac operator h is diagonal in the energy representation:

$$h(\omega^2)\,|\,\lambda_\omega\,\rangle = \mathrm{i}\omega\,|\,\lambda_\omega\,\rangle \ . \tag{9}$$

The lowest (valence) orbit is obtained as a solution of Eq. (9) for $\mathrm{i}\omega \to \varepsilon_0$:

$$\left[\frac{\boldsymbol{\alpha}\cdot\boldsymbol{\nabla}}{\mathrm{i}} + \beta\,\mathrm{e}^{\frac{1}{2\Lambda^2}({\varepsilon_0}^2+\boldsymbol{\nabla}^2)}\left(S(\boldsymbol{r}) + \mathrm{i}\gamma_5\tau_a P_a(\boldsymbol{r})\right)\mathrm{e}^{\frac{1}{2\Lambda^2}({\varepsilon_0}^2+\boldsymbol{\nabla}^2)}\right]|\lambda_0\rangle = \varepsilon_0|\lambda_0\rangle\,. \tag{10}$$

The solution of Eq. (10) exists for all M beyond M_{cr} as shown in Fig. 1.

The soliton is sought iteratively by assuming a hedgehog shape for the chiral fields. Starting from an initial guess for $S(\boldsymbol{r})$ and $P_a(\boldsymbol{r})$ Eq. (9) is solved for the valence orbit, $|\lambda_0\rangle$, and the sea orbits, $|\lambda_\omega\rangle$. In the next iterations new values for the chiral fields are obtained using the Euler-Lagrange equations:

$$\begin{aligned}
\frac{S(\boldsymbol{r})}{G} &= N_c \mathrm{z}_0 \langle\lambda_0|r|\boldsymbol{r}\rangle\beta\langle\boldsymbol{r}|r|\lambda_0\rangle + \int\frac{d\omega}{2\pi}\sum_{\lambda_\omega}\frac{\langle\lambda_\omega|r|\boldsymbol{r}\rangle\beta\langle\boldsymbol{r}|r|\lambda_\omega\rangle}{\mathrm{i}\omega + e_\lambda\,(\omega^2)}\,,\\
\frac{P_a(\boldsymbol{r})}{G} &= N_c \mathrm{z}_0 \langle\lambda_0|r|\boldsymbol{r}\rangle\mathrm{i}\beta\gamma_5\tau_a\langle\boldsymbol{r}|r|\lambda_0\rangle + \int\frac{d\omega}{2\pi}\sum_{\lambda_\omega}\frac{\langle\lambda_\omega|r|\boldsymbol{r}\rangle\mathrm{i}\beta\gamma_5\tau_a\langle\boldsymbol{r}|r|\lambda_\omega\rangle}{\mathrm{i}\omega + e_\lambda\,(\omega^2)}\,.
\end{aligned}$$

Here z_0 is the residue factor: $\mathrm{z}_0 = \left(1 - \mathrm{i}\;\mathrm{d}\varepsilon_0(\omega)/\mathrm{d}\omega|_{\omega=\mathrm{i}\varepsilon_0}\right)^{-1}$. The convergence is obtained for all values of M beyond a critical value M_{cr}.

In this approach the soliton acquires the correct baryon number [1].

4 Properties of the Soliton

Figure 1 (a) shows the soliton energy as a function of M for three regularization schemes: (i) the non-local model described in Sect. 3 yields stable solutions above $M_{\mathrm{cr}} = 280$ MeV which are also energetically stable since the energy of three free on-shell quarks is always higher than the soliton energy, (ii) the model with a non-local regulator of the Gaussian shape that depends only on the 3-momentum $\boldsymbol{k}$ possesses stable solutions above $M \approx 325$ MeV; however, beyond $M \approx 600$ MeV the solution is again unstable since it becomes energetically favorable for two quarks in the soliton to acquire high momenta, lower their effective mass, and escape from the third quark (Such an awkward behavior is a consequence of breaking the Lorentz invariance and does not show up in the model with the 4-momentum regulator.) [6], (iii) the model using the proper time regularization and the chiral-circle constraint has stable solutions above $M = 340$ MeV that are energetically stable only beyond $M = 400$ MeV.

The model with the 4-momentum non-local regulator supports another type of solutions with one or two quarks polarizing the sea (Fig. 1). The solution with three valence quarks is stable against disintegrating into solitons with a lower number of valence quarks. This does not hold for a soliton with four or more valence quarks since in such a case the additional quarks would have to fill the grand spin 1 orbit which is energetically less favorable.

The calculation of observables in the model with non-local regulator is a non-trivial task since the Noether currents acquire additional terms originating from the momentum dependent regulator [2]. Fortunately, their contribution remains small for physically interesting values of M below 400 MeV. The calculated properties reflect a relatively large size of the soliton compared to solutions using e.g. the proper time regularization as well as to the experimental values. In our opinion this is a consequence of the mean-field treatment rather than a serious deficiency of the model. Any improvement of the approximation such as elimination of spurious center-of-mass motion or projection onto subspace with good spin and isospin may considerably reduce the size [5].

5 Implications for Other Models

The solitons constructed in the chiral model with non-local regulators share several features found in more phenomenological models. From the model it is possible to derive a version of the linear σ-model that approximates well the full model [7].

The effective quark mass (i.e the square root of $S^2 + P^2$ shown in Fig. 2) drops almost to 0 in the center of the soliton for M between 300 MeV and 350 MeV similarly as in models based on restoration of the chiral symmetry inside the baryon like different versions of the bag model.

The Goldstone-boson exchange constituent quark model also finds a qualitative support in the non-local model. The solitons consisting of only one valence quark (see Fig. 1) can be interpreted as constituent quarks. Above $M \approx 310$ MeV these solutions are dressed in a cloud of chiral mesons that generate attraction and a 3-quark soliton is formed. The interaction between

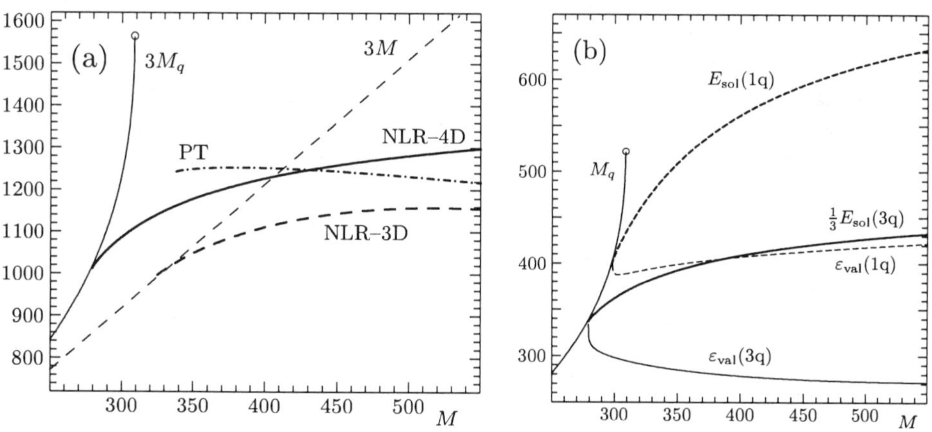

Figure 1. (a) The energy of the soliton for different regularization schemes plotted as functions of the parameter M using the 4-dimensional non-local Gaussian regulator (solid line), the 3-dimensional Gaussian regulator (dashed line), and the proper-time regularization (dashed-dotted line). (b) The energy per quark of the soliton with three valence quarks (solid line) and with one valence quark (dashed line) plotted as functions of the parameter M. Also shown are the corresponding valence energies. M_q is the free-space on-shell quark mass. All quantities in MeV. For each case the values of f_π and m_π have been fixed to their physical values.

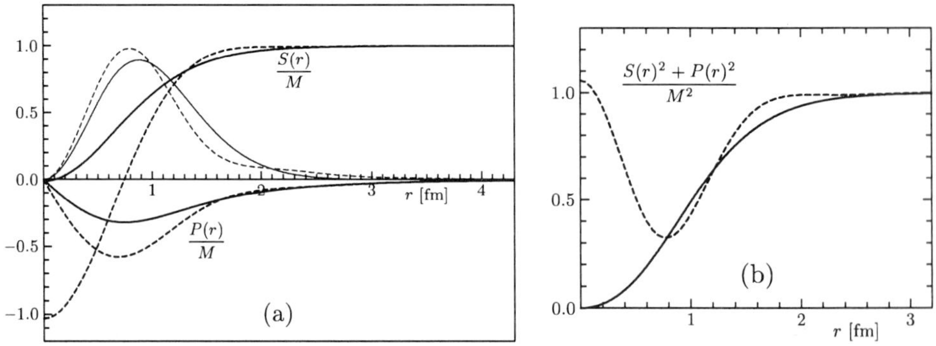

Figure 2. (a) Self consistently determined fields and baryon densities (multiplied by $4\pi r^2$); (b) effective squared quark mass for $M = 325$ MeV (solid line) and $M = 750$ MeV (dashed line), plotted as functions of the radial coordinate r.

such objects calculated in the linear σ-model [8] shows a typical behavior of the potential used in the Goldstone-meson exchange constituent quark model.

Acknowledgment. This work was supported in part by the Scientific and Technological Cooperation Joint Project between Poland and Slovenia, financed by the Ministry of Science of Slovenia and the Polish State Committee for Scientific Research, and by the Polish State Committee for Scientific Research, grant number 2 P03 09419.

References

1. B. Golli, W. Broniowski, and G. Ripka: Phys. Lett. **B437**, 24 (1998)
2. W. Broniowski, B. Golli, and G. Ripka: Nucl. Phys. **A703**, 667 (2002)
3. G. Ripka: *Quarks Bound by Chiral Fields.* Oxford: Clarendon Press 1997; C.V. Christov et al.: Prog. Part. Nucl. Phys. **37**, 1 (1996); R. Alkofer, H. Reinhardt, and H. Weigel: Phys. Rep. **265**, 139 (1996)
4. E. Shuryak: Nucl. Phys. **B203**, 93,116,140 (1982); D.I. Diakonov and V.Y. Petrov: Nucl. Phys. **B272**, 457 (1986)
5. G. Ripka and B. Golli: In: *Hadron Physics: Effective theories of low-energy QCD*, (AIP Conference Proceedings, vol. 508), p.3. AIP, Melville, New York 1999; hep-ph/9910479
6. B. Golli, talk presented at the Mini-Workshop on *Hadrons as Solitons*, Bled, Slovenia, 6-17 July 1999; hep-ph/0107115
7. W. Broniowski and B. Golli: Nucl. Phys. **A** (in print); hep-ph/0210200
8. B. Golli and M. Rosina: Phys. Lett. **B393**, 161 (1997)

Few-Body Systems Suppl. 14, 7–12 (2003)

Few-Body Systems

QCD Confinement and the Meson Spectrum

P. González[1], A. Valcarce[2], H. Garcilazo[3], J. Vijande[2]

[1] Dpto. de Física Teórica and IFIC, Universidad de Valencia-CSIC, E-46100 Burjassot, Valencia, Spain

[2] Grupo de Física Nuclear, Universidad de Salamanca, E-37008 Salamanca, Spain

[3] E.S.F.M., Instituto Politécnico Nacional, Edificio 9, 07738 México D.F., Mexico

Abstract. From QCD and lattice calculations two specific forms of quark confining potential, a strict linear and a screened linear confinement, come out. Both forms of the potential, implemented by the one gluon exchange interaction, are applied to the description of heavy quarkonia: $c\overline{c}$ and $b\overline{b}$. Applications to light hadrons, mesons and baryons, are also commented.

1 Introduction

The problem of hadron structure is closely related to the current impossibility to solve the bound state, or Bethe-Salpeter, equation in QCD. For a $q\overline{q}$ bound sate this equation reads:

$$(p_q \partial - m_q)\Psi_{q\overline{q}}(p; P)(p_{\overline{q}}\ \partial - m_{\overline{q}}) = \frac{1}{(2\pi)^4} \int K(p, p'; P)\Psi_{q\overline{q}}(p'; P) d^4p' \qquad (1)$$

where p_q and m_q ($p_{\overline{q}}$ and $m_{\overline{q}}$) stand for the quark (antiquark) momentum and mass, respectively. P and p are the total and relative momentum of the $q\overline{q}$ system, and $\Psi_{q\overline{q}}$ its probability amplitude. K, the so-called kernel, contains the $q\overline{q}$ interaction as provided by QCD. In fact K corresponds to a sum of an infinite number of diagrams as the ones drawn in Fig.1.

At the hadron energy scale perturbative QCD is not a reasonable approach and there seems to be no way to obtain a closed form for the kernel to be used in Eq. (1). Instead one can try to approximate such a closed form by making use of simplifying assumptions. A very drastic simplification stems from neglecting the time dependence in K and assuming the form of a non-relativistic potential: $K(\boldsymbol{p}-\boldsymbol{p'})$. Then if the relativistic kinematics is maintained one has the Salpeter equation, if non-relativistic kinematics is considered one gets the Schrödinger

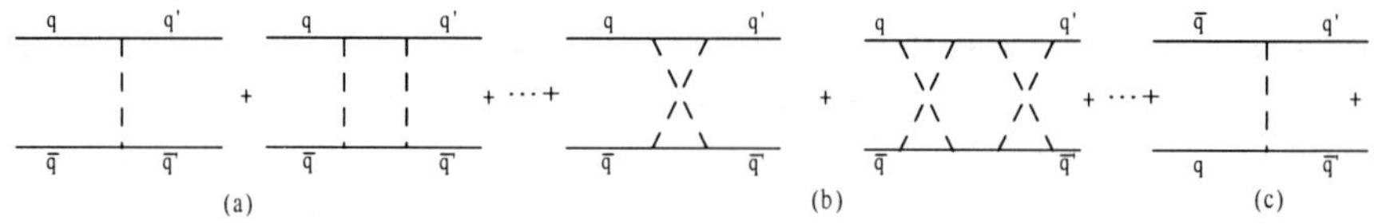

Figure 1. Diagrams contributing to the kernel K.

equation. These two equations constitute the basic formulation of the quark potential model of hadron structure.

2 The QCD Potential

If we consider a non-relativistic tree-level kernel, Fig. 1(a), the scattering amplitude generates, up to order $1/c$, a coulombic potential $V(r) \propto -1/r$, where r is the interquark distance. Proceeding further in the expansion other relativistic correction terms: spin-independent, tensor, spin-orbit and spin-spin, are generated [1]. In the spirit of a quark model the strength parameters should be considered as effective. Quite obviously, the more non-relativistic the system the less effective the parameters and the closer to their QCD values. This makes in principle heavy quarkonia the ideal experimental systems to test this QCD-inspired potential. Unfortunately, such a perturbatively derived interaction is not sufficient to get a physical description of the hadron spectrum since non-perturbative contributions (as well as radiative ones), in particular those related to confinement, cannot be skipped over.

Though not rigorously proven in QCD confinement can be inferred from the running of the quark-gluon coupling constant α_s. It has been shown, through a solution of the Schwinger-Dyson equation, that when decreasing the energy scale, i. e. at low Q^2, α_s does not increase continuously but saturates to a constant value [2]. Specifically:

$$\alpha_s(Q^2) = \frac{4\pi}{\beta_0 \ln\left(\frac{Q^2+4M_g^2(Q^2)}{\Lambda^2}\right)} \tag{2}$$

where $\beta_0 = (33 - 2n_f)/3$, being n_f the number of flavors with mass much smaller than Q, Λ is the QCD scale parameter for which we shall take 300 MeV hereforth and M_g is an effective gluon mass given by:

$$M_g^2(Q^2) = m_g^2 \left(\frac{\ln\left(\frac{Q^2+4m_g^2}{\Lambda^2}\right)}{\ln\left(\frac{4m_g^2}{\Lambda^2}\right)} \right) \tag{3}$$

with $m_g \sim 300$ MeV. Hence the effective gluon mass runs from 0 at $Q^2 \to \infty$ to m_g at $Q^2 \to 0$. Correspondingly α_s runs from 0 at $Q^2 \to \infty$ (asymptotic freedom) to ~ 0.8 at $Q^2 \to 0$.

This translates, at tree level and for $M_g = \Lambda$, in a heavy quark potential with a linear behavior at long distance and a coulombic one at short distance [3]. Lattice calculations corroborate this form of the potential [4]. In the quenched approximation, i.e. only valence quarks considered, a funnel (linear+coulomb) heavy quark interacting potential comes out. For a $q\overline{q}$ meson system it can be written as:

$$V(r) = ar - \frac{\alpha}{r} \tag{4}$$

where $\alpha = 4\alpha_s/3$ and a, the string tension, has a value $a \sim 900$ MeV fm.

Furthermore when sea quarks are also included (unquenched approximation), i.e. when the spontaneous creation of light-quark pairs in between the heavy quarks is taken into account, a breakup of the color flux tube takes place giving rise to a screened confinement [5]:

$$\overline{V}(r) = \left(\overline{a}r - \frac{\overline{\alpha}}{r}\right)\left(\frac{1 - e^{-\mu r}}{\mu r}\right) \tag{5}$$

where μ plays the role of an inverse screening length and the hat over the constants indicates that their values may not be the same than in the non-screened case.

This potential is essentially a coulombian one at short distances, $\overline{V}(r \to 0) \sim -\overline{\alpha}/r$, whereas it goes to a constant at long range, $\overline{V}(r \to \infty) \sim \overline{a}/\mu$. Regarding the inverse screening length μ one should realize that when $\mu \to 0$ the non-screened form is recovered. Besides μ enters in the potential in a way that reminds us the effect of the mass of an exchanged particle. So we propose the identification $\mu \sim \Lambda - M_g$. As a consequence μ should run with Q^2: $\mu(Q^2 = 0) < \mu(Q^2) < \mu(Q^2 = \infty)$ or $0 < \mu(Q^2) < 1.52$ fm^{-1}, a range quite compatible with the fitted value from lattice for a high Q^2 scale: $\mu = 0.9 - 1.4$ fm^{-1}.

3 Meson Spectrum

3.1 Strict Linear Confinement

Application of the (linear+coulomb) potential to charmonium and bottomium was done in the early 80's [6] by solving the Schrödinger equation. A successful description of the low-lying spin-triplet $c\overline{c}$ spectrum above the charm threshold was obtained as well as of the ratio of the leptonic decay widths of $\psi(2S)$ and J/ψ. The values of the parameters used were $m_c = 1840$ MeV, $\overline{a} = 925.5$ MeV fm^{-1}, and $\overline{\alpha}$ =102.61 MeV fm. The potential was also applied to $b\overline{b}$ with m_b =5170 MeV and the same $\overline{a}$ and $\overline{\alpha}$, accounting for the position of the $2S$ and $3S$ levels and for the leptonic decay widths of Υ and Υ' relative to that of J/ψ. Furthermore the model was extended with reasonable results to mesons composed of one heavy and one light quarks.

In the mid 80's a thorough study of masses and electromagnetic, weak and strong decays of mesons was carried out [7]. The funnel potential was implemented by spin-spin, tensor and spin-orbit interactions and the Salpeter equation (relativistic kinematics) was used. The running of α_s was also incorporated phenomenologically. In this manner a very successful universal description of all mesons, from π to Υ was attained. The confinement parameter used was $\overline{a}$ =912.2 MeV fm^{-1} and the quark masses: $(m_u + m_d)/2$ =220 MeV, m_s =419 MeV, m_c =1628 MeV, m_b =4977 MeV. However since a linear potential gives rise to an infinite discrete spectrum an explanation for the experimental absence of such a proliferation of states, specially above the current experimental upper thresholds, was needed. The calculation of the coupling of these absent states provided a partial explanation by realizing that many of them would most probably not be seen experimentally. Notwithstanding it some of them should clearly appear if dedicated experiments to look for them were done. These states were called missing resonances (experimental programs to look for missing baryons are at work at JLAB [8]).

3.2 Screened Confinement

In order to apply the screened potential to heavy quarkonia, via the Scrödinger equation, we should do first a reasonable guess of the value of the parameters. For the sake of simplicity let us try to get a unified description of $c\overline{c}$ and $b\overline{b}$. For this purpose we choose an intermediate scale Q between m_c and m_b and we take tentatively $\mu = \Lambda - M_g(Q^2)$ and $\overline{\alpha} = 4\alpha_s(Q^2)/3$ as provided by QCD. By taking as a reference the running quark masses given by the Particle Data Group [9] we may use for instance Q =2.7 GeV. Then μ =0.71 fm^{-1} and $\overline{\alpha}$ = 83.404 MeV fm. Regarding $\overline{a}$ it is considered as a free parameter chosen to fit the first $b\overline{b}$ excitation energy. The potential model masses $\overline{m_c}$ and $\overline{m_b}$ are fixed from the fitting of the ground states of $c\overline{c}$ and $b\overline{b}$ respectively. Thus we get $\overline{a}$ =1470 MeV fm^{-1}, $\overline{m_b}$ =4663.5 MeV and $\overline{m_c}$ =1238.5 MeV.

The results obtained for the masses as well as for the ratios of the leptonic decay widths are shown in Table 1. We marked by a $*$ those states used to fix the parameters of our calculation and by a † the experimental states whit different possibilities for the orbital angular momentum, it could be $L = 0$ or 2. We denote by Γ_r the ratio $\Gamma_{e^+e^-}/\Gamma_{e^+e^-}(1S)$. As can be seen, we successfully identify the ground and low-lying excited states. Again they appear more predicted states than experimentally found but it is important to emphasize that the spectra are finite with upper thresholds given by $\overline{m_b} + \overline{m_{\overline{b}}} + \overline{a}/\mu$ = 11397 MeV and $\overline{m_c} + \overline{m_{\overline{c}}} + \overline{a}/\mu$ = 4547 MeV. Certainly one cannot argue about the existence or not of missing states before a study of the couplings of these states is carried out. Furthermore the calculated thresholds cannot be taken for granted. Nevertheless the existence of thresholds imposes a severe restriction to possible three-quark bound states. As has been proposed [10] the experimental identification of resonances above thresholds could be a signature of exotics or glueballs.

Concerning light mesons let us assume that the same form of the potential may be applied in the Schrödinger equation. If we choose some scale Q between 0 and 1 GeV we have $0 < \mu < 0.5$ fm^{-1}. This value of μ gives rise to a screening factor bigger than in the heavy quark case. Then if $\overline{a}$ were kept constant there would be more screening for light than for heavy quarks. This

Table 1. Masses and ratios of the leptonic decay widths.

	$b\bar{b}$				$c\bar{c}$			
	Mass	Exp.	Γ_r	Exp.	Mass	Exp.	Γ_r	Exp.
1S	9460*	9460	1	1	3097*	3097	1	1
2S	10023*	10023	0.42	0.45±0.09	3679	3686	0.42	0.38±0.05
1d	10159				3794	3770		
3S	10350	10355	0.28	0.32±0.07	4023	4040	0.24	
2d	10436				4093			
4S	10588	10580	0.20		4248	4160†		
3d	10649				4294			
5S	10773				4397			
4d	10819				4427	4415†		
6S	10921	10860†			4494			
5d	10956							
7S	11040	11020						
1p0	9909	9860			3517	3417		
1p1		9893				3510		
1p2		9913				3556		
2p0	10262	10232			3914			
2p1		10255						
2p2		10269						
3p	10518				4174			
4p	10715				4348			
5p	10873				4463			
6p	11001							
	Thresholds							
	11397				4547			

sounds quite unphysical since one expects more screening in the heavy quark case according to more created intermediate pairs. Therefore $\overline{a}$ should also run with Q^2. As a criterium we shall assume that the splitting energy $\overline{a}/\mu$ keeps approximately constant all over the meson spectra. One should also realize that for light hadrons contributions to the potential coming from spontaneous chiral symmetry breaking are going to be present. As we do not consider them explicitly we shall assume that somehow they can be, at least in part, included through the effective value of $\overline{a}$. By proceeding in this manner it is easy to check that the triplet states corresponding to ρ mesons are well reproduced by taking $\overline{a} \sim 360$ MeV fm and $m_u = m_d = 350$ MeV.

It is worth to note that the effect of decreasing $\overline{a}$ in the potential is just the opposite to the effect of decreasing μ. Therefore both variations going from heavy to light quarks compensate to some extent. This gives rise to a kind of universality: the same potential (except for the running of $\overline{\alpha}$) used in the heavy quark sector may be applied to reproduce the excitation energies (though not the ground state) in the light quark spectrum.

Let us say finally that the same kind of potential with different values of the parameters has been applied to light baryons, the results showing an almost perfect parallelism between the number of predicted three-quark bound states and the number of experimentally found resonances [11].

4 Summary

We have shown that a screened form of the interquark potential as suggested by unquenched lattice calculations may account for the heavy quarkonia spectra. Furthermore approximate universality arguments can be used for its extension to light hadrons. The finite character of the spectra and the corresponding thresholds are of great importance to determine the highest energy three-quark bound states and the possible identification of exotics and glueballs.

Acknowledgement. This work has been partially funded by Ministerio de Ciencia y Tecnología under Contract No. BFM2001-3563, by Junta de Castilla y León under Contract No. SA-109/01, and by EC-RTN (Network ESOP) under Contract No. HPRN-CT-2000-00130.

References

1. A. de Rújula, H. Georgi, S.L. Glashow: Phys. Rev. **D12**, 147 (1975)
2. A.C. Aguilar et al.: Preprint hep-ph/0109223; J.M. Cornwall, J. Papavassiliou, Phys. Rev. **D40**, 3474 (1989); Phys. Rev. **D44**, 1285 (1991)
3. F.J. Ynduráin: Preprint hep-ph/9910399
4. G.S. Bali: Phys. Rep. **343**, 1 (2001)
5. K.D. Born et al.: Phys. Rev. **D40**, 1653 (1989)
6. E. Eichten et al.: Phys. Rev. **D17**, 3090 (1978); Phys. Rev. **D21**, 203 (1980)
7. S. Godfrey, N. Isgur: Phys. Rev. **D32**, 189 (1985)
8. M. Ripani: Few Body Syst. Supp. **11**, 284 (1999)
9. D.E. Groom et al.: Eur. Phys. J. **C15**, 1 (2000)
10. M.M. Brisudová et al.: Phys. Rev. **D61**, 054013 (2000)
11. P. González et al.: Proceedings of Baryons 02 (in print)

Few-Body Systems Suppl. 14, 13–18 (2003)

Few-Body Systems

Properties of Light and Strange Baryons within Constituent Quark Models

W. Plessas

Institut für Theoretische Physik, Universität Graz, Universitätsplatz 5, A-8010 Graz, Austria

Abstract. We discuss the performance of modern constituent quark models in the description of light and strange baryons. In particular we address the issues of excitation spectra, electroweak structure, and mesonic resonance decays. Special attention is paid to the implementation of essential features of low-energy quantum chromodynamics and to the role of relativity.

1 Introductory Remarks

Understanding the properties of hadrons is of eminent interest from many aspects. This is especially true for the properties of baryons at low and intermediate energies. Here again it is the light-flavour sector that attracts most attention. There one can investigate a number of important issues such as:

- What are the properties of quantum chromodynamics (QCD) prevailing in light and strange baryons
- Which degrees of freedom are active in low energy baryons?
- What is a suitable framework for describing baryon phenomena in a comprehensive manner?
- Which role is played by relativity?
- How can a working baryon model be implemented into nuclear physics?
- What are the possibilities of deducing nuclear forces from a microscopic theory of hadrons?

At a conference on few-body physics, naturally, the question of describing baryons as three-quark systems will be in the foreground. This approach has a long tradition, and the corresponding quantum-mechanical theory is amenable to rigorous solutions by few-body methods[1]. Over the recent years we have witnessed a vivid development of constituent quark models (CQMs), relevant especially for the physics of baryons in the light- and strange-flavour sectors. In addition to the traditional CQM with a hyperfine Q-Q interaction deriving from one-gluon exchange (OGE), other types of CQMs have been constructed, such as the ones based on Goldstone-boson exchange (GBE) dynamics or on instanton-induced (II) forces. In the following we shall discuss their different features along with results from typical representatives of these CQMs. In particular, we shall consider the OGE CQM after Bhaduri, Cohler, and Nogami (BCN) [1] (in the relativistic version as parametrized in ref. [2]), the GBE CQM by the Graz group [3, 4], and the II CQM of the Bonn group [5]. Due to limited space, we shall concentrate specifically on light- and strange-baryon spectroscopy, nucleon electroweak form factors, and some results on resonance decays.

2 Baryon Spectroscopy

The baryon spectra as provided by CQMs have been discussed from many viewpoints, especially also because of the big interest from the experimental side. While the description was rather qualitative in the early stages of CQMs, it is clear that considerable improvements have been achieved in the past years. Finally one has met the demand of describing all the light- and strange baryon spectra within one and the same framework. Due to the very nature of these spectra it is immediately evident that one needs flavour-dependent forces. As a result the OGE CQM, being flavour-indpendent, encounters difficulties in describing simultaneously the N and Λ spectra, in addition to the notorious shortcomings in the level orderings of positive- and negative-parity states in the N (and tentatively Δ) spectrum. The reasons have already been discussed in much detail in the literature (see, e.g., ref. [4]). Here we only recall the typical behaviour through a comparison of the OGE, II, and GBE CQMs for the N and Λ spectra in Figs. 1 and 2 (the complete spectra can be found in the original refs. [2], [5], and [3], respectively). This comparison actually shows the (most striking) current problems still prevailing in the various types of CQMs. Notably, all of them encounter severe shortcomings in the Λ spectrum, especially with regard to the $\Lambda(1405)$, the first excitation above the ground state. Furthermore the OGE and the II CQMs are left with the wrong level orderings of the first positive- and negative-parity excitations above the nucleon ground state. Only the GBE CQM can reproduce both the $N^*(1440)\ \frac{1}{2}^+$ and the $N^*(1535)\ \frac{1}{2}^-$ states in the right places, in accordance with experiment.

Obviously a correct quantitative description of all light and strange baryon spectra in a unified framework is of importance for applications of the CQM

[1] At least this is true at first sight. We will see later on that an increasingly refined description including additional degrees of freedom, beyond $\{QQQ\}$ configurations, and observing relativistic covariance constitutes respectable demands even for the most elaborate few-body methods presently available.

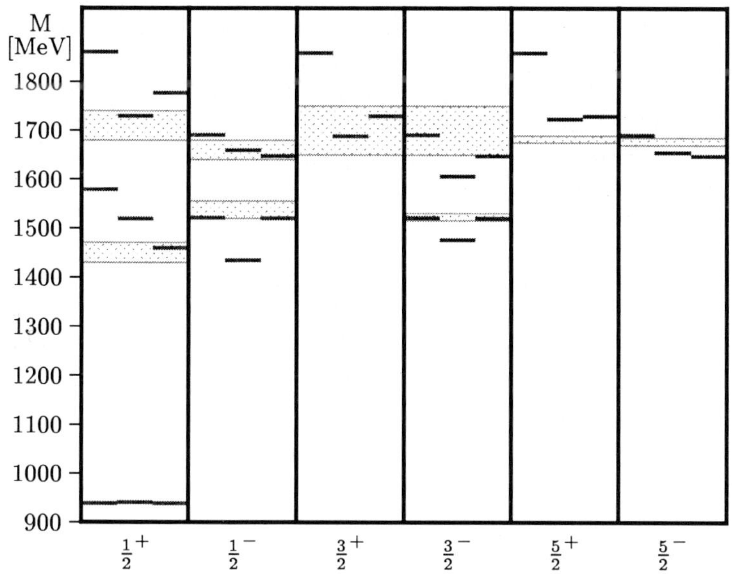

Figure 1. Nucleon excitation spectra of three different types of relativistic CQMs. In each column the left horizontal lines represent the results of the Bhaduri-Cohler-Nogami OGE CQM, as parametrized in ref. [2], the middle one of the instanton-induced CQM (Version A) [5], and the right one of the Goldstone-boson-exchange CQM [3]. The shadowed boxes give the experimental data with their uncertainties after the latest compilation of the PDG [6].

wave functions in dynamical processes involving resonances. Certain constraints are, of course, imposed by the spectra on the spin-flavour symmetry of the Q-Q interaction, and the various CQMs may not yet be complete in these respects[2]. However, one must also bear in mind that the description of resonances in terms of $\{QQQ\}$ bound states (without explicit couplings to decay channels) might be totally inadequate (see Sec. 4).

3 Electroweak Nucleon/Baryon Structure

The particular CQMs considered in the previous section have already been tested with regard to nucleon form factors and, partly, also for electric radii and magnetic moments of the light and strange baryon ground states. The corresponding comparisons are given in refs. [8] and [9]. In this context, the OGE and GBE CQMs were treated in point-form relativistic quantum mechanics whereas the predictions of the II CQM were obtained in a Bethe-Salpeter approach [10]. In general, all the relativistic results appear quite reasonable in comparison to experiments, regarding both electromagnetic and axial form fac-

[2]See, e.g., the attempts of extending the GBE CQM to include further force components [7].

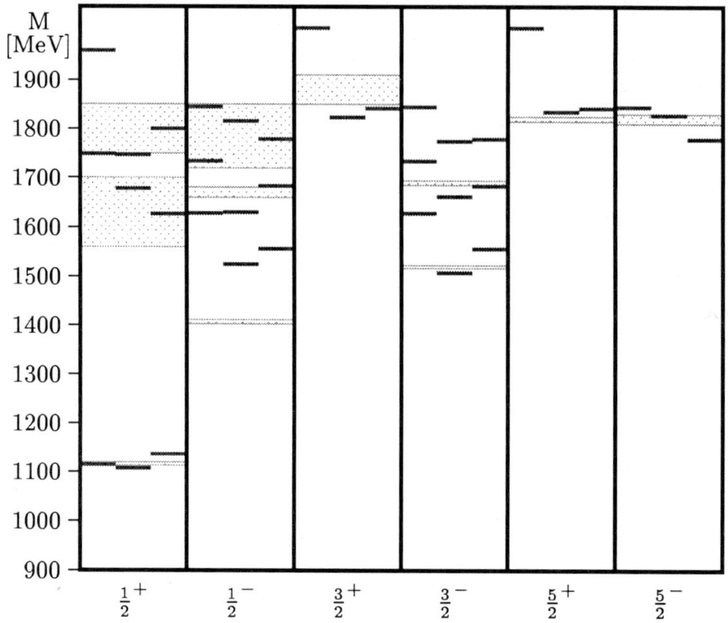

Figure 2. Λ excitation spectra. Same caption as in Fig. 1.

tors, even though only one-body models for the current operators were used. Of course, one must employ realistic nucleon wave functions, notably with mixed-symmetry spatial components included, in order to describe especially also the neutron[3]. From the comparisons of the observables on the electroweak nucleon structure as predicted by different CQMs it appears also remarkable that the dynamical ingredients in the wave functions are of secondary importance while relativistic (boost) effects are very essential. This is true for the results in point-form quantum mechanics as well as for the ones from the Bethe-Salpeter approach. It has become quite evident that any nonrelativistic theory for electroweak form factors is bound to fail and must be discarded [10, 11, 12].

While the final insight has certainly not yet been gained, in view of the conceptual incompleteness of the calculations managed so far, the description of the nucleon (ground state) properties within CQMs nevertheless appears promising. A number of investigations, e.g., employing two- and three-body currents, will be necessary in order to further manifest the theoretical understanding of the nucleon structure at low momentum transfers[4].

[3]With a spatially symmetric wave function, such as the one from a confinement potential alone, one would fail in reproducing, in particular, the neutron electric form factor [8].

[4]One should also mention that studies along relativistic aproaches other than the ones addressed here have generally led to congruent findings even though the one-body approximation to the current operators has produced quantitatively different results; see, e.g., the contribution by E. Pace et al. to this Conference [13].

4 Baryon Resonance Decays

Once having considered dynamical reactions involving the ground state(s), it is natural to proceed to the baryon resonances in order to further test the CQMs. Despite a number of investigations available in the literature, unfortunately, not many evidences that can really be considered reliable have so far been gained. In view of the experiences made with nucleon form factors and the enormous relativistic effects observed there, one must take any nonrelativistic study of resonances with considerable doubt. In this connection it is certainly not surprising that, notably, the mesonic baryon resonance decays are by far not understood. Theoretical predictions obtained hitherto with genuine CQM wave functions (i.e. without introducing additional phenomenological ingredients) compare quite unfavourably with existing experiments (see, e.g., ref. [2]).

Following the studies of nucleon form factors, the Graz group has recently started first exploratory calculations of (pionic) decay widths of N and Δ resonances with CQMs in relativistic point-form quantum mechanics. A first report thereof is given in ref. [14]. There one has employed a decay operator according to a relativistic generalization of the elementary emission model. Relativistic boost effects have been included exactly in the calculation of matrix elements with two types of CQM wave functions, namely, the ones from the OGE and the GBE CQMs. Again it has been found that relativistic effects are of considerable importance, with their magnitudes exceeding the influences of dynamical ingredients in the wave functions. Still, at this instance, one has not yet arrived at a satisfactory explanation of the decay widths. The problem appears much more intricate than anticipated. In order to reach a reasonable description of resonances a considerable extension of the concept of CQMs, going beyond $\{QQQ\}$ configurations, and/or further refinements in the decay operator will tentatively be needed.

5 Conclusions

In the talk given at the Conference and in this contribution we could address only very limited aspects of the many facets of CQMs. Nevertheless, from the evidences highlighted above, already a number of important conclusions can be drawn that will be important to observe in future research on CQMs and, more generally, on the physics of low- and intermediate-energy hadron phenomena.

Regarding spectroscopy a further improvement of CQMs is most desirable. One might be content with the description of ground states (as $\{QQQ\}$ bound states) but the treatment of resonances must certainly be made more realistic. In this respect it also not clear how far the concept of CQMs can be pushed up in excitation energy, even under the inclusion of further, say mesonic, degrees of freedom. The assumption of the constituent quark itself, as a quasiparticle of low-energy QCD, could soon break down in higher excitations, namely, when chiral symmetry gets restored.

From dealing with hadron reactions in CQMs it has become fairly evident that a relativistic formulation has to be employed. Any nonrelativistic treatment must be considered as inadequate. Still, it appears viable to work with a finite number of degrees of freedom. This paves the way for approaches following Poincaré-invariant quantum mechanics. It seems they provide the right framework of fulfilling Lorentz covariance and allow at the same time the application of established few-body techniques to solve a variety of few-quark problems. In any case, studies in these directions will bring along new exciting flavours for the traditional field of few-body physics.

References

1. R.K. Bhaduri, L.E. Cohler, Y. Nogami: Nuovo Cim. **A65**, 376 (1981)
2. L. Theußl, R.F. Wagenbrunn, B. Desplanques, and W. Plessas: Eur. Phys. J. **A12**, 91 (2001)
3. L.Ya. Glozman, W. Plessas, K. Varga, and R.F. Wagenbrunn: Phys. Rev. **D58**, 094030 (1998)
4. L.Ya. Glozman, Z. Papp, W. Plessas, K. Varga, and R.F. Wagenbrunn: Phys. Rev. **C57**, 3406 (1998)
5. U. Löring, B.C. Metsch, and H.R. Petry: Eur. Phys. J. **A10**, 395 (2001); ibid. **A10**, 447 (2001)
6. K. Hagiwara et al.: Phys. Rev. **D66**, 010001 (2002)
7. K. Glantschnig, R. Kainhofer, W. Plessas, B. Sengl, and R.F. Wagenbrunn: Contribution to these Proceedings
8. R.F. Wagenbrunn, S. Boffi, L.Ya. Glozman, W. Klink, W. Plessas, and M. Radici: Contribution to these Proceedings; hep-ph/0212190
9. K. Berger, W. Plessas, R.F. Wagenbrunn: Prog. Part. Nucl. Phys., in print
10. D. Merten, U. Löring, K. Kretzschmar, B. Metsch, and H.R. Petry: Eur. Phys. J. **A14**, 477 (2002)
11. R.F. Wagenbrunn, S. Boffi, W. Klink, W. Plessas, and M. Radici: Phys. Lett. **B511**, 33 (2001); L.Ya. Glozman, M. Radici, R.F. Wagenbrunn, S. Boffi, W. Klink, and W. Plessas: ibid. **B516**, 183 (2001); S. Boffi, L.Ya. Glozman, W. Klink, W. Plessas, M. Radici, and R.F. Wagenbrunn: Eur. Phys. J. **A14**, 17 (2002)
12. K. Berger, W. Plessas, and R.F. Wagenbrunn: Contribution to these Proceedings
13. E. Pace, G. Salmè, A. Molochkov: Contribution to these Proceedings
14. T. Melde, W. Plessas, and R.F. Wagenbrunn: Contribution to these Proceedings; hep-ph/0211356

Few-Body Systems Suppl. 14, 19–24 (2003)

Few-Body Systems

Pion Decay of Heavy-Light Mesons

D.-O. Riska

Helsinki Institute of Physics and Department of Physical Sciences, 00014 University of Helsinki, Finland

Abstract. As single pions do not couple to heavy flavor quarks, the pion decay rates of excited heavy-light flavor mesons to their ground states provide direct information on the coupling of pions to light-flavor constituent quarks. The present empirical widths rates of excited D-mesons are satisfactorily described by the chiral quark model, where the pion coupling is determined by the axial coupling of the light flavor quarks. The reaction $D_s^* \to D_s\pi^0$ provides information on the η-meson coupling to strange constituent quarks, as it is mediated by $\pi^0 - \eta$ mixing.

1 The Chiral Quark Model

The chiral quark model extends the chiral coupling of pions to hadrons to the constituent quarks, which form the hadrons. The pion coupling to a constituent quark is, accordingly,

$$\mathcal{L}_{\pi qq} = i\frac{g_A^q}{2f_\pi}\bar{\psi}_q\gamma_5\gamma_\mu\partial_\mu\boldsymbol{\pi}\cdot\boldsymbol{\tau}\psi_q. \tag{1}$$

Here ψ_q is the isospinor for the constituent quark and $\boldsymbol{\pi}$ the isovector pion field, f_π the pion decay constant ($f_\pi \simeq 93$ MeV), and g_A^q the axial coupling constant of the light flavor constituent quark. The value of g_A^q is of the order 0.87 [1] (typical values fall in the range 0.75-1.0 [2, 3]).

The coupling strength in (1) may be reexpressed in terms of a π-quark pseudovector coupling $f_{\pi qq}$ constant by setting $f_{\pi qq}/m_\pi = g_A^q/2f_\pi$. For $g_A^q = 0.87$ it follows that $f_{\pi qq} = 0.65$, which compares well to the static quark model value $f_{\pi qq} = 3/5 f_{\pi NN} \simeq 0.6$.

The static quark model value for $f_{\pi qq}$ is, however, well known to lead to an under-prediction of the $\pi N\Delta$ transition coupling $f_{\pi N\Delta}$. The quark model value for this is (with $g_A^q = 0.87$):

$$f_{\pi N\Delta} = 2\sqrt{2} f_{\pi qq} \simeq 1.84, \tag{2}$$

a value, which is $\sim$ 16% smaller than the value $f_{\pi N \Delta} = 2.2 \pm 0.4$, that is obtained from the empirical decay width of the Δ resonance.

This under-prediction of the width of the Δ resonance may however be quite natural, as shown by the following argument. As pions do not couple to strange quarks, the ratios of the widths for the transitions $\Delta \to N\pi$, $\Sigma^* \to \Sigma\pi$ and $\Xi^* \to \Xi\pi$ should be as 9:4:1. Empirically the ratios of these widths is, however, 12:4:1. That suggests that while the Σ^* and Ξ^* are three-quark states (sqq and ssq respectively, where q denotes a light-flavor quark), the Δ has a considerable 5 quark ($qqq(q\bar{q})$) (or "pion cloud") component. This argument applies not only if the pion coupling is to a single quark, but also if the pion couples through a seagull type interaction mechanism to two quarks. As the static quark model only takes into account the qqq component of the baryons, it is only natural that it under-predicts the decay rates of baryon resonances with large sea quark admixtures.

2 The Spectra of Heavy-Light Mesons

The simplest model for charm mesons are that they are $q\bar{c}$ (or $\bar{q}c$) systems, where q is a u or d quark and c a charm quark. Because of the small mass and high velocity of the confined light flavor quark, these systems call for a relativistic description. A convenient covariant approach is to describe the confined $q\bar{c}$ system by the Blankenbecler-Sugar equation [4].

This equation may be written in the form of Schrödinger equation

$$\left(\frac{\boldsymbol{p}^2}{2\mu} + V(r)\right)\psi(\boldsymbol{r}) = \epsilon\psi(\boldsymbol{r}), \tag{3}$$

where μ is the relative mass of the light and heavy quarks. The energy of the two-quark system is obtained from the eigenvalue ϵ as

$$E = \sqrt{M^2 + m^2 + 4\mu\epsilon + 2\sqrt{M^2m^2 + 2\mu(M^2 + m^2)\epsilon + 4\mu^2\epsilon^2}}. \tag{4}$$

In the large mass (non-relativistic) limit this expression reduces to $E = M + m + \epsilon$, which is the result for the Schrödinger equation.

A satisfactory description of the spectra of the D and D_s mesons may be achieved by assuming that the interaction between the heavy and light quarks is formed by the sum of a scalar linear confining interaction and a one-gluon exchange interaction. This interaction model describes heavy quarkonia well, including their small spin-orbit splittings. The small spin-orbit splittings are the result of a near cancellation between the spin-orbit components of the scalar confining interaction and the gluon exchange interaction.

To maintain this near cancellation and the correct spin-orbit splitting in heavy-light mesons, the momentum dependence of the running color coupling α_s has to be taken into account [5], as the static gluon exchange interaction would otherwise cause inversion of the spin-orbit splittings of the $L = 1$ charm mesons [6, 7], in conflict with experiment.

Table 1. Calculated pionic widths (in MeV) for the D^* for $g_A^q = 0.87$. The (relativistic) single quark approximation is denoted SQA, and the result with inclusion of the two-quark contribution is denoted FULL. The complete result is also given for $g_A^q = 1$ [8]. The experimental results are from [9, 10].

Decay	SQA	FULL	$g_A^q = 1$	Expt
$D^{*\pm} \to D\pi^0$	0.026	0.029	0.038	<0.04
$D^{*\pm} \to D^0\pi^\pm$	0.056	0.064	0.084	<0.09
$D^{*0} \to D^0\pi^0$	0.036	0.041	0.54	<1.3
$D^{*\pm} \to D^\pm\pi^0 + D^0\pi^\pm$	0.082	0.093	0.12	$0.096 \pm 0.04 \pm 0.022$

3 The $D^* \to D\pi$ Decay Rates

In the single quark approximation based on the quark-pion coupling (1) the pionic decays of the D^* to $D\pi$ are mediated by the spatial part of the axial current of the quark. With the standard values of g_A^q the single quark coupling leads to decay rates that are smaller than the present empirical values (Table 1).

The axial current of the $D-$mesons also obtains two-quark contributions that are associated with the effective interaction between the constituent quarks. The most obvious of these are the "seagull" or "pair" current contributions, which arise in the elimination of explicit negative energy components from the wave functions. While the analogous two-quark contributions to the axial current of nucleon pairs are known to be insignificantly small [11], these contributions are not negligible in the case of quark pairs, because of the small constituent mass of the light flavor quarks. As an example, the axial two-quark pair current that is associated with the scalar confining interaction leads to an enhancement of about 30 % of the calculated pion decay rates, as also shown in Table 1. It is apparent, that only with inclusion of this exchange current contribution, the chiral quark model gives a satisfactory description of the pion decay rates of the charmed vector meson D^*.

4 Pion Decay of the $D_1(2420)$ and $D_2^*(2460)$

The total widths of two of the $D-$meson resonances with $L = 1$ have been measured:

$$\begin{aligned} \Gamma(D_1(2420)) &= 18.9^{+4.6}_{-3.5}\,\text{MeV}, \\ \Gamma(D_2^*(2460)) &= 25^{+8}_{-7}\quad\text{MeV}. \end{aligned} \tag{5}$$

These widths are mainly due single π decay to the D and D^*, although $\pi\pi$ decay also contributes significantly to the widths [12].

For these $\Delta L = 1$ transitions the axial charge component of the coupling (1) also gives a contribution to the decay rate in addition to the axial current. In the single quark approximation the axial charge operator leads to a single

Table 2. π and $\pi\pi$ decay widths (in MeV) of the $D_1(2420)$ and $D_2^*(2460)$ mesons for $g_A^q = 1$. The complete result is also given for $g_A^q = 0.87$ [8, 12]. The experimental results are from [9, 10].

Decay	π	$\pi + \pi\pi$	$g_A^q = 0.87$	Expt
$D_2^*(2460)$	15.76	18.8	13.7	25^{+8}_{-7}
$D_1(2420)$	13.6	14.9	11.1	$18.9^{+4.6}_{-3.5}$

pion decay width of the $D_1(2420)$ meson, which is of the order 60 MeV, and therefore exceeds the empirical width by about a factor 4 [8].

The axial charge operator of the light flavor quark is, however, strongly suppressed by the confining interaction, if that has the form of a scalar interaction. The simplest way to see how this suppression arises is to note, that scalar interaction potentials are additive with the mass operators of the constituents. This gives rise to a shift of the constituent mass in the axial charge operator $-g_A^q \boldsymbol{\sigma} \cdot \boldsymbol{P}/m$, so that

$$m \to m + cr - b, \tag{6}$$

if the confining interaction is $cr - b$, where c is the string tension and b a constant. If $c = 1120$ MeV/fm and $b = 320$ MeV as in [5], this mass shift reduces the matrix element of the axial charge operator by about 45 %.

An alternate way of deriving the suppression of the axial charge operator from the scalar confining interaction, is to derive it as a two-quark operator, by elimination of the negative energy components from the wavefunction in the same way as employed for the axial current operator above. This latter method is more general, as it may be applied to the gluon exchange hyperfine interaction as well. The vector coupled gluon exchange contribution also causes a suppression of the matrix element of the axial charge operator of the light flavor constituent quark in $D-$mesons, which is similar to, if somewhat weaker than, the suppression caused by the confining interaction.

Once the suppression of the axial charge operator has been taken into account, reasonable values for the single pion decay widths of the excited positive parity charm mesons obtain (Table 2). The branching ratios for single pion decay also agree with the empirical values:

$$\begin{aligned} \frac{\Gamma(D_2^* \to D\pi)}{\Gamma(D_2^* \to D^*\pi)} &= 2.1 \quad \text{Expt}: \quad 2.3 \pm 0.6 \\ \frac{\Gamma(D_2^* \to D\pi + D^*\pi)}{\Gamma(D_1 \to D^*\pi)} &= 1.5 \quad \text{Expt}: \quad 1.5 \end{aligned} \tag{7}$$

The two-pion decay widths of the $L = 1$ charm mesons may be calculated from the Born terms that are associated with the two different time ordered iterations of the chiral coupling (1). The model is completed by the Weinberg-

Tomozawa coupling

$$\mathcal{L}_{\mathrm{WT}} = -\frac{i}{4f_\pi^2}\bar{\psi}_q\gamma_\mu\boldsymbol{\tau}\cdot\boldsymbol{\pi}\times\partial\boldsymbol{\pi}\psi_q. \tag{8}$$

With this model two-pion decay was found to contribute ~ 2 MeV to the decay width of the $D_2^*(2460)$ and ~ 1 MeV to the decay width of the $D_1(2420)$ [12] (Table 2). In the case of the two-pion exchange amplitude, there are exchange current contributions that are associated with the scalar confining and gluon exchange hyperfine interactions to the vector current of the quark, which appears in the Weinberg-Tomozawa coupling term. These reduce the value of the transition amplitude, which is obtained in the single quark approximation by about 25 %.

The net results for the decay widths of the $L=1$ charm mesons are closer to the empirical values if $g_A^q = 1$ than if $g_A^q = 0.87$. This result is obviously dependent on the model Hamiltonian model for the charm mesons. The degree of this model dependence is revealed by the fact that in ref.[2], who employed a similar, but static, interaction model and the Gross quasipotential equation, obtained good results with $g_A^q = 0.75$. The qualitatively significant result is nevertheless, that the chiral quark model is able to describe the pionic decays of the charm meson resonances.

5 Pion Decay of the D_s^*

The isospin forbidden π^0 decay of strange charm vector meson $D_s^* \to D_s\pi^0$ is interesting, because it indicates the presence of pion coupling to both the strange and the charm quark, which is mediated by the η meson through $\pi^0\eta$ mixing [13]. If the $\pi^0\eta$ mixing angle is known, this decay can provide information on the strength of the $\eta-$meson coupling to both the strange and the charm quarks. Direct evidence for the latter coupling is provided by the empirically known decay branching ratio for $\psi' \to J/\psi\,\eta$.

A phenomenological analysis of $D_s^* \to D_s\pi^0$ shows that the η coupling to the strange quark is likely to be only about 60% of the value that the $SU(3)$ flavor extended chiral quark model value ($f_{\eta ss} = -m_\eta g_A^q/\sqrt{3}f_\eta$) [14]. If the quark model relation for baryon couplings can be relied upon, this would suggest that the η couplings to nucleons should be small.

6 Conclusion

The chiral quark model is able to provide a qualitatively satisfactory description of the pionic decays of the charm meson resonances, with standard values for the parameters in the model. The role of the two-quark contributions to both the axial current and charge operators of the $D-$meson systems is significant. The two-quark contributions to the axial charge causes a strong suppression of the net matrix element of the axial charge operator, which is necessary for realistic values for the decay width of the $D_1(2420)$ meson.

Acknowledgement. Research supported in part by the Academy of Finland under grant 54038.

References

1. S. Weinberg: Phys. Rev. Lett. **67**, 3473 (1991)
2. J.L. Goity and W. Roberts: Phys. Rev. **D60**, 034001 (1995)
3. S. Weinberg: Phys. Rev. Lett. **65**, 1181 (1990)
4. R. Blankenbecler and R. Sugar: Phys. Rev. **142**, 1051 (1966)
5. T.A. Lähde, C.J. Nyfält and D.O. Riska: Nucl. Phys. **A674**, 141 (2000)
6. H. Schnitzer: Phys. Lett. **B76**, 461 (1978)
7. N. Isgur: Phys. Rev. **D57**, 4041 (1998)
8. K.O.E. Henriksson et al.: Nucl. Phys. **A686**, 355 (2001)
9. D.E. Groom et al.: Eur. Phys. J. **C15**, 1 (2000)
10. S. Ahmed et al.: Phys. Rev. Lett. **87**, 251801 (2001)
11. K. Tsushima and D. O. Riska: Nucl. Phys. **A549**, 313 (1992)
12. T.A. Lähde and D. O. Riska: Nucl. Phys. **A693**, 755 (2001)
13. P. Cho and M. Wise: Phys. Rev. **D49**, 6228 (1994)
14. T.A. Lähde and D.O. Riska: hep-ph/0204230

Few-Body Systems Suppl. 14, 25–28 (2003)

Few-Body Systems

The Chances to Produce and Detect the $bb\bar{u}\bar{d}$ Tetraquark at LHC

D. Janc[1,*], M. Rosina[1,2], D. Treleani[3], A. Del Fabbro[3]

[1] J. Stefan Institute, 1000 Ljubljana, Slovenia

[2] Faculty of Mathematics and Physics, University of Ljubljana, 1000 Ljubljana, Slovenia

[3] Universita di Trieste; Dipartimento di Fisica Teorica, Strada Costiera 11, Miramare-Grignano, and INFN, Sezione di Trieste, I-34014 Trieste, Italy

Abstract. In the LHC collider a significant rate of events with double parton scattering is expected. This will be the leading mechanism for production of two $b\bar{b}$ pairs. We estimate the probability of binding two b quarks into a diquark and the probability of dressing this diquark into a $bb\bar{u}\bar{d}$ ($IS^P = 01^+$) tetraquark. Calculations shows that that this bound state of two B mesons is stable against the strong interaction and has a life time of the order of ps. We estimate that the production rate at luminosity L = 0.1 events per second will be about 6 tetraquarks per hour or more.

1 Introduction

In the new large hadron collider at CERN on can expect a sizable rate of events where two heavy b quarks and two heavy $\bar{b}$ antiquarks are simultaneously produced. This makes it possible that two heavy quarks or antiquarks combine to create a double heavy hadron after they dress with light quarks.

There are two mechanisms for creation of two heavy quark-antiquark pairs. The first is by a single parton scattering where initially only one heavy quark-antiquark pair is produced and the second pair is then created by the fragmentation of one of initial quarks. The second mechanism is a double parton scattering. Calculations show [4] that the double scattering although power suppressed gives at the center of the mass energy of 14 TeV a four times larger cross section for the production of two b-quark pairs. For the creation of a heavy diquark one requires that the two quarks are close together in momentum space. Since in the fragmentation the two b quarks tend to fly in opposite

* *E-mail address:* damijan.janc@ijs.si

directions this also confirms that the dominant mechanism would be two parton scattering. The creation of the double heavy hadron is then a two step process. First two heavy quarks which are close together in momentum space bind into a diquark. We can estimate that the probability for this is proportional to the overlap of the wave functions of the quarks which are considered to be free with the wave function of the diquark. The second step is then dressing of this diquark with the light quark into a doubly heavy baryon or dressing with the two light quarks into a heavy dimeson.

2 Phenomenological Estimate

Let me first give a very simple phenomenological estimate why a BB dimeson should exist with a binding energy of about 100 MeV and why a similar DD dimeson is unstable against strong interaction [1]. In this estimation we consider the heavy dimeson to be a bound state of two heavy b-quarks combined in a diquark with the spin 1 and two light antiquarks with the quantum numbers spin 0, isospin 0. The starting point is that the light antiquarks around the heavy diquark behave as in Λ_{b}. This would be exactly true in the limit where the mass of the b quark goes to infinity and the heavy diquark is point-like. This correspond to assumption that we can neglect the size of the heavy diquark in dimeson. The second assumption is that the interaction between heavy b quarks in a colour antitriplet state is half as strong as the interaction between heavy quark and antiquark in heavy meson. With this assumption we can estimate the binding energy of heavy dimeson. Since the spectroscopy of heavy mesons is quite well described with nonrelativistic potential models we see that the binding energy for a system with half as strong interaction V_{QQ} is half the binding energy for a system with original interaction $V_{\mathrm{Q\bar{Q}}}$ and twice lighter particles ($m_{\mathrm{Q}} \rightarrow m_{\mathrm{Q}}/2$).

$$\left[\frac{p^2}{m_{\mathrm{Q}}} + V_{\mathrm{QQ}}\right]\psi = \frac{1}{2}\left[\frac{p^2}{m_{\mathrm{Q}}/2} + V_{\mathrm{Q\bar{Q}}}\right]\psi = \frac{1}{2}E(m_{\mathrm{Q}}/2)\psi$$

We obtain $E(m_{\mathrm{Q}}/2)$ by plotting the binding energy of heavy mesons as a function of the reduced mass of the system [1]. We do that for different choices of constituent quark masses, to estimate the uncertainty due to quark masses. The plot is very smooth and we estimate the binding energy of the heavy bb diquark $E(m_{\mathrm{b}}/2)/2$ to be -390 ± 15 MeV. If we combine this with the masses of heavy mesons and Λ_b baryon we can obtain the phenomenological estimate for binding energy of the heavy dimeson ΔT_{bb}

$$\Delta T_{\mathrm{bb}} = m_{\Lambda_{\mathrm{b}}} + \left[m_{\Upsilon} - E(m_{\mathrm{b}}) + E(m_b/2)\right]/2 - m_{\mathrm{B}} - m_{\mathrm{B}^*} = -130 \pm 15\,\mathrm{MeV}$$

while the heavy dimeson with the two c quarks is unbound. This agrees well with some previous four body calculations [2] [3] in the constituent quark model.

3 b-quark Pair Production

Since at LHC there will be a large parton luminosity we can hope that the production rate of double heavy dimesons will be high enough to make detection possible. The dominant contribution to b-quark pairs production will be double parton scattering - the gluon fusion $(\mathrm{gg} \to \mathrm{b\bar{b}})^2$. To calculate this cross section one needs to make an assumption about the two body distribution function Γ

$$\Gamma(x_1, x_2, d) = G(x_1)G(x_2)F(d), \tag{1}$$

where $G(x)$ is the one-body parton function while the parton pair density is described by the normalized function $F(d)$. The distance of the two partons belonging to the same proton is d and the x_1 and x_2 are their fractional momenta. Doing so, we neglect the correlations in momenta. Then the cross section σ_D for the double gluon fusion has a form

$$\sigma_D(\mathrm{b\bar{b}b\bar{b}}) = \frac{\sigma_s(\mathrm{b\bar{b}})^2}{2\sigma_{\mathrm{eff}}}, \quad \sigma_{\mathrm{eff}}^{-1} = \int \mathrm{d}^2 b F(b)^2, \tag{2}$$

where $\sigma_s(b\bar{b})$ is cross section for a fusion of two gluons into a b quark-antiquark pair. For σ_{eff} we take the value measured in the experimental study of double parton collision by CDF at Fermilab in the reaction $\mathrm{p\bar{p}} \to \gamma/\pi^0 + 3\,\mathrm{jets} + \mathrm{X}$ [5]. This experimental value is 14.5 mb. Doing this we assume that also some correlation between quarks in the same proton are taken into account which were neglected in (1). There is also same uncertainty about the single scattering cross section σ_s since the higher order corrections are large and they enter squared into the cross section for double scattering (2). For details see [4]. The cross section calculated then from (2), taking into account that only the pseudorapidity region $1.8 < \eta < 4.9$ will be covered by forward LHCb detector, is 85 nb. For two b quarks to combine into a diquark it is very important that they are close together in momentum space. Calculations gives that the cross section is approximately proportional to momentum volume $\mathrm{d}\sigma/\mathrm{d}^3p \approx 0.4\,\mathrm{nb/GeV}^3$ up to $p \leq 2\mathrm{GeV}$.

4 Formation of the Diquark and Dressing

We assume that the production of the two heavy b quarks in double parton scattering is uniform inside the protons. Furthermore we take an instantaneous approximation so that the amplitude $M(p)$ for the formation of the diquark is simply given by the overlap $M(p) = \langle\phi(p)|\psi\rangle$ of the wave function of the two b quarks $\phi(p)$ with the wave function of the diquark ψ. This wave function is calculated in constituent quark model used also in [2] and [3]. So the production cross section is given by

$$\sigma = \int \mathrm{d}^3p \frac{\mathrm{d}\sigma}{\mathrm{d}^3p} M^2(p) = 0.15\,\mathrm{nb} \tag{3}$$

Dressing of a heavy diquark is governed by the quantum chromodynamics. But since fragmentation of diquark into a hadron is impossible to calculate using perturbative QCD methods, we estimate the probability from experimental data obtained at Fermilab and at LEP experiment [6]. Their results show that the probability that the b quark will be dressed with u or d antiquark is 0.38 ± 0.02, with s antiquark 0.16 ± 0.03 and the probability for dressing with light quarks to form Λ_b is 0.10 ± 0.03. In our estimation we compared the heavy dimeson with the Λ_{b} baryon where the role of diquark in the first case is played by the heavy b quark in the second. So we assume that the diquark will get dressed into a dimeson with the same probability as heavy quark is dressed into a lambda b baryon — that is 0.10 ± 0.03. The cross section for the heavy dimeson production is thus

$$\sigma = 0.015\,\mathrm{nb}$$

which will give at the expected luminosity the production rate of 5-6 events per hour.

5 Decay and Detection

We expect that the $\mathrm{bb\bar{u}\bar{d}}$ tetraquark will be stable against strong and electromagnetic decay and will decay only weakly. The main channel would be the decay of one of the b quarks into c quark followed by the weak decay of the second b quark. But this channel will be difficult to distinguish from the channel of two unbound B mesons which will be the main background contribution. The more characteristic decay channel would be the the formation of the Υ and light meson through $\mathrm{b} \to \bar{\mathrm{b}}$ oscillation inspired by $\mathrm{B}^0 \to \bar{\mathrm{B}}^0$ oscillations. But this oscillations are here negligible since the tetraquark is not degenerate with final state of two unbound mesons. We are still searching for new ideas for the detection of doubly b tetraquarks.

References

1. D. Janc, M. Rosina: Few-Body Systems **31**, 1 (2001)
2. B. Silvestre-Brac, C. Semay: Z. Phys. **C57**, 273 (1993)
3. D.M. Brink, Fl. Stancu: Phys. Rev **D57**, 6778 (1998)
4. A. Del Fabbro and D. Treleani: Phys. Rev. **D63**, 057901 (2001)
5. F. Abe et al. (CDF Collaboration): Phys. Rev. **D56**, 3811 (1997)
6. F. Abe et al. (CDF Collaboration): Phys. Rev. Lett. **84**, 1663 (2000)

Few-Body Systems Suppl. 14, 29–32 (2003)

Few-Body Systems

Schwinger-Dyson Approach to Light Pseudoscalars

D. Klabučar[1], D. Kekez[2]

[1] Physics Department, Faculty of Science, Zagreb University, Bijenička c. 32, HR-10000 Zagreb, Croatia

[2] Rudjer Bošković Institute, P.O.B. 180, HR-10002 Zagreb, Croatia

Abstract. We review briefly some of the successes of Schwinger-Dyson (SD) approach to the physics of quarks and hadrons, primarily light pseudoscalar mesons including η and η'. The main purpose of this paper is to point out that SD results on η and η' mesons can be formulated and given also in the two-mixing-angle scheme.

Schwinger-Dyson (SD) equations provide a non-perturbative, covariant and chirally well-behaved approach to quark-hadron physics. This approach [1–3] can be formulated so that it has strong and clear connections with QCD. Admittedly, coupled integral SD equations for Green's functions of QCD cannot be solved exactly since some truncation of the infinite system of SD equations is unavoidable. However, the numerous advances [1–3] achieved in this field within the last dozen or so years also include increased understanding of how to perform this truncation so that some important known behaviors are preserved. This includes the known perturbative QCD behavior and dynamical chiral symmetry breaking (DχSB). The latter is crucial in low-energy QCD, especially for the description of the light-quark sector. The correct chiral behavior due to DχSB can be achieved through the coupling of SD and Bethe-Salpeter (BS) equations in a consistent approximation. That is, one usually solves

$$S_q^{-1}(p) = \not{p} - m_q - ig^2 C_F \int \frac{d^4k}{(2\pi)^4} \gamma^\mu S_q(k) \gamma^\nu G_{\mu\nu}(p-k) \ , \qquad (1)$$

the (rainbow-)ladder-approximated SD equation for "dressed" propagators, $S_q(p)$, of the light quarks $q = u, d, s$. The strong coupling constant is denoted by g and the color factor $C_F = 4/3$. The interaction $G_{\mu\nu}$ stands for an effective gluon propagator; it is known in the perturbative QCD regime, but has to be modeled in the non-perturbative regime, for low momenta, in order to be phenomenologically successful. The same "dressed" gluon exchange interaction $G_{\mu\nu}$ and the resulting solutions of Eq. (1) for light-quark propagators S_q, are then employed in

$$S_q^{-1}(k+\frac{P}{2})\chi_{q\bar{q}'}(k,P)S_{q'}^{-1}(k-\frac{P}{2}) = ig^2C_F\int\frac{d^4k'}{(2\pi)^4}\gamma^\mu\chi_{q\bar{q}'}(k',P)\gamma^\nu G_{\mu\nu}(k-k')\,, \tag{2}$$

the ladder-approximated BS equations for various quark-antiquark ($q\bar{q}'$) BS amplitudes $\chi_{q\bar{q}'}$. This defines a chirally well-behaved SD-BS approach, in which light pseudoscalar meson $q\bar{q}$ bound states ($\pi^{0,\pm}, K^{0,\pm}, \eta$) in the chiral limit (and *close* to it) **simultaneously** manifest themselves also as (*quasi-*)Goldstone bosons of DχSB. This enables one to work with the mesons as explicit $q\bar{q}$ bound states while reproducing analytically – in the chiral limit – the famous results of the Abelian axial anomaly for the light pseudoscalar mesons, namely the amplitudes for $\pi^0 \to \gamma\gamma$ and $\gamma \to \pi^+\pi^0\pi^-$ [4]. This is unique among the bound state approaches – *e.g.*, see refs. [2–5] and references therein. Nevertheless, one keeps the advantage of bound state approaches that from the $q\bar{q}$ substructure one can calculate many important quantities such as meson decay constants (*e.g.*, f_π), which are just parameters in most of other chiral approaches to the light-quark sector. Also, one can depart far from the chiral limit, soft point and on-shell limit in the anomaly-related processes $\pi^0, \eta, \eta' \leftrightarrow \gamma\gamma$ and $\gamma \to \pi^+\pi^0\pi^-$, for example in refs. [5–12] where the phenomenologically successful Munczek-Jain Ansatz [13] for the effective gluon propagator $G_{\mu\nu}$ was used. (This Ansatz is very similar to the interactions adopted, with small variations, in almost all recent phenomenologically most successful SD studies [1–3].) In this way obtained description of η–η' complex [7,9–11,14] is especially noteworthy, as it is very successful in spite of the limitations of the SD-BS approach in the ladder approximation, because of which Eqs. (1)-(2) cannot include the effects of gluon anomaly. They give only non-anomalous contributions to $q\bar{q}$ BS solutions and masses. Fortunately, gluon anomaly is suppressed as $1/N_c$ in the expansion in number of colors N_c, so that the procedure of refs. [7,9,10,15], which includes gluon anomaly only through parameterizing the anomalous shift of η_0 mass, leads to very satisfactory results. For example, our η–η' mass matrix is in agreement with phenomenology and lattice results. It can be seen it leads to a very good description of η–η' mixing, although it is formulated (except in Appendix of ref. [9]) in terms of one *state* mixing angle, for example ϕ,

$$|\eta\rangle = \cos\phi|\eta_{NS}\rangle - \sin\phi|\eta_S\rangle\,, \quad |\eta'\rangle = \sin\phi|\eta_{NS}\rangle + \cos\phi|\eta_S\rangle\,, \tag{3}$$

rotating the nonstrange-strange (*NS-S*) basis states η_{NS} and η_S to the mass eigenstates η and η'. In terms of the quark-antiquark states $|q\bar{q}\rangle$ ($q = u, d, s$), or the (effective, flavor-symmetry-broken) singlet-octet states η_8 and η_0,

$$|\eta_{NS}\rangle = \frac{1}{\sqrt{2}}(|u\bar{u}\rangle + |d\bar{d}\rangle) = \frac{1}{\sqrt{3}}|\eta_8\rangle + \sqrt{\frac{2}{3}}|\eta_0\rangle\,, \tag{4}$$

$$|\eta_S\rangle = |s\bar{s}\rangle = -\sqrt{\frac{2}{3}}|\eta_8\rangle + \frac{1}{\sqrt{3}}|\eta_0\rangle\,. \tag{5}$$

Although mathematically equivalent to the (*effective*) octet-singlet η_8–η_0 basis (with its state mixing angle $\theta = \phi - \arctan\sqrt{2}$), the *NS–S* mixing basis and its angle ϕ have the advantage that they offer the quickest way to show the consistency of our procedures and the corresponding results obtained using just one state mixing angle (be it θ or ϕ), with the **two-mixing-angle scheme**. This scheme (reviewed in, *e.g.*, ref. [16]) is defined with respect to the mixing of the four decay constants f_η^8, $f_{\eta'}^8$, f_η^0, and $f_{\eta'}^0$, and results are parameterized in terms of two auxiliary decay constants f_0, f_8, and two angles θ_0, θ_8. However, phenomenology seems to justify the central assumption of Feldmann, Kroll and Stech (FKS) [17] that in the *NS–S* basis, the decay constants follow (in a good approximation) the pattern of particle state mixing. As a consequence, θ_8, θ_0, f_8 and f_0 can be expressed [16–18] through the single *NS–S* state mixing angle ϕ and f_{NS}, f_S, the respective decay constants of η_{NS}, η_S:

$$\theta_8 = \phi - \arctan\left(\sqrt{2} f_S/f_{NS}\right), \qquad \theta_0 = \phi - \arctan\left(\sqrt{2} f_{NS}/f_S\right), \tag{6}$$

$$f_8 = \sqrt{\frac{1}{3} f_{NS}^2 + \frac{2}{3} f_S^2}, \qquad f_0 = \sqrt{\frac{2}{3} f_{NS}^2 + \frac{1}{3} f_S^2}, \tag{7}$$

The relations (6)-(7) are applicable also in our SD-BS approach, as shown in detail in Appendix of ref. [9]. This reference confirmed our earlier results [7] on the mixing angle, finding that the preferred *NS–S* state mixing angle in our SD-BS approach is $\phi \approx 42^\circ$, in good agreement with FKS results quoted in refs. [16–18]. The values we find [9] for f_0, f_8, θ_0 and θ_8 are also similar to theirs. This is the consequence of our values of f_{NS} and f_S, which are calculated directly from $q\bar{q}$ substructure, *i.e.*, our SD and BS solutions. The ratio $y = f_{NS}/f_S$ is suitable for giving the extent of the SU(3) breaking. Due to Goldberger–Treiman relation for constituent quarks, it is close to $\mathcal{M}_u/\mathcal{M}_s$, the ratio of the masses of the nonstrange and strange constituent quarks.

We get $f_{NS} \equiv f_\pi$ (*theoretical* FKS analysis [16, 17] assumes this), while our numerical calculation yields $f_S = 1.451 f_\pi$, just 3% above the theoretical FKS prediction [16–18]. This leads to $f_8 = 1.318 f_\pi$ and $f_0 = 1.170 f_\pi$. (Interestingly, this is practically equal to our SD-BS model values $f_{\eta_8} = 1.31 f_\pi$ and $f_{\eta_0} = 1.16 f_\pi$ [7] for the octet and singlet axial-current decay constants f_{η_8} and f_{η_0} defined in the standard way through the matrix elements $\langle 0|A^{a\mu}|\eta_a\rangle$, $(a = 8, 0)$, so that f_{η_8} and f_{η_0} are straightforwardly expressed through f_{NS} and f_S by the relations $f_{\eta_8} = \frac{1}{3} f_{NS} + \frac{2}{3} f_S$ and $f_{\eta_0} = \frac{2}{3} f_{NS} + \frac{1}{3} f_S$. We thus note that the quadratic relations (7) for differently defined octet and singlet constants f_8 and f_0, lead to similar values as the linear relations for f_{η_8} and f_{η_0}.)

Using in Eqs. (6)-(7) our model predictions $\phi = 42^\circ$ and $y = f_{NS}/f_S = 0.689$, we get the following two decay-constant-mixing angles: $\theta_8 = -22^\circ$ and $\theta_0 = -2.3^\circ$, close to the theoretical FKS results [17, 18]. See also Table 1 in ref. [16], line "FKS scheme & theory", giving $\theta_8 = -21.0^\circ$ and $\theta_0 = -2.7^\circ$, while the line "FKS scheme & phenomenology" in this table has only somewhat more negative θ_0 but larger f_0/f_π. Up to corrections of order $(1 - y)^2$, the "FKS scheme & theory" implies the *effective* η_8-η_0 state-mixing angle $\theta \approx (\theta_8 + \theta_0)/2 \approx -12^\circ$, in agreement with our results [9]. Notably, $\theta \approx -12^\circ$ agrees well with the mixing angle we get from our η–η' mass matrix, which agrees well with the *NS-S* mass matrices following from phenomenology and lattice calculations, as commented in the pedagogical exposition [15]. Reference [15] also explains how our approach practically ensures the reproduction of the empirical *NS-S* mass matrix once SD-BS calculations successfully reproduce pion and kaon masses and pion decay constant, as only one more parameter is then needed to fit η and η' masses satisfactorily.

Acknowledgment: D. Klabučar thanks the organizers, B. Golli, R. Krivec, M. Rosina and S. Širca, for their hospitality and partial support enabling his participation at XVIIIth European Conference on Few-Body Problems in Physics.

References

1. For recent reviews and numerous references, see, *e.g.*, refs. [2, 3]
2. R. Alkofer and L. von Smekal: Phys. Rept. **353**, 281 (2001)
3. C. D. Roberts and S. M. Schmidt: Prog. Part. Nucl. Phys. **45**, S1 (2000)
4. R. Alkofer and C. D. Roberts: Phys. Lett. **B369**, 101 (1996)
5. D. Kekez, B. Bistrović and D. Klabučar: Int. J. Mod. Phys. **A14**, 161 (1999)
6. D. Kekez and D. Klabučar: Phys. Lett. **B387**, 14 (1996)
7. D. Klabučar and D. Kekez: Phys. Rev. **D58**, 096003 (1998)
8. D. Kekez and D. Klabučar: Phys. Lett. **B457**, 359 (1999); hep-ph/9905251; hep-ph/9907537; hep-ph/9909273
9. D. Kekez, D. Klabučar and M. D. Scadron: J. Phys. **G26**, 1335 (2000)
10. D. Klabučar, D. Kekez and M. D. Scadron: hep-ph/0012267; published in "Rostock 2000/Trento 2001, Exploring quark matter", p. 145
11. D. Kekez and D. Klabučar: Phys. Rev. **D65**, 057901 (2002)
12. B. Bistrović and D. Klabučar: Phys. Lett. **B478**, 127 (2000); B. Bistrović and D. Klabučar: Phys. Rev. **D61**, 033006 (2000)
13. P. Jain and H. J. Munczek: Phys. Rev. **D48**, 5403 (1993)
14. D. Kekez, D. Klabučar and M. D. Scadron: J. Phys. **G27**, 1775 (2001)
15. D. Kekez and D. Klabučar: In: Bled Workshops in Physics, Vol.3, No.3. (Proceedings of the Mini-Workshop "Quarks and hadrons", Bled, Slovenia July 7-14, 2002, also available at http://www-f1.ijs.si/Bled2002/)
16. Th. Feldmann: Int. J. Mod. Phys. **A15**, 159 (2000)
17. Th. Feldmann, P. Kroll and B. Stech: Phys. Rev. **D58** 114006 (1998)
18. Th. Feldmann, P. Kroll and B. Stech: Phys. Lett. **B449**, 229 (1999)

Few-Body Systems Suppl. 14, 33–36 (2003)

Few-Body Systems

Magnetic Dipole Decays of Heavy and Heavy-Light Mesons

T.A. Lähde*, D.O. Riska†

Helsinki Institute of Physics and Department of Physics,
PL 64, University of Helsinki, 00014 Finland

Abstract. The widths of M1 transitions in heavy quarkonia are shown to be sensitive to two-quark exchange current operators that depend explicitly on the Lorentz structure of the quark-antiquark interaction. A possible solution to the long-standing overprediction by a factor ~ 3 of the width for $J/\psi \to \eta_c\gamma$ obtains, if the effective confining interaction has scalar Lorentz structure. The nonrelativistic predictions for the M1 widths of the $Q\bar{q}$ mesons are shown to be unrealistic, as the confined light constituent quark requires a relativistic treatment. However, as the form of the $Q\bar{q}$ interaction is uncertain, the results should be viewed as suggestive.

1 Introduction

Theoretical predictions for the M1 transitions in charmonium have remained unsatisfactory [1], as the width for $J/\psi \to \eta_c\gamma$ has typically been overpredicted by a factor ~ 3. Calculations of M1 widths within the Schrödinger framework [2] have demonstrated that M1 transitions are sensitive, not only to the relativistic aspects of the spin-flip operator, but also to the Lorentz structure of the effective $Q\bar{Q}$ interaction. The results of refs. [2] and [3] suggest that a scalar confining interaction may explain the observed width of ~ 1 keV for $J/\psi \to \eta_c\gamma$, provided that an unapproximated expression for the single quark spin-flip operator is used. This conclusion is supported by the calculation of ref. [4], which is based on the instantaneous approximation to the Bethe-Salpeter equation.

As the velocity of the light constituent quark in the $Q\bar{q}$ systems is close to that of light, the relativistic corrections to both single quark and two-quark operators will *a priori* be large. It is shown in ref. [5] that the large relativistic corrections to the single quark spin-flip operators, that yield unfavorable results for the M1 widths of $Q\bar{q}$ mesons, are in general counteracted by the

* *E-mail address:* talahde@pcu.helsinki.fi
† *E-mail address:* riska@pcu.helsinki.fi

two-quark operators associated with an interaction Hamiltonian that consists of one-gluon exchange (OGE) + scalar confinement components, thus allowing for better agreement with experiment. It is also argued that the instanton induced interaction for $Q\bar{q}$ systems suggested by ref. [6] may have a favorable effect on the predictions for the M1 widths of the D^* mesons.

2 Two-Quark Magnetic Moment Operators

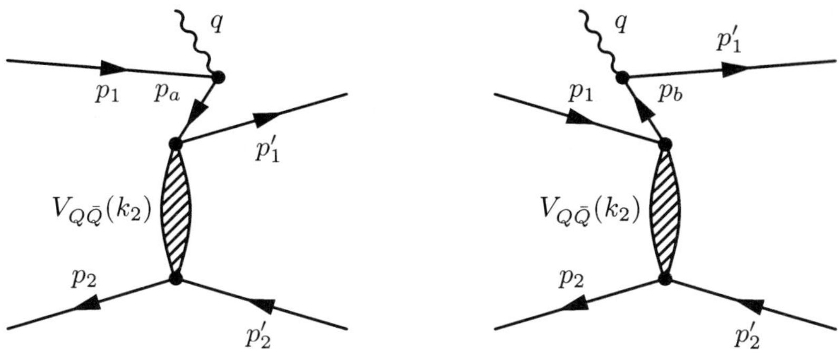

Figure 1. Relativistic Born diagrams for photon emission by a heavy constituent quark, which describe the negative energy components of the intermediate quark propagators p_a and p_b. The exchange current operators that correspond to the above diagrams have been calculated, for different Lorentz coupling structures of the interaction V, in ref. [7].

The spin dependent part of the (nonrelativistic) two-quark current operator associated with the scalar confining interaction, as calculated from the above Feynman diagrams, is of the form

$$\mathbf{j}_{\mathrm{ex}}^{\mathrm{c}} = \frac{i}{2}(\boldsymbol{\sigma}_1 + \boldsymbol{\sigma}_2) \times \boldsymbol{q} \left[\frac{Q_1^*}{2m_1^2} + \frac{Q_2^*}{2m_2^2}\right] + \frac{i}{2}(\boldsymbol{\sigma}_1 - \boldsymbol{\sigma}_2) \times \boldsymbol{q} \left[\frac{Q_1^*}{2m_1^2} - \frac{Q_2^*}{2m_2^2}\right], \quad (1)$$

where the variables Q_1^* and Q_2^* are defined as $Q_1^* = -e\,V_c(\boldsymbol{k}_2)Q_1$ and $Q_2^* = -e\,V_c(\boldsymbol{k}_1)Q_2$. The variables Q_1 and Q_2 denote the electrical charges of the quarks, in multiples of e. For the OGE interaction

$$\mathbf{j}_{\mathrm{ex}}^{\mathrm{g}} = Q_1^* \left[\frac{i\boldsymbol{\sigma}_1 \times \boldsymbol{k}_2}{2m_1^2} + \frac{i\boldsymbol{\sigma}_2 \times \boldsymbol{k}_2}{2m_1 m_2}\right] + Q_2^* \left[\frac{i\boldsymbol{\sigma}_2 \times \boldsymbol{k}_1}{2m_2^2} + \frac{i\boldsymbol{\sigma}_1 \times \boldsymbol{k}_1}{2m_1 m_2}\right], \quad (2)$$

with $Q_1^* = -e\,V_g(\boldsymbol{k}_2)Q_1$ and $Q_2^* = -e\,V_g(\boldsymbol{k}_1)Q_2$. The corresponding two-quark spin-flip magnetic moment operators, for transitions between S-wave states, may be obtained as [3]

$$\boldsymbol{\mu}_{\mathrm{ex}}^{\mathrm{Conf}} = -\frac{eV_c(r)}{4}\left[\frac{Q_1}{m_1^2} - \frac{Q_2}{m_2^2}\right](\boldsymbol{\sigma}_1 - \boldsymbol{\sigma}_2) \quad (3)$$

for the scalar confining interaction, and

$$\mu_{\rm ex}^{\rm Oge} = -\frac{eV_g(r)}{8}\left[\frac{Q_1}{m_1^2} - \frac{Q_2}{m_2^2} - \frac{Q_1 - Q_2}{m_1 m_2}\right](\boldsymbol{\sigma}_1 - \boldsymbol{\sigma}_2) \tag{4}$$

for the OGE interaction. For equal constituent quark masses, the above expressions reduce to those given in ref. [7]. The OGE spin-flip operator (4) contributes to M1 transitions only if the constituent quark masses are unequal.

3 Numerical Results

Table 1. The M1 transitions between low-lying S-wave states in the charmonium ($c\bar{c}$), bottomonium ($b\bar{b}$), and $B_c^\pm$ ($c\bar{b}, \bar{c}b$) systems, from ref. [3]. Experimental data [8] is available only for $J/\psi \to \eta_c\gamma$ and $\psi' \to \eta_c\gamma$.

Transition	Matrix element [fm]		Width		
	RIA	Exch	NRIA	RIA	**RIA+Exch**
$J/\psi \to \eta_c\gamma$ q_γ : 116 MeV	$3.762 \cdot 10^{-2}$	$-8.724 \cdot 10^{-3}$	2.85	2.12	**1.25 keV** ex: 1.14 ± 0.39
$\psi' \to \eta_c\gamma$ q_γ : 639 MeV	$-5.14 \cdot 10^{-4}$	$2.826 \cdot 10^{-3}$	3.35	0.06	**1.13 keV** ex: 0.84 ± 0.24
$\psi' \to \eta_c'\gamma$ q_γ : 46 MeV	$3.735 \cdot 10^{-2}$	$-1.870 \cdot 10^{-2}$	0.18	0.13	**0.03 keV**
$\eta_c' \to J/\psi\,\gamma$ q_γ : 502 MeV	$-7.584 \cdot 10^{-3}$	$5.206 \cdot 10^{-3}$	5.89	18.6	**1.83 keV**
$\Upsilon \to \eta_b\gamma$ q_γ : 59 MeV	$-6.39 \cdot 10^{-3}$	$2.41 \cdot 10^{-4}$	9.2	8.3	**7.7 eV**
$\Upsilon' \to \eta_b\gamma$ q_γ : 603 MeV	$-1.44 \cdot 10^{-4}$	$-8.76 \cdot 10^{-5}$	31.8	4.3	**11.0 eV**
$\Upsilon' \to \eta_b'\,\gamma$ q_γ : 25 MeV	$-6.39 \cdot 10^{-3}$	$5.45 \cdot 10^{-4}$	0.70	0.64	**0.53 eV**
$\eta_b' \to \Upsilon\,\gamma$ q_γ : 530 MeV	$6.30 \cdot 10^{-4}$	$-1.31 \cdot 10^{-4}$	71.5	162	**102 eV**
$\Upsilon'' \to \eta_b''\,\gamma$ q_γ : 16 MeV	$-6.35 \cdot 10^{-3}$	$8.02 \cdot 10^{-4}$	0.18	0.16	**0.13 eV**
$\Upsilon'' \to \eta_b'\,\gamma$ q_γ : 350 MeV	$-1.11 \cdot 10^{-4}$	$-1.55 \cdot 10^{-4}$	5.3	0.5	**2.9 eV**
$\Upsilon'' \to \eta_b\,\gamma$ q_γ : 910 MeV	$-6.59 \cdot 10^{-5}$	$-3.77 \cdot 10^{-5}$	30.2	3.0	**7.3 eV**
$B_c^* \to B_c\gamma$ q_γ : 53 MeV	$1.496 \cdot 10^{-2}$	$0.326 \cdot 10^{-3}$	50.0	32.6	**34.0 eV**

Table 2. Numerical results for the M1 transition $D^{\pm *} \to D^{\pm}\gamma$ from ref. [5], for different values of the light constituent quark mass m_q and a charm quark mass of 1580 MeV. The two-quark contributions from the exchange magnetic moment operators eqs. (3) and (4) have been added in the rightmost column. The present experimental width is 1.5 ± 0.6 keV [8].

m_q	NRIA	RIA	**RIA + Exch**
450 MeV	0.58 keV	$9.4 \cdot 10^{-3}$ keV	**1.09 keV**
420 MeV	0.79 keV	$1.5 \cdot 10^{-2}$ keV	**1.43 keV**
390 MeV	1.07 keV	$2.2 \cdot 10^{-2}$ keV	**1.90 keV**

4 Conclusions

The results of refs. [3, 5] indicate that the M1 transitions in the $Q\bar{q}$ and $Q\bar{Q}$ systems may provide direct information on the Lorentz structure of the effective confining interaction. Whereas the M1 transitions in charmonium have been shown to be satisfactorily described by a pure scalar confining interaction, the situation for the D mesons is more uncertain because of the substantial exchange current contributions from the OGE or instanton induced interactions. Moreover, a relativistic treatment of the two-quark currents is desirable because of the low masses of the light constituent quarks.

References

1. A.V. Smilga and M.A. Shifman (ed.): *Vacuum Structure and QCD Sum Rules*, p. 490, North Holland (1992)
2. T.A. Lähde, C.J. Nyfält and D.O. Riska: Nucl. Phys. **A645**, 587 (1999); hep-ph/9808438
3. T.A. Lähde, submitted to Nucl. Phys. **A** (2002); hep-ph/0208110
4. J. Linde and H. Snellman: Nucl. Phys. **A619**, 346 (1997); hep-ph/9703383
5. T.A. Lähde and D.O. Riska: Nucl. Phys. **A710**, 99 (2002); hep-ph/0204230
6. S. Chernyshev, M.A. Nowak and I. Zahed: Phys. Rev. **D53**, 5176 (1996); hep-ph/9510326
7. K. Tsushima, D.O. Riska and P.G. Blunden: Nucl. Phys. **A559**, 543 (1993)
8. K. Hagiwara et al.: Phys. Rev. **D66**, 010001 (2002) (Particle Data Group, ed. 2002)

Few-Body Systems Suppl. 14, 37–40 (2003)

Few-Body Systems

Relativistic Study of Mesonic Baryon Resonance Decays

T. Melde, R.F. Wagenbrunn, W. Plessas

Institut für Theoretische Physik, Universität Graz, Universitätsplatz 5, A-8010 Graz, Austria

Abstract. Mesonic baryon resonance decays are calculated from constituent quark models along a Poincaré-invariant generalization of the elementary emission model. Covariant results of pionic decay widths are presented for the Goldstone-boson-exchange constituent quark model.

1 Introduction

Investigations of mesonic resonance decays have a long tradition, and in the focus of interest have been, notably, the performances of various constituent quark models (CQMs) as well as the adequacy of different decay operators for the mechanism of meson creation/emission. Despite considerable efforts invested one has still not yet arrived at a satisfactory explanation especially of the N^* and Δ resonance decays. Also complementary attempts beyond the CQM approach have not succeeded much better with hadron decays, and more generally, with providing a comprehensive working model of low-energy hadronic physics based on QCD. This situation is rather disappointing from the theoretical side, especially in view of the large amount of experimental data accumulated over the past years and the ongoing high-quality measurements at such facilities as JLAB, MAMI and others (for a an overview of the modern developments see the proceedings of the past N^* Workshop [1]).

In a recent CQM the interaction between two constituent quarks was based on Goldstone-boson exchange (GBE) [2] in addition to a linear confinement. The so-called GBE CQM at the same time fulfills the requirements of Poincaré invariance and thus it paves the way to a relativistic treatment of reactions involving baryons. Specifically, due to the usage of a relativistic kinetic-energy operator, the Hamiltonian of the GBE CQM leads to an invariant mass operator in relativistic quantum mechanics, where the quark-quark interactions are introduced via the Bakamjian-Thomas construction [3].

Up till now the GBE CQM has already been put to some tests in calculating mesonic decays of resonances of light and strange baryons in a non-relativistic framework [4, 5, 6]. Here, we calculate the mesonic decay widths in a covariant approach along a generalization of the elementary emission model (EEM) in point-form relativistic quantum mechanics [7]. We compare the predictions of the GBE CQM to analogous results from a CQM with a one-gluon-exchange (OGE) hyperfine interaction, namely the relativized Bhaduri-Cohler-Nogami CQM as parametrized in ref. [6].

2 Theory

Generally, the decay width of a particle is defined by the expression

$$\Gamma = 2\pi\rho_f \left|F\left(i \rightarrow f\right)\right|^2, \tag{1}$$

where $F\left(i \rightarrow f\right)$ is the transition amplitude and ρ_f is the phase-space factor. In Eq. (1) one has to average over the initial and to sum over the final spin-isospin projections. In non-relativistic calculations of baryon resonance decays one has usually made an arbitrary choice of the phase-space factor. In the rest frame of the decaying resonance, either a non-relativistic form, $\rho_f = 2\pi\frac{M_f M_\pi}{M_i}q$, or a

Table 1. Predictions for pionic decay widths by the GBE CQM [2] along the EEM in PFSA in comparison to experiment, an analogous calculation with the OGE CQM [6], and results from a non-relativistic EEM approach.

Decays	Experiment [8]	Rel. PFSA CQM		Nonrel. EEM GBE CQM	
		GBE	OGE	dir	dir+rec
$N^\star_{1440} \rightarrow \pi N_{939}$	$(227 \pm 18)^{+70}_{-59}$	30.3	37.1	4.85	6.16
$N^\star_{1520} \rightarrow \pi N_{939}$	$(66 \pm 6)^{+9}_{-5}$	16.9	16.2	22.0	38.3
$N^\star_{1535} \rightarrow \pi N_{939}$	$(67 \pm 15)^{+55}_{-17}$	93.2	122.8	24.3	574.3
$N^\star_{1650} \rightarrow \pi N_{939}$	$(109 \pm 26)^{+36}_{-3}$	28.8	38.3	11.3	160.3
$N^\star_{1675} \rightarrow \pi N_{939}$	$(68 \pm 8)^{+14}_{-4}$	5.98	6.20	7.65	15.1
$N^\star_{1700} \rightarrow \pi N_{939}$	$(10 \pm 5)^{+3}_{-3}$	0.91	1.19	1.43	2.87
$N^\star_{1710} \rightarrow \pi N_{939}$	$(15 \pm 5)^{+30}_{-5}$	4.06	2.28	23.4	5.95
$\Delta_{1232} \rightarrow \pi N_{939}$	$(119 \pm 1)^{+5}_{-5}$	33.7	32.1	59.1	81.2
$\Delta_{1600} \rightarrow \pi N_{939}$	$(61 \pm 26)^{+26}_{-10}$	0.116	0.503	74.2	55.7
$\Delta_{1620} \rightarrow \pi N_{939}$	$(38 \pm 8)^{+8}_{-6}$	10.4	14.6	4.82	74.8
$\Delta_{1700} \rightarrow \pi N_{939}$	$(45 \pm 15)^{+20}_{-10}$	2.92	3.10	7.12	14.4

relativistic form $\rho_f = 2\pi \frac{E_f E_\pi}{M_i} q$ has been used. Herein, M_i is the mass of the initial state and M_f, M_π as well as E_f, E_π are the masses and energies of the decay products, the nucleon and the pion, respectively. In the present work we follow a Poincaré-invariant description of the transition amplitude resulting in a unique choice of the phase-space factor.

For the actual calculation in the point form it is advantageous to introduce so-called velocity states

$$|v; \boldsymbol{k}_1, \boldsymbol{k}_2, \boldsymbol{k}_3; \mu_1, \mu_2, \mu_3\rangle = U_{B(v)} |k_1, k_2, k_3; \mu_1, \mu_2, \mu_3\rangle = \prod_{i=1}^{3} D^{\frac{1}{2}}_{\sigma_i \mu_i} [R_W(k_i, B(v))] |p_1, p_2, p_3; \sigma_1, \sigma_2, \sigma_3\rangle , \quad (2)$$

where $B(v)$ is a boost with four-velocity v and $U_{B(v)}$ its unitary representation. The boosted momenta are defined by $p_i = B(v)\, k_i$, where $k_i = (\omega_i, \mathbf{k}_i)$ and $\sum \boldsymbol{k}_i = 0$; the $D^{\frac{1}{2}}$ are the spin-$\frac{1}{2}$ representation matrices of Wigner rotations $R_W(k_i, B(v))$. The baryons are described by simultaneous eigenstates of the four-momentum operator (or equivalently the mass operator), the total-angular-momentum operator, and its z-component; we denote them by $|P, J, \Sigma\rangle$. The transition amplitude is then defined in a Poincaré-invariant fashion, under overall momentum conservation ($P' - P = Q_\pi$), by

$$\begin{aligned} F(i \to f) &= \langle P, J, \Sigma | \hat{\mathcal{D}}_\alpha | P', J', \Sigma' \rangle \\ &\sim \int d^3k_2 d^3k_3 d^3k'_2 d^3k'_3 \\ &\quad \Psi^\star_{MJ\Sigma}(\boldsymbol{k}_1, \boldsymbol{k}_2, \boldsymbol{k}_3; \mu_1, \mu_2, \mu_3)\, \Psi_{M'J'\Sigma'}(\boldsymbol{k}'_1, \boldsymbol{k}'_2, \boldsymbol{k}'_3; \mu'_1, \mu'_2, \mu'_3) \\ &\quad \prod_{\sigma_i} D^{\frac{1}{2}\star}_{\sigma_i \mu_i} [R_W(k_i, B(v_{in}))] \prod_{\sigma'_i} D^{\frac{1}{2}}_{\sigma'_i \mu'_i} [R_W(k'_i, B(v_f))] \\ &\quad \langle p_1, p_2, p_3; \sigma_1, \sigma_2, \sigma_3 | \hat{\mathcal{D}}_\alpha | p'_1, p'_2, p'_3; \sigma'_1, \sigma'_2, \sigma'_3 \rangle , \end{aligned} \quad (3)$$

where the baryon wave functions $\Psi^\star_{MJ\Sigma}$ and $\Psi_{M'J'\Sigma'}$ enter as velocity-state representations of the baryon states $\langle P, J, \Sigma|$ and $|P', J', \Sigma'\rangle$, respectively. In a first attempt, we investigate a decay operator, which can be interpreted as a generalization of the EEM. Namely, we assume that a pion is created on one of the quarks, while the other two quarks of the decaying baryon resonance are merely spectators. Consequently, the decay operator is taken in point-form spectator approximation (PFSA). Here, we assume a pseudo-vector coupling for point-like particles. This yields the decay operator in the form

$$\begin{aligned} &\langle p_1, p_2, p_3; \sigma_1, \sigma_2, \sigma_3 | \hat{\mathcal{D}}_\alpha | p'_1, p'_2, p'_3; \sigma'_1, \sigma'_2, \sigma'_3 \rangle \\ &\quad \sim 3 i g_{q\pi} \bar{u}(p_1, \sigma_1)\, \gamma^5 \gamma^\mu \boldsymbol{\lambda}^F u(p'_1, \sigma'_1) \\ &\quad\quad \times 2p'^0_2 \delta(\boldsymbol{p}_2 - \boldsymbol{p}'_2)\, 2p'^0_3 \delta(\boldsymbol{p}_3 - \boldsymbol{p}'_3)\, \delta_{\sigma_2 \sigma'_2} \delta_{\sigma_3 \sigma'_3} Q_\mu , \end{aligned} \quad (4)$$

where $g_{q\pi}$ is the pion-quark coupling constant and $\boldsymbol{\lambda}^F$ the flavour operator. It should be noted that in PFSA the impulse delivered to the quark that emits the pion is not equal to the impulse delivered to the baryon as a whole. The momentum transfer $\tilde{q}$ to this single quark is uniquely determined from the momentum Q transferred to the baryon and the two spectator conditions.

3 Results

In Table 1 we present the predictions of the GBE CQM for pionic decay widths of N^* and Δ resonances. They have been calculated directly from the CQM wave functions with the EEM decay operator in PFSA; no further parameterization has been introduced. As is immediately evident, only in a few instances a reasonable agreement with experiment is found. In most cases the decay widths remain far from the data, with the theoretical values always being smaller than the measured ones. These characteristics of the relativistic decay widths are in contrast to the ones calculated along the EEM in a non-relativistic approach plus first-order relativistic (recoil) corrections [4, 5]. On the other hand, the OGE and GBE CQMs produce quite similar results indicating that dynamical effects from the CQM are of lesser importance than relativistic effects.

In any case, we find the relativistic results to exhibit a completely new behaviour. This might largely be due to Lorentz boost effects, which have here been included exactly. Certainly, the predictions for the pionic decay widths are by no means satisfactory. This hints to persisting deficiencies in the decay operator and/or the CQM wave functions. The latter, relying on $\{QQQ\}$ configurations only, are tentatively not realistic enough especially for the decaying resonance state. Further work will have to clarify these shortcomings.

Acknowledgement. This work was supported by the Austrian Science Fund (Project P14806).

References

1. D. Drechsel and L. Tiator (Eds.): *NSTAR2001* (Proc. of the Workshop on the Physics of Excited Nucleons, Mainz). Singapore: World Scientific 2001
2. L.Ya. Glozman, W. Plessas, K. Varga, and R.F. Wagenbrunn: Phys. Rev. **D58**, 094030 (1998)
3. B.D. Keister and W.N. Polyzou: Adv. Nucl. Phys. **20**, 225 (1991)
4. A. Krassnigg, W. Plessas, L. Theußl, K. Varga, and R.F. Wagenbrunn: Few Body Syst. Suppl. **10**, 391 (1999)
5. W. Plessas, A. Krassnigg, L. Theußl, K. Varga, and R.F. Wagenbrunn: Few Body Syst. Suppl. **11**, 29 (1999)
6. L. Theussl, R.F. Wagenbrunn, B. Desplanques, and W. Plessas: Eur. Phys. J. **A12**, 91 (2001)
7. W.H. Klink: Phys. Rev. **C58**, 3587 (1998)
8. K. Hagiwara et al.: Phys. Rev. **D66**, 010001 (2002)

Few-Body Systems Suppl. 14, 41–44 (2003)

Few-Body Systems

A Dressed Bag Study of the $d(\gamma, \pi)2N$ Reaction

I.T. Obukhovsky[1], M.M. Kaskulov[2], P. Grabmayr[2,*]

[1] Institute of Nuclear Physics, Moscow State University, Vorobievy Gory, 119899 Moscow, Russia

[2] Physikalisches Institut, Universität Tübingen, Auf der Morgenstelle 14, D-72076 Tuebingen, Germany

Up to now, there has been no detailed work done to investigate the role of the $N\Delta$ final state interaction (FSI) in pion photoproduction from nuclei. Recall that in the Δ-resonance region $E_\gamma \approx$ 250-400 MeV the quasi-free mechanism overestimates considerably (in 1.5-2 times) the forward-angle $d\gamma \to \pi^0 pn$ cross section [1], and thus the FSI should be taken into consideration. According the recent calculations [2] the pn FSI is most important in this region. However, it is not excluded that the 'direct' $N\Delta$ FSI plays also a certain role, and if so there will be a possibility to study the $N\Delta$ interaction (and its off-shell and short-range behaviour) in photoproduction processes on nuclei. In this work we evaluate the contribution of short-range $N\Delta$ FSI to the pion photoproduction off deuteron starting from the constituent quark model (CQM).

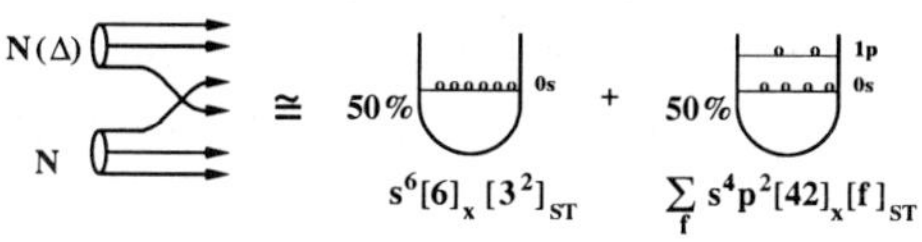

Figure 1.

In the quark approach the short-range $N\Delta$ and NN interactions are very similar each other. The microscopic evaluation [3] made in the framework of RGM leads to the following picture (Fig. 1) of the configuration mixing in the overlap region: in both $NN(ST{=}10)$ and $N\Delta(ST{=}21)$ channels the overlap consists of a fully symmetric configuration $s^6\,[6]_X$ to the extent of $\approx$50% and of a remainder being a superposition of states of the configuration s^4p^2, which are characterized by a non-trivial coordinate Young tableau $[42]_X$. The s^6 and s^4p^2 states differ in their inner structure: (i)the fully symmetric configuration s^6 is a bag-like state with the charge-spin structure strictly constrained by the Pauli exclusion principle (in both channels, $NN(ST{=}10)$ and $N\Delta(ST{=}21)$, the Young tableau in the product of spin and isospin spaces $[3^2]_{ST}$ is fixed by the condition to

*This work was sponsored by DFG grants No Fa67/20-1, GRK683, Gr1084/3, He2171/1 and RFBR grants No 01-02-04015, 02-02-16612

be conjugated to the color Young tableau $[2^3]_C$); (ii) in excited quark-shell-model states s^4p^2 the non-trivial coordinate Young tableau $[42]_X$ allows all the possible tableaux $[f_{ST}]$ in the inner product of spin and isospin spaces for both $NN(ST{=}10)$ and $N\Delta(ST{=}21)$ channels.

This is a sufficiently large basis to represent the main features of NN or $N\Delta$ channels at short range. In ref. [3] it was shown that some linear combinations of the s^4p^2 basis states are genuine NN (or $N\Delta$) cluster-like states, which can be referred to as short-range wave functions $\{\Psi_{NN}\}_{short\ range}$. (A similar picture can be also seen in the lowest odd partial wave $L = 1$, where the configuration $s^5p[51]_X$ is also strictly constrained by the Pauli exclusion principle.)

In the recently developed Moscow-Tuebingen dressed bag model [4] it was proposed that the non-nucleonic bag-like states with trivial coordinate symmetry $[6]_X$ and $[51]_X$ have got their own meson cloud which is not a trivial sum of meson clouds of initial nucleons (baryons). If the scalar-isoscalar (σ-meson) field is mostly enhanced in the presence of the six-quark bag, the following dynamical model for the short-range NN (or $N\Delta$) interaction could be proposed:

(*i*) At short range there is an enhanced σ field originated from transitions between the "proper" NN cluster-like state and the bag-like one $d_0(s^6)$: $\{\Psi_{NN}\}_{short\ range} \leftrightarrows d_0(s^6) + \sigma$ (or $\{\Psi_{N\Delta}\}_{short\ range} \leftrightarrows D_0(s^6) + \sigma$ in the $N\Delta$ channel).

(*ii*) The short-range NN dynamics is determined by coupled-channel equations for NN and $d_0 + \sigma$ components.

(*iii*) An effective separable and energy-dependent NN (or $N\Delta$) potential

$$V_{NqN}(r,r';E) = g^2_{N\sigma q} I(E) | \varphi_{2S}(r') \rangle \langle \varphi_{2S}(r) | \tag{1}$$

emerges [4] after excluding of the inner states with standard methods.

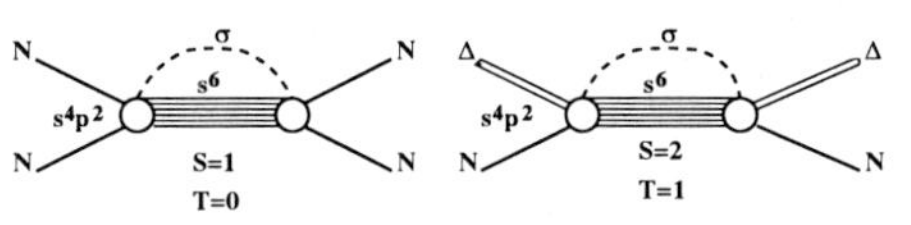

Figure 2.

These potentials are related to s-channel loop diagrams (Fig. 2), the contribution of which to the short-range force can be calculated in the $^3P_0$ model [4]. The effective potential for the short-range $N\Delta$ interaction $V_{\Delta q\Delta}(r,r';E)$ could be obtained from $V_{NqN}(r,r';E)$, and vice versa, by replacement of 'algebraic coupling constants' $g^2_{N\sigma q} \leftrightarrows g^2_{\Delta\sigma q}$.

In the π^0 and π^- photoproduction off deuteron there are at least two different mechanisms of short-range $N\Delta$ FSI shown in Fig. 3 and 4 correspondingly. In both cases there is the same intermediate $N\Delta$ state $^5S_2$ generated by the isovector $M1$ $\gamma N \to \Delta$ transition in the deuteron, but the final $\pi^0 pn$ and $\pi^- pp$ systems correspond to two different NN isospin states. The lowest partial wave $^3S_1$ in the np final state is allowed, but in the final pp state ($T{=}1$) the analogous $^1S_0$ wave is forbidden, and the lowest allowed partial pp wave is only the $^1D_2$ as it is shown in Fig. 3(b). However, the $D_0(s^6)$ bag-like state in the $N\Delta$ overlap can bypass the transition to the final $N\Delta$ state allowing the direct transition to the final $\pi^- pp$ state with $^3P_2$ pp partial wave (with the S-wave pion), as it is shown in Fig. 4. Hence the dressed bag $d_1(s^5p) + \sigma$ could be seen in the π^-

photoproduction as a correlated pp pair in the $^3P_2 - ^3F_2$ state originated from the decay of the dressed bag $D_0(s^6)$ into $\pi^-(l_\pi{=}0) + pp$ channel.

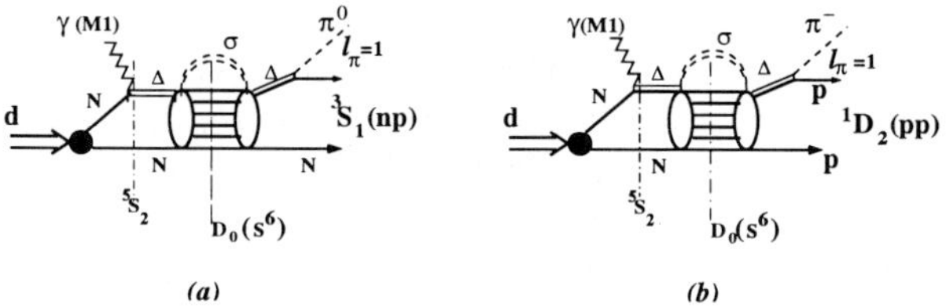

Figure 3. The $N\Delta$ type of FSI (P-wave pion).

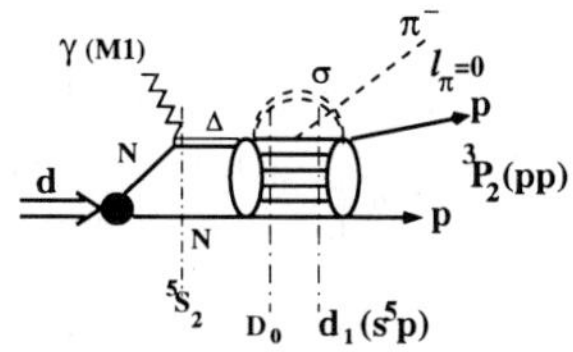

Figure 4. The 'dibaryon' type of FSI (S-wave pion).

Such a correlated pp state has to accept the recoil pion momentum as a unit. A possible dibaryon resonance with the quantum numbers of $d_1(s^5p) + \sigma$ system could enhance the role of such a correlated pp state, and thus any polarization experiments taking into consideration the $^3P_2 - ^3F_2$ final pp state would be of advantage. But in general (i.e. without a dibaryon) the role of $N\Delta$ FSI interaction in the π^- photoproduction should be considerably smaller then in the π^0 case.

First we test our approach in processes of pion photoproduction from free nucleons. For that the bare baryons of QCM should be dressed by a pion field. Then the Δ isobar acquires a finite decay width $\Gamma_\Delta \approx 120$ MeV and the propagator of Δ acquires a non-trivial off-shell behavior with the mass M_R and the width Γ_R dependent on the energy of a virtual Δ isobar (e.g. in the $N\Delta$ system) $W_\Delta = E - E_N$. The amplitude of pion photoproduction going through an intermediate Δ

$$\frac{v_\Delta(\mathbf{k},\lambda) w^\dagger_\Delta(\mathbf{q},\varkappa)}{W_\Delta - M_R(W_\Delta) + \frac{i}{2}\Gamma_R(W_\Delta)} \tag{2}$$

includes the e.-m. vertex $w^\dagger_\Delta(\mathbf{q},\varkappa)$ for the M1 transition $N + \gamma \to \Delta$ ($\varkappa{=}\pm 1$)

$$w^\dagger_\Delta(\mathbf{q},\varkappa) = 3\langle N|H^{(3)}_{\gamma qq}(\mathbf{q},\varkappa)|\Delta\rangle = -i\left(\frac{\sqrt{2}e}{2m_q}\right) e^{-q^2b^2/6}[\mathbf{S}^\dagger \times \mathbf{q}]\cdot \boldsymbol{\epsilon}^{(\varkappa)}\, T_z \tag{3}$$

and the $\pi^\lambda N\Delta$ vertex function ($\lambda{=}0,\pm 1$)

$$v_\Delta(\mathbf{k},\lambda) = 3\langle N|H^{(3)}_{\pi qq}(\mathbf{k},\lambda)|\Delta\rangle = i\left(\frac{2\sqrt{2}f_{\pi qq}}{m_\pi}\right) e^{-k^2b^2/6}\,\mathbf{S}\cdot\mathbf{k}\,T_\lambda, \tag{4}$$

calculated with elementary quark-photon $H^{(n)}_{\gamma qq}(\mathbf{q},\varkappa)$ and $^3P_0$-model quark-pion $H^{(n)}_{\pi qq}(\mathbf{k},\lambda)$ coupling operators correspondingly (for the n-th quark). Additional contributions to the amplitude are the nucleon and π-meson pole terms with the seagull term added for the current conservation. The calculated cross sections for pion photoproduction off nucleon are very close to the conventional model

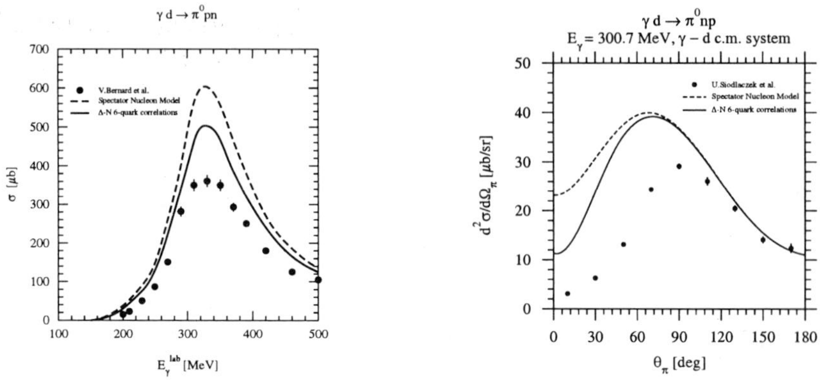

Figure 5. The total (left) and differential (right) cross sections for π^0 photoproduction off deuteron calculated in the quark model. The data from Ref. [5]. Quasi-free photoproduction (dash line) and quasi-free $+N\Delta$ FSI (solid line).

results. Results for pion photoproduction off deuteron obtained without $N\Delta$ FSI are also very similar to the conventional model results [1]. But including into consideration the short-range $N\Delta$ FSI leads to an additional suppression of the forward-angle π^0 photoproduction (see Figs. 5). One can see that including the $N\Delta$ FSI in terms of the six-quark-bag diagram of Fig. 3(a) leads to a detectable effect for the differential and integral cross section of π^0 production. However, for the π^- photoproduction the $N\Delta$ FSI is not so important as for the π^0.

Acknowledgement. The authors greatly acknowledge the contribution of all the collaborators of the Moscow-Tuebingen group [4] to this talk.

References

1. R. Schmidt, H. Arenhövel, and P. Wilhelm: Z. Phys. **A355**, 421 (1996)
2. M.I. Levchuk, M. Schumacher, and F. Wissman: Nucl. Phys. **A675**, 621 (2000); *ibid*, nucl-th/0011041
3. A.M. Kusainov, V.G. Neudatchin, and I.T. Obukhovsky: Phys. Rev. **C44**, 2343 (1991); I.T. Obukhovsky, Prog. Part. Nucl. Phys. **36**, 359 (1996)
4. V.I. Kukulin, I.T. Obukhovsky, V.N. Pomerantsev, and Amand Faessler: Int. J. Mod. Phys. **E11**, 1 (2002); I.T. Obukhovsky, M.M. Kaskulov, V.I. Kukulin, P. Grabmayr, and A. Faessler: to be published (2003)
5. U. Siodlaczek et al.: Eur. Phys. J. **A10**, 365 (2001); B. Krusche et al.: Eur. Phys. J. **A6**, 309 (1999)

Few-Body Systems Suppl. 14, 45–48 (2003)

Few-Body Systems

A Unified Meson-Baryon Potential

C. Semay[1,*,†], F. Brau[1,**,††], B. Silvestre-Brac[2,***]

[1] Université de Mons-Hainaut, Place du Parc, 20, B-7000 Mons, Belgium
[2] Institut des Sciences Nucléaires, 53, Av. des Martyrs, F-38026 Grenoble-Cedex, France

Abstract. We study the spectra of mesons and baryons, composed of light quarks, in the framework of a semirelativistic potential model including instanton induced forces. We show how a simple modification of the instanton interaction in the baryon sector allows a good description of the meson and the baryon spectra using an interaction characterized by a unique set of parameters.

The description of the meson and baryon spectra in the framework of nonrelativistic or semirelativistic potential models appears to be a rather successful approach. The meson properties or the baryon properties are generally investigated in disconnected works. There exist only few complete studies dealing consistently with both meson and baryon spectra. Even if some encouraging results have already been obtained, none is really satisfying.

In pioneer works, Isgur *et al.* have proposed a semirelativistic model relying on a confinement supplemented by spin-spin, spin-orbit, and tensor interactions [1] (denoted CGI model hereafter). Despite the use of a complex model, the authors need to choose different values for the slope of the confinement for mesons and for baryons to obtain good results (some other parameters have also been slightly modified).

Another approach was developed by Blask *et al.* [2] (denoted BBHMP model hereafter). The authors have proposed a nonrelativistic model relying on a confinement supplemented by an instanton induced interaction. The meson and baryon ground states are well described but the use of a nonrelativistic kinematics leads to a bad description of radial excitations of mesons and baryons.

In a previous work [3], we have tried such a description using a semirelativistic model relying on a Cornell potential supplemented by an instanton

*FNRS Research Associate
† *E-mail address:* claude.semay@umh.ac.be
**FNRS Postdoctoral Researcher
†† *E-mail address:* fabian.brau@umh.ac.be
*** *E-mail address:* silvestre@isn.in2p3.fr

induced interaction. No satisfying result was obtained, but this work and a previous one [4] have clearly shown that a separate description of meson and baryon properties was possible within this simple semirelativistic model. So, a natural question arises: "which simple modification of our model could lead to a consistent description of both meson and baryon spectra?". In this work, we show that, at least one such a modification exists (see ref. [5] for more details).

The model used in this work is similar to the one introduced in refs. [3, 4]. The Hamiltonian is written

$$H = \sum_{i=1}^{N} \sqrt{\boldsymbol{p}_i^2 + m_i^2} + \sum_{i<j=1}^{N} V_{ij} \quad (N = 2 \text{ or } 3), \tag{1}$$

with m_i the constituent mass of the ith quark ($m_u = m_d = m_n$), and V_{ij} the interaction between quarks (or antiquarks) i and j. The interaction contains the Cornell potential and the instanton induced interaction. The Cornell potential, which depends only on the distance r between two quarks, is given by

$$V_{\mathrm{C}}(r) = -\frac{3}{4}\frac{\lambda_i \cdot \lambda_j}{4}\left[-\frac{\kappa}{r} + a\,r + C_{\mathrm{M}} + C_{\mathrm{B}}\delta_{N3}\right]. \tag{2}$$

C_{M} is the constant for the meson spectra, while C_{B} is the constant that we need to add to C_{M} to obtain the correct absolute value of the energy levels of the baryon spectra.

The instanton induced interaction between one quark and one antiquark depends on the isospin of the $q\bar{q}$ pair via two dimensioned coupling constants g and g' [4]. Between two quarks in a baryon, this interaction is written [3]

$$V_{\mathrm{I}}(r) = -4\left(gP^{[nn]} + g'P^{[ns]}\right)P^{S=0}\delta(\boldsymbol{r}), \tag{3}$$

where $P^{S=0}$ is the projector on spin 0 and $P^{[qq']}$ is the projector on antisymmetrical flavor state qq'.

In our model, a quark is not a pure point-like particle, but an effective degree of freedom which is dressed by the gluon and quark-antiquark pair clouds. The form that we retain for the color charge density of a quark is a Gaussian function of size γ_n or γ_s, depending on the flavor of the quark.

We have shown in ref. [3] that it is necessary to use two different sets of parameters to get a correct description of hadron masses. So, we have performed of series of minimizations, starting with new ranges of parameters and studying only some classes of hadrons. We have then remarked that it was systematically possible to reproduce the masses of all the mesons, as well as the masses of the baryons for which the instanton induced interaction does not act. Consequently, we tried to modify in different ways $V_{\mathrm{I}}(r)$ in the baryon sector. We propose below the simplest form that we found, which gives good results:

$$V_{\mathrm{I}}(r) \to V_{\mathrm{I}}(r) + C_{\mathrm{I}}\left(P^{[nn]} + P^{[ns]}\right)P^{S=0}P^{L=0}. \tag{4}$$

In this formula, C_{I} is a new constant. The status of this supplementary term is up to now purely phenomenological.

In Table 1, we present the optimal values found for the parameters of our model.

Table 1. List of parameters of the model.

m_n	0.236 GeV	$C_{\rm M}$	−0.691 GeV	g	2.906 $\rm GeV^{-2}$
m_s	0.472 GeV	$C_{\rm B}$	−0.033 GeV	g'	1.710 $\rm GeV^{-2}$
a	0.210 $\rm GeV^2$	$C_{\rm I}$	−0.065 GeV	γ_n	0.779 $\rm GeV^{-1}$
κ	0.525			γ_s	0.566 $\rm GeV^{-1}$

A comparison between the meson spectra of π and ρ families obtained with the model proposed here, the data, and the results obtained with the CGI and the BBHMP models is shown in Fig. 1. The spectra obtained with our model are as good as those obtained in our previous work [4]. Our energy levels present the same characteristics than those found with the CGI model, but they are clearly better than those obtained with the nonrelativistic BBHMP model compared to experimental data. It is worth noting that the pseudoscalar mesons π-K-η-η' are very well described with the instanton induced interaction, while in the CGI model a purely phenomenological annihilation interaction was introduced.

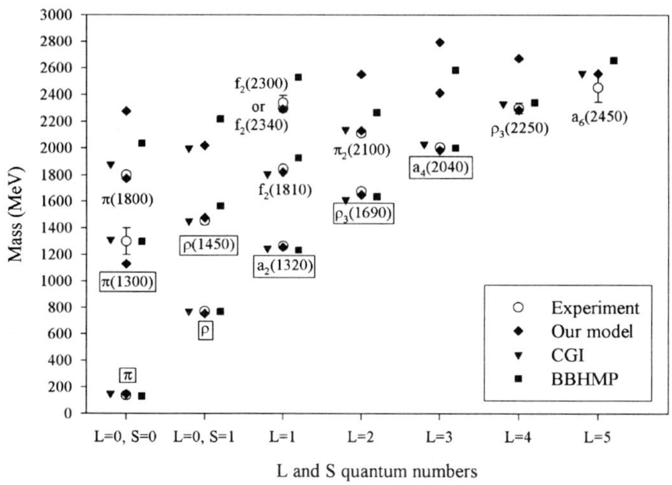

Figure 1. Masses of $n\bar{n}$ mesons as a function of total orbital angular momentum and total spin. Framed names indicate centers of gravity of multiplets of these families used in the fit of parameters.

The same analysis is presented in Fig. 2, but concerning baryon spectra of N and Δ families. Here also, the spectra obtained with our model are as good as the ones obtained in our previous work [3]. Globally, we have an improvement of the positive-parity states as compared to the corresponding ones obtained with the BBHMP model, while the negative-parity states are close in both models. Conversely, we have an improvement of the negative-parity states as compared to the those obtained with the CGI model, while the positive-parity states are close in both models. Although a semirelativistic kinetic energy operator substantially improves the various Roper resonances, there still remains a noticeable disagreement with experimental data.

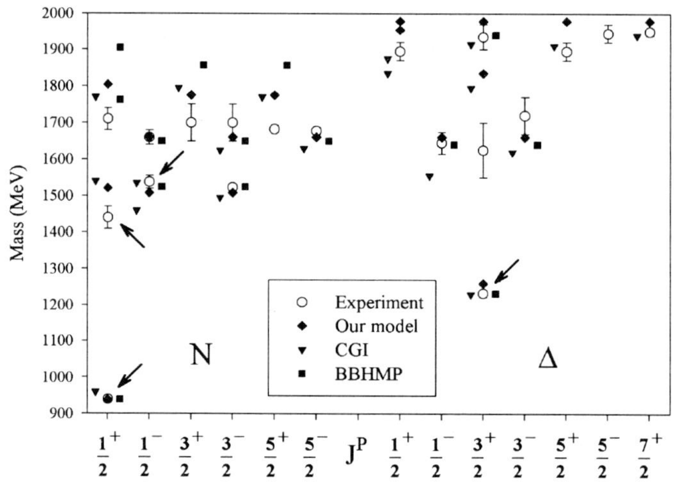

Figure 2. Masses of the N and Δ baryons (status ⋆⋆⋆⋆ and ⋆⋆⋆) as a function of total angular momentum and parity J^P. The arrows indicate baryons of these families used in the fit of parameters.

In this work, we showed that the addition of a new interaction with one free parameter to the instanton induced interaction in the baryon sector solely allows a good consistent description of both meson and baryon spectra.

References

1. S. Godfrey and N. Isgur: Phys. Rev. **D32**, 189 (1985); S. Capstick and N. Isgur: Phys. Rev. **D34**, 2809 (1986)

2. W.H. Blask, U. Bohn, M.G. Huber, B.C. Metsch, and H.R. Petry: Z. Phys. **A337**, 327 (1990)

3. C. Semay, F. Brau, and B. Silvestre-Brac: Phys. Rev. **C64**, 055202 (2001)

4. F. Brau and C. Semay: Phys. Rev. **D58**, 034015 (1998)

5. F. Brau, C. Semay, and B. Silvestre-Brac: to appear in Phys. Rev. **C**

Few-Body Systems Suppl. 14, 49–52 (2003)

Few-Body Systems

Effects of an Electromagnetic Quark Form Factor on Meson Properties

B. Silvestre-Brac*

Institut des Sciences Nucléaires, 53 Avenue des martyrs, 38026 Grenoble Cedex, France

Abstract. A form factor is introduced in the quark electromagnetic current. Two different expressions for the form factors, and two different types of quark potential are tested on meson form factors and leptonic decays. In any case the introduction of a quark form factor greatly improves the agreement with experimental data.

1 Introduction

In the non relativistic quark model (NRQM), the degrees of freedom (constituent quarks) are no longer the basic degrees of freedom of QCD; they are dressed by gluon clouds. This gluonic dressing modifies the quark potentials. In the same spirit the electromagnetic current should be modified and, in the expression $j^\mu(x) = e\bar{q}(x)\gamma^\mu q(x)$, the term γ^μ could be replaced by a quark form factor $\Gamma^\mu = \left[F_1(q^2)\,\gamma^\mu + \frac{\kappa}{2m}F_2(q^2)\,i\sigma^{\mu\nu}q_\nu\right]$ where q is the four momentum of the photon at the quark-photon vertex. By simplicity I assume $F_1(q^2) = F_2(q^2)$, and two forms are proposed : the monopole form $F_1(q^2) = 1/(1 - q^2/\Lambda^2)$ and the dipole form $F_1(q^2) = 1/(1 - q^2/\Lambda^2)^2$. Here I investigate the effect of such a form factor on two observables : the form factors of pseudoscalar mesons and the leptonic decays of vector mesons, as well as some of kinematical relativistic corrections.

2 Meson Potentials

The meson wave function $|\Psi\rangle$ follows from the usual dynamical equation $H\,|\Psi\rangle = E\,|\Psi\rangle$ where the hamiltonian H is composed of a kinetic energy operator T and a potential term V. One can use either a non relativistic kinetic energy $T = \mathbf{p}^2/2\mu$ or a relativistic expression $T = \sqrt{\mathbf{p}^2 + m_1^2} + \sqrt{\mathbf{p}^2 + m_2^2} - m_1 - m_2$.

* *E-mail address:* silvestre@isn.in2p3.fr

The potential that I consider is the convolution of bare potentials with a gaussian gluonic density. Precisely it is composed of 3 terms : $V = V_c + V_{hyp} + V_{inst}$. The central part V_c originates from the ordinary funnel potential, V_{hyp} is a hyperfine term chosen as a sum of 2 gaussians and V_{inst} is an instanton induced interaction. The parameters of these potentials depend on which type of kinetic energy is considered; for a non relativistic one they lead to the DNR potential, and for a relativistic one to the DSR potential. The values of these parameters can be found in [1]. They both give very good spectra.

3 Meson Form Factors

Due to the fact that the meson wave function is essentially non relativistic and time independent, the usual covariant definition of the form factor $F_M(\mathbf{Q}^2)$ cannot be invoked. I replace it by the 3-momentum equivalent : $\langle \Psi_{\mathbf{K}'} | \, \mathbf{j}(0) \, | \Psi_{\mathbf{K}} \rangle = e\, N\, (\mathbf{K}' + \mathbf{K})\, F_M(\mathbf{Q}^2)$ with $\mathbf{Q} = \mathbf{K}' - \mathbf{K}$. N is a normalization factor such that $F_M(0)$ is identified to the meson charge. In a complete treatment, one has just to suppose that $K \ll m_\alpha$ while the non relativistic approximation supposes in addition $E_p(\alpha) \approx m_\alpha$. It is convenient to introduce a renormalized mass $\widetilde{m}_\alpha$ and a renormalized charge $\widetilde{e}_\alpha$ by

$$\frac{1}{\widetilde{m}_\alpha} = \int d\mathbf{p}\, \frac{\Psi^*(\mathbf{p})\Psi(\mathbf{p})}{E_p(\alpha)} \, ; \, \widetilde{e}_\alpha = e_\alpha \frac{m_\alpha}{\widetilde{m}_\alpha} \, ; \, c_\alpha = \frac{m_\beta}{m_\alpha + m_\beta} \; ; \; c_\beta = -\frac{m_\alpha}{m_\alpha + m_\beta} \tag{1}$$

Then the meson form factor writes

$$F_M(\mathbf{Q}^2) = \frac{e_\alpha + e_\beta}{\widetilde{e}_\alpha + \widetilde{e}_\beta} \left[\widetilde{e}_\alpha F_1^\alpha(-\mathbf{Q}^2) J^\alpha(\mathbf{Q}^2) + \widetilde{e}_\beta F_1^\beta(-\mathbf{Q}^2) J^\beta(\mathbf{Q}^2) \right] \tag{2}$$

with a contribution from quark α and antiquark β and the dynamical integrals are defined by:

$$J^\rho(\mathbf{Q}^2) = \widetilde{m}_\rho \int d\mathbf{y}\, \frac{\Psi(\mathbf{y} - (c_\rho/2)\mathbf{Q})\, \Psi(\mathbf{y} + (c_\rho/2)\mathbf{Q})}{E_y(\rho)} \; ; \; \rho = \alpha, \beta \tag{3}$$

It is remarkable that only the F_1 form factors appear so that meson form factor is useful to determine the cut off Λ only ($\Lambda_u = \Lambda_d$ from the pion, Λ_s from the kaon). If, in these formulae, one replaces E by m, one recovers the usual NR approximation. Moreover, if we impose $F_1 = 1$, we get the situation concerning a pure Dirac particle. For a monopole form, $\Lambda_\alpha \approx 3m_\alpha$ in the light sector, $\Lambda_\alpha \approx 10m_\alpha$ in the heavy sector, while for a dipole form the cut off are slightly higher than those values.

In Fig. 1, I plot the pion form factor obtained with DNR wave function, a monopole form and for 4 different types of calculation explained in the legend. The relativistic corrections have a weak effect, although going in the right direction, while the presence of a quark form factor is a crucial ingredient to reproduce the data. The situation obtained with DSR is worse in bare NR, but even better for a complete calculation. A dipole form gives a fit slightly less good, due to the fact that the asymptotic behaviour is probably too rapid.

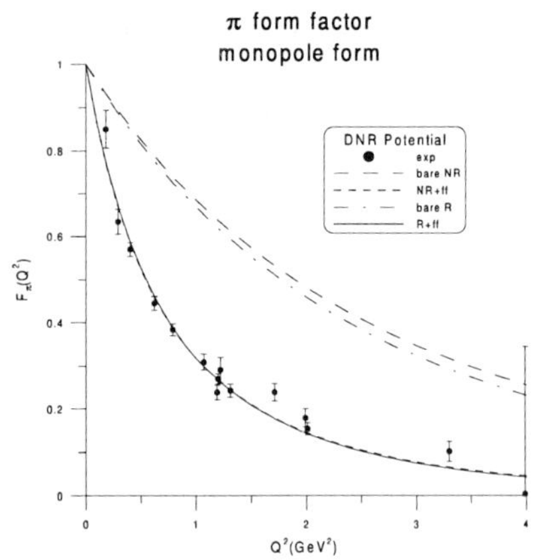

Figure 1. Pion form factor obtained with DNR wave function and a monopole form for the quark form factor. Long dashed line corresponds to bare NR, dash-dot line to bare R, short dashed line to NR approximation with presence of a monopole quark form factor, and solid line to R approximation with a quark form factor.

4 Leptonic Decays

The decay of a vector meson into a lepton-antilepton pair $V \to l\bar{l}$ is governed essentially by creation and annihilation of a virtual photon. The elementary amplitude is calculated with the usual Feynman rules. The only difference concerns the quark-photon vertex where the term γ^μ should be replaced by $\Gamma^\mu(q^2)$. Moreover the initial state is a meson bound state and the elementary amplitude has to be weighted with the probability that, in the meson, the quarks have a definite set of momenta. The calculation of the amplitude and the width is straightforward. The final expression for the leptonic width is given by (α is the fine structure constant):

$$\Gamma(V \to l\bar{l}) = \delta_{S,1}\delta_{J,1}16\pi\frac{\alpha^2}{M_V^2}\left(1+\frac{2m_l^2}{M_V^2}\right)\left(1-\frac{4m_l^2}{M_V^2}\right)^{1/2}\left(\sum_f c_f\, e_f\, \widetilde{\Psi}_V^f(0)\right)^2 \qquad (4)$$

where the summation on f is the summation of various flavours with their mixing coefficient c_f; m_l and M_V are respectively the lepton mass and the vector meson mass. $\widetilde{\Psi}_V(0)$ is a renormalized value of the wave function at the origin $\Psi_V(0)$ defined by :

$$\widetilde{\Psi}_V(0) = \delta_{L,0}\, A_\alpha\, \Psi_V(0) + \frac{\sqrt{2(2L+1)}}{6\pi}\,\langle L\,0\,1\,0|\,1\,0\rangle\; I_V^\alpha \qquad (5)$$

In addition to $\Psi_V(0)$, the meson wave function also participates to the dynamical integral :

$$I_V^\alpha = \int_0^\infty dp\, p^4\, T_\alpha(p)\,\varphi_V(p)\,;\; T_\alpha(p) = \frac{2B_\alpha}{E_p(\alpha)} - \frac{A_\alpha}{E_p(\alpha)[E_p(\alpha)+m_\alpha]} \qquad (6)$$

The A and B quantities are related to the quark form factors by :

$$A_\alpha = F_1^\alpha(M_V^2) + \kappa_\alpha F_2^\alpha(M_V^2)\,;\; B_\alpha = \frac{\kappa_\alpha}{2m_\alpha}F_2^\alpha(M_V^2) \qquad (7)$$

The Van-Royen (VR) approximation [2] consists in using $\Psi_V(0)$ instead of $\widetilde{\Psi}_V(0)$. Contrary to the VR approximation which allows only ${}^3S_1$ decays, ${}^3D_1$

decays are also allowed in the formalism. The anomalous magnetic moments κ can be deduced from the experimental data. The leptonic width is very sensitive to these parameters. In any case, I found them negative, around -0.5 in the light sector, around -0.2 for κ_c and around -0.1 for κ_b.

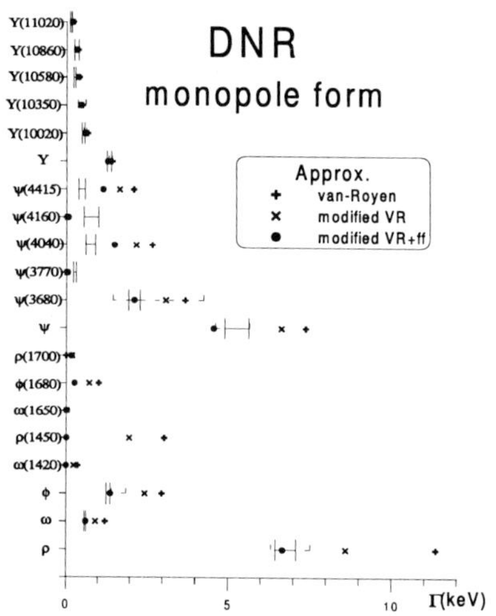

Figure 2. Leptonic decay width (in KeV) obtained with DNR wave function and a monopole form for quark form factor. The experimental values for e^+e^- decays are symbolized by solid error bars, and for $\mu^+\mu^-$ by dashed error bars. The theoretical values obtained with VR approximation are denoted with a $+$ sign, the m-VR with a $\times$ sign (relativistic corrections but no quark form factor), and the exact value with quark form factor with a circle.

The widths for all known transitions are shown in Fig. 2 for DNR and a monopole form. The VR approximation is too large, the effect of relativistic corrections is important although still too weak, and the introduction of a quark form factor does a pretty good job. The situation is very similar for DSR potential, or with a dipole quark form factor.

5 Conclusions

Starting with good nonrelativistic meson wave functions, we showed that the non relativistic approximation overestimates both the meson form factor and the leptonic decays. The relativistic corrections always go in the right direction, but with a too weak effect. The introduction of quark form factors improves both observables, without recourse to boost effects. These conclusions are the same whatever the dynamics in use, and a monopole form is favored, specially for describing meson form factors.

References

1. C. Semay, B. Silvestre-Brac: Nucl. Phys **A618**, 455 (1997)

2. R. van Royen, V.F. Weisskopf: Nuovo Cimento **A50**, 617 (1967)

Few-Body Systems Suppl. 14, 53–54 (2003)

Few-Body Systems

Baryon Electric Radii and Magnetic Moments from the GBE Constituent Quark Model

K. Berger, W. Plessas, R.F. Wagenbrunn

Institut für Theoretische Physik, Universität Graz, Universitätsplatz 5, A-8010 Graz, Austria

We have investigated electric radii and magnetic moments of all octet and decuplet baryon ground states within the constituent quark model (CQM) based on Goldstone-boson-exchange (GBE) dynamics [1]. For this purpose we have extended the previous study of nucleon electromagnetic form factors [2] also to baryons with strangeness.

The calculations have been performed in a relativistically covariant manner along point-form Hamiltonian dynamics [3]. This has allowed to include Lorentz-boost effects exactly in the evaluation of the current-operator matrix elements. With regard to the electromagnetic current the same model as in ref. [2] has been used, i.e. the point-form spectator approximation (PFSA).

The results for electric radii (Table 1) and magnetic moments (Table 2) obtained from the GBE CQM in PFSA fall rather close to the available experimental data. Typically the same characteristics are found as before in the cases of the nucleon electromagnetic form factors at low and moderate Q^2 [2]. The present findings further support the consistent picture that has already been seen with regard to all of the other observables of the electroweak nucleon structure [4]. It should be emphasized that only the quark model wave functions have been used as input into the calculations and no further parameterization has been introduced at this stage.

In order to demonstrate the influences of boost effects we have performed a calculation with the relativistic current operator but no Lorentz boosts applied (fourth column in the tables). The corresponding results differ grossly from the PFSA predictions in all instances, indicating the big importance of boost effects. For comparison we also give the results obtained in nonrelativistic impulse approximation (NRIA). While their failure is obvious with regard to the electric radii, it appears that the NRIA could be accepted for the magnetic moments. However, in this case the seemingly good results are only accidental as the relativistic effects from the current operator just cancel the ones from the boosts (cf. the figures in the last two columns in Table 2). Consequently a fully relativistic treatment, using both a relativistic current operator and including boost effects, is mandatory even though the electric radii and magnetic moments are observables in the limit $Q^2 \to 0$. In summary one learns that a realistic CQM as well as a fully relativistic formulation is required to achieve a reasonable description.

Table 1. Electric radii (in fm^2) of the proton, neutron, and Σ^- baryon as predicted by the GBE CQM.

Baryon	Experiment [5]	PFSA	RC w/o Boosts	NRIA
p	0.757 ± 0.014	0.8176	0.1829	0.1016
n	-0.1161 ± 0.0022	-0.1332	-0.0129	-0.0086
Σ^-	0.61 ± 0.15	0.4946	0.1558	0.0902

Table 2. Magnetic moments (in nuclear magnetons) of the octet baryons as well as Δ^+, Δ^{++}, and Ω^- as predicted by the GBE CQM.

Baryon	Experiment [5], [6]	PFSA	RC w/o Boosts	NRIA
p	2.793	2.6980	1.3056	2.7391
n	-1.913	-1.7004	-0.8521	-1.8185
Λ	-0.613	-0.5894	-0.3722	-0.6103
Σ^0		0.7001	0.3935	0.8059
Σ^+	2.458	2.3372	1.2457	2.6261
Σ^-	-1.160	-0.9371	-0.4588	-1.0143
Ξ^0	-1.250	-1.2744	-0.7871	-1.4041
Ξ^-	-0.6507	-0.6663	-0.3838	-0.5331
Δ^+	2.7 ± 2.8	2.0827	1.5139	2.7618
Δ^{++}	$3.7 - 7.5$	4.1654	3.0279	5.5235
Ω^-	-2.0200	-1.5907	-1.2924	-1.8780

Acknowledgement. This work was supported by the Austrian Science Fund, project no. P14806-TPH.

References

1. L.Ya. Glozman et al.: Phys. Rev. **D58**, 094030 (1998)
2. R.F. Wagenbrunn et al.: Phys. Lett. **B511**, 33 (2001)
3. W.H. Klink: Phys. Rev. **C58**, 3587 (1998)
4. S. Boffi et al.: Eur. Phys. J. **A14**, 17 (2002)
5. K. Hagiwara et al.: Phys. Rev. **D66**, 010001 (2002)
6. M. Kotulla et al.: Acta Phys. Polon. **B33**, 957 (2002)

Few-Body Systems Suppl. 14, 55–56 (2003)

Few-Body Systems

Goldstone-Boson-Exchange Dynamics for Constituent Quarks in Light and Strange Baryons

K. Glantschnig, R. Kainhofer, W. Plessas, B. Sengl, R.F. Wagenbrunn

Institute for Theoretical Physics, University of Graz, Universitätsplatz 5, A-8010 Graz, Austria

Recently the Goldstone-boson-exchange constituent quark model (GBE CQM) [1] has been quite successful in describing the spectra of all light and strange baryons in a unified framework [1, 2]. Furthermore it has been found to yield a remarkably consistent picture of the electroweak nucleon structure [3] (see also the contribution by Wagenbrunn et al. in these proceedings). The version of the GBE CQM of ref. [1] relies only on the spin-spin component of the pseudoscalar GBE for the hyperfine interaction of linearly confined constituent quarks.

Here we report from a further study of the GBE dynamics as effective hyperfine interaction between constituent quarks. In particular, we have extended the concept of GBE to include all force components as well as vector and scalar exchanges. The latter are to be viewed as effective description of multiple Goldstone-boson exchanges [4]. The corresponding version of the GBE CQM can be considered as more complete with regard to the quark dynamics since it contains not only spin-spin but also central, tensor, and spin-orbit forces.

In the extended version of the GBE CQM we use a linear confinement and the central, spin-spin, and tensor forces as produced by the pseudoscalar, vector, and scalar exchanges. The quark-meson coupling constants are deduced from meson-nucleon phenomenology. All the quark and meson masses are fixed to the standard values. The spin-orbit forces are treated with an independent (fitted) strength. In addition, the model involves five open cut-off parameters for the quark-meson vertices.

In the figure we show the light and strange baryon spectra as predicted by this type of the extended GBE CQM. It is seen that the specific features of the GBE CQM are, of course, preserved with regard to the level orderings of the positive- and negative-parity excitations; this is the consequence of the spin-flavor dependence induced by the GBE. Furthermore the near degeneracy of the levels in the higher multiplets is brought about even though tensor and spin-orbit forces are present. In particular, the accurately measured $\frac{5}{2}^+$-$\frac{5}{2}^-$

nucleon excitations are now well reproduced. For achieving such a result it turns out that the tensor-force effects from pseudoscalar and vector exchanges should practically cancel each other, while the (overall) spin-orbit forces must be kept at a moderate strength. Major shortcomings still persist only with the regard to the first excited states in the Λ spectrum. This, however, is known as a notorious difficulty of constituent quark models relying on three-quark configurations only.

Acknowledgement. This work has been supported by the Austrian Science Fund, project no. P14806-TPH.

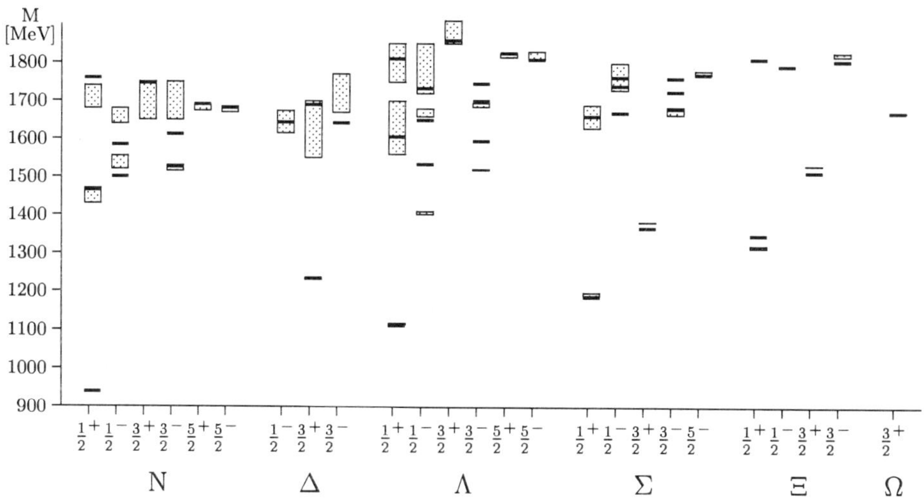

Figure 1. Low-lying light and strange baryon states with total angular momentum and parity J^P as predicted by the extended GBE CQM (solid horizontal bars). The shadowed boxes represent the experimental data with their uncertainties [5].

References

1. L.Ya. Glozman, W. Plessas, K. Varga, and R.F. Wagenbrunn: Phys. Rev. **D58**, 094030 (1998)

2. L.Ya. Glozman, Z. Papp, W. Plessas, K. Varga, and R.F. Wagenbrunn: Phys. Rev. **C57**, 3406 (1998)

3. S. Boffi, L.Ya. Glozman, W. Klink, W. Plessas, M. Radici, and R.F. Wagenbrunn: Eur. Phys. J. **A14**, 17 (2002)

4. G. Brown and D.O. Riska: Nucl. Phys. **A653**, 251 (1999)

5. K. Hagiwara et al.: Phys. Rev. **D66**, 010001 (2002)

Few-Body Systems Suppl. 14, 57–58 (2003)

Few-Body Systems

Baryon Magnetic Moments in a QCD String Model

J. Weda[1], Yu.A. Simonov[2], J.A. Tjon[1,3]

[1] Institute for Theoretical Physics, Leuvenlaan 4, 3584 CE Utrecht, The Netherlands

[2] Institute of Theoretical and Experimental Physics, 117118, B. Cheremushkinskaya 25, Moscow, Russia

[3] Kernfysisch Versneller Instituut, Zernikelaan 25, 9747 AA Groningen, The Netherlands

Abstract. In the field correlator method an effective interaction for the light quarks is obtained from QCD. Using this formalism the magnetic moments of the nucleons are calculated and reasonable agreement with experiment is obtained. Results of higher order corrections due to mesonic excitations are also shown. The method can easily be extended to the whole baryon multiplet.

A general effective quark Lagrangian can be obtained from the standard QCD Lagrangian by using the field correlator method (FCM) and integrating out the gluonic degrees of freedom. Keeping only the Gaussian (bilocal) correlator and exploiting the large N_c limit a nonlinear equation showing both confinement and chiral symmetry breaking (CSB) can be derived for a light quark in the field of a heavy quark [1, 2]. The method is quite general and can also be used to study the $q\bar{q}$ and $3q$ systems [3, 4]. Considering the baryon Green's function, neglecting gluon and meson exchanges, in lowest order of the approximation scheme the baryon wavefunction is given by a product of 3 independent single quark orbitals. As a result the Hamiltonian can be written as a sum of three quark terms, where the single quark solutions satisfy the Dyson-Schwinger equation with a nonlocal kernel.

The nonlinear equation for the single quark orbital (attached to the string in a gauge–invariant way) has been solved in the Gaussian correlator approximation. The resulting 3-quark wave function has been used to determine the magnetic moments of the baryons with the string tension σ the only parameter and without introduction of constituent quark masses [4].

Because of the coupling of the quark to the effective mesonic degrees of freedom, one- and two-body current contributions to the magnetic moments of the baryons exist from virtual excitations of the mesons. We have estimated the pionic and kaonic one-loop and one pion exchange contributions to the magnetic moment. The single quark corrections from pionic and kaonic loops are found to be of the order of 10%, whereas the total effect of 2-body current contributions are predicted to be small. The results for the nucleons are shown in Table 1. We find, that the predicted magnetic moment of the nucleon is improved substantially once we choose a string tension to give a reasonable nucleon mass, $M_B = \sum_{i=1}^{3} \epsilon_{n_i}^{(i)}$. The calculation can easily be extended to include the entire baryon multiplet, results are shown in ref. [4]. The mesonic contributions improve the predictions somewhat. Comparing the predictions we find that the magnetic moments are in close overall agreement with the experiment for a string tension of $\sigma = 0.10$ GeV^2.

Table 1. For different values of the string tension σ the single quark current contribution $\mu_N^{(1)}$ to the magnetic moment in units of nuclear magneton, together with the correction $\delta\mu_N$ arising from the pion and kaon one-loop and pion exchange diagrams are shown. Our total combined prediction and the experimental results are listed in the last two columns. The ground state energy ϵ_0 of the orbitals is also shown.

σ	$\epsilon_0(u,d)$	N	$\mu_N^{(1)}$	$\delta\mu_N$	μ_N^{tot}	exp
0.09 GeV^2	297 MeV	p	2.81	0.15	2.95	2.79
		n	-1.87	-0.16	-2.02	-1.91
0.12 GeV^2	342 MeV	p	2.44	0.14	2.59	2.79
		n	-1.63	-0.14	-1.78	-1.91
0.15 GeV^2	380 MeV	p	2.20	0.12	2.34	2.79
		n	-1.46	-0.13	-1.61	-1.91

We have assumed here that the baryon wave function can be described as a product of single quark orbitals, i.e. neglecting correlation effects. Our results in this approximation for the magnetic moments of baryons are encouraging, but are in need of including higher order corrections such as the one gluon exchange. This induces correlations in the three-quark wave function and its magnitude may give us insight whether our basic description in this contribution in terms of simply single quark orbitals is a reasonable one.

References

1. Yu.A. Simonov: Phys. At. Nucl. **60**, 2069 (1997); Few Body Syst. **25**, 45 (1998)
2. Yu.A. Simonov and J.A. Tjon: Phys. Rev. **D62**, 014501 (2000); Phys. Rev. **D62**, 094511 (2000)
3. Yu.A. Simonov: Phys. At. Nucl. **62**, 1932 (1999)
4. Yu.A. Simonov, J.A. Tjon and J. Weda: Phys. Rev. **D65**, 094013 (2002)

Few-Body Systems Suppl. 14, 59–64 (2003)

Few-Body Systems

Nucleon-Nucleon Interaction Models and Non-Locality

B. Desplanques[1,*], A. Amghar[2]

[1] Institut des Sciences Nucléaires (UMR CNRS/IN2P3–UJF), F-38026 Grenoble Cedex, France

[2] Faculté des Sciences, Université de Boumerdes, 35000 Boumerdes, Algeria

Abstract. The effect of non-locality in the NN interaction models is examined. It is shown that this feature can explain differences in predictions made from models evidencing a difference with this respect. This is done for both static and dynamical observables, taking into account that a non-local term can be transformed away by performing a unitary transformation. Some results for the deuteron form factors, the $A(Q^2)$ structure function and the $T_{20}(Q^2)$ tensor polarization, are given as an example. A few cases where discrepancies cannot be explained are also considered. They point to differences in the models as for the deuteron asymptotic normalizations, A_S and A_D, which are not affected by the present analysis.

1 Introduction

Apart from a recent work by Doleschall and Borbély [1], non-locality in the NN interaction has not been the object of dedicated studies for a long time. It is not absent however in models where it appears most often as a by-product of some prejudice in their construction. In the Paris model [2], for instance, it was realized that an energy dependence could help in fitting NN scattering data. The transformation of this energy dependence into a p^2/M dependence provides a non-local component. Later on, the Bonn group produced a model, field-theory motivated, taking into account the coupling of the NN channel to $NN\pi$, NΔ, $\Delta\Delta, \cdots$ channels [3]. It contains both a spatial and a time-non-locality. Moreover, the improvement consisting in introducing the Dirac spinors to describe 1/2-spin particles also provides non-locality. One could add other examples that have not been concretized in a high accuracy model. Taking into account the substructure of nucleons and mesons in terms of quarks most often leads to a non-locality [4], which is better expressed in configuration space than in momentum one, contrarily to the above sources of non-locality.

* *E-mail address:* desplanq@isn.in2p3.fr

A double question may be raised about this non-locality. Does it help in explaining NN scattering data and how this could be evidenced? On the other hand, taking into account that a non-locality can be transformed away by a unitary transformation (wave by wave), one can wonder whether the different models on the market are independent of each other?

Concerning the first question, some answer is obtained by examining models such as the versions Nij1 (non-local) and Nij2 (local) of the Nijmegen group [5]. They equally fit the scattering phase shifts (χ^2 per datum $= 1.03$), but in the first case, this is achieved with 41 parameters while 47 are required in the other one. The slightly smaller number in the former case perhaps provides indication that the introduction of some non-locality is beneficial.

The second question is the main object of the present paper. The plan will be as follows. In the second section, we present the different non-local terms which we are interested in. How they can be removed by a unitary transformation at the first order is given. The third section is devoted to a few selected results concerning the deuteron: static properties, form factors, structure function ($A(Q^2)$) and the tensor polarization ($T_{20}(Q^2)$). It involves a comparison of these quantities obtained with different models when the effect of the non-locality is taken into account. Section four contains the conclusion and a discussion. Due to a lack of space, we concentrate here on the essential points. Details and extended results could be found in refs. [6, 7, 8].

2 Transforming Away Non-Local Terms

The interaction of interest here may be written:

$$V = V_S + V_T + \{\frac{\mathbf{p}^2}{M}, \tilde{W}_S\} + \{\frac{\mathbf{p}^2}{M}, \tilde{W}_T\} + [\frac{\mathbf{p}^2}{M}, i\,U]. \tag{1}$$

where the non-local terms take the form of an anticommutator or a commutator. Another term of same order could be considered but those retained here are the only ones appearing in the pion-exchange contribution when this one is expanded up to order $1/M^4$. Due to its long range, it a priori provides larger contributions. Moreover, they are theoretically well identified while shorter range contributions are likely to have some effective character. The last term in the above equation has been studied at length in refs. [6, 7, 9]. Though it has a different origin, it has a strong similarity with a term arising from the difference of pseudo-scalar and pseudo-vector πNN couplings, considered in an earlier work by Friar [10]. As for the anticommutator terms, only rough estimates were made in the past. They are considered more completely here.

In principle, if two models are unitary equivalent, the corresponding Hamiltonians, H and H', should fulfill the following relation:

$$H = \frac{p^2}{M} + V + V_{NL} = e^{-S}\; H'\; e^S = e^{-S}\; (\frac{p^2}{M} + V')\; e^S. \tag{2}$$

At the first order in the interaction, the quantity, S, appearing in the unitary transformation, $\exp(S)$, can be determined by requiring:

$$V_{NL} + [S, \frac{\mathbf{p^2}}{M}] = \Delta V^0 \,, \tag{3}$$

where ΔV^0 has to be local. This equation is fulfilled as follows:

$$\begin{aligned} S &= iU + \frac{i}{2}\Bigg(\boldsymbol{p}{\cdot}\boldsymbol{r} \ \Big(V_0(r) + S_{12}(\hat{r})\, V_1(r)\Big) \\ &+ \Big(\boldsymbol{\sigma_1}{\cdot}\boldsymbol{p}\, \boldsymbol{\sigma_2}{\cdot}\boldsymbol{r} + \boldsymbol{\sigma_2}{\cdot}\boldsymbol{p}\, \boldsymbol{\sigma_1}{\cdot}\boldsymbol{r} - \frac{2}{3}\, \boldsymbol{\sigma_1}{\cdot}\boldsymbol{\sigma_2}\, \boldsymbol{p}{\cdot}\boldsymbol{r} \Big)\, V_2(r) + h.c. \Bigg), \end{aligned} \tag{4}$$

with

$$\begin{aligned} V_0(r) &= -\frac{1}{r}\int_r^\infty W_S(r')dr', \\ V_2(r) &= -\frac{1}{r}\int_r^\infty dr' \int_{r'}^\infty \frac{W_T(r'')}{r''}dr'', \\ V_1(r) &= -V_2(r) - \int_r^\infty \frac{W_T(r')}{r'}dr'. \end{aligned} \tag{5}$$

It is noticed that the difference of V and V' in Eq. (2) involves two-body but also many-body terms. Similarly, if a one-body current is introduced in one representation, the other one contains two, ... body currents to ensure the unitary equivalence. The role of the two-body part is examined in the following section. Applications involve the interaction models: Nij2 [5], Argonne V18 [11], Reid93 [5] which are local ones, Nij1 [5], Nij93 [5], Paris [2] which have a linearly p^2/M dependence, and the Bonn-QB, Bonn-CD ones [3].

3 Results for Static and Dynamical Quantities

Concerning static properties, a quantity of interest is the deuteron D-state probability, often referred to characterize different models. The difference between the Paris and Bonn-QB model is 0.78%. As the two models have quite close values for the mixing parameters, ϵ_1, their comparison is largely free of bias with this respect. Taking into account the effect of the non-local term, V_{NL}, explains 0.74% (0.60% and 0.14% for the commutator and anticommutator parts respectively). Notice that the change in the deuteron D-state probability just reflects the fact that this quantity is not an observable one. Similar results were reached by von Geramb et al. [12], using the inverse scattering problem methods. Another interesting quantity is the ratio $Q_D/(A_S\ A_D)$, which, contrary to the above one, is observable. The difference is 0.23 fm^3 while V_{NL} explains 0.20 fm^3. It is also instructive to look at quantities that only depend on the scalar anticommutator part of V_{NL}. In this order, we compare the squared charge radius for the models Nij1 and Nij2. The models differ by an amount of 0.004 fm^2 while the effect of the non-locality is estimated to be around 0.001 fm^2. The apparent failure to explain the discrepancy in this case actually points to the difference of the models for the asymptotic normalization A_S, which as is well known, governs the size of the charge radius. This factor is not affected by the present analysis of non-local effects.

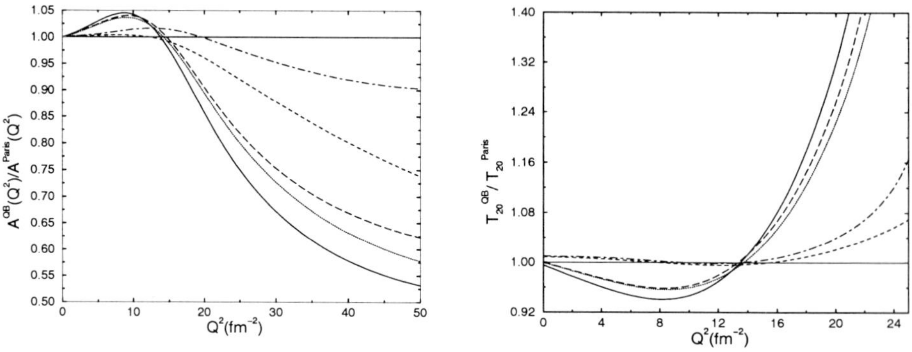

Figure 1. Ratio of predictions for the Bonn-QB and Paris models: effect of non-local terms in the interaction (see text for explanations)

Ratios of form factors and related quantities calculated in different representations of the interaction should go to 1 when the non-locality effect is accounted for and provided that the unitary equivalence is ensured. It is not so in practice because the unitary transformation is treated at first order. In each case, there are thus two sets of results, depending on whether corrections are added to predictions made with one model or removed from the other ones.

Ratios of predictions made from the Bonn-QB and Paris models for the $A(Q^2)$ and $T_{20}(Q^2)$ observables are shown in Fig. 1. As it can be seen, the ratio of bare predictions (continuous line) tends to 1 when the effect of the anticommutator (dashed and dotted lines) and commutator terms (small-dash and dashed-dotted lines) is considered. A large part of the effect is due to the tensor part of V_{NL}. Notice that a slight departure from 1 appears for $T_{20}(Q^2)$ around $Q^2 = 0$. The effect of the scalar part of the anticommutator, which is seen in Fig. 2 (left part), shows features similar to the previous ones. It involves the S-wave function close to the origin, which is generally suppressed in local models compared to non-local ones. The right part of the same figure emphasizes a case where an agreement between two models turns into a disagreement. In fact, this one is consistent with what is expected from the comparison of the asymptotic normalization, unaffected by the present analysis. In Fig. 3, all predictions corrected for the effect of a non-local term are compared to the Paris ones. At low Q^2, it is seen that some discrepancy is still present while the initial motivation of the work was rather to explain it by non-locality effects. Actually, for both $A(Q^2)$ and $T_{20}(Q^2)$, the discrepancy reflects a sensitivity to the A_D/A_S ratio in the first case and A_S in the second one. Around $Q^2 = 20$ fm^{-2}, the ratio becomes very close to 1, while in absence of corrections departures up to 10% and 20% respectively could be observed. The decrease of the uncertainty has motivated Schiavilla and Sick in using the quadrupole form factor to derive the neutron charge form factor [13].

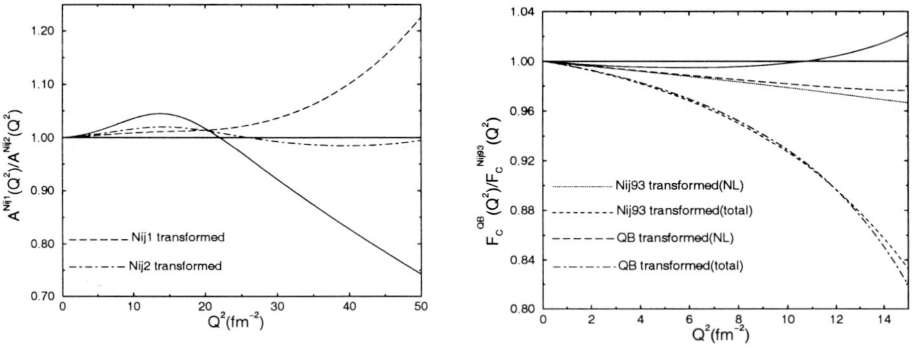

Figure 2. Examples showing a decrease of discrepancies between models due a scalar non-local effect (left part) and the appearance, on the contrary, of a discrepancy (right part); see text for comments

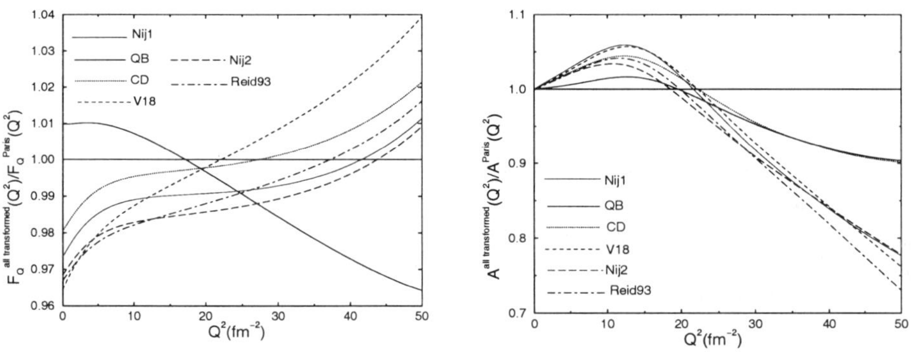

Figure 3. Comparison of predictions of different models after incorporating non-locality corrections (see comments in the text)

4 Conclusion

Effects of non-local terms in NN interaction models have been considered. Roughly, they explain a large part of the differences that the comparison of various model predictions for electromagnetic observables evidences. The part with a tensor character has been found to be the dominant one. It is also the best determined. Similar conclusions hold in some cases for the scalar part but the effect is often masked by other effects related to the fact that models correspond to different values of the asymptotic normalization, A_S, which is an observable and remains unchanged in the analysis performed here.

While the original goal of present studies was rather to relate discrepancies between models to some non-locality, it appears that this is not always possible, especially at low Q^2. Interestingly however, accounting for this effect tends to restore some hierarchy of the results, as expected from simple models. Thus, in the above range, the structure function, $A(Q^2)$, and the quadrupole form factor, $F_Q(Q^2)$, evidence a direct sensitivity respectively to the asymptotic normalizations, A_S and A_D, which are unaffected by the present analysis of non-local effects.

The result of the analysis presented here was not a priori guaranteed. The fact that it points to a unique family of phase-equivalent models indicates that the sensitivity of the models to different parameterizations of the radial part for instance or to a different fit to experimental data is rather small. Thus, the availability of various models is not without interest. The remaining sensitivity, as for A_S or equivalently the scattering length, a_t, strongly calls for a more accurate determination of these quantities.

Throughout the present study, we compare together predictions of models for electromagnetic observables. Ultimately, a comparison to experiment should be done. In this respect, a model to be preferred, most probably non-local, is that one based on degrees of freedom of which effective character, unavoidable in any case, is as low as possible.

References

1. P. Doleschall and I. Borbély: Phys. Rev. **C62**, 054004 (2000)
2. M. Lacombe et al.: Phys. Rev. **C21**, 861 (1980)
3. R. Machleidt, K. Holinde and Ch. Elster: Phys. Rep. **149**, 1 (1987); R. Machleidt: Adv. in Nucl. Phys. **19**, 189 (1989); R. Machleidt: Phys. Rev. **C63**, 024001 (2001)
4. V. Kukulin (plenary talk, this conference)
5. V.G.J. Stoks et al.: Phys. Rev. **C49**, 2950 (1994)
6. B. Desplanques and A. Amghar: Z. Phys. **344**, 19 (1992)
7. A. Amghar and B. Desplanques: Nucl. Phys. **A585**, 657 (1995)
8. A. Amghar and B. Desplanques: nucl-th/0209058
9. J. Forest: Phys. Rev. **C61**, 034007 (2000)
10. J.L. Friar: Phys. Rev. **C22**, 796 (1980)
11. R.B. Wiringa, V.G.J. Stoks and R. Schiavilla: Phys. Rev. **C51**, 38 (1995)
12. H.V. von Geramb, Lectures notes in physics: Vol. **427**, (1994); B.F. Gibson, H. Kohlhoff and H.V. von Geramb: Phys. Rev. **C51**, R465 (1995)
13. R. Schiavilla and I. Sick: Phys. Rev. **C64**, 041002(R) (2001)

Few-Body Systems Suppl. 14, 65–70 (2003)

Few-Body Systems

Probing Few-Body Systems with Bremsstrahlung

N. Kalantar-Nayestanaki*

Kernfysisch Versneller Instituut (KVI), Zernikelaan 25, 9747 AA Groningen, The Netherlands

Abstract. A series of bremsstrahlung measurements have been performed with the superconducting cyclotron, AGOR, at KVI. These measurements, on the proton-proton and proton-deuteron systems, aim to investigate the nucleon-nucleon interaction far away from elastic channel. Cross sections and analyzing powers have been measured for all possible exit channels of both systems with an incident proton energy of 190 MeV, which is below the particle-production threshold. Some results of the real-photon production are presented for both systems and compared to theoretical predictions.

1 Introduction

Nucleon-nucleon bremsstrahlung is the most fundamental reaction used in studying the off-shell behavior of the nucleon-nucleon interaction as it involves the strong interaction between two nucleons, and the well-known electromagnetic interaction. In 1958 Low proved the soft-photon theorem, which states, that in a series expansion in the photon momentum, the first two terms of the scattering amplitude can be expressed exactly in terms of the elastic amplitude for the same system [1]. Current potential models describe the elastic data very accurately and therefore also soft-photon production. To investigate the behavior of the potentials beyond elastic scattering, one has to go beyond the validity regime of the Low expansion and thus investigate high-energy bremsstrahlung.

Recently, the study of nucleon-nucleon bremsstrahlung below the particle-production threshold has been receiving increasing attention both theoretically and experimentally. The reason for this is the success of modern potential-model calculations on the one hand and the advent of high-precision detection systems capable of processing high count rates on the other. Polarized-proton beams [2] and high-flux neutron beams have also delivered a boost to the experimental efforts. With even higher accuracies achievable nowadays, one can hope to study, in detail, the properties of the off-shell nucleon as it propagates in

* *E-mail address:* nasser@kvi.nl

an intermediate state as well as the non-nucleonic degrees of freedom (such as the Δ-excitation) which can even admix at energies below the pion-production threshold. In this contribution, after a brief review of nucleon-nucleon bremsstrahlung in section 2, the analysis method along with the first results will be discussed in section 3.

2 Nucleon-Nucleon Bremsstrahlung and the KVI Experiment

The dominant ingredients in nucleon-nucleon bremsstrahlung are radiation from external and internal nucleon lines. The higher-order effects which have been investigated in the past few years include meson-exchange currents (MEC), the delta contributions and relativistic contributions. The Δ-isobar contribution in the intermediate state and the MEC effects have been studied in detail by several groups [3, 4, 5, 6]. Even though all these contributions are relatively small, precise measurements and selected kinematics should distinguish between models with or without them. As a result of the absence of the electric dipole radiation and electric MECs in the proton-proton system, one is more sensitive to higher-order effects in this system.

The experimental situation on the proton-neutron bremsstrahlung is quite different than that of the proton-proton system. Generally speaking, the neutron beams lack the high quality and intensity required for precise measurements. Because of this, the cross sections presented in the literature have been inclusive. In order to investigate the proton-neutron bremsstrahlung, the proton-deuteron system was chosen in the past for which no good calculations exist. Any calculation attempting to describe the inclusive results which have been published in the literature [7] should treat all the possible outgoing channels of this reaction properly. These are: radiative capture, three- and four-body final states.

The KVI bremsstrahlung program with AGOR on few-body systems, set out to study and measure the cross sections and analyzing powers of all the above-mentioned reactions in a series of measurements specifically tuned for nucleon-nucleon bremsstrahlung. The measurements started with the proton-proton system and extended to the proton-deuteron system in which all the different final states were studied. In addition to coplanar geometry, non-coplanar geometries have been measured for a large number of angle combinations. Angular coverage of protons was from 6° to 26° and photons were measured from 60° to 170°. All the measurements were exclusive in the sense that all the outgoing particles were measured. Besides, all the measurements were carried out using polarized protons of 190 MeV. For a complete description of the experimental setup and some results, refer to [8, 9, 10, 11, 12, 13, 14, 15]. In this paper, some experimental results on the proton-proton system will be presented along with some recent theoretical calculations. In addition, first results on proton-deuteron system will be presented.

3 Results and Discussion

For the analysis of the experiments, the fact that most measured channels were kinematically overdetermined, was used for background rejection. For instance, in the proton-proton case, five measured angles, namely the polar and azimuthal angles, θ and ϕ, of the two outgoing protons and the polar angle of the photon, θ_γ, and the incident proton energy were used to reconstruct the other parameters. By comparing these parameters to their measured counterparts most of the background left was rejected. In this section, some results are presented for the $p{+}p$, $p{+}d$ and $p{+}n$ systems.

As was mentioned earlier, our detection system is capable of measuring a very large number of angle combinations of the outgoing particles. Here, we only show a few percent of the available real-bremsstrahlung phase space. In Fig. 1, the typical results of the coplanar cross sections and analyzing powers are presented as a function of the polar angle of one of the protons, while the polar angle of the other proton and the polar angle of the photon are fixed [13]. Systematic errors on the cross sections are around 5% for the present measurements. For comparison, a data point from an earlier TRIUMF experiment at 200 MeV is also shown in the figure [16].

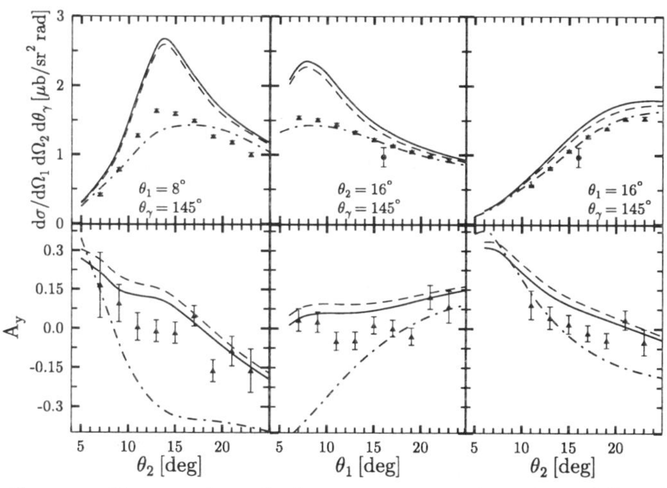

Figure 1. Cross sections and analyzing powers as a function of first (middle) and second (left and right) proton polar angles for $\theta_\gamma = 145°$. The other proton angle is fixed at the angles shown in the figure. The calculations are SPA (dash-dotted), the "nucleonic calculation" (dashed) and the full calculation (solid). The circle data point is from TRIUMF for E_p=200 MeV [16].

Two different types of calculations were performed for our kinematics. The first one uses the Soft-Photon Approximation (SPA) of Liou, Timmermans and Gibson [17]. This calculation is consistent with the low-energy theorem and is relativistic and gauge invariant. The second set of calculations is done by Martinus, Scholten and Tjon [6]. These relativistic, microscopic calculations are covariant, thereby including the negative-energy states. To see the higher order effects, a calculation called the "nucleonic calculation" is shown in which only the internal and external bremsstrahlung (rescattering) contributions are considered, see the dashed curve, and one which includes MEC, $N\bar{N}$ and Δ contributions as well (solid curve). As can be seen in the figure, there are sizable differences between the data and the calculations for the cross sections and analyzing powers. Note that Coulomb corrections are expected to be negligible for most of the kinematics considered here. The largest contribution is expected to be where the microscopic calculations have a peak (see, for example the top-left panel of Fig. 1). This is where the relative energy of the outgoing protons is smallest. Here, the discrepancy is expected to be reduced somewhat [15, 18]. Aside from our measurements, results from other laboratories have emerged on this system [19, 20]. The results for the virtual-photon production which, for the first time, was measured in parallel have already been published [21].

Next, the preliminary results for one of the proton-deuteron channels, namely the "coherent" channel consisting of a proton, a deuteron and a photon in the outgoing channel, is presented. Fig. 2 shows the cross sections and analyzing powers as a function of photon angle for a number of proton-deuteron angle combinations, indicated in the figure. The two data sets, taken at different times, are obtained with two different geometries of the photon detection system, TAPS, and are in good agreement with each other. The models shown are all inspired by low-energy theorems. The differences arise from the various methods of extrapolating to the on-shell points when calculating the matrix elements [22]. Even though one of the curves comes close to the data, the large variations in the calculations point to the fact that the dynamics of this simple system is rather complex and microscopic Faddeev calculations are called for.

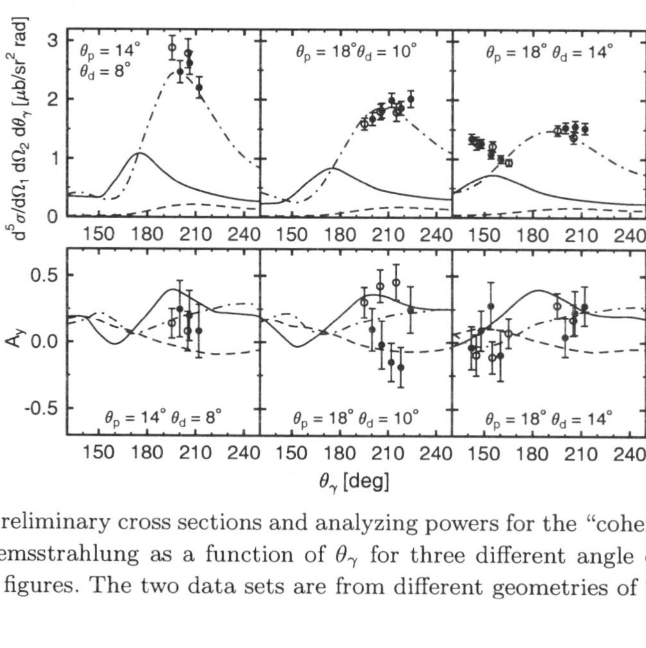

Figure 2. Preliminary cross sections and analyzing powers for the "coherent" proton-deuteron bremsstrahlung as a function of θ_γ for three different angle combinations given in the figures. The two data sets are from different geometries of TAPS.

Finally, the preliminary results for the *quasi-free* proton-neutron bremsstrahlung are shown in Fig. 3. In order to obtain these results, the quasi-free proton-proton bremsstrahlung data were compared to their free counter-part presented earlier. In this way, kinematics were chosen for which the bremsstrahlung took place from "quasi-free" particles. The curves are results of a fully microscopic calculation [23] shown by solid lines and calculations inspired by low-energy theorem [24] shown by dashed and dash-dotted lines. As expected, the cross sections are generally an order of magnitude larger than those of proton-proton bremsstrahlung shown in Fig. 1. Also the differences between the microscopic and the soft-photon calculations are much larger than those in the case of proton-proton bremsstrahlung. This might be due to the way the meson-exchanges, which are much larger for the $pn\gamma$ than $pp\gamma$, are taken into account.

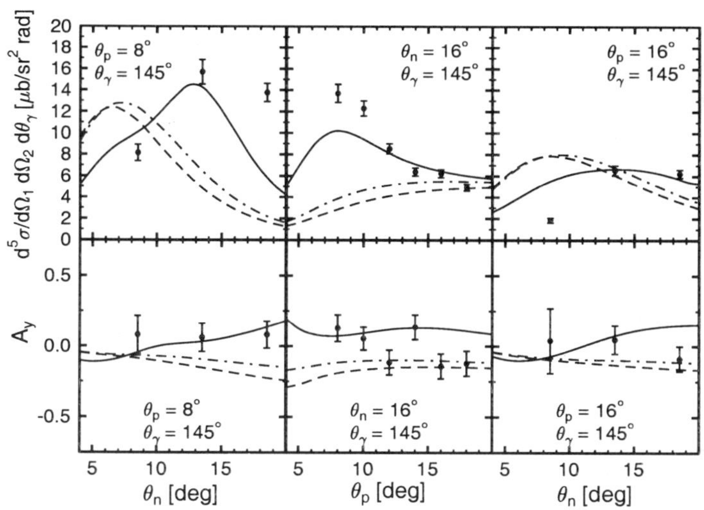

Figure 3. Preliminary cross sections and analyzing powers for the proton-neutron bremsstrahlung as a function of θ_γ for three different angle combinations given in the figures. See the text for the explanation of the curves.

There is still a large part of the phase space for each reaction which has been measured and for which the results are emerging and will be published in the near future. Data with high statistical accuracies from non-coplanar geometries have been looked at producing an extra handle on the systems under study and new observables, never measured before [14].

Acknowledgement. I would like to acknowledge the assistance of all those who helped build and commission SALAD, made TAPS operational at KVI, took part in all the preparations and data taking and were involved in the analysis. Specifically, I should thank my graduate students H. Huisman and M. Volkerts who are responsible for the analysis of the data presented here.

References

1. F.E. Low: Phys. Rev. **110**, 974 (1958) and E.M. Nyman: Phys. Lett. **B25**, 135 (1967); Phys. Rev. **170**, 1628 (1968)
2. K. Michaelian et al.: Phys. Rev. **D41**, 2689 (1990)
3. J.A. Eden and M.F. Gari: Phys. Lett. **B347**, 187 (1995); Phys. Rev. **C53**, 1102 (1996)
4. F. de Jong et al.: Phys. Lett. **B333**, 1 (1994); F. de Jong et al.: Phys. Rev. **C51**, 2334 (1995)
5. M. Jetter and H.W. Fearing: Phys. Rev. **C51**, 1666 (1995)
6. G. Martinus et al.: Phys. Lett. **B402**, 7 (1997); Phys. Rev. **C56**, 2945 (1997); Phys. Rev. **C58**, 686 (1998)
7. J.A. Pinston et al.: Phys. Lett. **B249**, 402 (1990) and J. Clayton et al.: Phys. Rev. **C45**, 1810 (1992)
8. N. Kalantar-Nayestanaki et al.: Nucl. Inst. and Meth. **A444**, 591 (2000)
9. M. Volkerts et al.: Nucl. Inst. and Meth. **A428**, 432 (1999)
10. S. Schadmand et al.: Nucl. Inst. and Meth. **A423**, 174 (1999)
11. N. Kalantar-Nayestanaki et al.: Nucl. Inst. and Meth. **A417**, 215 (1998)
12. O. Schwalb et al.: Nucl. Inst. and Meth. **A295**, 191 (1990); H. Ströher: Nucl. Phys. News Int. Vol. **6**, No. 1, 7 (1996)
13. H. Huisman et al.: Phys. Rev. Lett. **83**, 4017 (1999)
14. H. Huisman et al.: Phys. Lett. **B476**, 9 (2000)
15. H. Huisman et al.: Phys. Rev. **C65**, R031001 (2002)
16. J.G. Rogers et al.: Phys. Rev. **C22**, 2512 (1980)
17. M.K. Liou et al.: Phys. Lett. **B345**, 372 (1995); Phys. Rev. **C54**, 1574 (1996); R. Timmermans: Private Communication
18. D. Cozma et al.: Phys. Rev. **C65**, 024001 (2002)
19. R. Bilger et al.: Phys. Lett. **B429**, 195 (1998)
20. K. Yasuda et al.: Phys. Rev. Lett. **82**, 4775 (1999)
21. J.G. Messchendorp et al.: Phys. Rev. Lett. **82**, 2649 (1999); Phys. Rev. Lett. **83**, 2530 (1999); Phys. Rev. **C61**, 064007 (2000)
22. O. Scholten, A.Yu. Korchin: Phys. Rev. **C65**, 054004 (2002)
23. K. Nakayama et al.: Phys. Rev. **C43**, 394 (1991); Private Communication
24. O. Scholten: Private Communication

Few-Body Systems Suppl. 14, 71–76 (2003)

Dressed Dibaryon Production as a New Mechanism for Basic Nuclear Force and Meson-Exchange Currents

V.I. Kukulin*,†

Institute of Nuclear Physics, Moscow State University, 119992, Leninskie gory, Moscow, Russia

Abstract. A novel concept for the nuclear force is introduced. The approach assumes that a specific dressed six-quark state (dressed dibaryon) is generated in NN collision at intermediate and short ranges. The new dressing mechanism providing π-, σ-, ρ- and ω-meson clouds around the six quark core is discussed. These novel components produce both strong intermediate-range attraction and short-range repulsion in the NN sector due to the s-channel exchange mechanism which is supplemented with conventional π- and 2π- Yukawa exchanges at intermediate and long ranges. The model developed is demonstrated to lead to numerous new effects and contributions in many fields of nuclear physics.

Nowadays the traditional model for NN-interaction and basic nuclear force, which has been based on Yukawa's idea on meson-exchange in t-channel, works very well at large distances $r_{NN} \geq \lambda_\pi \simeq 1.4$ fm but there are some serious problems at intermediate ($r_{NN} \sim 1$ fm) and especially at short ranges ($r_{NN} \sim 0.4 \div 0.8$ fm). From the numerous difficulties peculiar to the traditional OBE-models we can list here only the most evident (more complete discussion of the difficulties and contradictions in currently used NN-potential models – see the refs. [1-3]).

(i) The values of cut-off parameters $\Lambda_{\pi NN}$, $\Lambda_{\rho NN}$, $\Lambda_{\pi N\Delta}$ etc. (in all OBE-models) are taken to be around $1.3 \div 2.0$ GeV, i.e. they are very high and in strong disagreement with all microscopic theoretical estimates and experimental fits, which gives the 'soft' values $\Lambda_{\pi NN}|_{\rm exp} \sim 0.5 \div 0.8$ GeV. Moreover, the $\Lambda_{\pi NN}$ and $\Lambda_{\rho NN}$ values, which fit the inelastic NN-data on π-meson production, like $pp \to pp\pi^0$ or $pn\pi^+$, are in good agreement just with the soft values of $\Lambda_{\pi NN} \simeq 0.5 \div 0.6$ GeV! However these values of Λ cannot fit the elastic NN scattering even at low energies and deuteron observables.

*This work was partially supported by the RFBR Grant. No. 02-02-16612 and RFBR-DFG Grant No. 01-02-04015

†*E-mail address:* kukulin@nucl-th.sinp.msu.ru

(ii) Quite unclear situation is with the quantitative description of intermediate-range attraction via σ-meson exchange in t-channel (*what is σ-meson in NN-force*). The traditional view to the σ-meson exchange in t-channel as an effective model representing the exchange by π-π s-wave correlated pair has recently been strongly criticized [4,5]. These studies established (independently from each other) that the above π-π exchange is able to give only *a short-range repulsion or a quite weak attraction* which is far insufficient to bind the deuteron. Complete account of $\pi-\pi$ continuum and $\pi\pi \rightarrow \sigma \rightarrow \pi\pi$ mode in NN-interaction, which is in agreement with recent experiments for $\pi - \pi$ scattering leads to *very weak intermediate range attraction* in the channel, or even a *repulsive contribution* instead of strong attraction needed to fit NN-phase shifts and deuteron data.

(iii) There is a long-standing problem with short-range tensor force in $^3\mathrm{S}_1 - {}^3\mathrm{D}_1$ channel. The current NN-models of second generation lead to too low values of deuteron quadrupole moment and ϵ_1 tensor mixing parameter at $E_{\mathrm{lab}} \geq 400$ MeV: $Q_d^{\mathrm{theor}} \simeq 0.270\,\mathrm{fm}^2$ vs. $Q_d^{\mathrm{exp}} \simeq 0.286\,\mathrm{fm}^2$ (the error $\Delta Q/Q >$ 5% which is too high to be explained by isoscalar MEC contribution).

(iv) The ωNN-coupling constant $(g^2_{\omega NN}/4\pi)|_{\mathrm{NN\,fit\,with\,OBE}} \simeq 13.6 \div 15$ found from the fits to NN phase shifts at $E_{\mathrm{lab}} \leq 300$ MeV with OBE models occurred in few times as large as the respective SU(6)-value: $(g^2_{\omega NN}/4\pi)|_{\mathrm{SU}_6} \simeq 5$ while all the other coupling constants were in a good agreement with the SU(6)-values.

However the main problem in OBE-description for the NN interaction at intermediate and short ranges is even more fundamental. In fact, the quark 'radius' of nucleon $r_N \simeq 0.5 \div 0.6$ fm, so that NN-distances less 0.8 fm are well inside the nucleon overlap range. On the other hand, the ranges of interaction which correspond to all heavy-meson exchanges (viz. ρ, ω etc.) with mass greater 700 MeV (and at $\Lambda_{\pi NN}$ and $\Lambda_{\rho NN} \simeq 1.5 \div 2.0$ GeV) are ca. 0.3 fm or even less and thus these interactions occur well inside the nucleon overlap field. Thus the heavy meson exchange occurs in the field of *all six quarks* (see the Fig. 1a,b). So that, one needs first *to justify* an employment of *free* meson-nucleon vertices when the two nucleons are in the overlap region (look e.g. to the Fig. 1a). Thus, it appears therefore, that one can reach a more consistent description if to assume that the σ- ρ- or ω-mesons are moving *simultaneously* in the field of two nucleon cores, or simply around six-quark bag, like shown in Fig. 1b. By this way we can come to the idea of dressed six-quark bag. However the problem can by no means be solved by simple reformulation of the three-quark cluster states to common six-quark degrees of freedom and should be posed on a *different physical ground.* We must incorporate into our consideration a few new concepts and features peculiar to high-density quark matter, e.g. restoration of (broken) chiral symmetry in multi-quark bag, non-linear mode of quark interaction with σ and ω-fields etc.

We start with a simple observation that there are two possible symmetries for 6q-wave functions in NN system viz.[6] and [42] for even and [51] and [33] for the odd partial waves respectively (assuming that the dominating space symmetry for 3q-configuration in the nucleon is [3]). These components

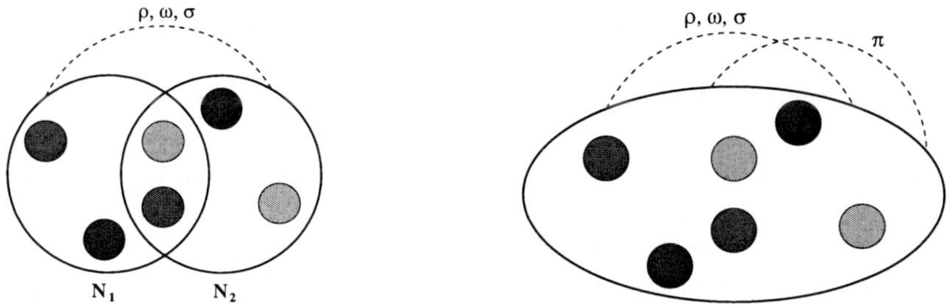

a b

Figure 1. The pass from the heavy-meson exchange (at intermediate and short ranges) (a) to a dressed six-quark bag picture (b).

can be identified with the following 6q shell-model configurations:$|s^6, L = 0\rangle$, $|s^4p^2, L = 0, 2\rangle$ and $|s^5p, L = 1\rangle$, $|s^3p^3, L = 1, 3\rangle$ for even and odd states respectively. In the states where there are the (excited) p-shell quarks they can deexcite (by transition to s-shell quarks — see the Fig. 2) by emission of one-or a few correlated pions all of which are in the s-wave respective to the six-quark core and thus these two pions can form easily the scalar σ-meson moving around the quark core. The two components with different symmetries (i.e. fully symmetric and of mixed symmetry) in the $^1S_0$ and $^3S_1 - {}^3D_1$ coupled channels can be mixed by quark-meson interactions: (i) by conventional t-channel π-exchange between quarks (the moderate contribution); (ii) by 2π-generation in s-channel. The latter process leads to a specific dressing mechanism which is illustrating in Figs. 2 and 3. There the σ-field is strongly enhanced near the

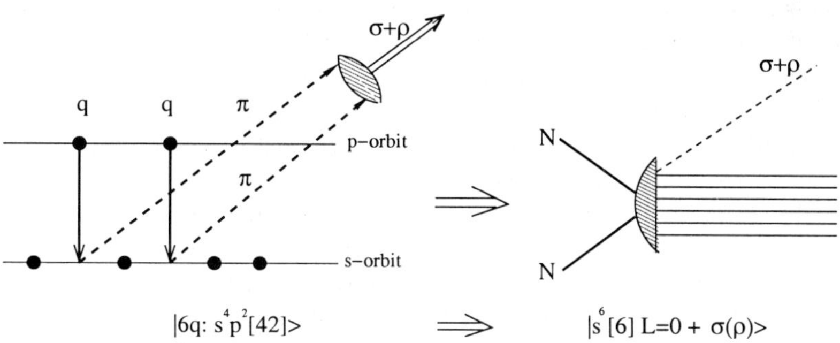

Figure 2. The 6q dressing mechanism.

dense six-quark core due to (partial) chiral symmetry restoration and highly non-linear effects (induced by instantons and diquark correlations). In fact, the fully symmetric six-quark bag has a highest possible density. This strong σ-field leads to shrinking the multi-quark bag due to strong 'pressure' of scalar field. The σ-field has mainly spherical symmetry due to the dominating $L_\sigma = 0$ component and high space symmetry ($s^6[6]L_q = 0$) of the bag, and thus the field pulls quarks towards the center of the bag. In other words, this collec-

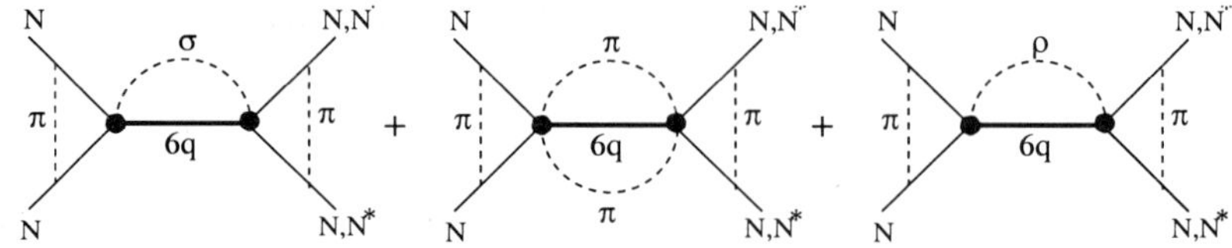

Figure 3. Series of diagrams illustrating the *s*-channel mechanism for NN interaction at intermediate and short ranges.

tive σ-field results in effective strong *attraction among all the six quarks* in the dressed bag state (DBS). As a net result of this inter-quark effective attraction there arises a strong attraction between two nucleons in NN-channel [1,3]. The above mechanism of the σ-field enhancement (and scalar s-channel $\pi\pi$ correlation) $2q\,(p\text{-wave}) \longrightarrow 2q\,(s\text{-wave}) + \sigma$ has been predicted [1] also for the Roper resonance: $|sp^2\rangle_{N^*} \overset{2\pi}{\Rightarrow} |s^3 + \sigma\rangle|_{L_\sigma=0} \to N^*(1440)$. Here non-shifted mass of the Roper resonance should be $M_0(R) \sim 2\hbar\omega \simeq 2$ GeV vs. $M_{exp}(R) = 1440$ MeV, i.e. one wins in this case $\Delta E_{\sigma-3q} \simeq 500$ MeV. Recently this our prediction for the Roper resonance structure has been fully confirmed in two independent studies[6,7]. In six-quark system with the strong σ-field the energy gain should be even higher, e.g. $\Delta E_{\sigma-6q} \simeq 600 \div 700$ MeV! As a result, the mass of the dressed six-quark bag occurs only ca. $1.8 \div 1.9$ GeV, i.e. near the 2N-threshold. So that, after the incorporation of the additional (to the DBS contribution) t-channel π and 2π-exchanges (they induce an additional moderate attraction in 2N-system) one gets eventually the bound deuteron state in $^3\mathrm{S}_1 - {}^3\mathrm{D}_1$ channel and $^1\mathrm{S}_0$-virtual state in $^1\mathrm{S}_0$-channel.

The chiral restoration predicted by the present mechanism is in a nice agreement with recent findings of Hatsuda and Kunihiro [8], and Cohen and Glozman [9] who suggested a chiral restoration at high excitation in baryons (as can be demonstrated e.g. by appearance of close parity doublets high in the spectra). In fact, $|s^4p^2[42]\rangle$ is $2\hbar\omega$ excited configuration in 6q-channel! Another important effect of the strong sigma field in the DBS is an appearance of new short-range NN tensor force. The σ-meson, being massive (m_σ in our model is ca. 400 MeV) may be rotating around the 6q-core and thus the dressed bag may have a non-zero angular momentum $L_\sigma > 0$ [1,3]. (The effect of the dressed-bag rotation might be even more important for the ρ-meson cloud). This rotation leads to a few important contributions: — to the deuteron quadrupole moment; — to the tensor mixing parameters at higher energies $E \geq 500$ MeV; — to the NN higher partial phase shifts, e.g. F-, G-, H- at higher energies; — to the tensor analyzing power of recoil deuterons in the reaction $d(e, e'd)$ etc. The attractive effect of the strong σ-field in the bag is balanced at very short ranges by ω-field of the bag. The collective vector ω-field, due to its higher mass, is localized *inside* the dressed bag (see Fig. 4) and thus provides repulsive contributions in 6q-bag. This inner repulsion will prevent the collapse of the dressed bag in the strong non-linear sigma field. Because this contribution will increase the total energy of the bag the net effect of this collective ω field in the NN sector is NN repulsion at short ranges. From this point of view, the inner part of the deuteron state consists mainly from *the above dressed dibaryon!* Our microscopic calculations confirm this very well. It turned out quite possible to describe almost perfectly with this model the NN lowest partial phase shifts in the energy range from zero until as high as 1 GeV, and also the deuteron structure [1]. The weight of the DBS data in deuteron occurred around 3.5%! Thus the physical picture of NN interaction at long, intermediate and short distances looks as shown in Fig. 4: This picture of baryon-baryon interaction is

$r_{NN} \sim \lambda_\pi \sim 1.4$ fm OBE interaction
($\pi + 2\pi$ exchange)

$r_{NN} \lesssim 1$ fm; DBS:
$|s^6[6] + (\pi\pi)_{corr}\rangle$

$r_{NN} \sim r_c(0.5)$ fm; DBS:
$|6q + \omega + \sigma\rangle$

$r_{NN} \sim \lambda_\pi$, OBE interaction

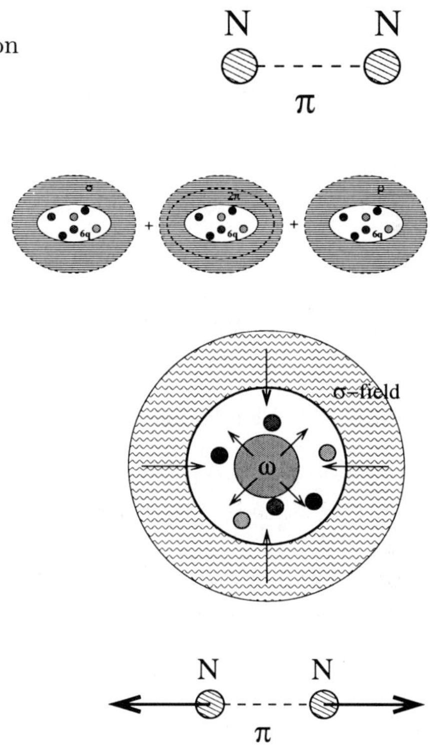

Figure 4. The physical pattern for NN interaction at long, intermediate and short ranges.

in a general agreement with the picture for e^+e^- interaction in Weinberg-Salam electroweak theory. At small momentum transfers in e^+e^- interaction the t-channel mechanism will be dominating, but at large momentum transfers the t-channel exchange will be of minor contribution while the s-channel exchanges should give the leading part of the interaction (e.g. in $e^+e^- \to \mu^+\mu^-$ etc. — see the Fig. 5). So that, if to complement the traditional t-channel NN interaction with s-channel DBS mechanism, there appears a rather close analogy between e^+e^- (or $\mu^+\mu^-$) electroweak interactions and NN (and $N\pi$, NK^+ etc.) strong interactions (compare Figs. 3 and 5)! If however to restrict ourselves by the traditional OBE t-channel exchanges only, any such analogy disappears!

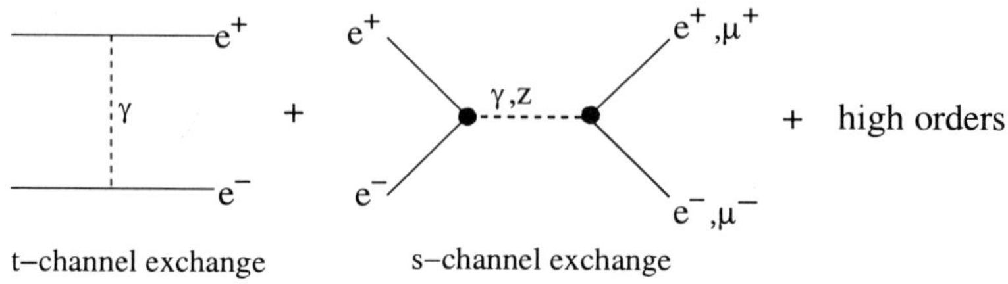

Figure 5. The basic diagrams for e^+e^- interaction in the electroweak theory.

I have described in the talk the basic ingredients of the new model of nuclear force and NN-interaction developed in recent years in a series of works of our group jointly with the Tuebingen university group. Due to limitation of space, I omitted in the above text all the mathematical and computational details of the model and focused only to physical aspects of the new concept referring reader to the original papers [1,3,6] for the details.

References

1. V.I. Kukulin, I.T. Obukhovsky, V.N. Pomerantsev and A. Faessler: Phys. At. Nucl. **64**, 1667 (2001); J. Phys. **G27**, 1851 (2001)
2. D. Plämper, J. Flender and M.F. Gari: Phys. Rev. **C49**, 2370 (1994)
3. V.I. Kukulin: Phys. At. Nucl. (to be published); V.I. Kukulin: Proceedings of the 5th Winter School on Theoretical Physics PIYaF, p.142: Gatchina, St. Petersburg 1999
4. M. Kaiser, R. Brockman and W. Weise: Nucl. Phys. **A625**, 758 (1997); M. Kaiser, S. Gerstendörfer and W. Weise: Nucl. Phys. **A637**, 395 (1998)
5. E. Oset, H. Toki, M. Mizobe and T.T. Takahashi: nucl-th/0011008
6. M. Kaskulov, V.I. Kukulin and P. Grabmayr: Phys. Rev. **C** (to be published)
7. O. Krell, C. Hanhart, S. Krevald and J. Speth: nucl-th/9911080
8. T. Hatsuda and T. Kunihiro: Phys. Rep. **247**, 221 (1994); T. Hatsuda, T. Kunihiro and H. Schimizu: Phys. Rev. Lett. **82**, 2840 (1999)
9. T.D. Cohen and L.Ya. Glozman: Int. J. Mod. Phys. **A17**, 1327 (2002); Phys. Rev. **D65**, 116006 (2002)

Few-Body Systems Suppl. 14, 77–82 (2003)

Few-Body Systems

The Two-Nucleon System beyond the Pion Threshold

M. Schwamb*, H. Arenhövel

Institut für Kernphysik, Johannes Gutenberg-Universität Mainz, D-55099 Mainz, Germany

Abstract. A model is developed for the hadronic and electromagnetic interaction in the two-nucleon system above pion threshold in the framework of meson, nucleon and Δ degrees of freedom. It is based on time-ordered perturbation theory and includes full meson retardation in potentials and exchange currents as well as loop contributions to the nucleonic one-body current. Results for some selected processes are discussed.

1 Introduction

At present, a very interesting topic in the field of medium energy physics is devoted to the role of effective degrees of freedom (d.o.f.) in hadronic systems in terms of nucleon, meson and isobar d.o.f. and their connection to the underlying quark-gluon dynamics of QCD. For the study of this basic question, the two-nucleon system provides an important test laboratory, because it is the simplest system for the study of the nucleon-nucleon interaction, the role of medium effects due to two-body operators and the relevance of off-shell effects, i.e. the change of single particle properties in the nuclear medium. Moreover, the deuteron is of specific relevance as an effective neutron target. Within the present effective description, one is able to obtain a reasonable, though not perfect description of the possible hadronic and electromagnetic reactions in the two-nucleon sector below π threshold. However, for energies beyond π threshold up to the Δ region additional complications occur. Present state-of-the-art models should incorporate not only a realistic NN interaction, but also a dynamical treatment of the Δ isobar. Moreover, in view of the requirement of unitarity, the various possible reactions should be described within one consistent framework, because they are linked by the optical theorem. In the past, we have started to realize this ambitious project within a *retarded* coupled channel $NN - N\Delta$ approach, based on a three-body scattering approach with nucleon, Δ and meson degrees of freedom [1, 2, 3, 4, 5].

* *E-mail address:* schwamb@kph.uni-mainz.de

2 The Model

For energies up to the Δ region, our model Hilbert space $\mathcal{H}^{[2]}$ consists of three orthogonal subspaces $\mathcal{H}^{[2]} = \mathcal{H}^{[2]}_{\bar{N}} \oplus \mathcal{H}^{[2]}_{\Delta} \oplus \mathcal{H}^{[2]}_{X}$, where $\mathcal{H}^{[2]}_{\bar{N}}$ contains two (bare) nucleons, $\mathcal{H}^{[2]}_{\Delta}$ one nucleon and one Δ resonance, and $\mathcal{H}^{[2]}_{X}$ two nucleons and one meson $X \in \{\pi, \rho, \sigma, \delta, \omega, \eta\}$.

Concerning the hadronic part, the basic interactions in our model are XNN and $\pi N\Delta$ vertices. Inserting these into the Lippmann-Schwinger equation, one obtains after some straightforward algebra [3] effective hadronic interactions acting in $\mathcal{H}^{[2]}_{\bar{N}} \oplus \mathcal{H}^{[2]}_{\Delta}$ which contain, apart from pion-nucleon loop contributions to the nucleonic self energy, *retarded* one-boson exchange (OBE) mechanisms describing the transitions $NN \to NN$, $NN \to N\Delta$ and $N\Delta \to N\Delta$. Due to the nonhermiticity and nonlocality of a retarded operator, which becomes moreover complex above pion threshold, in most applications the simple static approximation is used by neglecting the energy transfer by the exchanged meson. The corresponding static operators are much easier to handle, but one encounters on the other hand at least two serious problems. Above pion-threshold, unitarity is violated due to the absence of singularities in the static operators. Moreover, in the past it turned out that even the simplest photonuclear reaction, namely deuteron photodisintegration, cannot be described even qualitatively within a consistent static framework [6, 7].

In our explicit realization, we use for the retarded NN interaction the Elster potential [8] which takes into account in addition one-pion loop diagrams in order to fulfill unitarity above pion threshold. Therefore, one has to distinguish between bare and physical nucleons (see [3] for details). Concerning the transitions $NN \to N\Delta$ and $N\Delta \to N\Delta$, we take besides retarded pion exchange static ρ exchange into account. Moreover, the interaction of two nucleons in the deuteron channel in the presence of a spectator pion (the πd channel) is also considered. This mechanism is necessary for satisfying unitarity, because otherwise no asymptotic free πd state would exist. By a suitable box renormalization [9], we are able to obtain approximate phase equivalence between the Elster potential and our coupled channel approach below pion threshold.

Similarly, the basic electromagnetic interactions consist of baryonic and mesonic one-body currents as well as vertex and Kroll-Ruderman contributions [4, 5]. These currents are, together with the πNN vertex, the basic building blocks of the corresponding effective current operators. The latter contain beside the ordinary spin-, convection- and spin-orbit currents full retarded pionic meson exchange currents and electromagnetic loop contributions, where the latter can be interpreted as off-shell contributions to the baryonic one-body current [5]. Moreover, static ρ MEC as well as Δ MEC contributions are taken into account. It can be shown [5] that concerning the pionic part gauge invariance is fulfilled in leading order of $1/M_N$. This is a consequence of the fact that the hadronic pion-nucleon loop contributions, the retarded π-exchange contribution to the NN interaction, the electromagnetic loop contributions, and the π MEC are based on the same πNN vertex.

3 Results

The hadronic Δ parameters and the M1 $\gamma N\Delta$ coupling are simultaneously fitted to the $M_{1+}(3/2)$ multipole of pion photoproduction on the nucleon, to the P_{33} channel of pion-nucleon scattering and to the ${}^1D_2$ channel of nucleon-nucleon scattering [3, 4]. In Fig. 1, our results for the ${}^1D_2$ phase shift and inelasticity are depicted. We obtain a good description at least up to about $T_{lab} = 800$ MeV. This is of particular relevance for deuteron photodisintegration where the most important contribution to the unpolarized cross section in the Δ region comes from the ${}^1D_2$ channel. Concerning the other partial NN waves, the overall description is only fairly well [3]. Therefore, we have started to construct from scratch an interaction model whose parameters are fitted to the phase shifts and inelasticities of *all* relevant NN partial waves for energies up to about 1 GeV. However, as of yet we have not found a parameter set which would lead to a considerable improvement. One reason for this failure may be the fact that our effective baryon-baryon model is basically a one-boson exchange model which does not take into account more complicated mechanisms like the crossed two-pion-exchange. Moreover, the πN interaction has to be improved. Forthcoming studies on that topic are essential, especially for clarifying the role of final state interactions (FSI) in hadronic and electromagnetic breakup reactions on the deuteron for energies beyond the π threshold.

As next, we discuss very briefly deuteron photodisintegration. The starting point of our consideration is the static approach of Wilhelm *et al.* [7] which is based on the Bonn-OBEPR potential [10]. Similar to our present approach, there is no free parameter in the calculation of the photodisintegration process in [7]. As is evident form Fig. 2, Wilhelm *et al.* clearly fail in describing the data. One obtains a considerable underestimation of the cross section in the Δ peak. Moreover, a dip structure around 90° occurs at higher energies which is not present in the data. This feature is also present in other static approaches like the unitary three-body model of Tanabe and Otha [6]. On the other hand, these

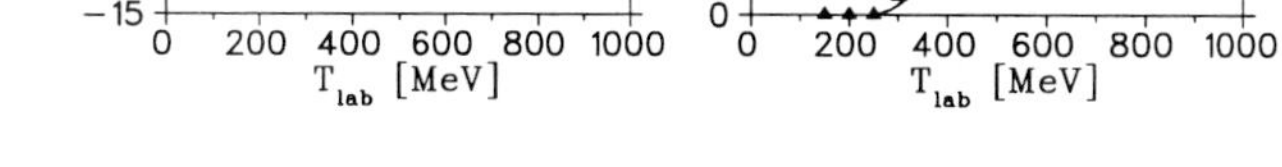

Figure 1. Phase shift δ and inelasticity ρ for the ${}^1D_2$ NN-channel in comparison with experiment (solution SM97 of Arndt et al. [11]) for two potential models: dash-dotted curve: static approach, based on the Bonn-OBEPR potential [10], full curve: retarded approach. See [3] for further details.

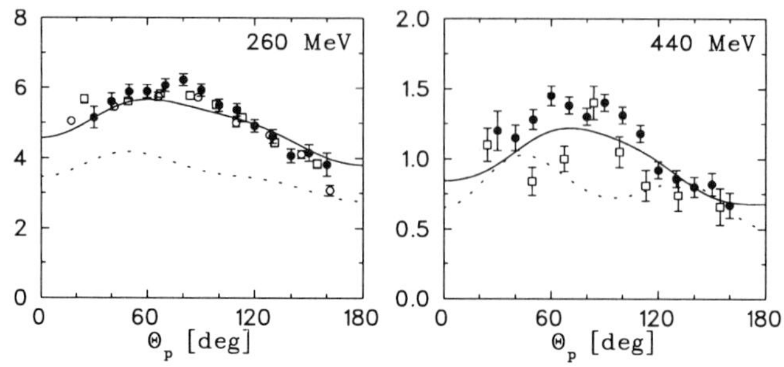

Figure 2. Differential cross section of deuteron photodisintegration for two photon energies k_{lab} as function of the c.m. proton angle θ_p: dotted curve: result of Wilhelm *et al.* in static approach [7], full curve: retarded approach of [4]. Off-shell contributions to the nucleonic one-body current are included, too [5]. Experimental data from [12] (•), [13] (open box) and [14] (○).

problems in the differential cross section vanish almost completely in a retarded approach [4, 5]. However, some discrepancies in polarization observables like the linear photon asymmetry Σ or the polarization $P_y(p)$ of the outgoing proton are still present [4, 5]. It is known that polarization observables may be sensitive to small reaction mechanisms which are suppressed in the unpolarized cross section. Therefore, we assume that the failure of our model in describing Σ and $P_y(p)$ may be attributed to the present treatment of the hadronic interaction which needs to be improved as has been outlined above.

A very interesting topic is the exploration of the spin asymmetry of the total cross section of the nucleon and light nuclei which determines the GDH-sum rule [15, 16]

$$I^{GDH} = \int dk \frac{\sigma^P(k) - \sigma^A(k)}{k} = 4\pi\kappa^2 \frac{e^2}{M^2} S \qquad (1)$$

and which is at present under investigation experimentally [17]. In (1), $\sigma^{P/A}(k)$ denotes the total photoabsorption cross section on a target with mass M, spin S and anomalous magnetic moment κ with the photon spin parallel (anti-parallel) to the target spin.

Due to the lack of a free neutron target, a measurement of the GDH-sum rule on the deuteron is of specific significance because it has been suggested in the past that one can measure the spin asymmetry of the total photoabsorption cross section of the neutron by using light nuclei like the deuteron. This would rest on the assumptions that (i) the polarized deuteron constitutes an effective polarized neutron target and (ii) one has to suppose that the spin asymmetry on the deuteron is dominated by the quasi free pion-production, so that binding and FSI effects from the presence of the spectator nucleon can be neglected, resulting in an incoherent sum of proton and neutron contributions.

However, both assumptions are questionable. First of all, due to the existence of the D state component in the deuteron wave function, even in a

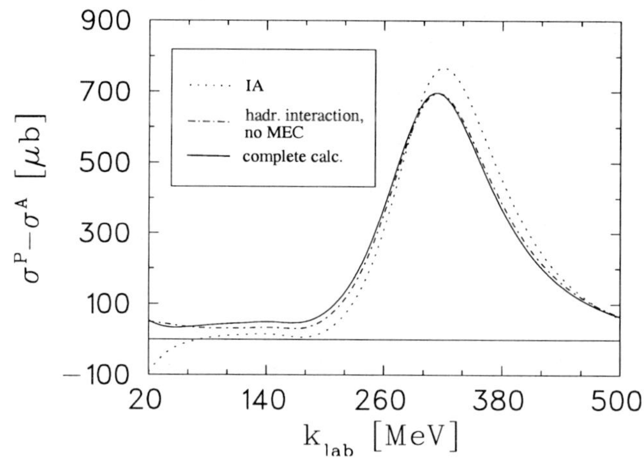

Figure 3. Preliminary result for the spin asymmetry on the deuteron within our retarded approach as a function of the photon energy k_{lab} in the laboratory frame. The dotted curve shows the result within a simple spectator approach. In the dashed-dotted curve, hadronic backscattering mechanisms are added. In the full curve, additional two-body meson exchange currents are taken into account. A part of the Born term contributions in pion production has been neglected.

completely polarized deuteron target, the neutron is not completely polarized. With respect to the second assumption, one has to recall that, in contrast to the nucleon case, for the GDH-integral on the *deuteron* also the energy region *below* π threshold has to be taken into account due to the existence of the breakup channel $\gamma d \to NN$ which is absent in the photoabsorption on a free nucleon. As has been pointed out in [18], there exists a strong anti-correlation between pion production and photodisintegration, because the latter reaction yields a very large *negative* contribution to the GDH-integral in (1) near the breakup threshold. This facts explains at least partially the large difference between the GDH-value on the deuteron ($I_d^{GDH} = 0.65$ μb) and the sum of the GDH-value on proton and neutron yielding 438 μb. Moreover, the effect of final state interactions in π production on the deuteron cannot at all be neglected [19] so that a simple spectator model picture is not appropriate for the evaluation of the spin asymmetry on the deuteron.

Due to these facts, it is obvious that an extraction of the neutron amplitude from the deuteron reaction cannot be performed in a model independent manner. Nevertheless, the experimental investigation of the GDH-sum rule on the deuteron is very important because the measurement of the spin asymmetry is for itself of great interest since it will give us much more detailed information on the underlying dynamics than just the unpolarized cross section. Therefore, it may serve in the future as a stringent test for existing models (see for example Fig. 3) in the two-nucleon sector.

4 Outlook

In the future, we plan to apply the present model to other reactions, especially electrodisintegration. Conceptually, we have to improve our hadronic interaction model. Moreover, additional degrees of freedom like the Roper, the D_{13} and the S_{11} resonances should be taken into account if one wants to consider higher energies.

References

1. M. Schwamb, H. Arenhövel, P. Wilhelm, Th. Wilbois: Phys. Lett. **B420**, 255 (1998)
2. M. Schwamb, H. Arenhövel, P. Wilhelm, Th. Wilbois: Nucl. Phys. **A631**, 583c (1998)
3. M. Schwamb, H. Arenhövel: Nucl. Phys. **A690**, 647 (2001)
4. M. Schwamb, H. Arenhövel: Nucl. Phys. **A690**, 682 (2001)
5. M. Schwamb, H. Arenhövel: Nucl. Phys. **A696**, 556 (2001)
6. H. Tanabe, K. Ohta: Phys. Rev. **C40**, 1905 (1989)
7. P. Wilhelm, H. Arenhövel: Phys. Lett. **B318**, 410 (1993)
8. Ch. Elster, W. Ferchländer, K. Holinde, D. Schütte, R. Machleidt: Phys. Rev. **C37**, 1647 (1988)
9. A. M. Green, M. E. Sainio: J. Phys. **G8**, 1337 (1982)
10. R. Machleidt, K. Holinde, Ch. Elster: Phys. Rep. **149**, 1 (1987)
11. R.A. Arndt et al.: program SAID
12. R. Crawford et al.: Nucl. Phys. **A603**, 303 (1996)
13. J. Arends et al.: Nucl. Phys. **A412**, 509 (1984)
14. G. Blanpied et al.: Phys. Rev. **C52**, R455 (1995); Phys. Rev. **C61**, 024604 (2000)
15. S. B. Gerasimov: Yad. Fiz. **2**, 598 (1965)
16. S.D. Drell, A.C. Hearn: Phys. Rev. Lett. **16**, 908 (1966)
17. J. Ahrens et al.: Phys. Rev. Lett. **87**, 022003 (2001)
18. H. Arenhövel, G. Kreß, R. Schmidt, P. Wilhelm: Phys. Lett. **B407**, 1 (1997)
19. E.M. Darwish, H. Arenhövel, M. Schwamb: contribution to these proceedings

Few-Body Systems Suppl. 14, 83–88 (2003)

Few-Body Systems

The Nucleon-Nucleon Problem in Quark Models

Fl. Stancu*

Institute of Physics B.5, University of Liège, Sart Tilman, B-4000 Liège 1, Belgium

Abstract. In the first part we summarize the status of the nucleon-nucleon (NN) problem in the context of Hamiltonian based constituent quark models and present results for the $\ell = 0$ phase shifts obtained from the Goldstone-boson exchange model by applying the resonating group method. The second part deals with the construction of local shallow and deep equivalent potentials based on a Superymmetric Quantum Mechanics approach.

1 Introduction

The strong interaction between two nucleons is the basic ingredient of nuclear physics. Since about 25 years there have been many efforts to derive the nuclear forces from the underlying theory of strong interactions, the Quantum Chromodynamics (QCD). As QCD cannot be solved exactly in the low energy regime, QCD-inspired models have been used both in baryon spectroscopy and the NN problem. We first briefly describe the situation in the framework of Hamiltonian based quark models. Next, we present recent progress obtained from using the Goldstone-boson exchange model.

After some pioneering work in the late '70 [1] a major breakthrough in the microscopical derivation of the NN interaction has been achieved at the beginning of the 80's with the work of Oka and Yazaki [2] who used the resonating group method to derive phase shifts from the one-gluon exchange model, the work of Harvey [3] on the symmetries of the most important six-quark states and the work of Golli, Rosina and collaborators [4] on local effective nucleon-nucleon interactions.

Presently there are several review papers available on the subject [5]. They show that the following steps are important: 1) the choice of the quark model, 2) the choice of the six-quark basis states, 3) the method to calculate the phase shifts. A challenge is to describe both the nucleon, as a three-quark system, and the NN interaction, as a six-quark problem, using the same quark model.

* *E-mail address:* fstancu@email.ulg.ac.be

2 Quark Models

The quark models used so far in the NN problem are mostly nonrelativistic. Studies based on the one-gluon exchange (OGE) model explained the short-range repulsion as due to the color-magnetic interaction combined with quark interchange between q^3 clusters. But in order to reproduce the scattering data some extra medium- and long-range repulsion was necessary. This was added at the baryon level. That is why more consistent models, called hybrid, have subsequently been introduced [6] where all interactions are introduced at the quark level. In such models the short-range repulsion is still attributed to the OGE interaction and the medium- and long-range attraction is due to scalar and pseudoscalar meson exchange. To our knowledge, these models have problems in fitting the nucleon resonances and the NN interaction with the same parameters.

Here we present results derived from the Goldstone-boson exchange (GBE) model which has been successful in describing the baryon properties [7]. The pseudoscalar meson exchange interaction between quarks, which is spin and flavour dependent, contains both a long and a short range part, appropriate for the NN problem. In this work we use the parameterization of ref. [8]. We employ the resonating group method to calculate the $\ell = 0$ phase shifts by using the technique proposed by Kamimura [9].

3 Phase Shifts in the GBE Model

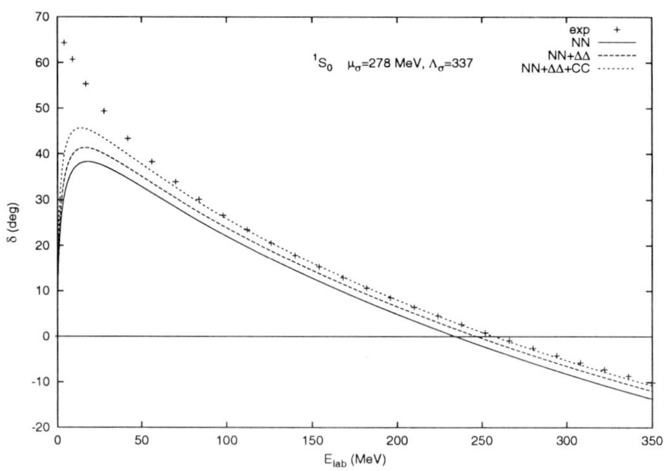

Figure 1. The $^1S_0$ NN scattering phase shift. The coupled channel calculations are from ref. [13]. The crosses are the Nijmegen fit [14] to the data.

The phase shifts obtained in ref. [10] from the GBE model also imply that the NN potential is strongly repulsive at short range. This means that the repulsion can equally be explained as due to a flavour-spin interaction combined with quark interchange. The repulsion obtained from the GBE model is somewhat stronger than that produced by the OGE model . However, to reproduce the experimental $^1S_0$ phase shift a certain amount of middle-range attraction was necessary. This has been provided by the addition of a scalar meson exchange interaction of the form [11]

$$V_\sigma = -\frac{g_{\sigma q}^2}{4\pi}\left(\frac{e^{-\mu_\sigma r}}{r} - \frac{e^{-\Lambda_\sigma r}}{r}\right) \tag{1}$$

where we chose $\frac{g_{\sigma q}^2}{4\pi} = \frac{g_{\pi q}^2}{4\pi} = 1.24$, $\mu_\sigma = 278$ MeV, $\Lambda_\sigma = 337$ MeV. Interestingly, the fit to the data favours $\mu_\sigma = 2m_\pi$, consistent with the findings of ref. [12]. Fig. 1 shows that the addition of (1) leads to a good agreement with the data over a large energy interval, the best result being obtained with three coupled channels NN+$\Delta\Delta$+CC. The quality of the baryon spectrum remains unchanged after the addition of (1).

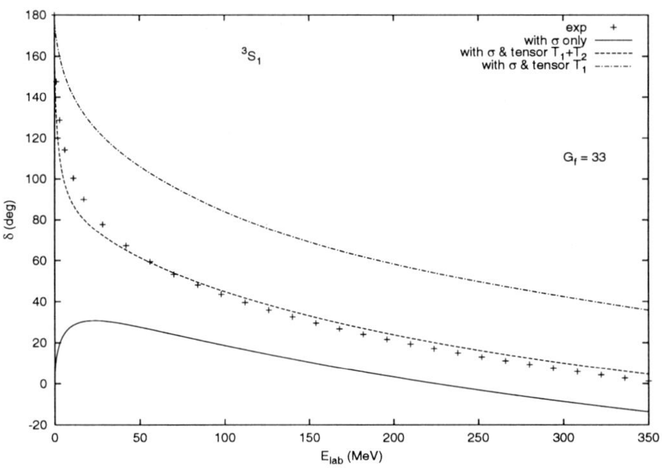

Figure 2. The $^3S_1$ NN scattering phase shift: with σ-meson exchange only (full line), with σ-meson exchange + full tensor interaction with (2) with $G_f = 33$ (dotted line), with σ-meson exchange + the first term of the tensor interaction (2) only, still with $G_f = 33$ (dot-dashed line). The crosses are the Nijmegen fit [14] to the data.

Besides a spin-spin, the pseudoscalar meson exchange gives rise to a tensor term as well

$$\begin{aligned} V_\gamma^T(r_{ij}) &= G_f\,\frac{g_\gamma^2}{4\pi}\frac{1}{12m_im_j}\times \\ &\left\{\mu_\gamma^2\left(1+\frac{3}{\mu_\gamma r}+\frac{3}{\mu_\gamma^2 r^2}\right)\frac{e^{-\mu_\gamma r}}{r} - \Lambda_\gamma^2\left(1+\frac{3}{\Lambda_\gamma r}+\frac{3}{\Lambda_\gamma^2 r^2}\right)\frac{e^{-\Lambda_\gamma r}}{r}\right\} \end{aligned} \tag{2}$$

where $\gamma = \pi, \eta$ and η'. It was found [13] that the tensor term used in baryon spectroscopy with $G_f = 1$ had to have $G_f = 33$ in order to fit the experimental $^3S_1$ phase shift. This is shown in Fig. 2. The tensor term (2) has a usual Yukawa type part, depending on the pseudoscalar meson masses μ_γ and a regularized part, containing cut-off parameters Λ_γ. If the latter term is removed, the attraction increases. Then a smaller factor, G_f=12, is required to reproduce the data. Further details of these studies can be found in ref. [13]. It would certainly be useful to search for alternative parameterizations of the GBE model which could better fit the NN phase shifts.

4 SUSY Approach to Local Phase Shift Equivalent Potentials

A microscopic derivation of the NN interaction, as above, leads to a nonlocal potential. This is a consequence of the complex structure of the interacting nucleons. As in the $\alpha - \alpha$ scattering [15], the wave function of this nonlocal potential presents a node in the $l = 0$ partial wave [10]. It means that if one would try to describe the interaction between two nucleons by an equivalent local potential, this potential would have an extra $l = 0$ state. This is an unphysical state known as *Pauli forbidden state.* The mathematical relation between such *deep* potentials and phenomenological *shallow* potentials (no bound state), as e. g. the Reid soft core potential [16], can interestingly and successfully be described through a Supersymmtric (SUSY) Quantum Mechanics approach. The procedure has been originated by Sukumar [17] and applied as a two-step SUSY transformation to the NN scattering by Sparenberg and Baye, see e.g. ref. [18].

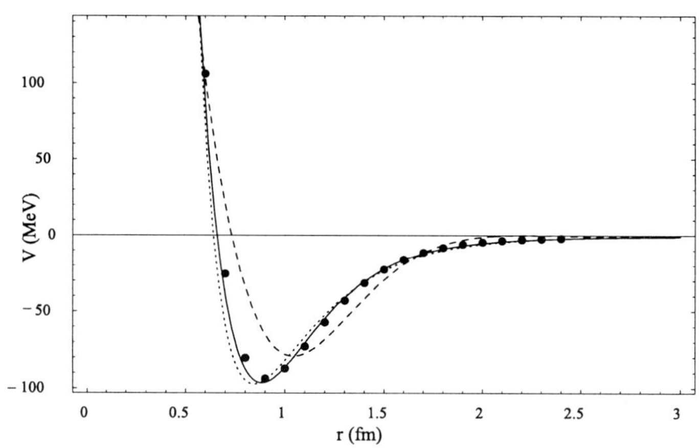

Figure 3. The shallow potential V_6 (solid line) compared to Reid68 [16] (dotted line), Baye & Sparenberg [18] (dashed line) and Reid93 [14] (dots) potentials.

Here we present an alternatively new procedure [19] based on phase equivalent chains of Darboux (or SUSY) transformations at fixed ℓ. These chains contain N successive transformations instead of two, as in ref. [18]. For poten-

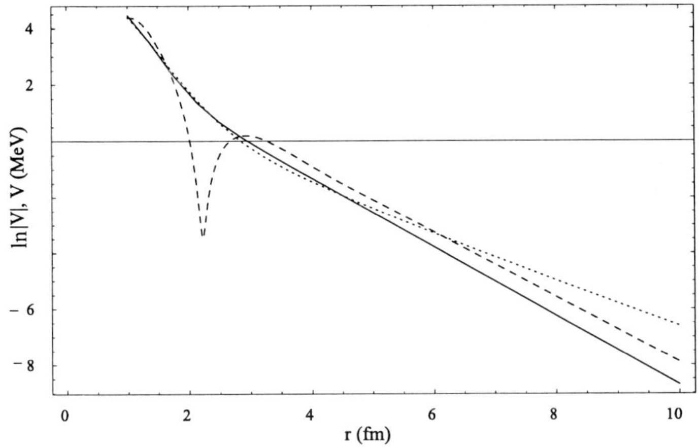

Figure 4. Asymptotic behaviour of the absolute value of V_6 (solid line), Reid68 [16] (dotted line), Baye & Sparenberg [18] (dashed line) in natural logarithmic scale.

tials related by a chain of transformations we derive an analytic expression for the Jost function and hence for the phase shift. They both contain N parameters, related to the N unphysical eigenvalues of a starting Hamiltonian, used to construct the chain. It is convenient to choose this Hamiltonian as the free particle Hamiltonian. These parameters are also poles of the S-matrix and can be fixed by a fit to the experimental phase shift. They can be distributed between poles and zeros of a Jost function and each distribution corresponds to a different potential. We applied this method to derive a shallow and a family of deep phase equivalent potentials for the $l = 0$ neutron-proton scattering. The $^1S_0$ phase shift has optimally been fitted with N = 6 S-matrix poles. The shallow potential, denoted by V_6, can be seen in Fig. 3. It is very close to the Reid soft core potential, denoted by Reid68, and also close to its updated version, called Reid93 [14]. The major improvement over the results of ref. [18] can be seen in Fig. 4. The unwanted oscillations in the potential tail [18] have disappeared so that the behaviour of V_6 is consistent with Yukawa's theory. A deep phase equivalent potential, accommodating a (Pauli forbidden) bound state, was found by adding two more poles. This is the supersymmetric partner of V_6. Varying the available parameters we brought this potential close to that of ref. [20], inspired from microscopic calculations as the sum of a Gaussian plus a Yukawa potential tail.

Thus the use of chains of Darboux (SUSY) transformations at fixed ℓ provides a powerful method for getting shallow and deep phase equivalent potentials for $\ell = 0$ partial waves. Studies of $l \neq 0$ phase equivalent potentials based on ℓ-changing Darboux transformations are underway.

Acknowledgement. I am grateful to my collaborators, Daniel Bartz and Boris Samsonov without whom this research could not have been accomplished.

References

1. D.A. Liberman: Phys. Rev. **D16**, 1542 (1977); C.E. DeTar: Phys. Rev. **D17**, 302 (1978); ibid. **D17**, 323 (1978)
2. M. Oka, K. Yazaki: Phys. Lett. **B90**, 41 (1980); Prog. Theor. Phys. **66**, 556 (1981); ibid. **66**, 572 (1981)
3. M. Harvey: Nucl. Phys. **A352**, 301 (1980); ibid. **A352**, 356 (1980)
4. M. Cvetič, B. Golli, N. Mankoč-Borštnik, M. Rosina: Nucl. Phys. **A395**, 349 (1983)
5. M. Oka, K. Yazaki: In: *Quarks and nuclei*, ed. W. Weise, p.469. Singapore: World Scientific 1984; F. Myhrer, J. Wroldsen: Rev. Mod. Phys. **60**, 629 (1988); K. Shimizu: Rep. Prog. Phys. **52**, 1 (1999); K. Shimizu, S. Takeuchi, A. Buchmann: Prog. Theor. Phys. Suppl. **137**, 43 (2000)
6. A.M. Kusainov, V.G. Neudatchin, I.T. Obukhovski: Phys. Rev. **C44**, 2343 (1991); Z. Zhang, A. Faessler, U. Straub, L.Ya. Glozman: Nucl. Phys. **A578**, 573 (1994); A. Valcarce, A. Buchmann, F. Fernandez, A. Faessler: Phys. Rev. **C50**, 2246 (1994)
7. W. Plessas: these proceedings
8. L.Ya. Glozman, Z. Papp, W. Plessas, K. Varga, R.F. Wagenbrunn: Nucl. Phys. **A623**, 90c (1997)
9. M. Kamimura: Prog. Theor. Phys. Suppl. **62**, 236 (1977)
10. D. Bartz, Fl. Stancu: Phys. Rev. **C63**, 034001 (2001)
11. D. Bartz, Fl. Stancu: Nucl. Phys. **A688**, 915 (2001)
12. B. Golli, M. Rosina: Phys. Lett. **B393**, 161 (1997)
13. D. Bartz: PhD Thesis, University of Liège, 2002 hep-ph/0205138
14. V.G.J. Stoks, R.A.M. Klomp, M.C.M. Rentmeester, J.J. de Swart: Phys. Rev. **C48**, 792 (1993); V.G.J. Stoks, R.A.M. Klomp, C.P.F. Terheggen, J.J. de Swart: Phys. Rev. **C49**, 2950 (1994) (cf. http://nn-online.sci.kun.nl/NN/)
15. S. Saito: Progr. Theor. Phys. **41**, 705 (1969)
16. R.V. Reid, Jr.: Ann. Phys. (N.Y.) **50**, 411 (1968)
17. C.V. Sukumar: J. Phys. **A18**, 2917 (1985); **A18**, 2937 (1985)
18. J.-M. Sparenberg, D. Baye: Phys. Rev. **C55**, 2175 (1997)
19. B.F. Samsonov, Fl. Stancu: Phys. Rev. **C66**, 034001 (2002)
20. V.I. Kukulin, V.M. Krasnopol'skii, V.N. Pomerantsev, P.B. Sazonov: Sov. J. Nucl. Phys. **43**, 355 (1986)

Few-Body Systems Suppl. 14, 89–92 (2003)

Few-Body Systems

Analyzing-Power Measurements for the $^2H(\boldsymbol{n},pn)n$ Reaction at 187 MeV

B.D. Anderson[1], W.M. Zhang[1], J. Sowinski[2], O. Osadchy[1], A.R. Baldwin[1], W. Franklin[1,2], J.W. Watson[1], S.W. Wissink[2], M. Wolanski[2], G. Yang[2], W. Glöckle[3], H. Witała[4]

[1] Kent State University, Kent, Ohio, USA

[2] Indiana University Cyclotron Facility, Indiana University, Bloomington, Indiana, USA

[3] Ruhr-Universität Bochum, Bochum, Germany

[4] Jagiellonian University, Cracow, Poland

Abstract. Analyzing-power measurements were performed for the $^2H(\boldsymbol{n},n)^2H$ and $^2H(\boldsymbol{n},pn)n$ reactions at 189 MeV at the Indiana University Cyclotron Facility. A polarized neutron beam was obtained using polarized protons incident upon a liquid deuterium target. Neutrons emitted at 10° were collimated into a beam 5 cm wide by 7 cm high, with a polarization of ~ 0.55 and an intensity of $\sim 10^7$/s. The neutron beam was incident upon a second liquid deuterium target to study the reactions of interest. Charged particles were detected in an array of plastic scintillator and wire chambers and neutrons were detected in an array of six segmented, mean-timed plastic scintillators, each 10 cm by 10 cm by 1 m, at a flight path of 2 m. The overall energy resolutions were about 10 to 15 MeV. The analyzing powers from the $^2H(\boldsymbol{n},n)^2H$ reaction agree well with measurements from $^2H(\boldsymbol{p},p)^2H$ at nearby energies, indicating no significant Coulomb effects in the latter at these energies. The $^2H(\boldsymbol{n},pn)n$ reaction was studied in two geometries; a quasifree scattering (QFS) geometry and a final-state interaction (FSI) geometry. The QFS geometry is not expected to be sensitive to 3-body forces while the FSI geometry is predicted to have a strong sensitivity to such forces. We find that the analyzing powers from the QFS geometry generally agree with those from free $^1H(\boldsymbol{n},n)^1H$ scattering as expected and that the analyzing powers from the FSI geometry may indicate the presence of 3-body forces in the nuclear interaction.

1 Introduction

It is generally acknowledged that three-body forces must exist in nature, including in the nuclear interaction. At present, the best evidence for such forces comes from the analyses of the binding energies of A = 3 nuclei, which are clearly underbound by modern nucleon-nucleon (NN) potentials. It is hoped that by measuring polarization observables in three-body reactions information can be provided that will allow for more detailed tests of present day 3NF models. Most 3N breakup measurements involve a proton beam and are influenced by the Coulomb force. The present theoretical description of these reactions based on the Faddeev formulation do not include the long-range Coulomb force. The measurements described here use a polarized neutron beam to breakup the deuteron and avoid problems with the Coulomb force in the theoretical interpretation. Of course, a neutron beam must be produced in a nuclear reaction and is a secondary beam with much reduced intensity. The results available from this study cannot approach the statistical accuracy of proton-beam experiments, but provide both an independent check of such experiments and a check on Coulomb-force contributions.

Two geometries were studied in this experiment. In one geometry the detected proton and the undetected neutron have equal momenta and interact strongly in the $^1S_0$ final state. For this configuration the Faddeev calculations predict strong sensitivity of the nucleon analyzing power to possible three-nucleon forces. This geometry allowed us also to measure the elastic-scattering reaction $^2H(\boldsymbol{n},n)^2H$, which can be compared with available analog $^2H(\boldsymbol{p},p)^2H$ analyzing-power measurements. The other geometry studied corresponds to quasifree scattering (QFS) kinematics and is predicted to not be sensitive to possible three-body forces. The predictions were obtained using exact, three-body Faddeev calculations based on the CD-Bonn NN interaction combined with the 2π-exchange Tucson-Melbourne 3-body force [1].

2 Experiment

The measurements were performed with the polarized neutron facility at the Indiana University Cyclotron Facility. 200-MeV polarized protons from the cyclotron were incident upon a liquid deuterium target, 5.7 cm thick. Neutrons emitted near 10° were collimated into a beam 5 cm by 7 cm using about 3 m of concrete and steel sheilding. The neutron beam had a polarization of ~ 0.55 and an intensity of $\sim 10^7$ neutrons/s. The neutron beam was incident upon a second liquid deuterium target in order to study the reactions of interest. The diagram of the experimental arrangement is shown in Fig. 1. The emitted charged particles were detected in a wire-chamber, scintillator array located about 1 m from the secondary target. This array consisted of two sets of "X" and "Y" multiwire drift chambers, a "START" scintillator, a "ΔE" scintillator, and two "E" scintillators. The array was able to locate charged particles to better than 1 mm in location and to measure the kinetic energies of the charged particles to ~ 5 MeV. Neutrons emitted from the secondary target were detected in an array of 6 segmented, mean-timed neutron detectors each 10 cm high by 10 cm deep by 1 m long, located 2 m from the target. The intrinsic time resolution of the neutron detectors was less 0.5 ns and the position location along the long length of each detector was good to ~2 cm. The neutron energy resolution was 5 to 12 MeV.

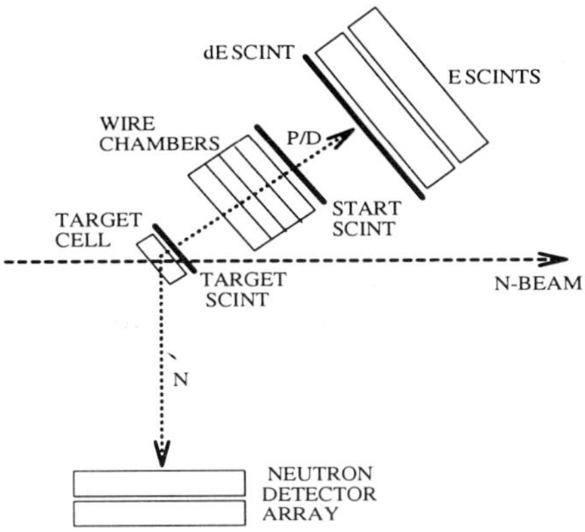

Figure 1. Experimental arrangement.

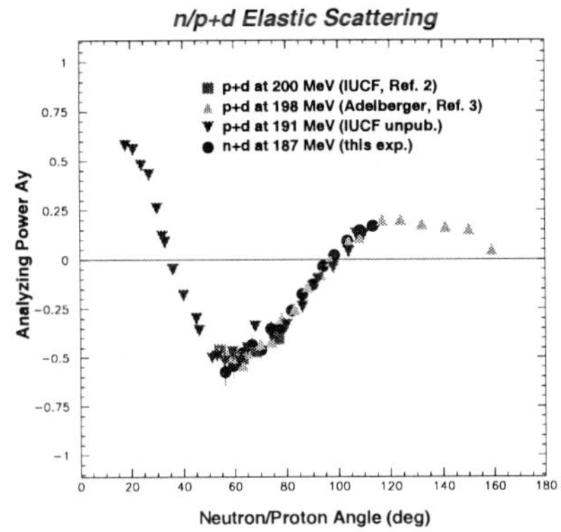

Figure 2. n-d and p-d elastic-scattering analyzing powers.

3 Data Analysis

Various requirements were imposed upon the data. First of all, a coincidence between the charged-particle array and a neutral particle into the neutron array was required in the hardware. The wire chambers in the charged particle array were used to require that the charged particle trajectory originated in the secondary target. Time-of-flight between the "START" detector near the target and the accelerator RF was used to pick out the correct beam burst. The position information from both the charged-particle drift chambers and the mean-timed, segmented neutron detectors was used to require co-planarity and correct opening angle between the two emitted particles. The charged particle energy was obtained by summing up the pulse-height signals from the scintillators, using known kinematics for calibration. The neutron kinetic energy was obtained by measuring the time-of-flight of the neutron to the neutron array, using the "START" detector as a fiducial mark.

Backgrounds were measured by performing empty-cell runs, using a hydrogen target, and gating an "accidental" spectrum on an adjacent beam burst. Calibrations were obtained using two-body kinematics for n-p elastic scattering from a CH_2 target and n-d elastic scattering from the LD_2 target.

4 Results

The results for the measurement of the analyzing power, A_y, for the elastic scattering of neutrons by deuterium are shown in Fig. 2, compared with the analogous measurements for the elastic scattering of protons by deuterium at nearby energies [2, 3]. The agreement between the neutron and proton analyzing powers is seen to be excellent, indicating that Coulomb effects are small for the proton scattering case at these energies and angles.

The results for the $^2H(\boldsymbol{n}, np)n$ breakup reaction in the quasifree scattering (QFS) geometry are shown in Fig. 3, compared with analyzing powers from free n-p scattering measured also in this experiment. These experimental results are compared with N-N phase-shift predictions from the SAID phase-shift code [4] and with the prediction for the breakup reaction from the 3-body, Faddeev calculations [1]. This prediction is practically unchanged if 3-body forces are included. Within the error bars, the agreement of the breakup reaction results with the free scattering results is good.

The results for the $^2H(\boldsymbol{n}, np)n$ breakup reaction to the $^1S_0$ final-state interaction (FSI) are presented in Fig. 4. The results are shown compared to the Faddeev calculations with and without the three-body force included. As seen, the calculations with the three-body force included describe the data better at the more forward and especially at the more backward angles, possibly indicating the significance of such forces in this reaction.

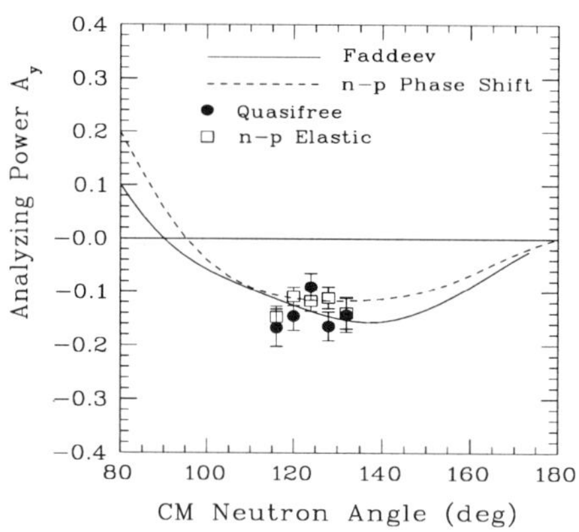

Figure 3. The $^2H(\boldsymbol{n}, np)n$ QFS analyzing powers.

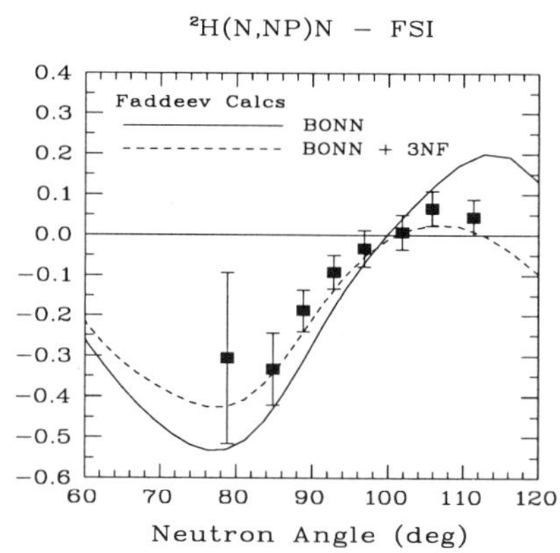

Figure 4. The $^2H(\boldsymbol{n}, np)n$ FSI analyzing powers.

Acknowledgement. This work was supported in part by the U.S. NSF.

References

1. H. Witała, W. Glö ckle, Th. Cornelius: Few-Body Systems **6**, 967 (1989)
2. S.P. Wells et al.: Nucl. Instr. Meth. **A325**, 205 (1993)
3. R.E. Adelberger, C.N. Brown: Phys. Rev. **D5**, 2139 (1972)
4. R.A. Arndt, L.D. Roper: SAID program, VPI, unpublished

Few-Body Systems Suppl. 14, 93–96 (2003)

Few-Body Systems

Very Low-Energy Hydrogen-Antihydrogen Scattering

E.A.G. Armour*, C.W. Chamberlain

School of Mathematical Sciences, University of Nottingham, Nottingham, NG7 2RD, U.K. Phone: +44 (0)115 9514922 Fax: +44 (0)115 9514951

Abstract. In view of current interest in the trapping of antihydrogen ($\bar{\text{H}}$) atoms at very low temperatures, we have carried out a calculation of s-wave hydrogen-antihydrogen scattering at very low energies, using the Kohn variational method, taking into account rearrangement scattering into the three channels that contain positronium in its ground state and lie closest to threshold. We find that our values for the elastic cross section are in good agreement with the the values obtained by Jonsell et al. [2001 Phys. Rev. A 64 052712] using a distorted wave approximation. However, our values for the total rearrangement cross section are much larger than their values and we predict that cooling of $\bar{\text{H}}$ by cold H would be considerably less efficient than was found to be the case by Jonsell et al..

1 Introduction

Experimental work with the aim of producing very cold,trapped antihydrogen ($\bar{\text{H}}$) [1, 2], is currently being carried out on two projects, ATHENA (ApparaTus for High precision Experiments on Neutral Antimatter) and ATRAP (Antihydrogen TRAP). This has generated increasing theoretical interest in the $\text{H} - \bar{\text{H}}$ system [3, 4, 5, 6, 7, 8, 9]. Very recently, experimentalists on the ATHENA project have succeeded in preparing about 50,000 $\bar{\text{H}}$ atoms, cooled to within 15 degrees of absolute zero [10].

In view of this, we have carried out detailed calculations of very low-energy H-$\bar{\text{H}}$ scattering using the Kohn variational method [11]. Some preliminary work [6] convinced us that the best way to obtain accurate results for both elastic and rearrangement scattering would be to do a Kohn calculation starting from the BO continuum function,

$$\phi_{BO} = \chi(R) \sum_{q=1}^{N} c_q(R)\varphi(\mathbf{r}_1, \mathbf{r}_2; \mathbf{R}), \tag{1}$$

E-mail address: eaga@maths.nott.ac.uk

where $\mathbf{r}_1$ and $\mathbf{r}_2$ are the position vectors of the electron and the positron, respectively, with respect to the nuclear centre of mass. The radial part of ϕ_{BO}, $\chi(R)$, makes allowance for the dominance of the Coulombic potential when the nuclei are close together and takes into account the van der Waals nature of the potential when H and $\bar{\mathrm{H}}$ are far apart.

We have shown that $0.8a_0$ is an upper bound to R_c [4] and a very recent calculation has obtained an improved upper bound value of $0.744a_0$ [12]. It is reasonable to assume that $R_c > 0.639a_0$, the value of R at which the nuclei cease to be able to bind an electron or a positron on its own.

Calculations of elastic scattering using the BO potential have the defect that the electron and positron are unbound for $R < R_c$ and separate off as Ps. As a consequence, the overall BO wave function cannot be said to represent purely elastic scattering. We allow for this by including rearrangement channels in our Kohn calculation.

Jonsell et al. [8] found that dominant contribution by far to the cross section for rearrangement was from the channel with Pn in its $N = 24$, $L = 0$ state and Ps in its ground state. To test this, we included rearrangement channels corresponding to protonium (Pn) in states with $N = 22$, 23 and 24, all with $L = 0$ and Ps in its ground state.

To carry out a Kohn calculation including these channels, we must have a trial function of the form $\{\Psi_i\}_{i=1}^4$, whose i^{th} component represents the scattering state associated with the ith entrance channel. We take $i = 1$ to be the state corresponding to the $\mathrm{H} + \bar{\mathrm{H}}$ entrance channel and $i = 2$, 3 and 4 to be the states with Pn in the entrance channel in its $N = 24$, $N = 23$ and $N = 22$ state, respectively.

We use both K-matrix and S-matrix boundary conditions. If we use K-matrix boundary conditions, the trial function for state i takes the form

$$\Psi_i = \phi_i^{(1)} + \sum_{q=1}^{4} K_{iq}^t \phi_q^{(2)} + \sum_{q=1}^{N} c_{iq} u_q \tag{2}$$

where for the H-$\bar{\mathrm{H}}$ entrance channel ($i = 1$)

$$\phi_1^{(1)} = \phi_{BO} - \tan\eta_{BO}\phi_1^{(2)} \tag{3}$$

and

$$\phi_1^{(2)} = \sqrt{\frac{\mu_p}{2\pi k_1}} \left(1 - e^{-\gamma_1 R}\right)^{t_1} \frac{\cos k_1 R}{R} \sum_{q=1}^{N} c_q(R)\varphi(\mathbf{r}_1, \mathbf{r}_2; \mathbf{R}) \tag{4}$$

where k_1 is the wave number associated with the relative motion in the $\mathrm{H}+\bar{\mathrm{H}}$ channel, η_{BO} is the s-wave BO phase shift and μ_p is the reduced mass of the proton and the antiproton.

$$\phi_j^{(1)} = \sqrt{\frac{\mu_r}{2\pi k_j}} \frac{\sin k_j \rho}{\rho} \Phi_{\mathrm{Pn}}^{N_j}(R)\phi_{\mathrm{Ps}}^{n_j}(r_{12}) \qquad (j = 2, 3, 4) \tag{5}$$

where $\Phi_{\mathrm{Pn}}^{N_j}(R)$ and $\phi_{\mathrm{Ps}}^{n_j}$ are the wave functions for the s-states of Pn and Ps with principal quantum numbers N_j and $n_j (= 1)$, respectively, μ_r is the reduced mass of the Pn and Ps and k_j is the wave number associated with the relative motion in the jth channel ($j = 2, 3, 4$). ρ is the distance of the centre of mass of the Ps from the centre of mass of the Pn and r_{12} is the distance between the electron and the positron. The second continuum function in the jth entrance channel is similar but has a shielded cosine factor instead of the sine factor. Appropriate linear combinations of $\phi_j^{(1)}$ and $\phi_j^{(2)}$ were used in calculations with trial functions which had S-matrix boundary conditions.

We have used two types of short-range correlation functions. The first are a set of the form

$$u_i = \sqrt{\frac{1}{4\pi^3}} R^{q_i} e^{-\zeta R} R^{m_i+n_i+j_i+k_i} \lambda_1^{m_i} \lambda_2^{n_i} \mu_1^{j_i} \mu_2^{k_i} r_{12}^{p_i} \times \\ \exp\left[-\alpha(\lambda_1+\lambda_2)+\beta(\mu_1-\mu_2)\right] (i = 1, ..., M), \tag{6}$$

where λ_i,μ_i and ϕ_i are prolate spheroidal coordinates for light particle i [4]. These functions are suitable for describing the $\mathrm{H} + \bar{\mathrm{H}}$ channel.

The second set are of the form

$$u_i = \sqrt{\frac{1}{4\pi}} s_i(\rho) \rho^{l_i} e^{-\xi\rho} \Phi_{\mathrm{Pn}}^{N_i}(R) \phi_{\mathrm{Ps}}^{n_i}(r_{12}) (i = M+1, ..., N) \tag{7}$$

where $l_i = -1, 0, 1, ...$ The factor $s_i(\rho)$ is of the form,

$$s_i(\rho) = \left(1 - e^{-\nu\rho}\right)^s. \tag{8}$$

These functions are suitable for describing the rearrangement channels. If $l_i = -1$ and ξ is small, u_i behaves asymptotically like a weakly bound state of Ps. The BO basis set for R values near or below R_c was improved by the inclusion of a basis function of this type, as in our earlier calculation [4]. Corrections to the BO approximation were included as far as possible.

For elastic $\mathrm{H} + \bar{\mathrm{H}}$ scattering, we obtain good agreement with the results of Jonsell et al. [8] over the range of energy values up to just below 10^{-3} a.u. covered in our calculation. In particular, we obtain the same structure in the cross section as they do in the vicinity of 10^{-4} a.u.. We obtain a low-energy limiting value of the elastic cross section of $908a_0^2$, which is quite close to the value $829a_0^2$, obtained by Jonsell et al.

The values we obtain for the total rearrangement cross sections, σ_{1j} ($j = 2, 3, 4$), obey Wigner's threshold law, which predicts that they should be proportional to $\epsilon_1^{-\frac{1}{2}}$, where $\epsilon_1 = (2\mu_p)^{-1}k_1^2$. Thus $\sigma_{12} = 0.21\epsilon_1^{-\frac{1}{2}}$, $\sigma_{13} = 0.45\epsilon_1^{-\frac{1}{2}}$ and $\sigma_{14} = 0.01\epsilon_1^{-\frac{1}{2}}$, all in units of a_0^2. Thus overall, $\sigma_{\mathrm{re}} = 0.67\epsilon_1^{-\frac{1}{2}}$.

Jonsell et al. [8] found that $\sigma_{\mathrm{re}} = 0.09\epsilon_1^{-\frac{1}{2}}$ and contributions from all other channels besides channel 2, i.e. $N = 24$, $n = 1$, were sufficiently small that they could be neglected. Thus, in contrast to Jonsell et al., we find that both channels 2 and 3, i.e. $N = 24$, $n = 1$ and $N = 23$, $n = 1$, make substantial contributions to σ_{re}, and not just channel 2. Our value for σ_{12} is over twice as large as their value and our value for σ_{13} is larger still by a factor of more than two. This is an important difference. A full explanation for this difference is given in ref. [11].

As expected from this difference, our calculations predict considerably more of the $\bar{H}$ is lost in cooling of $\bar{H}$ using ultra-cold H than is predicted by Jonsell et al. [8]. For example, they find that for an initial temperature of 10K, about 40% of the antihydrogen would still be remaining at a temperature of 0.5K, whereas we find that only about 14% would be left at this temperature.

Full details of our calculations will be published at a later date.

We are currently considering antihydrogen scattering by helium, which is a process that is likely to lead to the loss of trapped antihydrogen [6].

We wish to thank P. Froelich, J. W. Humberston, S. Jonsell and P. G. Van Reeth for valuable discussions and the University of Nottingham for support for this research.

References

1. M.H. Holzscheiter, M. Charlton: Rep. Prog. Phys. **62**, 1 (1999)
2. G. Gabrielse: Adv. At. Mol. Phys. **45**, 1 (2001)
3. W. Kołos, D.L. Morgan, D.M. Schrader, L. Wolniewicz: Phys. Rev. **A11**, 1792 (1975)
4. E.A.G. Armour, V. Zeman, J.M. Carr: J. Phys. **B31**, L679 (1998)
5. P. Froelich, S. Jonsell, A. Saenz, B. Zygelman, A. Dalgarno: Phys. Rev. Lett. **84**, 4577 (2000)
6. E.A.G. Armour, C.W. Chamberlain: *New Directions in Antimatter Chemistry and Physics*, C.M. Surko and F.A. Gianturco eds., Kluwer Academic Press, p. 53 (2001)
7. A.Yu. Voronin, J. Carbonell: Nuclear Physics **A689**, 529c (2001)
8. S. Jonsell, A. Saenz, P. Froelich, B. Zygelman, A. Dalgarno: Phys. Rev. **A64**, 052712 (2001)
9. E.A.G. Armour, C.W. Chamberlain: Few-Body Systems **31**, 101 (2002)
10. M. Amoretti et al.: Nature **419**, 456 (2002)
11. E.A.G. Armour, C.W. Chamberlain: J. Phys. B (2002), at press
12. K. Strasburger: J. Phys. **B35**, L435 (2002)

Few-Body Systems Suppl. 14, 97–100 (2003)

Few-Body Systems

Deuteron Two-Body Photodisintegration in the Quark-Hadron Picture

M. Mirazita*, for the CLAS Collaboration

I. N. F. N. – Laboratori Nazionali di Frascati

Abstract. Preliminary results of the new data taken with the CLAS large angle spectrometer of Hall B at TJNAF are shown. They determine a nearly complete angular distribution for photon energy between 0.5 and 3 GeV and show a persistent forward-backward asymmetry. They are well described by the recent calculations derived in the framework of the Quark Gluon String Model.

One of the primary goals of nuclear physics is the study of the interplay between hadronic and partonic degrees of freedom, and of the effectiveness of the traditional nucleon-nucleon theories or QCD inspired model in describing the data. For this purpose, the deuteron photodisintegration is an ideal reaction, because the deuteron is the simplest nucleus and photon interaction with quarks is well known.

Traditional models based on mesonic and baryonic degrees of freedom can be used for photon energy up to few hundred of MeV. However, in the Asymptotic Meson Exchange model [4] the dNN interaction vertex has been extended in the few GeV energy region. The calculation provides a scaling law for the cross section with an exponent depending on the scattering angle, but it fails to fully reproduce the available experimental data.

For high proton transverse momentum, pQCD predicts that the Constituent Counting Rule (CCR) holds [1]: for sufficiently high energy and fixed angle, the differential cross section should scale with the total energy as s^{2-n}, where n is the number of microscopic fields involved in the reaction. In our case, $n = 13$, thus $d\sigma/dt \propto s^{-11}$. However, the available experimental data for photon energy above 1 GeV [10] show such a scaling only at large proton angle, $\theta_{cm} = 69°$ and $89°$, while for more forward angles the situation is less clear. More sophisticated models based on pQCD have been proposed. The Reduced Nuclear Amplitude model [3] is basically a development of the CCR. It needs a separate normalization of the cross section at each angle, but it can't reproduce

* *E-mail address:* marco.mirazita@lnf.infn.it

satisfactory the experimental data, especially in the forward direction. In the Hard quark Rescattering Model [2], the scattering amplitude is a convolution of the deuteron wave function, the hard photo-quark interaction vertex and the *pn* scattering amplitude, but the poor knowledge of the *pn* amplitude generates large uncertainties in the cross section calculation.

At high energies, pQCD also predicts that the hadron helicity must be conserved (HHC) [5], because photon-quark and quark-gluon interactions conserve the chirality, if quark masses can be neglected. So far, no exclusive reactions satisfying HHC have been measured, but the number of studies is limited. For example, in *pp* elastic scattering the failure of HHC have been attributed to long distance phenomena in which there are three independent scattering of quarks in the beam proton with the quarks in the target proton [6, 7]. In photoreactions this mechanism is suppressed, thus HHC is expected to hold [8]. Based on HHC, predictions for many spin observables in deuteron photodisintegration can be made [7, 9]. Measurements exist only at 90° for proton polarization and for the Σ asymmetry [11]. Although the data are not fully inconsistent with HHC predictions [1], there are indications that HHC could be violated also in reactions involving real photons.

In the few GeV energy range, it is expected that non perturbative effects could become important. A non-perturbative approach for the description of binary reactions is the Quark Gluon String Model (QGSM) [12]. In this model, the reaction $\gamma d \to pn$ is described by the exchange of three valence quarks in the t-channel [13]. This picture corresponds to the formation and break up in the intermediate state of a quark-gluon string, that can be easily identified with the nucleon Regge trajectory. The model predicts a strong angular dependence of the cross section at fixed photon energy, with two peaks at forward and backward angles and forward-backward asymmetry.

In order to check the behaviour of the cross section predicted by the different calculations, data covering all proton angles are necessary in a wide energy region. These new data have been measured at TJNAF using tagged photons and the CLAS spectrometer. They provide the cross section for proton angles between 8° and 140° and photon energy between 0.5 and 3 GeV. Photodisintegration events are selected by detecting the initial photon in the tagger system and the produced, high energy proton in the spectrometer, while the neutron is identified by missing mass cut. The CLAS acceptance has been evaluated by Monte Carlo simulations of the photodisintegration reaction, and the background contribution has been computed by a fit to the experimental missing mass distributions and then subtracted from the measured yield. The preliminary results from about 30% of the accumulated statistic and with the additional cut $\theta_p^{CM} > 20°$ (which will be removed in the final analysis) are shown in the figure, where the differential cross section is reported as a function of the proton angle in the CM frame, for fixed photon energy above 0.9 GeV.

[1] The limit $\Sigma = -1$ in [9] is incorrect. Taking properly into account the symmetry properties of the isospin function of *pn* system, the correct HHC limit is $\Sigma = +1$.

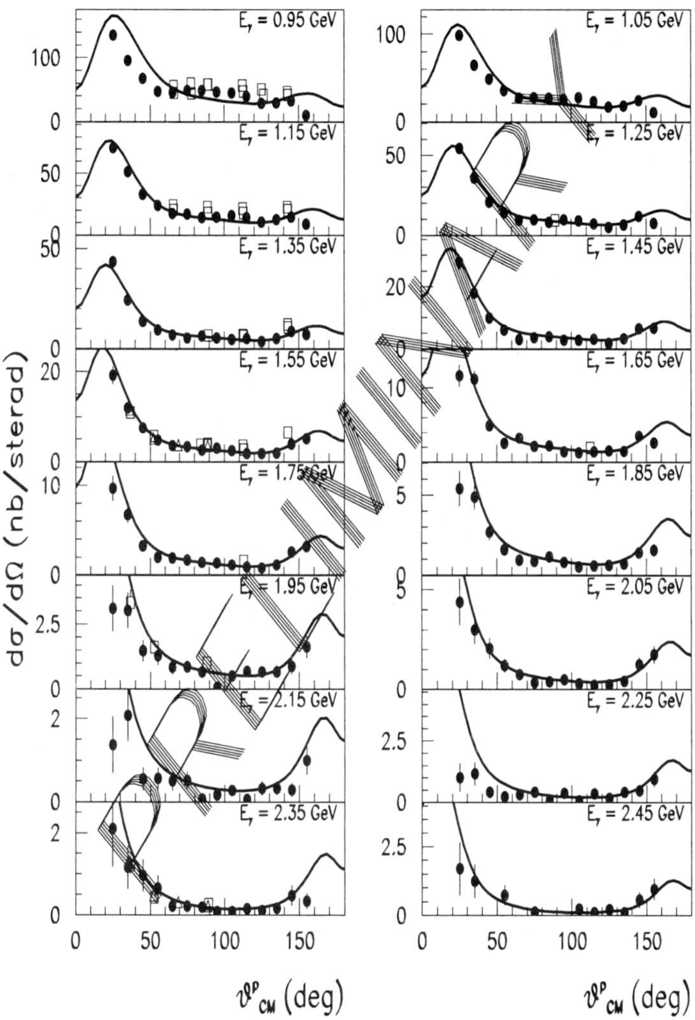

Figure 1. Preliminary results of the deuteron photodisintegration differential cross section measured in Hall B of TJNAF (black circles), compared with the published data from SLAC (open squares) and Jlab Hall C (open triangles). The curve is the QGSM calculation

In spite of the limited data set, they cover all proton angles with statistical error lower than 5% for $E_\gamma < 1.5$ GeV. When the whole data set will be analysed, the statistical error will be improved by a factor of 3 or more for photon energies up to 2.5 GeV. These angular distributions show a persistent forward-backward asymmetry. They also agree well with the few earlier published data.

The result of the QGSM calculation is also shown in the figure as full line. It is the only model describing well the measured angular dependence of the cross section and correctly predicting the forward-backward asymmetry showed by the data. It also predicts a decrease of the cross section for angles smaller than $10° - 20°$, which will be verified by the final analysis of our data.

Acknowledgement. I am pleased to thank L.A. Kondratyuk for pointing to my attention the wrong sign in the prediction for the Σ asymmetry in [9].

References

1. S.L. Brodsky, G.L. Farrar: Phys. Rev. Lett. **31**, 1153 (1973); V. Matveev et al.: Lett. Nuovo Cimento **7**, 719 (1973)
2. L.L. Frankfurt et al.: Phys. Rev. Lett. **84**, 3045 (2000); L.L. Frankfurt et al.: Nucl. Phys. **A663&664**, 349 (2000)
3. S.L. Brodsky, J.R. Hillier: Phys. Rev. **C28**, 475 (1983)
4. A.E.L. Dieperink, S.I. Nagorny: Phys. Lett. **B456**, 9 (1999)
5. G.P. Lepage, S.J. Brodsky: Phys. Rev. **D22**, 2157 (1980); S.J. Brodsky, G.P. Lepage: Phys. Rev. **D24**, 2848 (1981)
6. T. Gousset et al.: Phys. Rev. **D53**, 1202 (1996)
7. C. Carlson, M. Chachkhunashvili: Phys. Rev. **D45**, 2555 (1992)
8. A. Afanasev et al.: Phys. Rev. **D61**, 034014 (2000)
9. S.I. Nagornyi et al.: Sov. J. Nucl. Phys. **55**, 189 (1992)
10. R. Crawford et al.: Nucl. Phys. **A603**, 303 (1996); C. Bochna et al.: Phys. Rev. **C41**, 4576 (1998); E.C. Shulte et al.: Phys. Rev. Lett. **87**, 102302-1 (2001); S.J. Freedman et al.: Phys. Rev. **C48**, 1864 (1993); J.E. Beltz et al.: Phys. Rev. Lett. **74**, 646 (1995)
11. K. Wijesooriya et al.: Phys. Rev. Lett. **86**, 2975 (2001), and references therein; F. Adamian et al.: Eur. Phys. J. **A8**, 423 (2000)
12. A.B. Kaidalov: Z. Phys. **C12**, 63 (1982) 63; A.B. Kaidalov: Surv. High Energy Phys. **13**, 265 (1999)
13. V.Yu. Grishina et al.: Eur. Phys. J. **A10**, 355 (2001)

Few-Body Systems Suppl. 14, 101–102 (2003)

Few-Body Systems

New Types of Meson-Exchange Currents in Few-Nucleon Systems Based on Dressed-Bag Model for NN Interaction

M.M. Kaskulov[1,*], V.I. Kukulin[2], P. Grabmayr[1,†]

[1] Physikalisches Institut, Universität Tübingen, D-72076 Tübingen, Germany
[2] Institute of Nuclear Physics, Moscow State University, 119899 Moscow, Russia

A new model for basic NN-force has been developed in recent years by Moscow-Tübingen group [1]. The model assumes that two p-shell excited quarks in proper six-quark description for the NN-channel with a microscopic configuration $|s^4p^2[42]_xL = 0,2;ST\rangle$ deexcites via well known $^3P_0$-mechanism into two s-wave pions and then pass to the two s-shell quarks. This leads to production of the intermediate state in NN-interaction in the form of a dressed six-quark bag: $|s^6[6]_x + \sigma, L_\sigma = 0,2; S_\sigma = T_\sigma = 0\rangle$ and $|s^6[6]_x + 2\pi, L_{2\pi} = 0,2; S_{2\pi} = 0, T_{2\pi} = 1\rangle$ at intermediate and short-range region. In the transition $NN \to 6q + \sigma$ a strong scalar-isoscalar field φ_σ (represented by σ meson production) is generated due to partial restoration of chiral symmetry. This spherical symmetric σ-field around $6q$-bag results in strong effective "pressure" to six-quark bag from outside, which leads effectively to a strong attraction in the associated NN-channel. The external pressure to the six-quark core induced by σ-field generation in this approach is confronted with the so-called t-channel σ-exchange in traditional meson-exchange models. Similarly two meson-generation mechanism leads to the appearance of charged currents in the dressed bag (see Fig.1). These charged currents are different from traditional $\pi^\pm$-currents in NN-channel due to their s-channel nature [2], but having the similar to the traditional MEC tensor structure of current operators. This non-trivial similarity leads to the same "selection rules" in our approach as in traditional MEC-operators. As an important application of the new technique for the s-channel MEC induced by dressed-bag formation we studied the reaction $np \to d\gamma$ with polarised and unpolarised nucleons. In first

---

* *E-mail address:* kaskulov@pit.physik.uni-tuebingen.de
† This work was sponsored by DFG grants No Fa67/20-1, GRK683, Gr1084/3-3, He2171/1 and RFBR grants No 01-02-04015, 02-02-16612

case the ejected γ-quant should have a circular polarisation, P_γ, which could not be explained by any traditional NN-models and by also rather complete χPT- calculations.

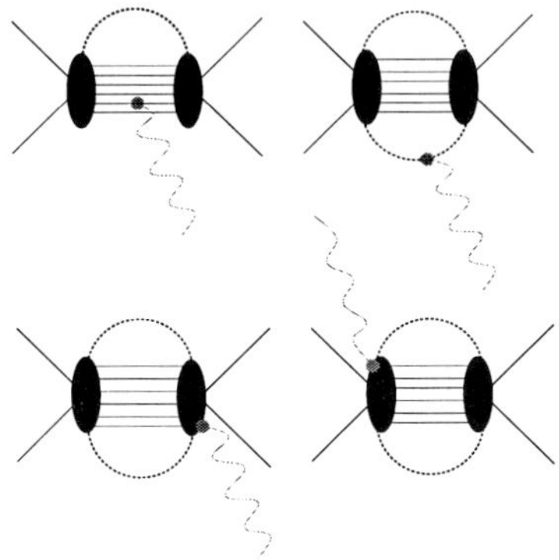

Figure 1. The "beetle" diagrams for the coupling of the γ-quanta with the six-quark dressed-bag.

For electro-magnetic properties of the NN-system we studied with the new model two types of currents *isovector* (i.e. those which are accompanying the isovector transition $NN(I = 1) \to NN(I = 0)$ or vice versa) and *isoscalar*, e.g. in ${}^3S_1 + {}^3D_1 \to {}^3S_1 + {}^3D_1$ transitions. The first type contributes mainly to cross-section of the $np \to d\gamma$ process with thermal neutrons while the second one gives rise to additional contribution to P_γ, and also deuteron quadrupole and magnetic moments. By consistent microscopic calculations we found quite significant contribution of the new MEC-operators to the $np \to d\gamma$ transition. The isoscalar currents give (in matrix elements) a 5 times smaller contribution to the matrix elements than the isovector currents, but still their size is sufficient to explain the experimentally observed cross section for $np \to d\gamma$ with thermal neutrons and the asymmetry of P_γ.

The theory for new currents developed in the present work should be crucially important in quantitative treatment of any e.-m. process with the deuteron and in other few-body systems at moderate and high momentum and energy transfer. E.g. the developed model gives a consistent and "smooth" transition using a unified microscopic description from very low up to very high momentum transfer in e.-m. properties of the few-nucleon systems including crossover to quark counting rules at $Q^2 > 10$ GeV^2.

References

1. V.I. Kukulin, I.T. Obukhovsky, V.N. Pomerantsev and A. Faessler: Int. J. Mod. Phys. **E11**, 1 (2002)

2. M.M. Kaskulov, V.I. Kukulin and P. Grabmayr: submitted for publication

Few-Body Systems Suppl. 14, 103–104 (2003)

Few-Body Systems

Virtual Bremsstrahlung in Proton-Proton Scattering

M. Kiš[1], J.C.S. Bacelar[1], R. Castelijns[1], R. Čaplar[2], K. Ermisch[1], I. Gašparić[2], M.N. Harakeh,[1] N. Kalantar-Nayestanaki[1], H. Löhner[1], M. Mahjəur-Shafiei[1]

[1] KVI, Zernikelaan 25, 9747 AA Groningen, The Netherlands

[2] IRB, Box 180, 10002 Zagreb, Croatia

The virtual bremsstrahlung process, in addition to the real photon bremsstrahlung, provides an elaborate tool for studying the nucleon-nucleon interaction. The additional degree of freedom which exists in the virtual bremsstrahlung production, a longitudinal polarization of the virtual photon, allows transversal, longitudinal, and interference terms to be studied. The squared transition amplitude for $p+p \to p+p+e^+ + e^-$ can be written [1, 2]:

$$
\begin{aligned}
|A|^2 \propto \frac{1}{M_\gamma^2} \Bigg\{ & W_{\mathrm{T}} \left(1 - \frac{2\mathbf{l}^2}{M_\gamma^2} \sin^2\theta_l\right) + W_{\mathrm{L}} \left(1 - \frac{4\mathbf{l}^2}{k_0^2} \cos^2\theta_l)\right) \\
& + \frac{2\mathbf{l}^2 \sin^2\theta_l}{M_\gamma^2} (W_{\mathrm{TT}} \cos 2\phi_l + W'_{\mathrm{TT}} \sin 2\phi_l) \\
& + \frac{2\mathbf{l}^2 \sin 2\theta_l}{k_0 M_\gamma} (W_{\mathrm{LT}} \cos\phi_l + W'_{\mathrm{LT}} \sin\phi_l) \Bigg\} ,
\end{aligned} \tag{1}
$$

where $\mathbf{l}$ is a relative momentum of dilepton pair and θ_l, ϕ_l are polar and azimuthal angle of $\mathbf{l}$ relative to a reaction plane. By measuring the momenta of the outgoing dilepton pair, all kinematic factors in Eq. (1) are determined and in principle the response functions (transversal, longitudinal, and interference terms) $W_i, i = \mathrm{T}, \mathrm{L}, \mathrm{TT}, \mathrm{TT}', \mathrm{LT}, \mathrm{LT}'$ can be extracted.

The measurement of virtual bremsstrahlung production in proton-proton scattering was carried out at KVI using the AGOR facility. We used a 190 MeV proton beam and a liquid hydrogen target. A coincidence setup of two detector systems was employed for an exclusive measurement. The Plastic Ball [1] detector [3], with approximately 550 $\Delta E - E$ modules was used for the detection of outgoing dilepton pairs, while the Small-Angle Large-Acceptance Detector (SALAD) [4] was employed for the detection of scattered protons.

[1]We gratefully acknowledge the loan of the Plastic Ball detector by GSI, Darmstadt.

The $\Delta E - E$ detector modules consist of a 4 mm CaF_2 layer in front of 35.6 cm plastic scintillator yielding a good particle identification for light charged particles. The high granularity (each detector module covers approximately 17.5 msr) results in polar and azimuthal angular resolution of approximately 5° and in an overall coverage of 77% of the 4π solid angle.

SALAD was specifically designed for the measurement of bremsstrahlung events. It consists of a MWPC and two cylindrically-shaped plastic scintillator layers. The MWPC provides polar and azimuthal angular resolution of approximately 0.5° with a polar angle coverage between 6° and 26°.

In Fig. 1 the extracted interference response functions are shown and compared to the LET calculation [2] which has detector response folded in with Monte Carlo technique. The results presented are preliminary. The calculations are normalized to the data using one arbitrary factor. Except that there seems to be a sign problem in extracting of W_{LT}, the overall agreement is reasonable. Presently, a comparison to a microscopic calculation by [5] is being studied.

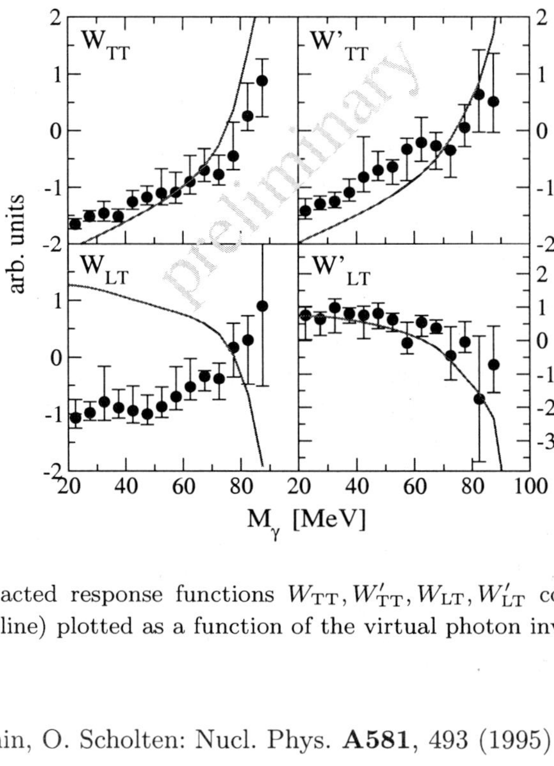

Figure 1. Extracted response functions $W_{\mathrm{TT}}, W'_{\mathrm{TT}}, W_{\mathrm{LT}}, W'_{\mathrm{LT}}$ compared to LET calculation (full line) plotted as a function of the virtual photon invariant mass M_γ.

References

1. A.Yu. Korchin, O. Scholten: Nucl. Phys. **A581**, 493 (1995) 493
2. A.Yu. Korchin, O. Scholten, D. van Neck: Nucl. Phys. **A602**, 423 (1996)
3. H.H. Gutbrod et al.: Nucl. Instr. and Meth. **A203**, 189 (1982)
4. N. Kalantar-Nayestanaki et al.: Nucl. Instr. and Meth. **A444**, 591 (2000)
5. G.H. Martinus et al.: Phys. Rev. **C56**, 2945 (1997)

Few-Body Systems Suppl. 14, 105–110 (2003)

Few-Body Systems

^{3,4}He(e,e′p) Experiments in Jefferson Lab's Hall A

D.W. Higinbotham, for the Jefferson Lab Hall A Collaboration

Thomas Jefferson National Accelerator Facility, Newport News, VA 23606

Abstract. Coincidence experiments have proven to be very useful tools in studying specific aspects of the nucleus. In particular the (e,e′p) reaction has been used not only to study the single-nucleon structure of nuclei but also to study the behavior of nucleons embedded in the nuclear medium. In this paper, the results of the Jefferson Lab ^{3}He(e,e′p) measurements will be presented along with a discussion of recent and upcoming ^{4}He(e,e′p) measurements.

1 Introduction

The 6 msr, 4 GeV/c high resolution spectrometer pair in Jefferson Lab's Hall A have proven to be an exceptional tool for making cross section and response function separation measurements of the (e,e′p) reaction at high momentum transfers [1, 2]. The first Hall A few-body (e,e′p) experiment was performed on ^{3}He [3]. The range of missing momentum for a fixed (ω, $\mathbf{q}$) is unprecedented and the complete set of data will allow the extraction of the in-plane response functions described below. Data for both the two-body ^{3}He(e,e′p)d reaction and the continuum ^{3}He(e,e′p)pn reaction were taken and preliminary results will be presented herein. In addition to ^{3}He, a focused set of cross section measurements have been performed on ^{4}He [4]. This experiment sought to observe the predicted dip in the ^{4}He(e,e′p)t momentum distribution. It is generally accepted that final state interactions have obscured this dip in previous measurements [5]. These data are presently being analyzed. There are future plans to obtain a set of ^{4}He data in a kinematic range similar to that of the ^{3}He experiment in order to allow the extraction of the in-plane response functions [6]. Taken together, these measurements on ^{3}He and ^{4}He will add to our understanding of relativistic effects in the few-body wave function, ground state short-range correlations, final state interactions, and non-nucleonic degrees of freedom.

2 Kinematics

The kinematics for the (e,e′p) reaction are shown in Fig. 1. The scattering plane is defined by the incoming electron, $e = (E_e, \mathbf{e})$, and the outgoing electron, $e' = (E'_e, \mathbf{e}')$. The four-momentum of the virtual photon is given by $q^\mu=(\omega, \mathbf{q})$ and the four-momentum of the outgoing proton is given by $p'^\mu = (E_p, \mathbf{p}')$. The four-momentum square, $Q^2 = q^2 - \omega^2$, is defined such that for electron scattering Q^2 is always positive. The missing momentum vector is defined as $\mathbf{p}_m = \mathbf{q} - \mathbf{p}'$ and represents the momentum of the recoiling system.

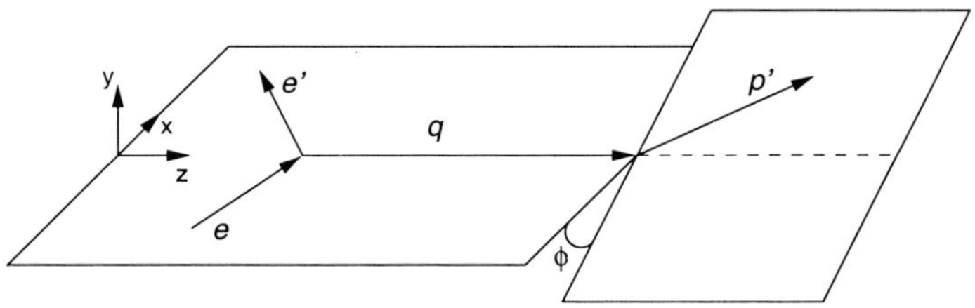

Figure 1. A schematic of the kinematics for the (e,e′p) reaction.

The form of the differential cross section for (e,e′p) reactions in the one-photon exchange approximation without polarization is:

$$\frac{d^6\sigma}{d\Omega_{e'}dE_{e'}d\Omega_{p'}dE_{p'}} = \frac{E_p p'}{(2\pi)^3}\sigma_{Mott}\Big[v_T R_T + v_L R_L + v_{TL} R_{TL}\cos\phi + v_{TT} R_{TT}\cos 2\phi\Big], \quad (1)$$

with ϕ the angle between the plane defined by $\mathbf{e}$ and $\mathbf{e}'$ and the plane defined by $\mathbf{p}'$ and $\mathbf{q}$, σ_{Mott} is the Mott cross section,

$$\sigma_{Mott} = \frac{4\alpha^2 E_{e'}^2}{Q^4}\cos^2\frac{\theta_e}{2}. \quad (2)$$

The kinematics factors v_L, v_T, v_{TL},and v_{TT} are:

$$v_L = \frac{Q^4}{q^4}, \quad (3)$$

$$v_T = \frac{Q^2}{2q^2} + \tan^2(\theta_e/2), \quad (4)$$

$$v_{TL} = \frac{Q^2}{q^2}\left[\frac{Q^2}{q^2} + \tan^2(\theta_e/2)\right]^{1/2}, \text{ and} \quad (5)$$

$$v_{TT} = \frac{Q^2}{2q^2}. \quad (6)$$

For the two body break-up channel, ^{3}He(e,e′p)d or ^{4}He(e,e′p)t, the proton energy and angle with respect to $\mathbf{q}$ are correlated because the missing energy is fixed. In this case the differential cross is written

$$\frac{d^5\sigma}{d\Omega_{e'}d\Omega_{p'}dE_{e'}} = \frac{E_p p'}{(2\pi)^3}\sigma_{Mott} f_{rec}^{-1}\Big[v_T R_T + v_L R_L + v_{TL} R_{TL}\cos\phi + v_{TT} R_{TT}\cos 2\phi \Big], \tag{7}$$

where f_{rec} is the recoil factor,

$$f_{rec} = \left[1 - \frac{E_{p'}}{E_t}\frac{\mathbf{p_m}\cdot\mathbf{p}'}{p'^2} \right]. \tag{8}$$

The response functions, R_L, R_T, R_{TL}, R_{TT} can be separated by a suitable choice of the kinematic parameters. In perpendicular in-plane kinematics with a fixed $\mathbf{q}$ and ω, one can separate R_T, R_{TL}, and a combination of the R_L and R_{TT} response functions, denoted as R_{L+TT}. One can also measure the cross section asymmetry A_{TL} for a given $\mathbf{q}$ and ω. This asymmetry is defined as:

$$A_{TL} = \frac{\sigma(\phi = 180) - \sigma(\phi = 0)}{\sigma(\phi = 180) + \sigma(\phi = 0)}. \tag{9}$$

In parallel and anti-parallel kinematics, i.e. when the out-going proton is in the direction of $\mathbf{q}$, one can separate the R_L and R_T response functions. In parallel kinematics p_m points in the opposite direction as $\mathbf{q}$ with $x_B < 1$ while in anti-parallel kinematics p_m points in the same direction as $\mathbf{q}$ with $x_B > 1$ where

$$x_B = \frac{Q^2}{2M\omega} \tag{10}$$

is the Bjorken scaling variable. For $x_B > 1$, the region in ω between the quasi-elastic peak and the elastic peak is being probed; while for $x_B < 1$, the region in ω towards the delta peak is being probed. The region in ω between the quasi-elastic peak and delta peak is often referred to as the dip region.

3 Results

The preliminary cross section results of the ^{3}He(e,e′p)d reaction for a fixed $\mathbf{q} = 1500$ MeV/c and $\omega = 840$ MeV for a beam energy of 4807 MeV are shown in Fig. 2. The agreement with the theoretical calculations of J.-M. Laget at missing momentum up to 750 MeV/c is striking. The cross section for p_m of up to 300 MeV/c follows the simple plane wave (PW) predictions. For higher p_m, the cross section becomes dominated by final state interactions, as indicated by the full calculations. J.-M. Laget's calculations use the Faddeev Paris two-body potential, the Hannover 18 three-body potential, and final state nucleon-nucleon interactions have been included using the eikonal approximation. There is as yet no clear indication as to what is causing the cross section at the largest missing momentum to be much greater than predicted.

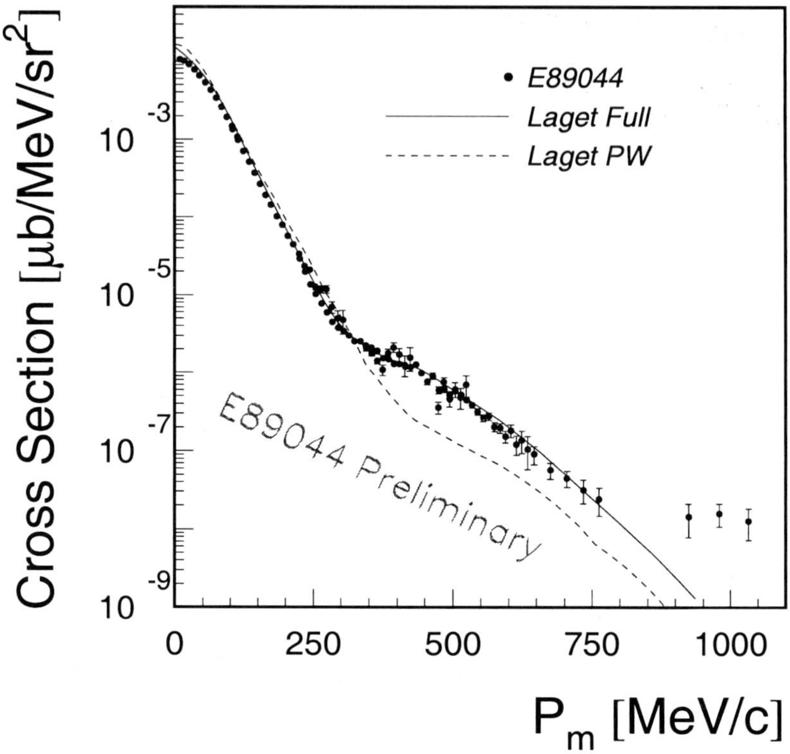

Figure 2. Shown are the preliminary cross section results for the reaction ^{3}He(e,e$'$p)d as a function of missing momentum with a beam energy of 4807 MeV and with a fixed $\mathbf{q}$ = 1500 MeV/c and ω = 840 MeV [1]. The theory curves show a plane wave impulse approximation calculation along with the latest full calculation of J.M. Laget. The enhancement in the cross section near 300 MeV/c and continuing to larger missing momentum is predominately due to final state interactions. There is no clear indication yet from theory what causes the enhancement of the cross section near 1000 MeV/c.

J. Udias is also working on making theoretical calculations on (e,e$'$p) reactions [9]. Udias has provided the collaboration with a rough A_{TL} calculation [10], shown in Fig. 3. Both the relativistic distorted wave calculation (RDWIA) and the relativistic plane wave calculation (RPWIA) were made using a mean field approximation.

From the preliminary cross section and A_{TL} asymmetry results, one can already see regions where the theoretical models clearly predicted the experimental results, such as the rise in the cross section at 300 MeV/c in Fig. 2 and the oscillations in Fig. 3. Also seen are the deficiencies in the theories, as indicated by much larger than expected cross section at extremely large missing momentum in Fig. 2 and by the phase and magnitude of the A_{TL} asymmetry.

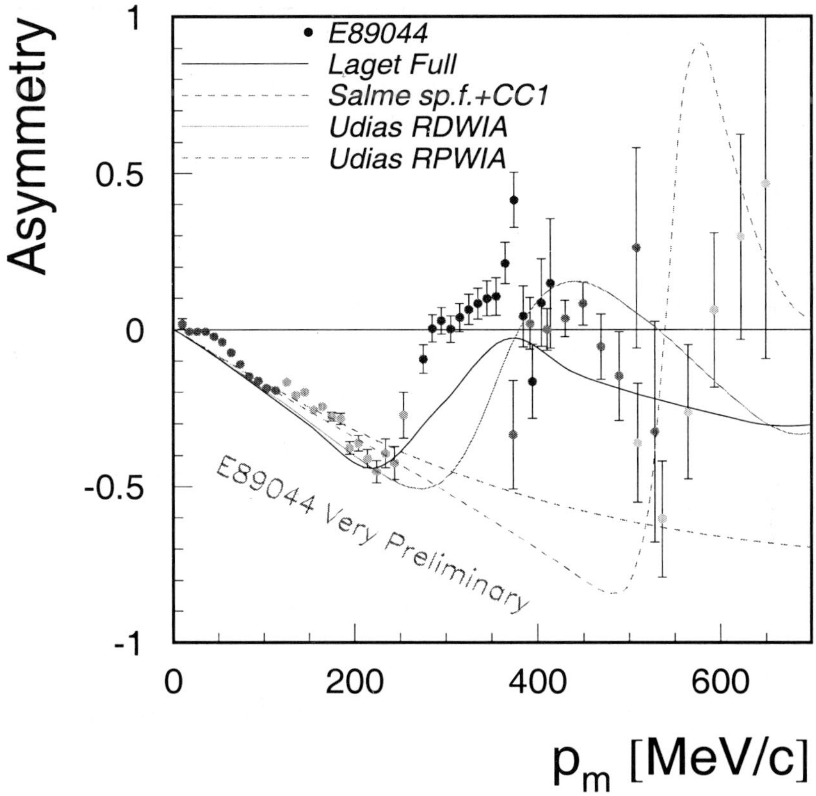

Figure 3. Shown is the preliminary A_{TL} data. The curves show the latest calculation of J.M. Laget along with a preliminary results of J. Udias. At lower missing momentum the theories are most sensitive to relativistic effects, while at the larger missing momentum the theories become sensitive to final state interactions effects.

Presently, the analysis of the separated response function separations is underway. Once these become available, it should become clear what effects are causing the discrepancies between theory and experiment and allow a significant improvement in our understanding of the three-body system. Also in the near future, the Hall A ^{4}He(e,e′p) cross section results will become available. In the coming years, the E01-108 ^{4}He(e,e′p) experiment will be run in Hall A and will provide a valuable investigation of the four-body system by measuring the in-plane response functions.

Acknowledgement. I would like to acknowledge the hard work and dedication of the doctoral students working on the analysis of the E89-044 experiment: Marat Rvachev with M.I.T., Emilie Penel-Nottaris with Grenoble, and Fatiha Benmokhtar with Rutgers. Also, I would also like to thank the experimental spokespersons, Marty Epstein, Arun Saha, and Eric Voutier, for their hard work and dedication along with the Hall A collaboration and the E89-044 collaboration. The Southeastern Universities Research Association (SURA) operates the Thomas Jefferson National Accelerator Facility under DOE Contract No. DE-AC05-84ER40150.

References

1. J. Gao et al.: Phys. Rev. Lett. **84**, 3265 (2000)
2. N. Liyanage et al.: Phys. Rev. Lett. **86**, 2975 (2001)
3. M.B. Epstein, A. Saha, E. Voutier (Spokespersons): Jefferson Lab Experiment E89-044, *Selected Studies of the ^{3}He Nuclei through Electrodisintegration at High Momentum Transfer*
4. J. Templon, J. Mitchell, B. Reitz (Spokespersons): Jefferson Lab Experiment E97-111, *Systematic Probe of Short-Range Correlations via the Reaction $^4He(e,e'p)^3H$*
5. J.J. van Leeuwe et al.: Phys. Rev. Lett. **80**, 2543 (1998)
6. K. Aniol, S. Gilad, D.W. Higinbotham, A. Saha (Spokespersons): Jefferson Lab Experiment E01-108, *Detailed Study of the ^{4}He Nuclei through Response Function Separations at High Momentum Transfers*
7. J.M. Laget: Phys. Lett. **B199**, 493 (1987)
8. J.M. Laget: Phys. Rev. **D579**, 333 (1994)
9. J.M. Udias, J.A. Caballero, E. Moya de Guerra, J.E. Amaro and T.W. Donnelly: Phys. Rev. Lett. **83**, 5451 (1999)
10. J. Udias: Private Communication

Few-Body Systems Suppl. 14, 111–116 (2003)

Few-Body Systems

Coulomb Effects in Three- and Four-Nucleon Systems

A. Kievsky, M. Viviani, L.E. Marcucci, S. Rosati

Istituto Nazionale di Fisica Nucleare, and Dipartimento di Fisica, Universita' di Pisa, Via Buonarroti 2, I-56100 Pisa, Italy

Abstract. Electromagnetic effects are studied in $p-d$ as well as in $p-{}^3$He scattering. Different observables are considered such as differential cross section and the vector and tensor analyzing powers. As an example calculations of the polarization-transfer coefficients $K_y^{y'}$ and $K_x^{x'}$ recently measured in $p-d$ and $n-d$ scattering are shown. Moreover the effect of the magnetic moment interaction is studied in the vector analyzing powers A_y and iT_{11} at different energies. The calculations have been done using the Argonne v_{18} two-nucleon plus the Urbana three-nucleon interaction. It is shown that Coulomb effects as well as magnetic moment effects are appreciable and cannot be neglected.

1 Introduction

The three- and four-nucleon systems are excellent testing grounds for the nuclear interaction. The last generation NN interactions can be used to calculate $3N$ and $4N$ bound and scattering states and, from a comparison to the experimental data, important conclusions about the capability of those interactions to reproduce the dynamics can be extracted. It is widely accepted that the potential energy in a few-nucleon system consists of a sum of the pairwise NN interaction and a term including a pure three-nucleon interaction (TNI). This last term is not very well known and, in general, its strength is fixed so as to reproduce the experimental $A = 3$ binding energy. With the recent advances in the solution of the $3N$ and $4N$ continuum, the possibility of using scattering data to improve our knowledge of the three-nucleon interaction is at present feasible. Because of this, a correct treatment of the Coulomb interaction in the description of few-nucleon reactions is required. In fact, Coulomb effects and three-nucleon force effects are very often of the same size. A rough estimate can be performed comparing the binding energy difference between ^{3}H and ^{3}He, which is 0.764 MeV. This is approximately the size of the underbinding obtained when the three-nucleon system is described only with a pure NN force.

2 Polarization-Transfer Coefficients

An interesting example where the size of Coulomb effects is comparable to that one obtained including TNI terms is represented by the polarization-transfer coefficients $K_y^{y'}$ and $K_x^{x'}$ recently measured in $p-d$ and $n-d$ scattering [1, 2]. Theoretical predictions for these observables at $E_{lab} = 19$ MeV are given in Fig. 1. The calculations have been made for $p-d$ scattering using the Argonne v_{18} (AV18) potential [3] (solid line) and including the Urbana (UR) [10] TNI (dotted line). The AV18+UR potential model has the property of reproducing the triton binding energy. The corresponding $n-d$ calculations using AV18 (dashed line) and AV18+UR (dotted-dashed line) are also shown.

The $p-d$ curve obtained with AV18+UR is very close to the $n-d$ curve calculated with AV18 showing that Coulomb and TNI effects are of the same size. Moreover the AV18+UR curve reproduces the $p-d$ data reasonable well. For $K_y^{y'}$ two $n-d$ data exist shown in Fig. 1 (squares). In the minimum the $n-d$ datum is above the corresponding $p-d$ datum. On the contrary the AV18 $n-d$ curve is below the AV18 $p-d$ curve. In order to clarify this conflict between theory and experiment a larger number of data points are required.

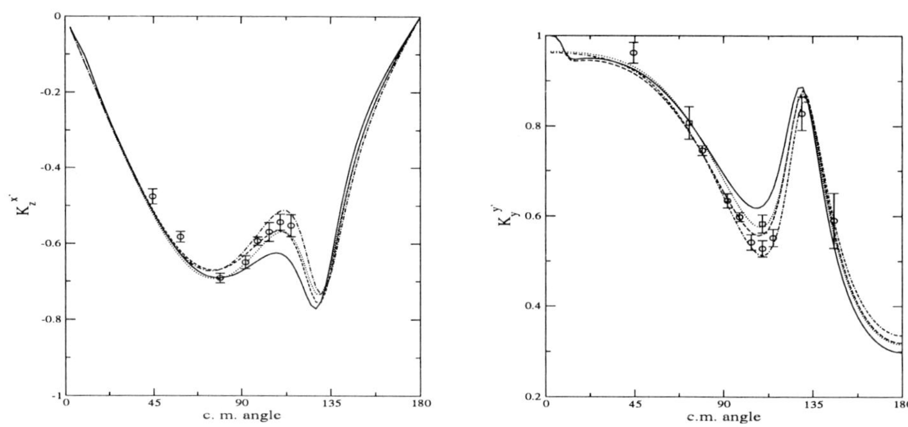

Figure 1. Polarization-transfer coefficients $K_x^{x'}$ and $K_y^{y'}$ at $E_{lab} = 19$ MeV calculated for $p-d$ and $n-d$ scattering using the AV18 (solid line and dashed line) and AV18+UR (dotted line and dotted-dashed line) respectively. Experimental points for $p-d$ scattering (circles) and for $n-d$ scattering (squares) are taken from refs [1,2], respectively.

3 Magnetic Moment Interaction

Following for example the notation used in the determination of the AV18 potential [3], all the modern NN potentials can be put in the general form

$$v(NN) = v^{EM}(NN) + v^{\pi}(NN) + v^{R}(NN) \ . \tag{1}$$

The short range part $v^R(NN)$ of all of these interactions includes a certain number of parameters (around 40), which are determined by a fitting procedure to the NN scattering data and the deuteron binding energy (BE), whereas the long range part is represented by the one-pion-exchange potential $v^{\pi}(NN)$ and an electromagnetic part $v^{EM}(NN)$.

For AV18, $v^{EM}(pp)$ consists of the one- and two-photon Coulomb terms plus the Darwin-Foldy term, vacuum polarization and magnetic moment (MM) interactions. The $v^{EM}(np)$ interaction includes a Coulomb term due to the neutron charge distribution in addition to the MM interaction. Finally, $v^{EM}(nn)$ is given by the MM interaction only. All these terms take into account the finite size of the nucleon charge distributions. Explicitly the NN magnetic moment interaction reads:

$$\begin{aligned}
v_{MM}(pp) &= -\frac{\alpha}{4M_p^2}\mu_p^2\left[\frac{2}{3}F_\delta(r)\boldsymbol{\sigma}_i\cdot\boldsymbol{\sigma}_j + \frac{F_t(r)}{r^3}S_{ij}\right] \\
&\quad -\frac{\alpha}{2M_p^2}(4\mu_p - 1)\frac{F_{ls}(r)}{r^3}\boldsymbol{L}\cdot\boldsymbol{S} \qquad (2) \\
v_{MM}(np) &= -\frac{\alpha}{4M_nM_p}\mu_n\mu_p\left[\frac{2}{3}F_\delta(r)\boldsymbol{\sigma}_i\cdot\boldsymbol{\sigma}_j + \frac{F_t(r)}{r^3}S_{ij}\right] \\
&\quad -\frac{\alpha}{2M_nM_r}\mu_n\frac{F_{ls}(r)}{r^3}(\boldsymbol{L}\cdot\boldsymbol{S} + \boldsymbol{L}\cdot\boldsymbol{A}) \qquad (3) \\
v_{MM}(nn) &= -\frac{\alpha}{4M_n^2}\mu_n^2\left[\frac{2}{3}F_\delta(r)\boldsymbol{\sigma}_i\cdot\boldsymbol{\sigma}_j + \frac{F_t(r)}{r^3}S_{ij}\right] \qquad (4)
\end{aligned}$$

In the above formula F_δ, F_t and F_{ls} describe the finite size of the nucleon charge distributions. M_p (M_n) is the proton (neutron) mass whereas M_r is the nucleon reduced mass. The MM interaction presents the usual r^{-3} behavior and has an operatorial structure with a spin-spin, a tensor and a spin-orbit term. In the np case, this last term includes an asymmetric force (proportional to $\boldsymbol{A} = (\boldsymbol{\sigma}_i - \boldsymbol{\sigma}_j)/2$) which mixes spin-singlet and spin-triplet states. This term is expected to have a very small contribution and has not been considered in the present analysis.

The inclusion of the MM interaction in the description of few nucleon systems has been limited at present to bound states [4]. Recently a detailed analysis of the contribution of the electromagnetic terms to the ^{3}He–^{3}H mass difference has been performed [5]. Here we would like to show the contribution of the MM interaction to $p-d$ and $p-^3$He scattering at low energies. A first analysis in $p-d$ scattering has been performed by Stoks [6] including the electromagnetic terms in Born approximation between a spin-1/2 particle and a spin-1 particle. In this approximation the main modifications have been obtained in A_y and iT_{11} at very forward angles. No modifications have been observed in other observables as differential cross section and tensor analyzing powers and in the maximum of A_y and iT_{11}. So the first conclusion was that the MM interaction does not help in the theoretical underprediction of the last two observables. However the Born approximation could not be correct in the description of low partial waves which govern the polarization observables at low energies. Using the Correlated Hyperspherical Harmonic basis [7, 8], we calculated the polarization observables at different energies using the Kohn Variational Principle in complex form [9] for $p-d$ and $p-{}^3$He scattering. We have considered the AV18+UR potential model to which we have added the complete MM interaction given above (except for the asymmetric force in the np case).

In Figs. 2 and 3 the $p-d$ vector analyzing powers A_y and iT_{11} are given for four potential models: AV18, AV18+UR, AV18+MM and AV18+UR+MM at $E_{lab}=3$ MeV and 650 keV respectively. The contribution of the MM interaction is evident mainly at low energies. Though the underprediction is still present, there is a reduction of the discrepancy in particular for iT_{11}. Moreover the MM interaction gives a much better description at forward angles as expected. A similar effect is observed in $p-{}^3$He elastic scattering. The corresponding A_y has been calculated at $E_{lab}=2.25$ MeV using AV18 and AV18+UR and, as it is shown in Fig. 4, both results are very close to each other and underpredict the analyzing power. The inclusion of the MM interaction improves the description of the observable.

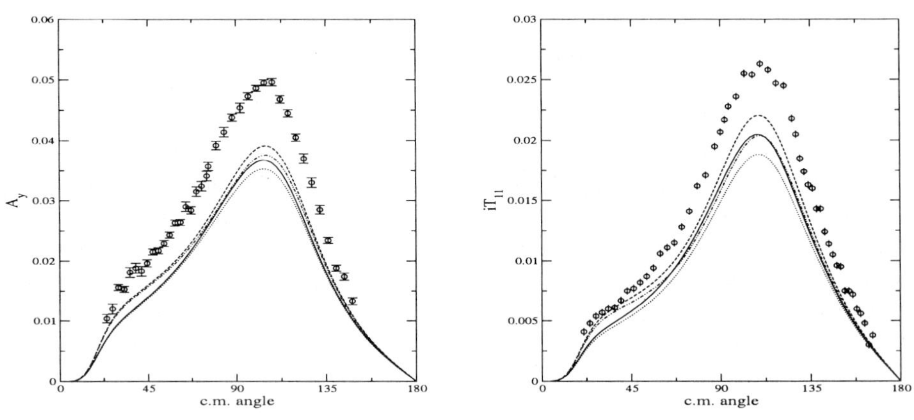

Figure 2. The vector analyzing powers A_y and iT_{11} calculated at $E_{lab}=3$ MeV using the AV18 (dotted line), AV18+UR (solid line), AV18+MM (dotted dashed line) and AV18+UR+MM (dashed line). Data points are from ref. [11].

An analysis of the discrepancy between theory and experiment for iT_{11} in $p-d$ scattering up to 10 MeV is given in Fig. 5. The difference at the maximum has been calculated using AV18 and AV18+MM. In the later case the discrepancy reduces to the level of 20% – 25%.

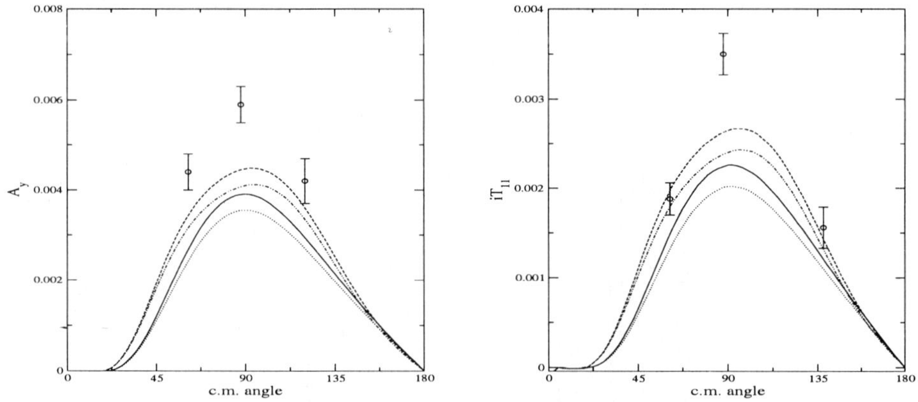

Figure 3. The vector analyzing powers A_y and iT_{11} calculated at $E_{lab} = 650$ keV using the AV18 (dotted line), AV18+UR (solid line), AV18+MM (dotted dashed line) and AV18+UR+MM (dashed line). Data points are from ref. [12].

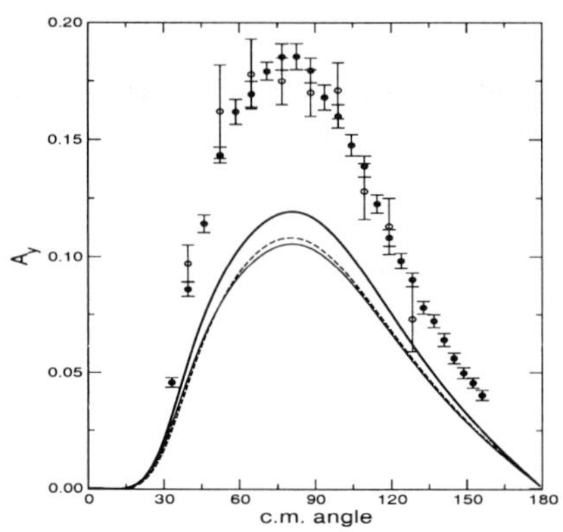

Figure 4. The $p-{}^3$He proton analyzing power A_y at $E_{lab} = 2.25$ MeV using the AV18 (solid line), AV18+UR (dashed line) and AV18+MM (thick solid line). Data points are from ref. [13] (open circles) and ref. [14] (filled circles).

References

1. L. Sydow et al.: Few-Body Syst. **25**, 133 (1998)
2. P. Hempen et al.: Phys. Rev. **C57**, 484 (1998)

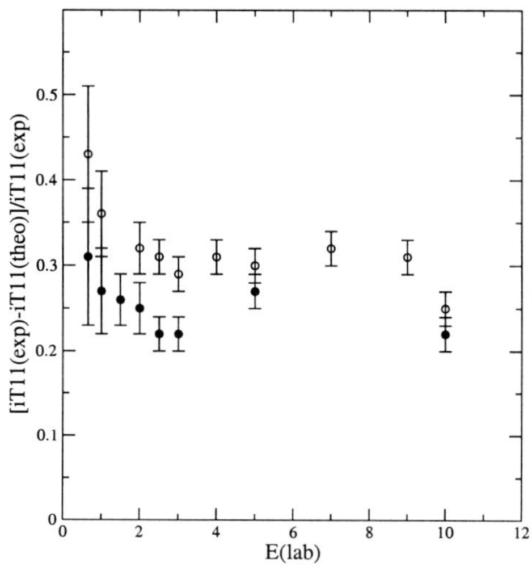

Figure 5. Difference in the maximum of iT_{11} between experiment and theory calculated using AV18 (open circles) and AV18+MM (filled circles).

3. R.B. Wiringa, V.G.J. Stoks, R. Schiavilla: Phys. Rev. **C51**, 38 (1995)
4. S.C. Pieper et al.: Phys. Rev. **C64**, 014001 (2001)
5. A. Nogga et al.: nucl-th/0202037
6. V.G.J. Stoks: Phys. Rev. **C57**, 445 (1998)
7. A. Kievsky, M. Viviani, S. Rosati: Nucl. Phys. **A577**, 511 (1994)
8. M. Viviani, S. Rosati, A. Kievsky: Phys. Rev. Lett. **81**, 1580 (1998)
9. A. Kievsky: Nucl. Phys. **A624**, 125 (1997)
10. B.S. Pudliner et al.: Phys. Rev. Lett. **74**, 4396 (1995)
11. S. Shimizu et al.: Phys. Rev. **C52**, 1193 (1995)
12. C.R. Brune et al.: Phys. Rev. **C63**, 044013 (2001)
13. M. Viviani et al.: Phys. Rev. Lett. **86**, 3739 (2001)
14. B. Fisher: Private Communication

Few-Body Systems Suppl. 14, 117–122 (2003)

Few-Body Systems

Testing the Nuclear Hamiltonian in Few-Nucleon Systems

H. Witała[1,*], E. Epelbaum[2], W. Glöckle[2], J. Golak[1,2], H. Kamada[3], J. Kuroś-Żołnierczuk[1], A. Nogga[4], R. Skibiński[1]

[1] M. Smoluchowski Institute of Physics, Jagiellonian University, PL-30059 Kraków, Poland

[2] Institut für Theoretische Physik II, Ruhr Universität Bochum, D-44780 Bochum, Germany

[3] Department of Physics, Faculty of Engineering, Kyushu Institute of Technology, 1-1 Sensuicho, Tobata, Kitakyushu 804-8550, Japan

[4] Department of Physics, University of Arizona, Tucson, Arizona 85721, USA

Abstract. Recent results on three-nucleon (3N) scattering are presented. The discrepancies between a theory based on numerical solutions of 3N Faddeev equations with NN interactions only and data point to the action of three-nucleon forces (3NF's). Also results of application of 3N bound and scattering states to photodisintegration of 3N bound states are shown.

1 Introduction

A major goal in nuclear physics is to establish the form of nuclear forces to be used in the nuclear Hamiltonian. The construction of such forces guided by meson theory led to a new generation of realistic nucleon-nucleon (NN) potentials which describe with high precision the NN data set. These potentials can be tested in nontrivial manner in 3N system. This system is also the most simple one which can give information on the significance and properties of 3N forces (3NF's) in the nuclear Hamiltonian. That effects of such interactions are non-negligible became evident with the advance of exact solutions for three- and four-nucleon bound states using standard integration and differentiation methods. In all cases realistic NN forces lead to a clear-cut underbinding. A natural idea to explain this underbinding is the consideration of 3N forces in the nuclear Hamiltonian.

* *E-mail address:* witala@if.uj.edu.pl

The advance in computational resources allowed to solve numerically exactly the 3N scattering equations using 2N and 3N forces in all their complexity. It provided the unique tool to study the 3N scattering processes with their rich set of spin observables for a more detailed investigation of 3NF properties than offered by bound states. In the next section a brief review of the 3N scattering formalism and the present state of 3N continuum calculations are given together with some examples where data are compared to various NN potential predictions only and combined with different 3NF's. These examples are chosen to show the importance of the 3NF in the 3N system.

The application of rigorous 3N bound and scattering states to photodisintegration of 3N bound states presented in section 3 shows that also this reaction is a valuable source of information on 3NF properties and electromagnetic current operator of the 3N system. The summary is given in section 4.

2 Three Nucleon Scattering

We solve the 3N Faddeev equation for an amplitude $T|\phi\rangle$ from which all observables for elastic scattering and the breakup process can be obtained [1, 2]. Using the modern NN forces: AV18 [3], CD Bonn [4], NijmI, II, and 93 [5] in general one gets predictions for 3N scattering observables which agree well with data at energies below $\approx$ 30 MeV. A fairly complete overview of those theoretical predictions in comparison to data is presented in [1]. At higher energies discrepancies develop. They are exemplified for the elastic scattering in Fig. 1. The large discrepancy in the minimum of the cross section using NN forces only is removed when 3NF's are included reproducing the experimental triton binding energy [2]. A similar behavior shows the deuteron vector analyzing power [2] (Fig. 1). But there are many spin observables for which large 3NF effects are predicted and where moreover the TM [6] and the Urbana IX [7] give different predictions [2]. This is also the case e.g. for the nucleon analyzing power A_y where in addition both 3NF's do not reproduce the data [2]. Similar happens for the tensor analyzing powers [2]. In both cases the data cannot be reproduced by pure 2N force predictions either.

The Nd breakup process in specific kinematically complete configurations is also very promising to provide information on the nuclear Hamiltonian and cross sections and some spin observables exhibit very large 3NF effects at higher energies, which moreover are different for the TM and the Urbana IX models [8]. Therefore observables for elastic Nd scattering and the breakup process have been identified, which are sensitive to the 3NF structure [2, 8]. With precise data on such spin observables it should be possible to reveal the proper spin structure of the 3NF.

3 Photodisintegration of 3N Bound States

The driving term of the Faddeev-like equation for the amplitude $|U_T\rangle$ defining the nuclear matrix element for photodisintegration process contains all of

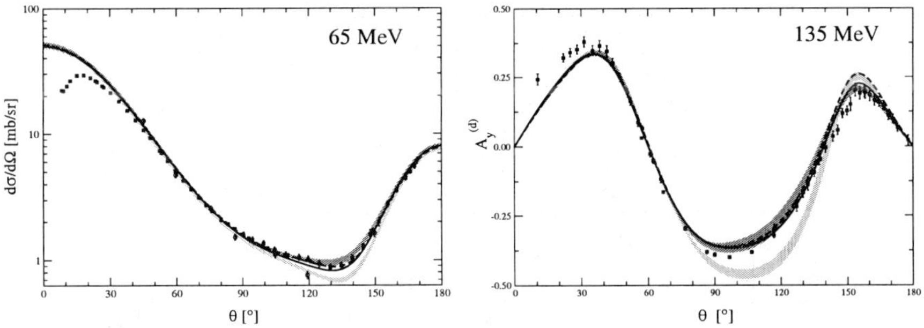

Figure 1. The cross section at E= 65 MeV and the deuteron vector analyzing power $A_y(d)$ at E= 135 MeV for elastic Nd scattering. For curves and bands see text. The data for E=65 MeV are from [13] (pd - filled squares) and from [14] (nd - open circles). The data for E= 135 MeV are from [15] and [16].

the dynamical input: the 3N bound state, the current operator, the free 3N propagator and the NN t-operator together with 3NF [9]. The nucleonic electromagnetic current operator acting on 3N bound state has single, two- and three-nucleon parts. We use the single nucleon current whose operator form is well established in terms of the electromagnetic form factors of proton and neutron. It is supplemented by the 2N current in the form of π and ρ exchange currents. As an alternative we use the Siegert approach. Some many-body currents are thereby included based on the continuity equation.

In Fig. 2 we show angular distributions for various energies without and with 3NF's and using the Siegert approach. We see that for $E_\gamma \leq 30$ MeV the maximum is decreased by adding a 3NF while it is increased above that energy. Also it is important to state that the results are fairly stable under the exchange of the dynamical input. The differential cross sections data taken at 103.05° shown in Fig. 3 seem to support the predicted 3NF effects. However, more and high precision data would be needed to challenge our choice of currents and 3NF effects.

The search over the photodisintegration three-body breakup phase space revealed a lot of interesting cases with large 3NF effects [10]. It also revealed that the choice of the current leads to differences in the predictions, which disturb the search for 3N force effects. The geometries can be found where the cross section is large, 3NF effects are big and the difference between "Siegert" and MEC's is small (example is shown in Fig. 4, where the five fold differential cross section is displayed along the kinematically allowed S-curve).

Based on the results obtained up to now we can say that photodisintegration of 3N bound states can very likely contribute significantly to identify 3NF effects and will be important to test the overall dynamical picture.

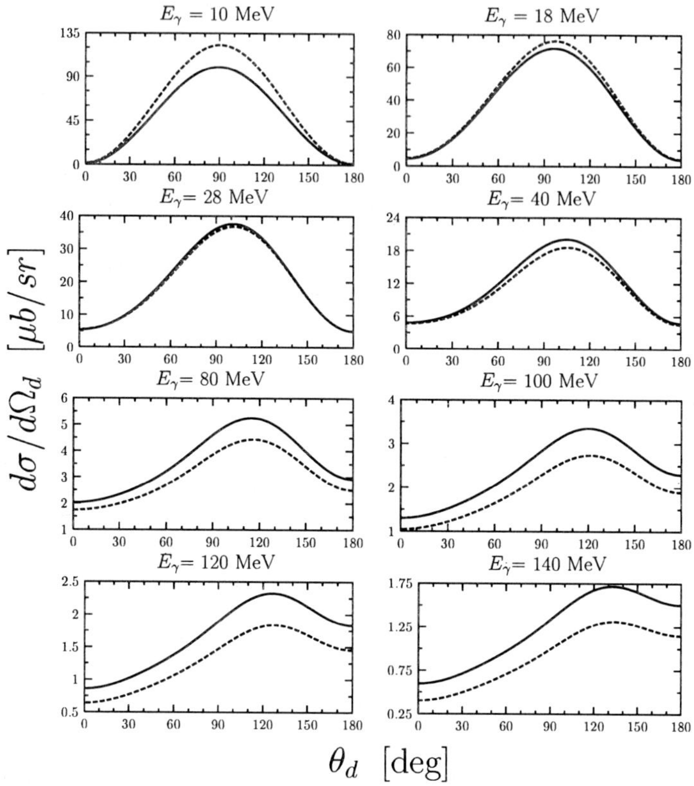

Figure 2. Deuteron laboratory angular distribution for the process ^{3}He$(\gamma, d)p$ at different photon laboratory energies E_γ. Curves show results of calculations with the AV18 NN and Urbana IX 3NF forces (solid) and with the AV18 NN force alone (dashed). The current is treated in the Siegert approach.

4 Summary

Solving 3N scattering exactly in a numerical sense up to energies below the pion production threshold allows to test 3N Hamiltonians based on modern NN potentials and 3NF's. At the higher energies for some observables large 3NF effects are predicted when using present day models such as TM and Urbana IX. Recent Nd elastic scattering cross sections and polarization data support some of these predictions. Others, however, indicate defects of present day 3NF's. It can be expected that a precise and rich data basis, comprising elastic scattering and breakup data, will enable to establish the proper spin structure of 3N forces.

A well founded 3N Hamiltonian is a prerequisite for a theoretical analysis of photodisintegration of 3N bound states. Exact 3N bound and scattering states

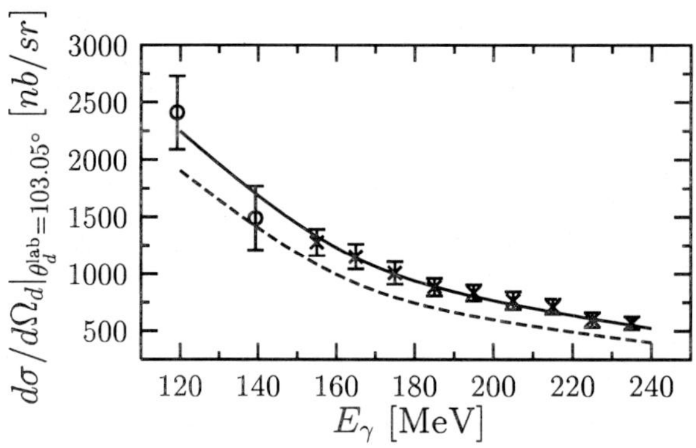

Figure 3. Deuteron angular distribution for the process $^3\text{He}(\gamma, d)p$ at given laboratory angle as a function of the photon laboratory energy E_γ. Curves show results of calculations with the AV18 NN and Urbana IX 3NF forces (solid) and with the AV18 NN force alone (dashed). Explicit π- and ρ-like MEC's are included in the current operator. Data are from [17] (x-symbols) and [18] (circles).

are mandatory in such an analysis to account for the final state interactions. First results of our calculations indicate that inclusion of a 3NF might be necessary in some kinematical regions. Consistent treatment of two- and three-nucleon forces together with electromagnetic current operator is called for. The chiral perturbation theory (χPT) seems to be promising in this respect [11, 12].

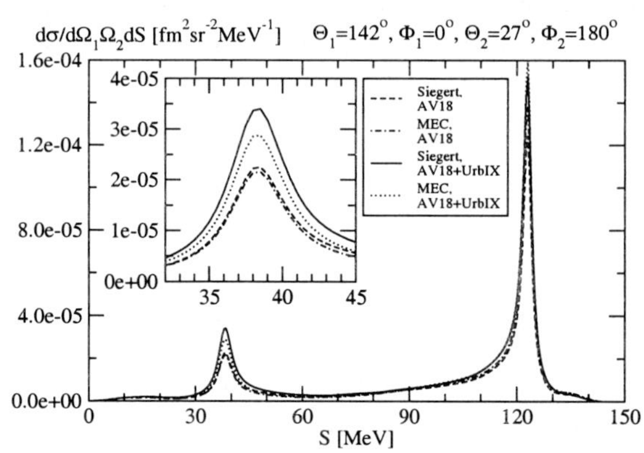

Figure 4. Five-fold differential cross section displayed along the kinematically allowed S-curve for the case, where the cross section is large, 3NF effects are big and the difference between "Siegert" and MEC's is small.

Acknowledgement. This work was supported by the Polish Committee for Scientific Research under Grants No. 2P03B02818 and 2P03B05622, by the Deutsche Forschungsgemeinschaft (J.G.),and by NFS grant # PHY0070858. R.S. is a holder of scholarship from the Foundation for Polish Science and acknowledges its financial support. The numerical calculations have been performed on the Cray T90 of the NIC in Jülich, Germany.

References

1. W. Glöckle et al.: Phys. Rep. **274**, 107 (1996)
2. H. Witała et al.: Phys. Rev. **C63**, 024007 (2001)
3. R.B. Wiringa et al.: Phys. Rev. **C51**, 38 (1995)
4. R. Machleidt et al.: Phys. Rev. **C53**, R1483 (1996)
5. V.G.J. Stoks et al.: Phys. Rev. **C49**, 2950 (1994)
6. S.A. Coon and W. Glöckle: Phys. Rev. **C23**, 1790 (1981)
7. B.S. Pudliner et al.: Phys. Rev. **C56**, 1720 (1997)
8. J. Kuroś-Żołnierczuk et al.: Phys. Rev. **C66**, 024003 (2002); Phys. Rev. **C66**, 024004 (2002)
9. J. Golak et al.: Nucl. Phys. **A707**, 365 (2002)
10. R. Skibiński: PhD Thesis, Jagiellonian University 2002; to appear in http://www.if.uj.edu.pl/ZFJ/prj/thFewbod.htm
11. E. Epelbaum et al.: nucl-th/0201064
12. E. Epelbaum et al.: nucl-th/0208023
13. S. Shimizu et al.: Phys. Rev. **C52**, 1193 (1995)
14. H. Rühl et al.: Nucl. Phys. **A524**, 377 (1991)
15. H. Sakai et al.: Phys. Rev. Lett. **84**, 5288 (2000)
16. N. Sakamoto et al.: Phys. Lett. **B367**, 60 (1996)
17. D.I. Sober et al.: Phys. Rev. **C28**, 2234 (1983)
18. N.M. O'Fallon et al.: Phys. Rev. **C5**, 1926 (1972)

Few-Body Systems Suppl. 14, 123–126 (2003)

Few-Body Systems

An Effective Field Theory for the Three-Body System

T. Barford, M.C. Birse

Theoretical Physics, Department of Physics and Astronomy, University of Manchester, Manchester, M13 9PL, UK.

Abstract. We study the scattering of a particle from a bound pair in an effective field theory using a distorted-wave renormalisation group method to find the power-counting for the three-body force terms. We find that three-body terms appear at lower orders than naively expected. They start with a marginal term that varies as a logarithm rather than a power of the energy scales in the problem. The marginal term has important implications for the three-body problem in nuclear physics.

1 Introduction

The need for a model-independent approach to the three-nucleon system is clear. The plethora of two-body potentials, each capable of describing two-body data extremely well, fail when applied to the three-body problem without an additional three-body force constructed to fit the data.

Effective Field Theories (EFTs) offer a systematic and model-independent treatment of nuclear and hadronic physics at low energies [1]. In an EFT we hope to take advantage of a seperation of scales by expanding observables in powers of the ratio Q/Λ_0, where Q denotes a generic low-energy scale and Λ_0 a typical scale of the underlying physics. The expansion will be useful if the seperation of scales is large enough to ensure quick convergence.

An EFT is defined by a Lagrangian containing all possible local terms consistent with the symmetries of the underlying theory. Although this will invariably lead to an infinite number of terms, we hope to organise them according to some "power-counting", related to the number of low-energy scales in each term. EFTs are well-developed and understood for two-body physics and now the three-body problem is an area of keen interest.

The mathematical tool that allows us to determine the power-counting is the renormalisation group (RG) [2]. If we are concerned with problems where the wavelengths of the particles are far longer than the range of the interaction, the scattering is insensitive to the detailed structure of that interaction. Replacing the interaction with contact interactions and forming an EFT Lagrangian leads to UV divergences from the high-momentum modes of loop diagrams. Accordingly we regulate the loop diagrams with a floating UV cut-off Λ. Since all observables should be independent of this cut-off we renormalise the theory by absorbing the resulting Λ-dependence into the EFT couplings. Solving the resulting RG equation provides a frame-work for constructing power-counting schemes. The resulting EFT absorbs the effects of the short-range physics and parameterises it on a power-series in Q.

The key to constructing power-counting schemes is fixed points. After rescaling all quantities in terms of Λ, we may look for solutions to the RG equation which are independent of Λ. Such solutions are referred to as fixed points. Since these have no scale attached to them solutions of the RG equation tend to one of them as $\Lambda \to 0$. Study of the scaling behaviour of solutions close to the fixed points leads to different power-counting schemes.

2 Two-Body EFT

The simplest EFT has only one field, the field of the asymptotic particles being considered. Such a theory is known as a "pionless" EFT in nuclear physics. For a two-body system the Lagrangian takes the form,

$$\mathcal{L}_{2B} = \psi^\dagger \left(i\partial_0 + \frac{\nabla^2}{2M} \right) + \frac{C_0}{4}|\psi|^4 + \frac{C_2}{4} \left(\psi^\dagger (\overrightarrow{\nabla} - \overleftarrow{\nabla})^2 \psi^\dagger \right) \psi^2 + \text{H.c.}) + \dots \quad (1)$$

Using the RG method outlined above one may construct two different power-counting schemes for such a system. The RG equation for the leading term C_0 is,

$$\Lambda \frac{\partial \hat{C}_0}{\partial \Lambda} = \hat{C}_0(1 + \hat{C}_0), \quad (2)$$

where the hat signifies a dimensionless rescaled coupling. There are clearly two fixed point solutions to this equation, $\hat{C}_0 = 0$ and $\hat{C}_0 = -1$. These lead to the two different schemes.

Perturbing about the fixed point $\hat{C}_0 = 0$ leads to Weinberg or naive counting in which the leading perturbation scales with Q. This counting system is useful for weakly interacting systems.

Perturbing about the second fixed point $\hat{C}_0 = -1$ leads to KSW counting, in which the leading perturbation scales with Q^{-1}. The fixed-point solution corresponds to a system with a bound-state at exactly zero energy, consequently this expansion is useful for strongly attractive systems with a shallow bound-state and is the expansion most useful in nuclear and atomic physics. The counting scheme is equivalent to the effective range expansion. The couplings, C_0 and C_2, can be directly related to the scattering length, a_2, and effective range, r_2, respectively.

Once a power-counting scheme is chosen the free parameters in the couplings can be fixed using experimental data. The resulting theory is then predictive for other processes.

3 Building a Three-Body EFT

Our aim is to extend the KSW EFT to describe three-body interactions, by using the already determined two-body couplings and constructing a counting scheme for the three-body couplings,

$$\mathcal{L}_{3B} = \frac{D_0}{36}|\psi|^6 + \frac{D_2}{36}\left(\psi^\dagger(\overrightarrow{\nabla} - \overleftarrow{\nabla})^2\psi^\dagger\right)\psi^\dagger\psi^3 + \mathrm{H.c.}) + \ldots \tag{3}$$

In scattering of a third particle from a bound pair the two-body contact potential leads to a particle exchange force with a range $\sim a_2$. In nuclear physics in particular, the length scale set by the scattering length is far longer than any other scale in the problem. We hope to use this seperation of scales to construct an EFT for three-bodies.

We may treat the long-range physics, $Q \sim 1/a_2$ exactly by working in terms of the distorted waves (DWs) of the particle-exchange potential [3]. The tool for constructing the power-counting schemes is then the distorted wave renormalisation group (DWRG) in which the cut-off Λ is applied in the basis of the distorted waves. This method ensures that all non-analytic behaviour resulting from the long-range potential is factored out in terms constructed from the DWs.

To apply the DWRG method to the three-body problem we must find the three-body wavefunction for a system with a finite two-body scattering length and zero effective-range. This two-body interaction can be expressed simply as a boundary condition on the wavefunction at zero seperation. For the three-body wavefunction this takes the form,

$$\left[\frac{\partial\psi(r_{23}, r_1)}{\partial r_{23}}\right]_{r_{23}=0} = -\frac{1}{a_2}\left[\psi(r_{23}, r_1)\right]_{r_{23}=0}, \tag{4}$$

where we are using the three-body Jacobian coordinate system. When this is applied to the integral equation for the three-body wavefunction we obtain an integro-differential equation for the projection of the wavefunction, $\phi(r_1) = \psi(r_{23} = 0, r_1)$. This is equivalent to the equation given by Skorniakov and Ter-Martirosian [4]. Generally, the equations for ϕ and ψ must be solved numerically. In the zero-energy limit we can solve them analytically. These analytic solutions give the short-range asymptotic behaviour of the full solutions

$$\psi(r_{23}, r_1) = \mathcal{N}(ka_2) f(\theta) \sin(s_0 \ln(kR) + \eta(R_0)), \tag{5}$$

where R and θ are the hyperradius and hyperangle respectively, f is some known function, $k = \sqrt{4(a_2^2 + ME)/3}$, $s_0 \approx 1$, $\mathcal{N}$ is some numerically computed normalisation. Here, η is an arbitrary phase that must be fixed by applying a boundary condition at some R_0, $\eta(R_0) = -s_0 \ln(kR_0)$. This solution is precisely that found by Efimov [5].

Using the DWRG method, we may use this result to construct the power-counting for the three-body system. One finds that the DWRG equation has no true fixed point but a logarithmic evolution. Perturbations around this lead to an expansion of the form $\mathcal{D}_{n,m}p^{2n}a_2^{-m}\Lambda_0^{-2n-m}$, where p is the on-shell momentum, $n, m \geq 0$, $\mathcal{D}_{n,m}$ are free dimensionless parameters that should be of order one and Λ_0 is the scale at which the EFT breaks down. This expansion provides the power-counting for the system.

The leading term is marginal, i.e. it is dimensionless and does not scale with Q/Λ_0. As expected, such a marginal perturbation leads to logarithmic behaviour in Q which necessitates the introduction of a new scale. In this case the scale is in fact R_0, the scale appearing in the boundary condition on the three-body wavefunction.

It is important to notice that since the leading three-body term is marginal, it occurs at next-to-leading order in the expansion (two-body leading term occurs at Q^{-1}) rather than the naively expected next-to-next-to-leading order.

The final result may be expressed as a DW effective range expansion,

$$\mathcal{N}(pa_2)\cot(\delta_{3B}) = -\mathcal{M}(pa_2) + \sum_{n,m} \mathcal{D}_{n,m}p^{2n}a_2^{-m}\Lambda_0^{-2n-2m}. \tag{6}$$

δ_{3B} is the correction to the phaseshift due to the three-body force and $\mathcal{N}$ and $\mathcal{M}$ are numerically computed functions.

4 Conclusion

This analysis shows the importance of the three-body force in this system. The marginal behaviour of this force means that it must be included in the low-energy EFT. It, or equivalently the boundary condition on the three-body wave function at short distances, is required to resolve the ambiguities in the three-body system with zero-range two-body forces. In nuclear physics, its role can be seen from the fact that different phenomenological nucleon-nucleon potentials lead to quite different predictions for three-nucleon binding energies and scattering lengths. However, these predictions all lie on a single "Phillips line" [6] showing that their differences can be expressed in a single effective parameter R_0, the scale appearing the marginal term of the three-body potential.

References

1. S.R. Beane, P.F. Bedaque, W.C. Haxton, D.R. Phillips and M.J. Savage: nucl-th/0008064; P.F. Bedaque and U. van Kolk: nucl-th/0203055
2. M.C. Birse, J.A. McGovern and K.G. Richardson: Phys. Lett. **B464**, 169 (1999); hep-ph/9807302
3. T. Barford and M.C. Birse: hep-ph/0206146
4. G.V. Skorniakov and K.A. Ter-Martirosian: Sov. Phys. JETP **4**, 648 (1957)
5. V.N. Efimov: Sov. J. Nucl. Phys. **12**, 589 (1971)
6. A.C. Phillips: Nucl. Phys. **A107**, 209 (1968)

Few-Body Systems Suppl. 14, 127–130 (2003)

Few-Body Systems

Nucleon-Deuteron Scattering with Δ-Isobar Excitation: Chebyshev Expansion of Two-Baryon Transition Matrix

A. Deltuva, K. Chmielewski, P.U. Sauer

Institut für Theoretische Physik, Universität Hannover, D-30167 Hannover, Germany

Abstract. A new technique for solving three-particle scattering equations is developed. It is based on the two-dimensional Chebyshev expansion of the two-baryon transition matrix. Its validity and its effectiveness is demonstrated. The dynamics of the examples is based on a two-baryon potential which allows for the excitation of a nucleon to a Δ-isobar; the coupled-channel potential yields an effective three-nucleon force in the three-nucleon system. The purely nucleonic reference potential is the charge-dependent Bonn potential.

We embarked on the description of nucleon-deuteron scattering with Δ-isobar excitation. Reference [1] developed and used a coupled-channel formulation within the framework of nonrelativistic quantum mechanics: Three-nucleon channels are coupled to those in which one nucleon is turned into a single Δ-isobar. The Δ-isobar is considered a stable baryon with spin and isospin $\frac{3}{2}$. The description applies to scattering energies well below the pion-production threshold. The virtual excitation of the Δ-isobar yields an effective three-nucleon force, besides other Δ-isobar effects. First results of that coupled-channel description of nucleon-deuteron scattering are given for elastic scattering in ref. [1] and for break-up in ref. [2]. Compared to refs. [1, 2], the underlying purely nucleonic reference potential is a more modern one, the charge-dependent CD Bonn potential [3]. Partial waves up to total two-baryon angular momentum $I = 5$ and up to total three-baryon angular momentum $\mathcal{J} = \frac{27}{2}$ with total isospin $\mathcal{T} = \frac{1}{2}$ and $\frac{3}{2}$ are taken into account.

Till now, we have solved the three-particle scattering equations in momentum space by a separable expansion of the two-baryon transition matrix. Though the validity of the separable expansion is tested in ref. [1] and confirmed there to be quite reliable, this contribution radically improves the numerical technique. Instead of using separable expansions as dynamic input, the two-baryon transition matrix $T_\alpha(Z)$ is calculated exactly, but for further applications its momentum dependence is represented and effectively interpolated with the help of Chebyshev polynomials, i.e.,

$$T_\alpha(Z) \approx \sum_{\nu'\nu} \int q^2 dq \int p'^2 dp' \int p^2 dp$$
$$\times\, |p'q\nu'(I'j')\rangle_\alpha \sum_{i',i=0}^{n_c-1} t^{i'}_{L'}(p')\, \delta_{\chi'\chi}\, T^{i'i}_{\eta'\eta}(\chi q, Z)\, t^i_L(p)\, {}_\alpha\langle pq\nu(Ij)|. \quad (1)$$

The $|pq\nu(Ij)\rangle_\alpha$ are the three-particle basis states as used in ref. [1], η and χ are subsets of the whole set of discrete three-particle quantum numbers. The employed momentum functions

$$t^i_L(p) = \frac{p^L}{(p^2 + a_L^2)^{L/2}} T_i(x_c(p)) \quad (2)$$

are related to the Chebyshev polynomials $T_i(x) = \cos(i \arccos(x))$, defined in the interval $[-1, 1]$. $x_c(p) = (p^2 - a^2)/(p^2 + a^2)$ is the function which maps the interval $[0, \infty)$ of the physical values of momentum p to the interval $[-1, 1]$. The form of the mapping function $x_c(p)$ and the parameters a and a_L are chosen beforehand by experience and are taken to be the same for all polynomials. Separating out factors of type $p^L/(p^2 + a_L^2)^{L/2}$, L being the pair orbital angular momentum, makes the remaining function, which is to be represented by Chebyshev polynomials, smoother and ensures correct asymptotic behavior of the expansion for small momenta of the interacting pair. The expansion parameters are the Chebyshev coefficients $T^{i'i}_{\eta'\eta}(\chi q, Z)$. The three-particle scattering equations are then solved without any further approximation.

Figure 1 studies the convergence of sample physics observables with the number of Chebyshev polynomials employed. The convergence is impressively rapid. Understandably, it is faster for lower energies. As an alternative to the Chebyshev expansion, favored by us and to be used by us from now on, also spline interpolation is used, as usually adopted in few-body numerics. The results derived from both interpolation schemes are indistinguishable for a sufficiently large number of interpolating functions; however, spline interpolation reaches the same quality of results only with a considerably larger number of functions than the corresponding Chebyshev expansion. The results of Fig. 1 make us conclude: *The Chebyshev expansion is systematic and efficient and thereby superior to spline interpolation.*

As sample application we revisit the problem of neutron analyzing power $A_y(n)$ of elastic neutron-deuteron scattering at 10 MeV neutron lab energy. This observable is haunted by a persistent discrepancy with theoretical predictions, called the A_y-puzzle. Calculations based on realistic two-nucleon potentials and complemented by a traditional three-nucleon force, either by an irreducible one or by an effective one as due to Δ-isobar excitation, are unable to account for the experimental height of the peak. Reference [4] discusses a three-nucleon force as possible remedy which has a phenomenological spin-orbit component with rather long range.

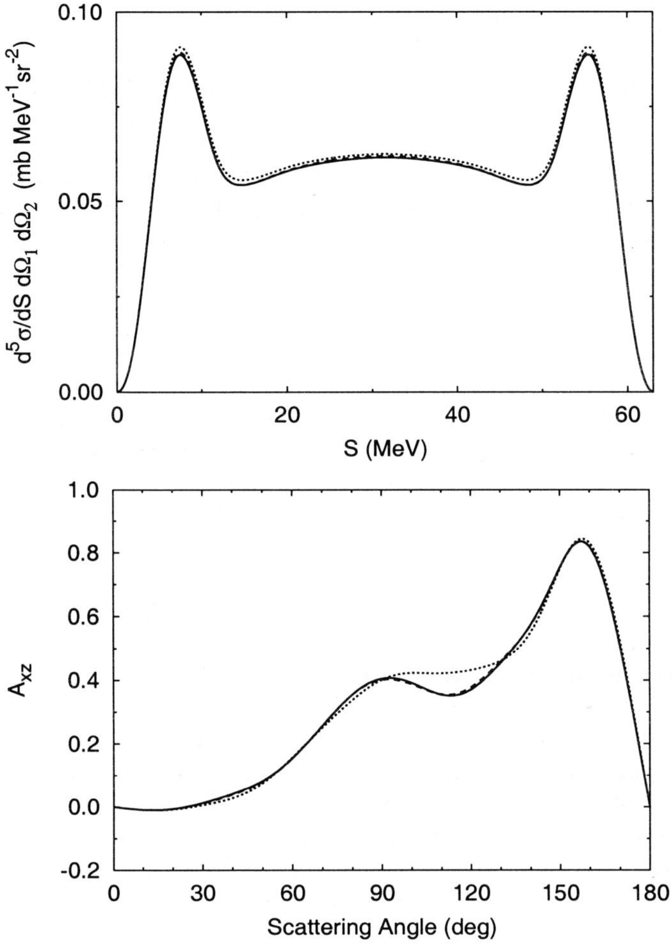

Figure 1. Fivefold differential cross section for nucleon-deuteron break-up at 65 MeV nucleon lab energy in the space star configuration $(54.0°, 54.0°, 120.0°)$ and deuteron analyzing power A_{xz} for elastic nucleon-deuteron scattering at 135 MeV nucleon lab energy. The underlying dynamics is the CD Bonn potential extended for single Δ-isobar excitation. The convergence of the Chebyshev expansion is studied. The dotted and dashed curves are results obtained with 8 and 12 Chebyshev polynomials, respectively, the solid curve corresponds to the indistinguishable results obtained with 16, 24 and 48 Chebyshev polynomials and with 48 spline functions; always more than 24 spline functions are needed to reproduce the solid curve very well.

We therefore test an effective three-nucleon force which obtains a microscopically motivated spin-orbit component arising from the spin-orbit part of the ρ-meson exchange mediating single Δ-isobar excitation; that spin-orbit component is of rather short range, its strength and range being predetermined by the ρ-parameters used in the transition potential from two-nucleon to nucleon-Δ states. The obtained results, displayed in Fig. 2, are disappointing: The inclusion of the spin-orbit mechanism of ρ-exchange does not significantly decrease the long-standing discrepancy. Although the three-nucleon force contribution is quite significant, it is canceled in part by two-nucleon dispersion, leaving the full Δ-isobar effect small.

The inability of resolving the A_y-puzzle by a microscopically motivated three-nucleon force with spin-orbit contribution is interesting, but also disappointed. However, our novel numerical technique for solving the three-particle scattering equations is promising for future physics applications.

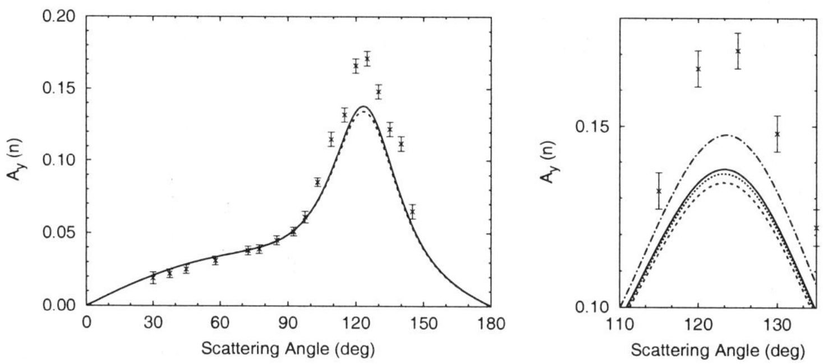

Figure 2. Neutron analyzing power $A_y(n)$ as function of the c.m. scattering angle for elastic neutron-deuteron scattering at 10 MeV nucleon lab energy. Results of the coupled-channel potential with Δ-isobar excitation including spin-orbit interaction (solid curves) are compared with results of the CD Bonn potential (dashed curves). On the right side the peak results are shown in finer resolution; there also the three-nucleon force effect (dash-dotted) and the result for Δ-isobar excitation without spin-orbit interaction (dotted) are shown. The experimental data are from ref. [5].

References

1. S. Nemoto et al.: Few-Body Systems **24**, 213 (1998); Few-Body Systems **24**, 241 (1998); Phys. Rev. **C58**, 2599 (1998)
2. K. Chmielewski et al.: submitted to Phys. Rev. **C**
3. R. Machleidt: Phys. Rev. **C63**, 024001 (2001)
4. A. Kievsky: Phys. Rev. **C60**, 034001 (1999)
5. C. R. Howell et al.: Few-Body Systems **2**, 19 (1987)

Few-Body Systems Suppl. 14, 131–134 (2003)

Few-Body Systems

Three-Nucleon Bound States with Phenomenological Nonlocal Nucleon-Nucleon Interactions

P. Doleschall[1], I. Borbély[1], Z. Papp[2], W. Plessas[3]

[1] Research Institute for Particle and Nuclear Physics, H-1525 Budapest P.O. Box 49, Hungary

[2] Institute of Nuclear Research of the Hungarian Academy of Sciences, Bem tér 18/c. P.O. Box 51, H-4001 Debrecen, Hungary

[3] Institute for Theoretical Physics, University of Graz, Universitätsplatz 5, A-8010 Graz, Austria

Abstract. Longer and shorter range nonlocal nucleon-nucleon (NN) interactions are constructed which fit the NN phase shifts with high accuracy, take into account the charge symmetry and charge independence breaking, and reproduce the triton and ^3He binding energies. The nonlocal potential model gives a better simultaneous agreement of the triton and ^3He binding energies than the conventional NN interaction models with a properly normalized 3N force.

This paper is the continuation of an investigation reported earlier [1]. The nonlocal INOY (Inside Nonlocal, Outside Yukawa tail) interactions determined there have a relatively long range. The local Yukawa tail was cut-off in the $2-4$ fm region, i. e. below 2 fm there is no local potential at all while around 4 fm the local potential becomes the intact Yukawa tail. The internal nonlocal potential was subsequently fitted to the NN data with the local potential being kept fixed. The overlap region of the fixed local and the fitted nonlocal interactions is practically the same $2-4$ fm interval as the cut-off, although the nonlocal interaction was rather small above 3 fm.

In the meantime a successful attempt was made to fit nonlocal interactions in the presence of a Yukawa tail cut-off in the $1-3$ fm region (they are denoted by IS). The overlap region between the local and nonlocal parts of the interaction became the same $1-3$ fm interval. Analogously to the earlier interactions, the nonlocal part decreases rather fast above $2-2.5$ fm. Its bulk is within the 1.5 fm sphere, which satisfies a more rigorous expectation [2] for the range of the nonlocality.

With an earlier simplification, namely that the nonlocal part of the nn $^1S_0$ interaction was derived from the nonlocal part of the np $^1S_0$, one produced a relatively poor fit to the pp phase shifts. In contrast, the present nn interactions are obtained from fitted pp interactions by switching off the Coulomb potential. To improve the earlier nonlocal NN interactions [1],a new longer range nonlocal interaction (the cut-off of the Yukawa tail is in the $2-4\,\mathrm{fm}$ region) was also defined (they are denoted by IL). The only charge-symmetry-breaking is the adjustment of the nn singlet scattering length around the $-18.5\,\mathrm{fm}$ value by a small change of a strength parameter in order to have similar nn singlet scattering lengths for all of the compared potentials. Without this adjustment the nn singlet scattering length would be $-16.809\,\mathrm{fm}$ for the shorter range (IS) and $-16.839\,\mathrm{fm}$ for the longer range (IL) nonlocal interactions.

Two approximations are also applied: i) only non-relativistic Coulomb interaction was taken into account for the proton pair with distributed charge [3], ii) all of the nucleon masses are considered to be equal to the average nucleon mass. These approximations were also applied for the Argonne v_{18} potential, the neglected electromagnetic terms (see Ref. [4]) and the effect of the equal neutron and proton masses on the scattering lengths and on the deuteron binding energy were compensated by small changes of the nuclear part. This modified Argonne v_{18} is labeled by ARG mod., while the original Argonne v_{18} potential [4] is denoted by ARG.

The shape of the nonlocal part of the NN interaction used in this paper is the same as that of ref. [1] except that the earlier term

$$e^{-(a_{ll'}F(y,y_{ll'})(x-x'))^2}$$

is omitted. It can be interpreted as an $a_{ll'}=0$ parameter choice. The aim of this modification is to exclude the sharp change in the $r\approx 0, r'\approx 2\,\mathrm{fm}$ region caused by this term. In any case, the present new parametrization makes this term obsolete.

In the fitting procedure a minimal requirement was to fit the NN phase shifts [5] and the deuteron properties. The ${}^1S_0$ scattering lengths and effective ranges are nearly the same for all interactions. The only exception is the nn scattering length. Using the results of ref. [1], the deuteron D-state probability for both the IL and IS tensor forces was chosen to be 3.60%. This low deuteron D-state probability is an essential feature of the nonlocal ${}^3SD_1$ tensor force which make possible to reproduce the 3N binding energies. The difference between local and nonlocal NN interactions is that the local ones cannot produce D-state probabilities lower than 5% [6] if other deuteron properties and the experimental phase shifts are reproduced. The deuteron and low energy ${}^3S_1$ properties of the used NN interactions are shown in Table 1. The lower deuteron D-state probability seems to be connected with a higher asymptotic normalization constant A_D/A_S (see ref. [1] and Table 1).

The Coulomb interaction in a 3N bound state problem poses only technical difficulties since the $1/r$ long range behaviour of the Coulomb interaction loses its significance because of the bound state wavefunction exponentially decreases. The technique of Z. Papp and W. Plessas [7] was used to include the Coulomb interaction in the ${}^3He$ calculations.

The fine tuning of the interactions was performed by changing the S-state wavefunctions in the small distance region. In the first step the ^3He binding energy was reproduced, because the triton binding energy depends at least on one more free parameter: on the neutron-neutron (nn) scattering length, which is not too well known. If the nn scattering length is chosen to be the same as that of the Argonne v_{18} potential, the resulting triton is underbound with about 5 keV for the IL and IS set of interactions. Therefore the nn scattering lengths were chosen to be −18.601 fm in order to reproduce the triton binding energy.

Table 1. Deuteron properties, triplet scattering lengths and effective ranges given by different interactions.

	ARG	ARG mod.	IL	IS
$\varepsilon_D(MeV)$	−2.224575	−2.224575	−2.224574	−2.224582
$P_D(\%)$	5.76	5.764	3.600	3.600
$Q_D(\text{fm}^2)$	0.270	0.2699	0.2749	0.2751
$A_S(\text{fm}^{-1/2})$	0.8846	0.8851	0.8850	0.8850
A_D/A_S	0.0250	0.02509	0.02706	0.02697
$r_{rms}(\text{fm})$	1.967	1.96735	1.96487	1.96514
$a_t(\text{fm})$	5.419	5.4192	5.4190	5.4190
$r_t(\text{fm})$	1.753	1.7532	1.7531	1.7531

A calculation was performed using the modified Argonne v_{18} potential (ARG mod.), in two other cases the IL and IS nonlocal interactions were used for the $^1S_0$ and $^3SD_1$ states (Table 2).

The original and the modified Argonne potentials produce nearly the same 3N binding energies (Table 2), in spite of the fact that the n-p mass difference and the electromagnetic terms (except the nonrelativistic distributed charge Coulomb interaction) were substituted by small corrections to the $^1S_0$ and $^3SD_1$ nuclear forces. Because these corrections were chosen to reproduce the low-energy scattering data and the deuteron binding energy, it is quite plausible that these low energy data (which are practically defined by the pole terms of the $^1S_0$ and $^3SD_1$ interactions) are the dominant factors in producing the 3N binding energies.

Contrary to the presented INOY interactions, the Argonne+3N force model produces around 48 keV deficit for the triton binding energy. The CD-Bonn+3N force model seems to perform better, because the triton is underbound only with 20 keV [8]. However, the CD-Bonn potential has already a larger nn scattering length (−18.968 fm). In order to produce a correct triton binding energy for a 3N force adjusted to the ^3He binding energy, the singlet nn scattering length has to be adjusted to the value −19.33 fm [8], which seems to be too large.

The fact that the IL and IS forces reproduce the 3N binding energies is

Table 2. The ^3H, ^3He binding energies and the ΔE energy shifts.

Interactions	$E(^3H)(MeV)$	$E(^3He)(MeV)$	$\Delta E(MeV)$
ARG [8]	7.628	6.917	0.711
ARG mod.	7.6268	6.9167	0.7101
IL	8.4818	7.7181	0.7637
IS	8.4818	7.7181	0.7637
Experiment	8.48182	7.71806	0.76376

just a new example that with nonlocal NN interactions one can avoid using 3N forces. Since the used nonlocal interactions are purely phenomenological ones, one cannot state that they are the real NN interactions. Especially since different variants can be constructed, although they are rather similar if the 3N binding energies are reproduced. But the low D-state probability for the deuteron and the higher asymptotic normalization constant A_D/A_S connected to it may have an effect in scattering processes especially at lower energies. On the other hand, the strong nonlocal feature of the INOY interactions at short distance may have effect at higher energies.

Acknowledgement. The work has been supported by OTKA under contract No. T034334.

References

1. P. Doleschall, I. Borbély: Phys. Rev. **C62**, 054004 (2000)
2. J. Siemens, A. P. Vischer: Ann. Phys. (N. Y.) **238**, 167 (1995)
3. N. Auerbach, J. Hüfner, A. K. Kerman, C. M. Shakin: Rev. Mod. Phys. **44**, 48 (1972)
4. R.B. Wiringa, V.G.J. Stoks, R. Schiavilla: Phys. Rev. **C51**, 38 (1995)
5. V.G.J. Stoks, R.A.M. Klomp, M.C.M. Rentmeester, J.J. Swart: Phys. Rev. **C48**, 792 (1993)
6. B.F. Gibson, H. Kohlhoff, H.V. von Geramb: Phys. Rev. **C51**, R465 (1995); J.L. Forest: Phys. Rev. **C61**, 034007 (2000)
7. Z. Papp, W. Plessas: Phys. Rev. **C54**, 50 (1996)
8. A. Nogga, H. Kamada, W. Glöckle: Phys. Rev. Lett. **85**, 944 (2000); A. Nogga, H. Kamada, W. Glöckle, B. R. Barett: Phys. Rev. **C65**, 054003 (2002); A. Nogga, A. Kievsky, H. Kamada, W. Glöckle, L. E. Marcucci, S. Rosati, M. Viviani: nucl-th/0202037

Few-Body Systems Suppl. 14, 135–138 (2003)

Few-Body Systems

Polarization Effects of the $3/2^+$ Resonance on the ^{3}He$(d,p)^4$He Fusion Reaction

S. Gojuki*,†, S. Oryu**

Department of Physics, Faculty of Science and Technology, Frontier Research Center for Computational Science, Tokyo University of Science, Noda, Chiba 278-8510, Japan

Abstract. The nearest resonance state above the ^{3}He-d threshold is investigated by using the three-body Faddeev formalism in the n-p-^{3}He system. The AV14 NN potential in the $^1S_0$, $^3S_1$-$^3D_1$, $^1P_1$, $^3P_0$, $^3P_1$, $^3P_2$-$^3F_2$, $^1D_2$, and $^3D_2$ partial waves is adopted, while the p-^{3}He, and n-^{3}He effective interactions are constructed by using the well known Resonating Group Method (RGM), but only the ^{1}S$_0$ and ^{3}S$_1$ states for the N-^{3}He interaction are used in our first trial. We found a $3/2^+$ resonance state of ^{5}Li which appears at around E$_{\rm lab}$= 400keV above the d-^{3}He threshold, which is slightly lower than the 450 keV experimental value. The low energy unpolarized ^{3}He$(d,p)^4$He reaction cross section and polarized $^3\overrightarrow{\rm He}(\overrightarrow{d},p)^4$He cross section are compared in the resonance energy region. Coulomb effects are taken into account in the initial and the final states.

1 Introduction

^{3}He has, to a large extent, a number of similarities with the neutron, except of course for the charge and mass, and thus it can be considered as a potential candidate for use instead of a neutron. That is, ^{3}He could be a good polarized target substitute for the neutron [1]. Moreover, its reactions with other light nuclei exhibit a much richer physics than, for example the ^{3}He+d reaction, due to its structure and the presence of resonances which can enhance the fusion reaction rates. The two-body states describing the p-^{3}He and n-^{3}He interactions are almost the same as those for the NN cases; however, the potential for p-^{3}He is completely different from that for n-^{3}He due to the Pauli principle.

* *Alternative address:* SGI Japan, Ltd. Yebisu Garden Place Tower 4-20-3 Ebisu Shibuya-ku, Tokyo 150-6031 Japan
† *E-mail address:* gojuki@sgi.co.jp
** *E-mail address:* oryu@ph.noda.tus.ac.jp

As a result, the three-body Faddeev equations become 3×3 coupled channel equations.

We have performed three-cluster Faddeev calculations for the n-p-^{3}He system at very low energies. For the NN-interaction we adopted the realistic AV14 potential [2], while for the p-^{3}He and n-^{3}He potentials we constructed new effective inter-cluster potentials called the "RGM-OCM-LS potential" [3] based on the Resonating Group Method, the use the Orthogonal Condition Model to remove the Pauli forbidden states, and on the inclusion of LS-forces between clusters for p-^{3}He but for the n-^{3}He interaction is constructed from a genuine RGM-OCM potential with LS-forces included via the NN interaction of Kaneko et al. [4]. All the states included are shown in Table 1. The separable form factors for the p-^{3}He and n-^{3}He effective interactions are illustrated in Fig. 1.

Table 1. Two-body states.

System	States
p-n (AV14)	$^1S_0$, $^3S_1$-$^3D_1$
	$^1P_1$, $^3P_0$, $^3P_1$, $^3P_2$-$^3F_2$
	$^1D_2$, $^3D_2$ (all states rank=1)
p-^{3}He (RGM-OCM-LS)	$^1S_0$, $^3S_1$ (rank=1)
n-^{3}He (RGM-OCM-LS)	$^1S_0$ (rank=1), $^3S_1$(rank=3)

2 Results and Discussion

We calculated the total cross section for total spin and parity of $J^\pi = 1/2^\pm, \cdots, 9/2^\pm$. Results for the unpolarized reaction ^{3}He$(d,p)^4$He are shown in Fig. 2 (top left), while the polarized results are illustrated in Fig. 2 (top right). The $3/2^+$ state enhances the cross section at around E_{lab}= 390keV, and the $1/2^+$ state enhances it at around 450keV. However, the polarized cross section is enhanced very much by the $3/2^+$ state but the $1/2^+$ state contribution disappears. This means that such an alignment of spins could enhance the reaction rate via this $3/2^+$ resonance near the threshold. By considering parallel spin up for ^{3}He and the deuteron, we obtain, using spin algebra, the $3/2^+$ resonance and predict that the reaction rate, at an incident deuteron energy of $\mathrm{E_{lab}}$ =400 keV, should exhibit a factor of 1.5 enhancement, because the $3/2^+$ state is open but the $1/2^+$ state is closed in the resonance energy region. In other words, it is easily seen from the unpolarized cross section $(\mathrm{d}\sigma/\mathrm{d}\Omega)^{\mathrm{unpol}}$ vs. $(\mathrm{d}\sigma/\mathrm{d}\Omega)^{\mathrm{pol}}$, and with $\boldsymbol{S} = \boldsymbol{S}_{\mathrm{d}} + \boldsymbol{S}_{\mathrm{h}}$ ('h' stands for ^{3}He), i.e.

$$\left(\frac{\mathrm{d}\sigma}{\mathrm{d}\Omega}\right)^{\mathrm{unpol}} = \frac{1}{(2\mathrm{S_d}+1)(2\mathrm{S_h}+1)} \sum_{\mathrm{S}=1/2,3/2} (2\mathrm{S}+1)\left(\frac{\mathrm{d}\sigma}{\mathrm{d}\Omega}\right)^{(\mathrm{S})} = \frac{4}{6}\left(\frac{\mathrm{d}\sigma}{\mathrm{d}\Omega}\right)^{(\frac{3}{2})} \tag{1}$$

$$\left(\frac{\mathrm{d}\sigma}{\mathrm{d}\Omega}\right)^{\mathrm{pol}} = \left(\frac{\mathrm{d}\sigma}{\mathrm{d}\Omega}\right)^{(\frac{3}{2})} \tag{2}$$

where we assumed $(\mathrm{d}\sigma/\mathrm{d}\Omega)^{(\frac{1}{2})} = 0$ at the resonance energy of the $3/2^+$ state.

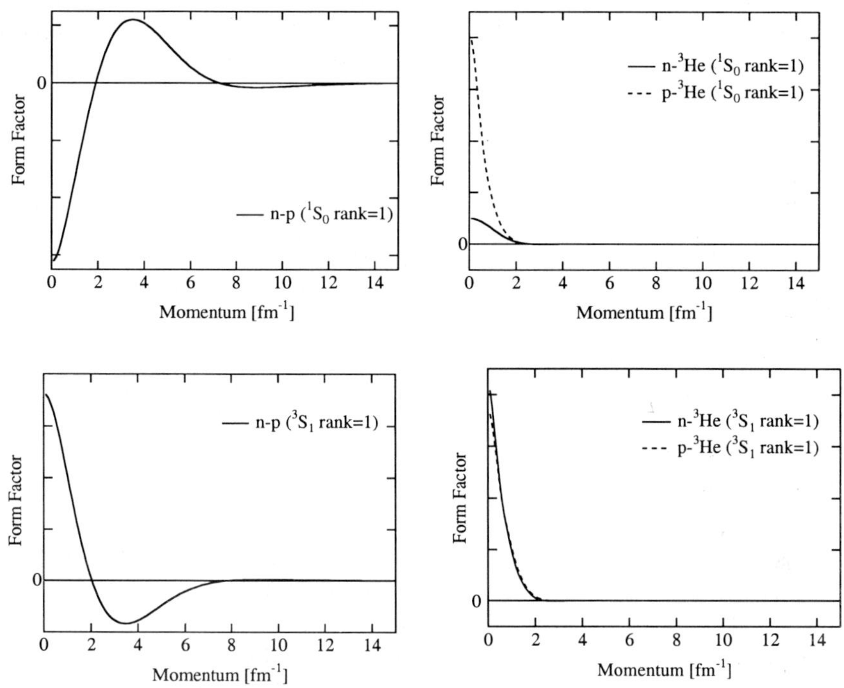

Figure 1. The form factors of the p-^{3}He, and n-^{3}He potentials. The vertical axis is in arbitrary units.

The results obtained for the polarized and unpolarized total cross section are shown in Fig. 2 (bottom left) and for the ratio $\sigma^{\mathrm{unpol}}/\sigma^{\mathrm{pol}}$ in Fig. 2 (bottom right). It is seen that there is a resonance at $E_{\mathrm{lab}} \sim 390\,\mathrm{keV}$ which enhances the ratio $\sigma^{\mathrm{unpol}}/\sigma^{\mathrm{pol}}$ by a factor of 3.4. This is much larger than the value of 1.5 obtained via spin algebra. The partial cross section $(\mathrm{d}\sigma/\mathrm{d}\Omega)^{(\frac{3}{2})}$ in Eq. (2) is enhanced more than two times by itself. The reason could be understood that the constituent neutron of deuteron is strongly dissociated by the unsatisfied shell cluster ^{3}He. The small difference between this value and the experimental resonance energy $E_{\mathrm{lab}} = 450$ keV could be attributed to Coulomb repulsion and higher partial wave effects [5].

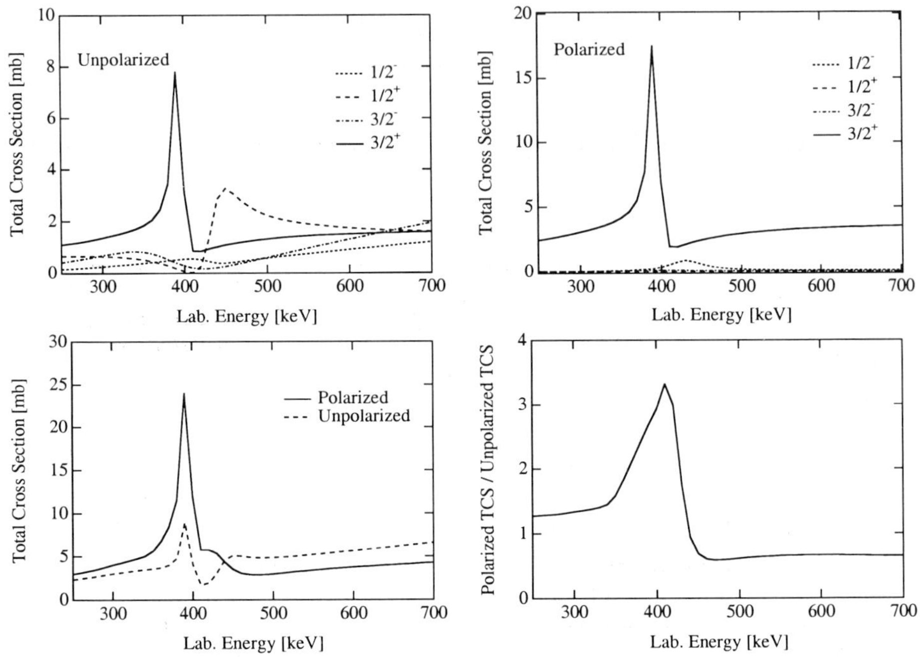

Figure 2. Polarized and unpolarized total cross sections and the ratio between polarized and unpolarized total cross sections at very low energy for the ^{3}He$(d,p)^4$He reactions.

References

1. T. Uesaka: Private Communication; Doctoral Dissertation, Tokyo University, 1997
2. R.B. Wiringa et al.: Phys. Rev. **C29**, 1207 (1984)
3. S. Gojuki, S. Oryu, S. A. Sofianos: nucl-th/0112020
4. T. Kaneko: Private Communication
5. S. Gojuki, S. Oryu: In: Proceedings of the Asian-Pacific Few-Body Conference 2002, Shanghai, China, to be published in Mod. Phys. Lett. **A** (2002)

Few-Body Systems Suppl. 14, 139–140 (2003)

Few-Body Systems

Indications for the Nonexistence of Three-Neutron Resonances Near the Physical Region

A. Hemmdan*,†, W. Glöckle**

Institut für Theoretische Physik II, Ruhr Universität Bochum, D-44780 Bochum, Germany

This work presents the calculations of three-neutron($3n$) resonance trajectories by solving Faddeev equations in momentum space for the states $\frac{3}{2}^{\pm}$ and $\frac{1}{2}^{\pm}$. The neutron-neutron (nn) forces for the states $^1S_0$, $^3P_0$, $^3P_2$ and $^1D_2$ are of rank 2 [1]. This implies Faddeev calculations including up to 10 channels. To calculate the resonance positions the kernel of the Faddeev equations has to be analytically continued into the second energy sheet below the positive real energy axis by contour deformation. There one hits singularities arising from the form factors of the potential and the free propagator. We follow ref.[2] to avoid the singularities. The resonance trajectories are generated by changing an artificial enhancement factor of the nn forces from values larger than 1 towards 1, the physical case. The final resonance positions for three-neutrons are in the four studied states far away from the real energy axis as shown in Fig. 1. If the resonance positions are located near the real energy axis, the Padé method can be used to predict their energies using the knowledge of the $3n$ bound state energies for different enhancement factors [3]. This requires, however, a high precision of the "artificial" bound state energies. One example is displayed in Table 1 for the state $\frac{3}{2}^{-}$. For that state it was sufficient to enhance only the nn force component in the state $^3P_2$ and we could keep all other force components at their physical values. The resonance positions calculated directly by contour deformation are compared to those predicted by the Padé approximation in Table 2. We see that the agreement among the two approaches is quite good. A more detailed description of our results together with references for previous works can be found in [4]. Since the final resonance positions are far away from the positive real energy axis we expect that replacing the model forces by realistic nn forces will not change the outcome qualitatively and there is little hope that low-lying $3n$ resonances exist in nature. A recent investigation [5]

* *On leave from:* Faculty of science, South Valley University, Aswan, Egypt
† *E-mail address:* amel.hemdan@tp2.ruhr-uni-bochum.de
** *E-mail address:* walter.gloeckle@tp2.ruhr-uni-bochum.de

using modern realistic nn forces came to the same tentative conclusion. In that work the artificial enhancement factors could not be, however, fully reduced to the physical value 1 because of numerical instabilities. Nevertheless the widths of the pole positions established there were already quite large.

Table 1. The $3n$ bound state energies in MeV for various enhancement factors in the state $J^\pi = 3/2^-$.

$\lambda_{^3P_2}$	E_b^{3n}
3.29	-1.38338
3.28	-1.13457
3.27	-0.89220
3.26	-0.65666
3.25	-0.42848
3.24	-0.20849
3.2300873	-1.1×10^{-8}

Table 2. The directly calculated resonance positions E_{res}^{3n} via Faddeev compared to Padé, all in MeV.

$\lambda_{^3P_2}$	Padé [3,3]	E_{res}^{3n}
3.22	$0.20 - \mathrm{i}\,0.003$	$0.20 - \mathrm{i}\,0.003$
3.21	$0.39 - \mathrm{i}\,0.012$	$0.39 - \mathrm{i}\,0.014$
3.20	$0.56 - \mathrm{i}\,0.03$	$0.57 - \mathrm{i}\,0.03$
3.19	$0.74 - \mathrm{i}\,0.05$	$0.74 - \mathrm{i}\,0.06$
3.18	$0.90 - \mathrm{i}\,0.08$	$0.91 - \mathrm{i}\,0.09$
3.17	$1.06 - \mathrm{i}\,0.12$	$1.08 - \mathrm{i}\,0.13$
3.16	$1.21 - \mathrm{i}\,0.16$	$1.24 - \mathrm{i}\,0.18$
3.15	$1.35 - \mathrm{i}\,0.22$	$1.40 - \mathrm{i}\,0.23$

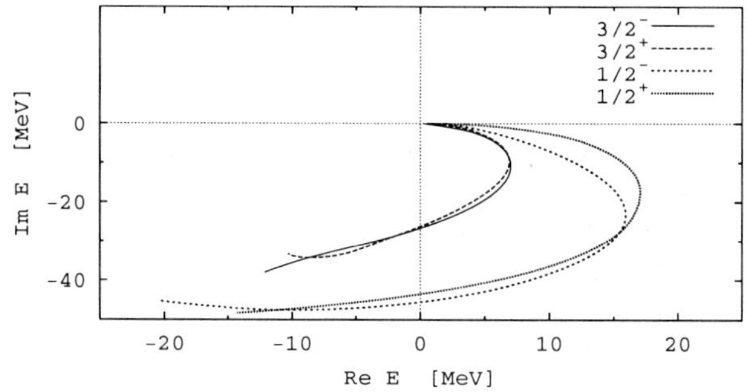

Figure 1. $3n$ resonance trajectories for the states $\frac{3}{2}^{\pm}$ and $\frac{1}{2}^{\pm}$.

References

1. W. Schweiger, W. Plessas, L.P. Kok and H. van Haeringen: Phys. Rev. **C27**, 515 (1983)

2. W. Glöckle: Phys. Rev. **C18**, 564 (1978)

3. N. Tanaka, Y. Suzuki, K. Varga, R.G. Lovas: Phys. Rev. **C59**, 1391 (1999)

4. A. Hemmdan, W. Glöckle, H. Kamada: Phys. Rev. **C66**, 054001 (2002)

5. H. Witala and W. Glöckle: Phys. Rev. **C60**, 024002 (1999)

Few-Body Systems Suppl. 14, 141–142 (2003)

Few-Body Systems

Three-Nucleon Dynamics Studied via $^1\mathrm{H}(\boldsymbol{d},pp)n$ Breakup at 130 MeV

St. Kistryn[1], R. Bieber[3], A. Biegun[2], K. Bodek[1], K. Ermisch[3], W. Glöckle[4], J. Golak[1], M.N. Harake[3], N. Kalantar-Nayestanaki[3], H. Kamada[4], J. Kuroś-Żołnierczuk[1], M. Kiš[3], A. Micherdzińska[2], A. Nogga[4], M. Shafiei[3], R. Skibiński[1], E. Stephan[2], H. Witała[1], J. Zejma[1], W. Zipper[2]

[1] Institute of Physics, Jagiellonian University, PL-30059 Kraków, Poland

[2] Institute of Physics, University of Silesia, PL-40007 Katowice, Poland

[3] Kernfysisch Versneller Instituut (KVI), NL-9747 AA Groningen, The Netherlands

[4] Institut für Theoretische Physik II, Ruhr Universität Bochum, D-44780 Bochum, Germany

Abstract. An experiment to measure cross sections and both, vector and tensor analyzing powers, for the $^1\mathrm{H}(\boldsymbol{d},pp)n$ breakup reaction at 130 MeV deuteron beam energy is presented. Employing a position-sensitive detection system covering a large solid angle enables probing of a significant part of the whole phase space. First results are presented and compared with the rigorous Faddeev calculations using modern NN and 3NF potentials.

The three-nucleon (3N) system is the simplest non-trivial environment in which quality of modern, realistic nucleon-nucleon (NN) interaction models can be tested. Rigorous calculations for the elastic nucleon-deuteron (Nd) scattering show that inclusion of additional dynamics, represented in the form of genuine 3N potential, improves the agreement between the predictions and the cross section data [1]. It is therefore natural to extend the search for three-nucleon force (3NF) effects to other observables in the 3-body system.

Cross sections and polarization observables for the Nd breakup reaction contain exceptionally rich physical information on the interaction dynamics. A systematic scan of the whole phase space performed in the framework of the rigorous Faddeev calculations [2] shows that the predicted effects of the 3NF in numerous configurations are very large, up to 20%, and that their magnitude is quite strongly dependent on the used 3NF model. In order to test the details of the structure of the 3N force it is strongly desirable to conduct the experimental search covering possibly large phase-space regions.

Along these guidelines we have started an experiment to measure cross sections and both, vector and tensor analyzing powers for the $^1\mathrm{H}(\boldsymbol{d},pp)n$ breakup reaction at 130 MeV deuteron beam energy. The experiment is performed at KVI, Groningen, using a polarized deuteron beam focused on a liquid hydrogen target. The charged reaction products are registered in a position sensitive detection system, covering laboratory polar angles ϑ between 10° and 40° and full range of azimuthal angles φ. It consists of a three-plane MWPC and a two-layer scintillator hodoscope, used for trigger definition and particle identification.

Events originating in the elastic dp-scattering are used for normalization and provide additional information allowing very accurate checks of the whole system geometry. Analysis of data taken in special runs with energy degradation leads to precise energy calibration of the scintillation counters. More information about the setup and on the data analysis procedure can be found in ref. [3].

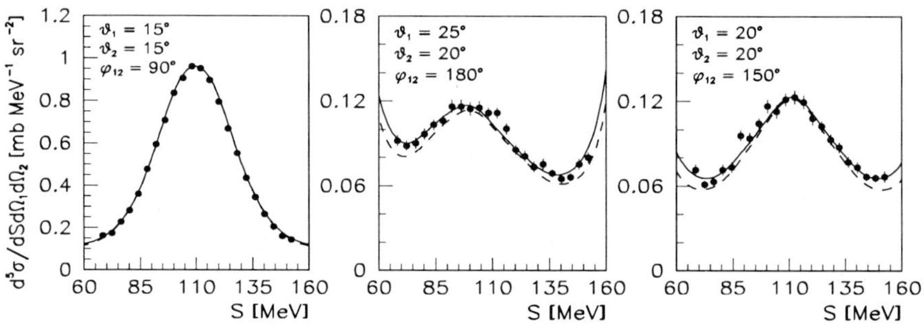

Figure 1. Cross sections for three kinematical configurations of the dp breakup reaction at 130 MeV. The dashed lines show the results of theoretical calculations with the CD-Bonn NN potential only, while the predictions represented by the solid lines include also the Tucson-Melbourne 3NF.

The first data taking run provided nearly sufficient amount of data to reach the required accuracy (1%) for cross sections determination. Examples of the cross section results for three configurations of the two outgoing protons and their comparison with the rigorous Faddeev calculations are shown in Fig. 1. From the results for all configurations analysed up to now a preference of the full (including 3NF) predictions can be deduced.

References

1. D. Hüber et al.: Acta Phys. Pol. **28**, 1677 (1997)
2. J. Kuroś-Żołnierczuk et al.: Phys. Rev. **C66**, 024003 (2002)
3. St. Kistryn et al.: Nucl. Phys. **A689**, 345c (2001); A. Micherdzińska et al.: KVI Annual Report, Groningen 2001

Few-Body Systems Suppl. 14, 143–144 (2003)

Few-Body Systems

Scattering of the $(p - {}^3\mathrm{H})$ System with the Lorentz Integral Transform Method

M.A. Marchisio[1], N. Barnea[2], W. Leidemann[1], G. Orlandini[1]

[1] Dipartimento di Fisica, Università di Trento and INFN (Gruppo collegato di Trento), I-38050 Povo, Italy

[2] The Racah Institute of Physics, The Hebrew University, 91904 Jerusalem, Israel

Integral transform methods can be applied to calculate the T matrix of a scattering process [1]. In this contribution we show how the Lorentz Integral Transform method (LIT) [2], which in the recent years has revealed to be a powerful tool in few-body calculations, can be applied to calculate the T matrix in $(p - {}^3\mathrm{H})$ scattering also for energies above the three-body breakup threshold. The T matrix elements of the considered reaction are given by

$$T_{\mathrm{fi}}(E_{\mathrm{p}}) = \langle \psi^{\mathrm{f}}(\mathbf{k}')|\hat{V}|\phi^{\mathrm{i}}(\mathbf{k})\rangle\,, \tag{1}$$

where E_{p} is the proton energy and $\hat{V}$ is the NN potential. The initial unperturbed state $\phi^{\mathrm{i}}(\mathbf{k})$ of the system is realized by the antisymmetrized product of the proton and ${}^3\mathrm{H}$ wave functions times the Coulomb functions. The full final state wave function of the system $\psi^{\mathrm{f}}(\mathbf{k}')$ has the form

$$\langle \psi^{\mathrm{f}}(\mathbf{k}')| = \langle \phi^{\mathrm{f}}(\mathbf{k}')| + \langle \phi^{\mathrm{f}}(\mathbf{k}')|\hat{V}\frac{1}{E_{\mathrm{p}} - \hat{H} + i\epsilon}\,, \tag{2}$$

where $\phi^{\mathrm{f}}(\mathbf{k}')$ is the final unperturbed state of the system. Using Eq.(2) the matrix elements in Eq.(1) can be written as the sum of two separate parts

$$T_{\mathrm{fi}}(E_{\mathrm{p}}) = \langle \phi^{\mathrm{f}}(\mathbf{k}')|\hat{V}|\phi^{\mathrm{i}}(\mathbf{k})\rangle + \langle \phi^{\mathrm{f}}(\mathbf{k}')|\hat{V}\frac{1}{E - \hat{H} + i\epsilon}\hat{V}|\phi^{\mathrm{i}}(\mathbf{k})\rangle\,, \tag{3}$$

where the first part is the Born term and the second one, through the total Hamiltonian operator $\hat{H}$ of the system, contains the information about the complete final state interaction (FSI term). The FSI term is obtained from the relation

$$T_{\mathrm{fi}}^{\mathrm{FSI}}(E_{\mathrm{p}}) = -i\pi F_{\mathrm{fi}}(E_{\mathrm{p}}) + \mathcal{P}\int_{E_0}^{\infty} dE \frac{F_{\mathrm{fi}}(E)}{E_{\mathrm{p}} - E} , \tag{4}$$

where $F_{\mathrm{fi}}(E)$ is the response function, defined as

$$F_{\mathrm{fi}}(E) = \int d\gamma \langle \phi^{\mathrm{f}}(\mathbf{k}')|\hat{V}|\psi_\gamma\rangle\langle\psi_\gamma|\hat{V}|\phi^{\mathrm{i}}(\mathbf{k})\rangle \delta(E - E_\gamma) \tag{5}$$

and $|\psi_\gamma\rangle$ represent the Hamiltonian eigenstates. With the LIT method one calculates $F_{\mathrm{fi}}(E)$ by the inversion of the corresponding Lorentz Integral Transform

$$L(\sigma = \sigma_{\mathrm{R}} + i\sigma_{\mathrm{I}}) = \int_{E_0}^{\infty} dE \frac{F_{\mathrm{fi}}(E)}{(E - \sigma_{\mathrm{R}})^2 + \sigma_{\mathrm{I}}^2} . \tag{6}$$

In the LIT method $L(\sigma)$ can be calculated directly without knowledge of $F_{\mathrm{fi}}(E)$. The transform $L(\sigma)$ turns out to be the scalar product of the complex functions $\tilde{\psi}_1$ and $\tilde{\psi}_2$. The latter are the solutions of the following differential equations

$$(\hat{H} - \sigma)|\tilde{\psi}_1\rangle = \hat{V}|\phi^{\mathrm{i}}(\mathbf{k})\rangle ; \quad (\hat{H} - \sigma)|\tilde{\psi}_2\rangle = \hat{V}|\phi^{\mathrm{f}}(\mathbf{k}')\rangle , \tag{7}$$

where $\sigma = \sigma_R + i\sigma_I$ is a complex parameter. Note that Eqs. (7) can be solved by using bound state techniques. Here we expand $\tilde{\psi}_1$ and $\tilde{\psi}_2$ on symmetrized hyperspherical harmonics [3].

For the calculation of the $T_{\mathrm{fi}}(E_{\mathrm{p}})$ of the $(p - {}^3\mathrm{H})$ scattering we use the semi-realistic MT-I/III potential [4] as NN interaction.

We employ the Effective Interaction Hyperspherical Harmonics [5] approach (see also contribution of N. Barnea in this proceedings) in order to accelerate the convergence of $|\tilde{\psi}_1\rangle$ and $|\tilde{\psi}_2\rangle$ states and make use of the Lanczos algorithm [6] to avoid the inversion of the Hamiltonian matrix in solving Eqs.(7) [7].

Preliminary results for the phase shift of various L, S, T channels look very promising but they have not yet undergone consistency checks and can not be presented here.

References

1. V.D. Efros: Sov. J. Nucl. Phys. **41**, 949 (1985); Phys. Atom. Nucl. **62**, 1833 (1999)
2. V.D. Efros, W. Leidemann, G. Orlandini: Phys. Lett. **B338**, 130 (1994)
3. N. Barnea, A. Novoselsky: Ann. Phys. (N.Y.) **256**, 192 (1997)
4. R.A. Malfliet, J. Tjon: Nucl. Phys. **A677**, 161 (1969)
5. N. Barnea, W. Leidemann, G. Orlandini: Phys. Rev. **C61**, 054001 (2000)
6. C. Lanczos: J. Res. Nat. Bur. Stand. **45**, 255 (1950)
7. M.A. Marchisio, N. Barnea, W. Leidemann, G. Orlandini: nucl-th/0202009

Few-Body Systems Suppl. 14, 145–146 (2003)

Few-Body Systems

Calculation of the α–Particle Ground State with the Hyperspherical Harmonic Basis

M. Viviani, A. Kievsky, S. Rosati

Istituto Nazionale di Fisica Nucleare, Via Buonarroti, 2, I-56100 Pisa (Italy), and Phys. Dept., Pisa University, Via Buonarroti, 2, I-56100 Pisa (Italy)

Abstract. We report on the application of the hyperspherical harmonic expansion method to the study of the α–particle. Preliminary results are presented for realistic hamiltonians including the three nucleon interaction.

In recent years the theoretical study of the ground state of the α particle has received great attention. At present, the calculations performed with different techniques (see, for example, ref. [1] and references therein), when the nuclear interaction is taken to be the sum of a (realistic) nucleon-nucleon (NN) potential, have reached a great accuracy. As an example, by adopting the Argonne $v8'$ potential, it has been recently shown that the α-particle binding energies B_α computed by different techniques differ at most by 0.1% [1].

On the other hand, when in the nuclear hamiltonian is included also a three nucleon force (3NF) so as to reproduce the ^{3}H binding energy, the situation is still not completely satisfactory. At present, only two techniques have been applied successfully to treat this problem, namely the Green Function Monte Carlo Method (GFMC) method [3] and the Faddeev-Yakubovsky (FY) integral equation method [4]. For example, when the Argonne $v18$ (AV18) NN potential plus the Urbana IX (UIX) 3NF [2] is considered, the calculated GFMC and FY binding energies are $B_\alpha = 28.3 \pm 0.2$ MeV and 28.53 MeV, respectively. The reason for this slight discrepancy could be due either to the difficulties of the GFMC in treating the L^2 and $(LS)^2$ components of the two–body potential or to an incomplete convergence of the partial wave expansion in the FY. It is therefore of interest to have new independent estimates of B_α.

In the hyperspherical harmonic (HH) method, the wave function is expanded in terms of a complete set of spin-isospin-HH functions [5]. Due to the large degenerancy of the HH basis, the main difficulty is the selection of a restricted and effective subset of basis states. We have grouped the states into 7 classes depending on the number of particles correlated, the value of the total isospin T and the value of $\mathcal{L} = \ell_1 + \ell_2 + \ell_3$ (ℓ_i being the orbital angular momentum quantum number associated to the Jacobi vector i). The convergence patterns are rather different for each class; the optimal choice of the classes results in a noticeable reduction of the number of HH functions to be included in the expansion.

Table 1. Binding energies B_α, mean kinetic energies T and percentages P_P, P_D, $P_{T=1}$, $P_{T=2}$ of various waves for the α–particle obtained by applying the HH expansion method are compared with the most accurate results reported in the literature, for two potential models. The HH results for the AV18/UIX are preliminary.

	AV18		AV18/UIX			Expt.
	FY [4]	HH	FY [4]	GFMC [3]	HH	
B_α [MeV]	24.23	24.21	28.53	28.30(2)	28.45	28.30
T [MeV]	97.80	97.79	113.21	112.1(8)	113.28	
P_P [%]	0.35	0.35	0.75		0.72	
P_D [%]	13.78	13.77	16.03		15.92	
$P_{T=1}$ [%]	0.003	0.0004				
$P_{T=2}$ [%]	0.005	0.005				

The results obtained using the HH expansion for the AV18 and AV18/UIX potentials are reported in Table 1. In the case of the AV18, we observe in Table 1 a good agreement with the results obtained by solving the FY equation [4]. The mean kinetic energy and the percentage of the P, D and $T = 2$ waves are also in good agreement. Only for the component with total isospin $T = 1$ (a very tiny quantity) we have found a discrepancy. In the case of the AV18/UIX only some preliminary results are reported. Again, as can be seen by inspecting Table 1, the various estimates obtained by means of the HH expansion are rather close to the FY results.

In conclusion, the HH method can be successfully applied to describe the four nucleon ground state, even when a 3NF is considered. The overprediction of about 0.2 MeV of B_α using the AV18/UIX potential model as first observed in Ref. [4] is confirmed. This could suggest the existence of different 3NF terms in the nuclear Hamiltonian. In addition, the HH technique can be extended to the calculation of the scattering states of the four nucleons, which is currently under way.

References

1. H. Kamada et al.: Phys. Rev. **C64**, 044001 (2001)
2. R.B. Wiringa, V.G.J. Stoks, and R. Schiavilla: Phys. Rev. **C51**, 38 (1995); B.S. Pudliner et al.: Phys. Rev. Lett. **74**, 4396 (1995)
3. R.B. Wiringa et al.: Phys. Rev. **C62**, 014001 (2000)
4. A. Nogga et al.: Phys. Rev. **C65**, 054003 (2002)
5. M. Fabre de la Ripelle, Ann. Phys. **147**, 281 (1983)

Few-Body Systems Suppl. 14, 147–152 (2003)

Few-Body Systems

Two-Electron Photoionization Cross Sections at High Energies

M.Ya. Amusia[1,2], R. Krivec[3], V.B. Mandelzweig[1]

[1] Racah Institute of Physics, The Hebrew University, Jerusalem 91904, Israel
[2] Ioffe Physico-Technical Institute, St. Petersburg, 194021, Russia
[3] J. Stefan Institute, PO Box 3000, 1001 Ljubljana, Slovenia

Abstract. Double and single electron photoionization cross sections and their ratios at high and ultra-relativistic energies are calculated for H^-, He and helium-like ions in ground and excited states including triplet states. The ratios contain shake-off and quasi-free terms. A high precision non-variational wave function is used. The quasi-free mechanism increases the ratios impressively: for He we get 0.0762 instead of 0.0164 in the non-relativistic case. Ratios are inversely proportional to Z^2, with a factor increasing from 0.094 in the non-relativistic to 0.595 in the ultra-relativistic limit.

1 Introduction and the Main Formulas

Elimination of two electrons by a single photon, double-electron photoionization (DPI), is entirely determined by the interelectron interaction. Although this process was predicted in the 1930's, it took more than 30 years before it was observed [1]. First DPI experiments were performed at high photon energies, $\omega \gg I^{++}$ (I^{++} is the two-electron ionization potential) for He. The results were explained [1] assuming that the second electron is eliminated due to the instant change of the field after the first electron leaves the atom by photon absorption, a mechanism was called *shake-off*. Its theory was developed in refs. [2,3] where it was demonstrated that one electron becomes fast with energy $\varepsilon_1 \approx \omega$, while the second is slow, $\varepsilon_2 \approx I^{++}$.

A little later the DPI mechanism was treated by perturbation theory [4] and calculated for the noble gas atoms (see refs. [5,6] and references therein). Simultaneously, cross sections in the near threshold of outer shell area were measured [7], stimulating the theoretical investigation [6]. The results of ref. [7] were reasonably well described by perturbation theory, giving the ratio of the double- to single ionization cross sections, $R(\omega) = \sigma^{++}(\omega)/\sigma^{+}(\omega)$, which rapidly increases from several percent in He to about 80% in Xe. It was demonstrated that at $\omega \gtrsim I^{++}$, up to $\omega = 3 - 4I^{++}$, the DPI cross-section can be presented as a product of single-electron ionization $\sigma^{+}(\omega)$ and the probability of the ionization of the ion $W(\varepsilon)$ by the photoelectron with the energy ε, $\sigma^{++}(\omega) = \sigma^{+}(\omega)W(\varepsilon)$ [8]. [1]

Lowest order perturbation theory in the $e - e$ interaction allowed to discover a new *quasi-free* mechanism of DPI, where two electrons share the photon energy almost equally and have almost the same momenta. This becomes particularly important at relativistic frequencies [9].

Until about 1992 the main features of DPI were clarified, but no new experiments were performed due to lack of proper radiation sources and detection techniques. The general belief was that at high ω, $\omega \gg I^{++}$, the shake-off mechanism is absolutely dominant so that even in a 1994 review [10] the quasi-free mechanism was not mentioned. Third generation synchrotron radiation sources and new detection methods (see e.g. ref. [11]) enabled new investigations of DPI, both experimental and theoretical (ref. [11] and references therein), most of them for He at quite high ω. It was demonstrated that from $\omega \approx 6$ keV for He the dominant mechanism of simultaneous elimination of two electrons is not photoionization but Compton scattering [12]. More complicated processes treated include three-electron ionization of Li [13] and DPI on multi-electron atoms.

DPI is determined by the $e - e$ interaction in the initial and final states of the ionized object. There exist four basic mechanisms:

a) *Shake-off* [1-3], for which

$$\sigma^{++}(\omega) = \sigma^{+}(\omega)W_{so}(\omega - I^{++}), \tag{1}$$

where the second electron shake-off probability $W_{so}(\omega - I^{++})$ behaves as

$$W_{so}(\omega - I^{++}) \to W_{so}, \quad \omega \to \infty. \tag{2}$$

b) *Ground state correlations* [4] cause DPI due to interaction between ionized electrons in the initial state. The expression for $\sigma^{++}(\omega)$ is similar to (1).

c) *Final state interaction* causes DPI by the interaction between the primary photoelectron, ejected by the absorption of the incoming photon, and the second electron. The higher the photoelectron velocity, the greater the probability of knocking out the second electron.

c) *Quasi-free absorption* [8], with the specific feature that while $\sigma^{+}_{qfa}(\omega) \approx 0$, the inequality

$$\sigma^{++}_{qfa}(\omega) > \sigma^{++}_{so}(\omega). \tag{3}$$

holds at $\omega \gtrsim c^2$.

It is seen from these expressions that in the high ω region the main contribution to $\sigma^{++}(\omega)$ and the ratio $R(\omega) = \sigma^{++}(\omega)/\sigma^{+}(\omega)$ comes from shake-off and the ground state correlations, and at $\omega \gtrsim c^2$ from the quasi-free absorption. The final state interaction contributes only at $\omega \sim I^{++}$.

---

[1]The atomic system of units is used: $e = m_e = \hbar = 1$.

In the high-energy region, the expressions for cross sections are greatly simplified since one or even both (for the quasi-free absorption) of the outgoing electrons are fast so they can be described by plane waves. Non-relativistically ($I^{++} \ll \omega \ll c^2$) $\sigma^{++}(\omega)$ is determined by shake-off and ground state correlations (see e.g. ref. [14]):

$$\sigma_i^{S++}(\omega) \approx \frac{32\sqrt{2}Z^2\pi^2}{3c\omega^{7/2}} \left\{ \int |\Psi_i^S(0,\mathbf{r})|^2 d\mathbf{r} - \sum_{nlm} \left| \int \Psi_i^S(0,\mathbf{r})\varphi_{nlm}(\mathbf{r}) d\mathbf{r} \right|^2 \right\} \quad (4)$$

where $\Psi_i^S(\mathbf{r}_1, \mathbf{r}_2)$ is the initial (i) state wave function of two electrons with coordinates $\mathbf{r}_1, \mathbf{r}_2$ in the field of the respective nuclei, i.e. a three-body wave function, Z is the nuclear charge, and $\varphi_{nlm}(\mathbf{r})$ is the single-particle wave function of the "slow" electron excited to the state (nlm). For helium-like ions it is a pure Coulomb wave function. The sum runs over all excited states of the second electron. S denotes the total spin of the initial state, $S = 1$ for singlet and $S = 3$ for triplet.

The sum of the cross section of ionization with excitation and $\sigma_i^+(\omega)$ is [14]

$$\sigma_i^{S+}(\omega) + \sigma_i^{S+*}(\omega) \approx \frac{32\sqrt{2}Z^2\pi^2}{3c\omega^{7/2}} \sum_{nlm} \left| \int \Psi_i^S(0,\mathbf{r})\varphi_{nlm}(\mathbf{r}) d\mathbf{r} \right|^2 \quad (5)\,.$$

Pure one-electron ionization is obtained from (5) by retaining only the term $(nlm) = i$. The ratio $R_i^S(\omega)$ is ω-independent: $R_i^S(\omega) = R_d^{iS}$.

In the relativistic limit $\omega \gtrsim c^2$ the ω dependence of $\sigma_i^+(\omega)$ becomes ω^{-1} instead of $\omega^{-7/2}$. Despite this, the ratio in the ultra-relativistic limit, R_i^{ur}, remains ω-independent and can be presented as the sum

$$R_i^{urS}(\omega) = R_d^{iS} + \frac{4}{Z^2}\mathcal{I}_i^S, \quad (6)$$

$$\mathcal{I}_i^S = \frac{1}{N_i^S} \int d\mathbf{r} \left|\Psi_i^S(\mathbf{r},\mathbf{r})\right|^2 = \frac{1}{N_i^S} \langle \delta(\mathbf{r}_{12}) \rangle_i^S, \quad (7)$$

$$N_i^S = \sum_{n=1}^{\infty} \left| \int d\mathbf{r} \Psi_i^S(\mathbf{r},0)\varphi_{n00}(\mathbf{r}) \right|^2. \quad (8)$$

In refs. [8,15] $R(\omega)$ was expanded in powers of $\omega/c^2 \ll 1$ using integrals (7):

$$R_i^{rS}(\omega) = R_d^{iS} + \frac{8\sqrt{24}}{5Z^2}\mathcal{I}_i^S \frac{\omega}{c^2}. \quad (9)$$

It is essential to note that since in the triplet states the electrons can not be at the same place, $\Psi_i^S(\mathbf{r},\mathbf{r}) = 0$ and $\mathcal{I}_i^1 = 0$, the relation

$$R_i^{ur1}(\omega) = R_d^{i1} \quad (10)$$

holds at any $\omega \gg I^{++}$, including $\omega \gtrsim c^2$. In deriving equations (6)-(10) it was assumed that Z is not too large: $Z \ll 137$.

We shall consider singlet and triplet initial states with total angular momentum $L = 0$ as well as excited states with $n = 2 - 5$, for He and helium-like ions with $Z = 2 - 10$. The extrapolation to $Z = \infty$ is compared with analytic calculations where available: for high Z refs. [8,9] predicted:

$$R_i^{r0}(\omega) \approx [0.094 + 0.141\omega/c^2]/Z^2, \tag{11}$$

$$R_i^{ur0}(\omega) \approx 0.59/Z^2. \tag{12}$$

It is seen that the relativistic non-dipole corrections (the term proportional to ω/c^2) becomes noticeable, of the order of 10%, at $\omega > 30$ keV.

2 Results of Calculations

To calculate $\sigma_i^{S+}(\omega)$, $\sigma_i^{S+*}(\omega), \sigma_i^{S++}(\omega)$ and their ratios

$$R_i^S = \left. \frac{\sigma_i^{S++}(\omega)}{\sigma_i^{S+}(\omega) + \sigma_i^{S+*}(\omega)} \right|_{\omega\to\infty}, \tag{13}$$

$$R_{1i}^S = \left. \frac{\sigma_i^{S+}(\omega) + \sigma_i^{S++}(\omega) + \sigma_i^{S+*}(\omega)}{\sigma_i^{S+}(\omega)} \right|_{\omega\to\infty} = (1 + R_i^S)(1 + R_i^{S*}) \tag{14}$$

we use $\Psi_i^S(\mathbf{r}_1, \mathbf{r}_2)$ calculated by the Correlation Function Hyperspherical Harmonic Method (CFHHM) (ref. [16] and references therein). In CFHHM, Ψ is decomposed as

$$\Psi = e^f \Phi , \tag{15}$$

where the "correlation function" f describes singularities of Ψ so Φ is smooth and can be expanded in hyperspherical harmonics (HH). f depends on interparticle distances, which is necessary and sufficient to take into account analytically the two- and three-body Coulomb singularities (cusps). Ψ is obtained by direct solution of the three-body Schrödinger equation, guaranteeing its local correctness [16] because of uniform convergence across the configuration space.

The singlet state calculations were published [17]; we concentrate here on the triplet states. The obtained values of the integral in Eq. (5), $I_{nn'}^3 = \left|\int \Psi_n^3(0, \mathbf{r})\varphi_{n'00}(\mathbf{r})d\mathbf{r}\right|^2$, and $\mathcal{I}_i^1$ show that the most probable process for the excited initial state is the transition with excitation, namely not $n \to n$, but $n \to n+1$, $n+2$. Only for considerable n the de-excitation transition $n \to n-1$ becomes noticeable compared with $n \to n+1$.

Table 1 presents our results for R_n^3 and the triplet/singlet ratio $\eta_n = R_n^3/R_n^1$.

For triplet states the cross section of ionization with excitation $\sigma_i^{3+*}(\omega)$ is larger than $\sigma_i^{3+}(\omega)$ and therefore R_{1i}^3 is much larger than 1 (see Table 2).

The relativistic ω/c^2 corrections to $R(\omega)$ are calculated in [18]. These results are closely connected, according to (6) and (9), to those for ultrarelativistic ω presented in [19].

Table 1. $R_n^3(Z) \times 10^3$ and $\eta_n(Z)$ for triplet states in the He isoelectronic sequence.

n	2		3		4		5	
Z	$R_2^{(3)}$	η_2	$R_3^{(3)}$	η_3	$R_4^{(3)}$	η_4	$R_5^{(3)}$	η_5
2	3.116	0.35	1.965	0.53	1.043	0.62	0.573	0.65
3	3.040	0.25	3.662	0.44	3.076	0.56	2.332	0.65
4	2.222	0.22	3.420	0.40	3.526	0.52	3.198	0.60
5	1.615	0.21	2.811	0.39	3.218	0.50	3.211	0.58
6	1.209	0.20	2.262	0.38	2.755	0.49	2.911	0.57
7	0.933	0.20	1.830	0.37	2.321	0.48	2.546	0.56
8	0.739	0.20	1.500	0.37	1.957	0.48	2.204	0.56
9	0.599	0.19	1.247	0.37	1.660	0.48	1.906	0.55
10	0.495	0.19	1.050	0.37	1.421	0.48	1.655	0.55

The use of an accurate initial state wave function modifies the ratios considerably. For instance, for He one has $R_1^{ur} = 0.07615$ while the ultra-relativistic formula (12) gives about twice as much: 0.1475. R_n^{ur}/R_n decreases rapidly with n: from $5-6$ (depending on Z) for $n = 1$ to $1 - 1.5$ for $n = 4$ [19].

Table 2. R_{1n}^3 values for triplet states in the He isoelectronic sequence.

n	2	3	4	5
Z				
2	29.618	115.017	292.485	597.101
3	17.889	63.421	155.538	310.760
4	14.608	49.868	120.384	238.421
5	13.086	43.752	104.703	206.395
6	12.211	40.291	95.890	188.441
7	11.644	38.071	90.259	177.000
8	11.246	36.527	86.357	169.085
9	10.953	35.392	83.495	163.284
10	10.727	34.523	81.307	158.859

3 Conclusion

The cross sections $\sigma_i^{S++}(\omega)$, $\sigma_i^{S+}(\omega)$ and their ratios $R(\omega)$ were calculated with high accuracy for He and helium-like ions in ground and S-excited singlet and triplet states. The non-variational wave function used in calculations proved to be very effective. Some of the considered ratios, particularly for the excited initial states, can reach very large values. It was shown that the quasi-free absorption plays a very important role already at relatively low energy of about 10 keV for He, contrary to the high-Z asymptotic estimate which gave a 10% correction at 30 keV.

Acknowledgement. The research was supported by the Hebrew University Intramural Fund (MYA), by the Bilateral Cooperation Program at the Ministry of education, science and sport of Slovenia (RK), and by the Israeli Science Foundation grant 131/00 (VBM).

References

1. T.A. Carlson and M.O. Krause: Phys. Rev. **140**, 1057 (1965)
2. F.W. Byron, Jr. and C.J. Joachain: Phys. Rev. **164**, 1 (1967)
3. T. Åberg: Phys. Rev. **A2**, 1726 (1970)
4. M.Ya. Amusia and M.P. Kazachkov: Phys. Lett. **A28**, 27 (1968)
5. T.N. Chang and R.T. Poe: Phys. Rev. **A12**, 1432 (1975)
6. S.L. Karter and H.P. Kelly: Phys. Rev. **A16**, 1525 (1977)
7. J.A.R. Samson and G.N. Haddad: Phys. Rev. Lett. **33**, 875 (1974)
8. M.Ya. Amusia, V.G. Gorshkov, E.G. Drukarev and M.P. Kazachkov: J. Phys. B: At. Mol. Phys. **8**, 1248 (1975)
9. E.G. Drukarev and F. Karpeshin: J. Phys. B: At. Mol. Phys. **9**, 399 (1976)
10. A. Dalgarno and H.R. Sadeghpour: Comm. At. Mol. Phys. **30**, 143 (1994)
11. J.H. McGuire et al.: J. Phys. B: At. Mol. Opt. Phys. **28**, 913 (1995)
12. J.A.R. Samson et al.: Phys. Rev. Lett. **71**, 201 (1993)
13. Y. Azuma: In: X-ray and inner-shell processes, AIP conference proceedings, vol. 506, p. 116 (2000)
14. M.Ya. Amusia: *Atomic Photoeffect.* New York, London: Plenum Press 1990
15. E.G. Drukarev: Phys. Rev. **A52**, 3910 (1995)
16. R. Krivec, V.B. Mandelzweig, K. Varga: Phys. Rev. **A61**, 062503 (2000)
17. R. Krivec, M.Ya. Amusia, V.B. Mandelzweig: Phys. Rev. **A63**, 052708 (2001)
18. R. Krivec et al.: Phys. Rev. **A64**, 012713 (2001)
19. M.Ya. Amusia, E.G. Drukarev, R. Krivec and V.B. Mandelzweig: Phys. Rev. **A**, in press (2002)

Few-Body Systems Suppl. 14, 153–158 (2003)

Few-Body Systems

Multi-Particle Break-up Near Threshold

T. Pattard*

Max Planck Institute for the Physics of Complex Systems, Nöthnitzer Str. 38, D-01187 Dresden, Germany

Abstract. A short overview on the subject of threshold laws for multiple break-up is given. In order to limit the length of the paper, emphasis is put on the basic physical ideas rather than on mathematical rigor. The threshold laws for short-range as well as long-range and dipolar potentials are summarized and put into perspective.

For various reasons, the derivation of threshold laws for few-particle break-up, i.e. the question of how the probability for a total break-up changes as a function of the excess energy E close to the threshold $E = 0$ where this process becomes possible, has attracted continuing interest for by now more than 50 years. Obviously, calculating this probability in principle requires a detailed (quantum mechanical) description not only of the bound states, but also of the continuum final states describing the outgoing particles and the interactions between them. As opposed to the high-energy limit, all particles are involved on an equal footing and have to be taken care of. Therefore, perturbative methods cannot be applied, which makes the treatment of the problem much more complicated. Furthermore, since the particles move very slowly in the final state (infinitely slowly in the limit $E \to 0$, $r_i \to \infty$), even weak interactions for large interparticle distances may play a crucial role, so that asymptotic properties of the final state wave functions become important. However, in fact it is just this dominance of asymptotic properties that considerably simplifies the underlying physical concepts and allows a treatment of certain aspects of the phenomenon.

Historically, Wigner [1] was the first to point out that the derivation of threshold laws could be considerably easier than obtaining complete cross sections for a wide range of energies. In his paper, he derived threshold laws for two-particle break-up processes for short-range (i.e. vanishing faster than $1/r^2$ asymptotically) as well as Coulomb interactions. The only two-body break-up law not given by Wigner is that for potentials behaving as $1/r^2$ asymptotically, which has been formulated much later by Gailitis and Damburg and O'Malley [2, 3].

* *E-mail address:* tpattard@mpipks-dresden.mpg.de

In order to gain some insight into the special role played by the dipole potential as the borderline case between two quite different regimes, let us start with the following consideration: The threshold break-up in which we are interested corresponds to a limit where the energy E goes to zero while the interparticle distances r_i approach infinity. The usual criterion for the validity of a quasi-classical ansatz given in textbooks (e.g. [4]) is connected to the rate of change of the deBroglie wavelength: $|d\lambda_{dB}/dr| \ll 1$. For long-range potentials ($\propto 1/r^n$ with $n < 2$) like the Coulomb interaction, $d\lambda_{dB}/dr \to 0$ for $E \to 0$ and $r \to \infty$, while for short-range potentials ($n > 2$) $d\lambda_{dB}/dr \to \infty$. Thus, we conclude that threshold break-up is an essentially classical process for long-range interactions, while the same is not true for short-range potentials. Moreover, we recall from basic scattering theory that scattering of a particle at a short-range potential leads to a well-defined phase-shift, while for Coulomb potentials the phase-shift shows a logarithmic dependence on r even asymptotically and does not converge towards a constant. In other words, (only) for short-range interactions the potential becomes negligible asymptotically and the particles behave like free particles.

From this last observation, we are led to the suggestion that for short-range interactions the threshold break-up law should not depend on the details of the potential, but be purely statistical. In this case, the threshold law would simply be determined by the number of available phase space cells the system can go into in the final state. It is then easy to show (see, e.g., the excellent review by Rau [5]) that the threshold law for short-range potentials is found to be

$$\sigma_{SR}^{(N)} \propto E^{(3N-5)/2}, \tag{1}$$

where N is the number of fragments in the final state. We will show later that this expression can also be derived from a more general consideration including dipolar potentials[1].

For charged particles in the final state, on the other hand, the situation becomes much more difficult. Due to the long-range nature of the Coulomb forces, the potential does not become negligible asymptotically, but will subtly influence the threshold law. This will become apparent in particular for break-up into more than two fragments. For a two-body break-up, the forces are always acting along the interparticle axis, and it is only the absolute strength of the interaction which varies. For this case, Wigner derived the break-up law

$$\sigma_C^{(2)} \propto \left\{ \begin{array}{ll} 1 & \quad attr. \\ \mathrm{e}^{-\alpha/\sqrt{E}} & \quad rep. \end{array} \right\}, \tag{2}$$

i.e. starting with a finite value at threshold for an attractive potential and being exponentially small for repulsive potentials. The latter is easily interpreted as a Gamov-like factor for tunneling through the potential barrier formed by the asymptotically dominating repulsive Coulomb interaction and the attractive

[1] Since angular momentum behaves as $1/r^2$, the above expression is valid for vanishing angular momentum only. The case $L \neq 0$ will be discussed below.

short-range potential which must be present in addition in order to allow for bound initial states, see also [6]. For more than two fragments, the potential "landscape" will depend sensitively on the charges and distances of all outgoing particles, hence a much richer behavior of the break-up process can be expected. Indeed, the first threshold law going beyond the two-particle break-up considered by Wigner was derived by Wannier [7], who was able to obtain a power law for the break-up into ion plus two electrons, i.e. electron impact ionization or photo-double ionization[2]. For a long time, this most simple case was the only theoretical result for Coulomb break-up into more than two fragments, apart from a very general half-page note by Wannier on the threshold law for break-up into ion plus $N-1$ electrons [8]. (In this note, Wannier argued that the threshold law should be a power law with an exponent "slightly larger (because of the electron-electron repulsion) than $N-2$ (which would be the result for independent electrons)", he did not, however, make any effort to derive precise numbers.) The first work deriving a precise numerical value for the exponent of a break-up power law for a different system, namely the four-body break-up into ion plus three electrons, was published more than twenty years after Wannier's paper [9], and generalization to asymmetric systems involving three different kinds of particles (i.e. ionization by positron or antiproton impact) came even later. The first general theory for break-up into N charged particles has been formulated by Kuchiev and Ostrovsky as late as 1998, i.e. fifty years after Wigner's original paper [10, 11]. Here, due to shortage of space, I will only outline the most important and fundamental physical idea, for a detailed derivation the reader is referred to [12] (and, of course, [10], although the derivation by Kuchiev and Ostrovsky starts from a somewhat different perspective which leads, however, to equivalent results).

What is the basic idea behind Wannier's analysis? Wannier was able to show that, within the framework of classical mechanics (the reader may recall the discussion about the validity of a classical treatment above!), the only configurations leading to a complete break-up at $E = 0$ are those asymptotically approaching the symmetrical situation where the two electrons recede from each other with the same velocity and the same distance to the nucleus. If one of the electrons is initially a little closer to the nucleus, e.g., this small asymmetry increases during the further evolution of the system and will finally result in this electron remaining bound (although in a highly excited state) while the other electron escapes with a finite positive energy, i.e. the system fails to reach the total break-up. Increasing the energy E will allow for increasing initial asymmetries in the motion of the outgoing electrons around this symmetrical configuration while still reaching complete break-up. It is this increase of phase space volume of "favorable" initial conditions that determines the fragmentation probability.

[2]While it has not been mentioned so far, it should be clear from the emphasis on the asymptotic properties of the potential that the energy dependence of the threshold law depends on the final state only, but not on the initial state and the way the final state is reached. This will of course influence the absolute value of the cross section, but not its functional form.

Trying to cast this physical picture into a more precise mathematical formalism, it is convenient (although by no means necessary) to introduce hyperspherical coordinates defined by $\mathcal{R} \equiv \sqrt{r_1^2 + r_2^2}$ and $\alpha \equiv \arctan\frac{r_2}{r_1}$, where r_1 and r_2 are the distances of the electrons from the nucleus. In general, hyperspherical coordinates are defined by a hyperradius $\mathcal{R}$ determining the "diameter" of the system and a number of angles and hyperangles — which will be called *internal* coordinates and collectively denoted by $\boldsymbol{\omega}$ in the following — describing the shape of the system, i.e. the relative positions of the particles. In these coordinates, it is possible (using mass-scaled coordinates) to write the kinetic energy as a sum of two terms where one only contains the momentum conjugate to the hyperradius while the other contains only internal coordinates and the corresponding momenta divided by $\mathcal{R}^2$,

$$H = \frac{p_{\mathcal{R}}^2}{2} + \frac{1}{2\mathcal{R}^2} D(\boldsymbol{\omega}, \boldsymbol{P}) + \frac{C(\boldsymbol{\omega})}{\mathcal{R}} = E.$$

(The possibility to write the potential in the separated form $C(\boldsymbol{\omega})/\mathcal{R}$ follows from the fact that the Coulomb interaction is a homogeneous function of the interparticle distances.) Obviously, fragmentation corresponds to $\mathcal{R} \to \infty$. At threshold, however, there is just enough energy available to reach this goal, so no energy must be "wasted". All available energy has to be used to change $\mathcal{R}$, and no energy is available to change the internal configuration (at least asymptotically for $\mathcal{R} \to \infty$). Mathematically, this means that the trajectory leading to fragmentation corresponds to a partial fixed point of the Hamiltonian equations of motion, where all internal coordinates are fixed and the corresponding momenta vanish. Therefore, as pointed out by Eckhardt [13], we should expect the instability of this fixed point to determine the threshold law. Indeed, a rather lengthy derivation involving, along with the stability analysis of the fixed point, an energy scaling argument akin to renormalization techniques familiar from the derivations of exponents in critical phenomena and phase transitions finally leads to the threshold law [10, 14]

$$\sigma_C^{(N>2)} \propto E^{\beta} \quad ; \quad \beta = \frac{\sum_{\lambda_i > 0} \lambda_i}{\sqrt{-2C^*}} \,. \tag{3}$$

Here, the λ_i are the Lyapunov exponents characterizing the stability of the fixed point of the internal coordinates[3], while C^* denotes the value of $C(\boldsymbol{\omega})$ at the fixed point. Since $E = 0$, the kinetic energy associated with the hyperradial momentum has to balance the potential asymptotically, so that there is a direct relation between the behavior of $p_{\mathcal{R}}$ and C^*. Equation (3) is therefore interpreted in the way that it is not only the instability of the fixed point (given by

[3]The Lyapunov exponents are obtained by introducing small deviations from the fixed point values of the coordinates and momenta and linearizing the equations of motion in these deviations. This leads to a matrix equation, the solution of which gives the "normal modes" around the fixed point. In general, some modes will oscillate around the fixed point (λ complex), some will converge towards it ($\lambda < 0$), and, since the fixed point corresponds to a saddle rather than a minimum of the potential, some will exponentially diverge from the fixed point ($\lambda > 0$).

the sum over the λ_i) that determines the threshold exponent. Rather, it is the ratio of the velocities of two competing processes, namely motion away from the fixed point (as determined by its Lyapunov exponents) and the outward motion (as determined by $p_{\mathcal{R}}$ or, equivalently, C^*). It should be noted that for break-up into more than three particles, there will in general be more than one single fixed point of the internal coordinates. The above arguments then apply to each individual fixed point, and the cross section is a sum of independent contributions from the different fixed points, with the one with the smallest β determining the threshold behavior.

Finally, what is left is the threshold law for potentials behaving as $1/r^2$ asymptotically. As stated before, for the case of two-body break-up it has been derived in [2, 3]. For the general case of N-body break-up it has been given in [15]. The derivation again makes use of the fact that in hyperspherical coordinates the potential can be written in the form $C(\boldsymbol{\omega})/\mathcal{R}^2$. With the substitution $\psi(\mathcal{R},\boldsymbol{\omega}) \equiv \mathcal{R}^{(D-1)/2}\tilde{\psi}(\mathcal{R},\boldsymbol{\omega})$, the Schrödinger equation (neglecting shorter-range potentials that become negligible compared to the dipole potential for $\mathcal{R} \to \infty$) becomes separable asymptotically, so that $\tilde{\psi}$ can be written in the form $\tilde{\psi} = \sum_j F_j(\mathcal{R})\phi_j(\boldsymbol{\omega})$. The solution of the "angular" equation then enters the radial equation only in the form of the corresponding eigenvalue λ_j, and the effective radial potential is found to be $V_{eff}(\mathcal{R}) = (\nu_j^2 - 1/4)/2\mathcal{R}^2$ with $\nu_j^2 \equiv (3N-5)^2/4 + \lambda_j$. Thus, the radial equation is of two-body form irrespective of the actual number N of fragments, which only enters in the value for ν, i.e. the strength of the effective potential. Hence, the general problem of N-body break-up has been reduced to the corresponding problem for two-body break-up, which has been solved by Gailitis and Damburg and O'Malley [2, 3]. The solution is given by

$$\sigma_D^{(N)} \propto \left\{ \begin{array}{lll} E^{\nu_j} & rep. & (\nu_j > 0) \\ (\sinh^2(\bar{\nu}_j\pi/2) + \cos^2(\bar{\nu}_j \ln\sqrt{2E} + \delta_j))^{-1} & attr. & (\nu_j = i\bar{\nu}_j) \end{array} \right\}, \tag{4}$$

i.e. a power law for asymptotically repulsive (and weakly attractive) potentials and oscillating, starting with a finite value at threshold, for attractive potentials. (The reader should be cautioned that it is misleading to assume a simple analogy between the power laws in Eq. (3) and Eq. (4). For dipole interactions, the power law arises for overall repulsive potentials, while in the Coulomb case the radial potential is attractive [$C < 0$].) As mentioned before, the statistical threshold law for short-range potentials can also be obtained from the preceding derivation. In this case $C(\boldsymbol{\omega}) = 0$, which leads to the solution $\lambda_0 = 0$ for the angular problem, and hence $\nu_0^2 = (3N-5)^2/4$ which leads exactly to the result given in Eq. (1). Of course, Wigner's two-body break-up law for short-range interactions but non-vanishing angular momentum also follows from the above, since then $\lambda_l = l(l+1)$.

In summary, a short overview on threshold laws for few-body break-up has been presented. The basic physical ideas have been summarized which underlie the various laws applicable for different potentials, and the laws for short-range as well as dipole and Coulomb interactions have been given. However, the reader should keep in mind that there are always exceptional cases where these general ideas do not apply and the threshold behavior will vary from the expected form.

Acknowledgement. Many helpful discussions with Jan-Michael Rost are gratefully acknowledged.

References

1. E.P. Wigner: Phys. Rev. **73**, 1002 (1948)
2. M. Gailitis and R. Damburg: Proc. Phys. Soc. **82**, 192 (1963)
3. T.F. O'Malley: Phys. Rev. **137**, A1668 (1965)
4. A. Messiah: *Quantenmechanik*, de Gruyter (1981)
5. A.R.P. Rau: Comments At. Mol. Phys. **14**, 285 (1984)
6. J.M. Rost: Phys. Rev. Lett. **82**, 1652 (1999)
7. G.H. Wannier: Phys. Rev. **90**, 817 (1953)
8. G.H. Wannier: Phys. Rev. **100**, 1180 (1955)
9. H. Klar and W. Schlecht: J. Phys. **B9**, 1699 (1976)
10. M.Y. Kuchiev and V.N. Ostrovsky: Phys. Rev. **A58**, 321 (1998)
11. V.N. Ostrovsky: Phys. Rev. **A64**, 022715 (2001)
12. T. Pattard and J.M. Rost: in preparation
13. B. Eckhardt: *Habilitationsschrift.* Marburg 1991 (unpublished)
14. T. Pattard and J.M. Rost: Phys. Rev. Lett. **80**, 5081 (1998); Erratum ibid. **81**, 2618 (1998); Comment ibid. **85**, 4409 (2000); Reply ibid. **85**, 4410 (2000)
15. T. Pattard and J.M. Rost: Found. Phys. **31**, 535 (2001)

Few-Body Systems Suppl. 14, 159–162 (2003)

Few-Body Systems

Effects of Breakup Thresholds in the Single-Particle Momentum Distributions of a Few-Body Continuum

R.O. Barrachina*,†

Centro Atómico Bariloche and Instituto Balseiro, 8400 S. C. de Bariloche, Río Negro, Argentina.

Abstract. We present an alternative approach for the study of a break-up process at threshold. Specifically, we demonstrate that the single differential cross section of one particle in a few-body system carries basic information about the breakup threshold of the remaining cluster. Thus, it represents an alternative approach for the study of few-body continua at very low energies. We exemplify these ideas by analyzing the particular case of a three-body system. In particular, we show that it is possible to extract electron-atom scattering parameters, such as the s-wave scattering length from ionization measurements.

Let us consider any given nuclear, atomic or molecular process leading to a N-body final state of mass M and energy E, whose center of mass moves with velocity $\boldsymbol{v}_{CM}$ in the laboratory reference frame. The energy E is measured from a zero corresponding to a complete breakup of the system in all its N constituents with velocities equal to $\boldsymbol{v}_{CM}$ at infinity. Let us now single out one fragment of mass m and characterize it by its Jacobi impulse $\mathbf{K}$ with respect to the center of mass of the other $N-1$ particles in the system. By energy conservation, the particle cross section single differential in this momentum K reads

$$\frac{\mathrm{d}\sigma}{\mathrm{d}K} = \frac{K}{\nu}\,\frac{\mathrm{d}\sigma}{\mathrm{d}\epsilon}\,, \tag{1}$$

where $\epsilon = E - K^2/2\nu$ is the excess energy of the remaining N-1 particles in the breakup channel, and $\nu = m(1 - m/M)$ is the corresponding reduced mass. Note that the particle momentum $\boldsymbol{P}$ in the laboratory reference frame is linearly related to its Jacobi impulse $\boldsymbol{K}$ by $\boldsymbol{P} = m\boldsymbol{v}_{CM} + \boldsymbol{K}$. Thus, the previous equation provides a dynamical constraint that relates the single differential cross section (SDCS) for one particle to the energy distribution of the

*Member of Comisión Nacional de Energía Atómica (CNEA) and Universidad Nacional de Cuyo (UNC), Argentina. Also member of the Consejo Nacional de Investigaciones Científicas y Técnicas (CONICET), Argentina.
†*E-mail address:* barrachina@ib.edu.ar, *Fax:* +54 2944 44-5102, *Phone:* +54 2944 44-5163

other remaining particles as a whole. Furthermore, due to energy conservation, the Jacobi momentum $\boldsymbol{K}$ of one particle reaches its maximum allotted value within the breakup channel whenever the energy ϵ of the remaining $N-1$ particles approaches zero. In essence, this means that at the limiting value $|\boldsymbol{P} - m\boldsymbol{v}_{CM}| \rightarrow \sqrt{2\nu E}$ the cross section of one particle carries information on the breakup threshold $\epsilon \rightarrow 0^+$ of the remaining cluster.

Here we want to exemplify the far-reaching consequences of these ideas by discussing the simplest case, i.e. a two-body breakup embedded in a three-body continuum. It represents one of the simplest systems which shows the main characteristics that we have been discussing so far. Let us consider a situation where the relative momentum $k = \sqrt{2\mu\epsilon}$ of the two-body system (of reduced mass μ) is very close to zero. In this case, the *Final-State Interaction* (FSI) theory [1] indicates that, under very general conditions [2, 3], the limiting value of the SDCS of the third particle is proportional to [1]

$$\frac{\mathrm{d}\sigma}{\mathrm{d}K} \propto \frac{K}{\nu}\frac{k}{|f_0(k)|^2} \qquad \text{for} \qquad k = \sqrt{2\mu E - \mu K^2/\nu} \rightarrow 0^+ \,. \tag{2}$$

Here $f_0(k)$ is the s-wave Jost function for the other two-body particles, which is defined by the corresponding normalized $\ell = 0$ radial wavefunction $\psi_{0,k}(r)$ evaluated at the origin $\psi_{0,k}(r) \underset{r\rightarrow 0}{\approx} kr/f_0(k)$. Note that we are not concerned with how this particular final-state configuration of $2+1$ particles was reached. In fact, how one particle behaves in this microscopic *ménage à trois* when its Jacobi momentum reaches the maximum allowed value does not depend on the details of the breakup process or its interaction with its other two partners, but only on how these two particles behave when they are close together in momentum space. For this reason Eqs. 1 and 2 provide an alternative approach for the study of a two-particle continuum at extremely low-energies that might not be easily accessible in direct scattering experiments.

Let us now review different possible threshold behaviors. According to the Wigner laws [4], the SDCS $\mathrm{d}\sigma/\mathrm{d}K$ should approach a constant value for an asymptotic Coulomb attraction between the other two particles. This behavior is related to the fact that the discrete spectrum of the two body system accumulates at zero energy, and -by continuity across the ionization limit- the SDCS does not vanish at threshold. On the other hand, for an asymptotically Coulomb repulsive interaction, it decreases exponentially as $\log(\mathrm{d}\sigma/\mathrm{d}\varepsilon) \propto -1/k$. These behaviors are shown in Figure 1 for the ionization of an atom by proton and antiproton impact. The SDCS $\mathrm{d}\sigma/\mathrm{d}K$ of the recoiling target-ion is proportional to the distortion factor $k/|f_0(k)|^2$ in Eq. 2, which is plotted as a function of $k = \sqrt{2\mu E - \mu K^2/\nu}$. Note that Wigner laws correctly describe the threshold behavior for both proton and antiproton impact. These results can be generalized to the case where the two-body interaction is such that $V(r) \approx -Z/r^\eta$ with $\eta < 2$ for $r \rightarrow \infty$. The SDCS in Eq. 2 displays a modified exponential fall-off if $Z < 0$, namely $\log \mathrm{d}\sigma/\mathrm{d}\varepsilon \propto -k^{1-2/\eta}$ [1], and approaches a constant value for $Z > 0$ as for a Coulomb potential. Finally, for a short range two-body interaction, the small-k behavior of the SDCS displays a characteristic shape

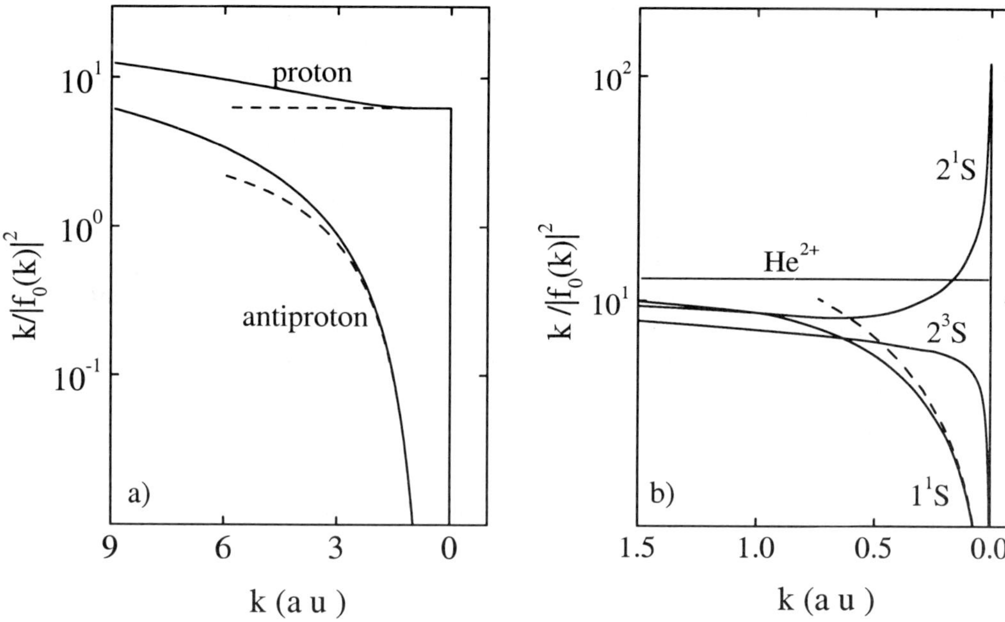

Figure 1. Distortion factors $k/|f_0(k)|^2$ at threshold for the interaction of electrons with a) protons and antiprotons and b) Helium in the 1^1S, 2^1S, 2^3S and ionic He^{2+} states [7]. Wigner threshold laws are also shown as dash lines.

$k/|f_0(k)|^2 \propto k\,a_o^2/(1+a_o^2\,k^2)$ [1], where the s-wave "scattering length" a_o is associated to the presence of a low-lying bound ($a_o > 0$) or virtual ($a_o < 0$) state. Whenever the energy of this s-wave bound or virtual state is far from zero, f_0 can be assumed to be constant over a small range of k, and the SDCS follows the Wigner law $\mathrm{d}\sigma/\mathrm{d}K \propto k$. However, if a_o is large enough, the cross section will deviate from the Wigner law at threshold, presenting a sharp structure. In particular, in the case of a "zero-energy resonance", a_o is infinite and $\mathrm{d}\sigma/\mathrm{d}K$ presents a $1/k$ divergence at threshold. These results are strictly valid for potentials that fall off faster than any power of $1/r$. In other cases, the Jost function might have a branch-point singularity at the origin, and some distortion of the previous shape might occur. For a two body interaction with a long $-\alpha/2r^4$ tail, the SDCS up to order k^2 reads [5]

$$\frac{k}{|f_0(k)|^2} \propto k\,a_o^2 \left\{1 - \frac{4}{3}(\beta k)^2 \log\left(\frac{\beta k}{4}\right) + \left(a_o^2 - \frac{2\,\pi}{3} a_o \beta\right) k^2 \right\}^{-1} \tag{3}$$

with $\beta^2 = \mu\alpha/\hbar^2$. Figure 1 shows this kind of behavior for the case of an electron interacting with a neutral Helium atom in its ground 1^1S or metastable 2^1S and 2^3S states [6]. The Wigner threshold law is reproduced only in the case of an e^- + He(1^1S) interaction. The e^- + He(2^3S) system presents an important deviation from this simple law, which can be ascribed to the proximity of a ^{2}S resonance. The deviation from the Wigner law is much stronger for the e^- + He(2^1S) system, due to the presence of a low-lying virtual state with a scattering length of about a_o = - 300 a.u. [6].

Equations 2 and 3 lead to the possibility of extracting experimental information on low-energy electron-excited atom collisions from coincidence measurements in ionization atomic collisions [5]. For instance, in 1997 Báder *et al.* [8] fitted their experimental data for 400 keV He + Ar → He + Ar^+ + e^- collisions and obtained the s-wave scattering lengths for the collision of electrons with 2^1S and 2^3S metastable Helium atoms, in agreement with theoretical estimates. Furthermore, in a recent work [9], an electron detachment 100 keV Li^- + He → Li + He + e^- collision was measured with a coincidence technique. A fitting of these data provides an s-wave scattering length of the $\mathrm{Li}(2p^2P) + \epsilon s$ that agrees with values obtained from photodetachment $h\nu + \mathrm{Li}^-(2s^2\ ^1S) \rightarrow \mathrm{Li}(2p^2P) + \epsilon s$ experiments [10, 11].

References

1. R.O. Barrachina: Nucl. Instr. Meth. **B124**, 198 (1997)
2. J. Gillespie: *Final State Interactions*, Holden-Day Inc., San Francisco (1964)
3. J.R. Taylor: *Scattering Theory: The Quantum Theory on Nonrelativitic Collisions*, John Wiley & Sons, New York (1972)
4. E.P. Wigner: Phys. Rev. **73**, 1002 (1948)
5. P. Macri and R.O. Barrachina: Phys. Rev. **A65**, 062718 (2002)
6. P. Macri and R.O. Barrachina: J. Phys. **B31**, 1303 (1998)
7. R.O. Barrachina, J. Fiol, V.D. Rodríguez and P. Macri: In: *The Physics of Electronic and Atomic Collisions*, Y. Itikawa et al. (eds.), p. 510, American Institute of Physics, New York (2000)
8. A. Báder, L. Sarkadi, L. Víkor, M. Kuzel, P.A. Závodszky, T. Jalowy, K.O. Groeneveld, P.A. Macri and R.O. Barrachina: Phys. Rev. **A55**, 14 (1997)
9. D.H. Lee, D.J. Pegg and D. Hanstorp: *Investigation of Shape and Virtual Resonances in Fast Negative Ion-Atom Collisions*, (Meeting of the Am. Phys. Soc., Division of Atomic, Molecular and Optical Physics, Storrs, Connecticut, USA, 14-17 June 2000)
10. Y.K. Bae and J.R. Peterson: Phys. Rev. **A32**, 1917 (1985)
11. G. Haeffler, D. Hanstorp, I. Kiyan, A.E. Klinkmüller, U. Ljungblad and D. Pegg: unpublished (2002)

Few-Body Systems Suppl. 14, 163–166 (2003)

Few-Body Systems

The Latest Determination of Antiproton Mass by Precision Three-Body Calculation of Antiprotonic Helium Atoms

Y. Kino[1], M. Kamimura[2], H. Kudo[1]

[1] Department of Chemistry, Tohoku University, Sendai 980-8578, Japan
[2] Department of Physics, Kyushu University, Fukuoka 812-8581, Japan

Abstract. The Particle Listings 2000 edition gave the recommended value of antiproton mass, for the first time, to be within 5×10^{-7} in the relative deviation from the proton mass; this can be a test of the CPT invariance. The value was derived by our precision 3-body calculation to analyze the laser spectroscopy data of metastable states in antiprotonic helium atoms taken at CERN. The 2002 edition reported an one-order smaller value, 6×10^{-8}, based on further development of both the experiment and our calculation.

In the Particle Listings 1998 edition, the Particle Data Group gave no recommended value of the mass of antiproton. Instead, they only cited several scattered values so far obtained. The reason why it is difficult to determine antiproton mass is as follows: The charge-to-mass ratio is determined very precisely, within error of 10^{-11}, by the periodic motion of antiproton in a magnetic field. But, one more relation is necessary between charge and mass. It is given by observing the energy of the X-ray decay in $\bar{p}$ atoms, but, the error of this observation is as large as 10^{-5} to 10^{-4}.

The antiproton mass is believed to be the same as the proton mass, but there had been no precise experimental information about itself before Particle Listings 2000 [1] in which a recommended information of antiproton mass is given for the first time; deviation of antiproton mass from proton mass is relatively within 5×10^{-7}. This can be a test of CPT invariance. This number was derived by the present authors [2, 3, 4] by analyzing the high-resolution laser experiment on the antiprotonic helium atoms done at CERN [5].

Antiprotons injected into matter annihilate within pico second. But, it was surprising that much delayed annihilation phenomena was observed at KEK in 1991: About 3 % of antiprotons injected into the helium target survived as long as a few microseconds. This phenomena was soon understood as production of metastable states of antiprotonic helium atoms ($\bar{p}\mathrm{He}^+ \equiv \mathrm{He}^{2+} + e^- + \bar{p}$). This exotic atom can be obtained by replacing one of the electrons by an antiproton. Therefore the principal quantum number n of antiproton is estimated to be $n \approx 38$, so highly excited states. About 3 % of the injected antiprotons are caught and located in nearly the circular orbitals ($l = n - 1$) with $n \approx 38$.

In the vicinity of circular states, the Auger transition in $\bar{p}He^+$ is strongly suppressed because the Auger electron needs here a large gap in angular momentum ($\Delta l_e \geq 4$). Thus, $\bar{p}He^+$ in metastable states are deexcited only by a slow radiative decay. More than a dozen of transition-frequencies between metastable states of $\bar{p}He^+$ were measured with fine laser spectroscopy in CERN (see Refs. [5, 7], and further references therein). Such long-lived antiprotonic atoms have stimulated spectroscopic studies from a viewpoint of antimatter science: CPT invariance and weak equivalence principle, interaction between matter and antimatter, etc.

The antiprotonic helium atom has two different faces of three-body systems at the same time. (i) From a viewpoint of interaction, the helium nucleus attracts the electron and the antiproton. This shows an atomic picture like a helium atom. (ii) From a viewpoint of dynamics, on the other hand, the electron moves around the two massive particles like a molecule. Calculational methods should describe these two pictures flexibly. Moreover, difficulty in calculation is that accuracy of 8 significant figures in transition frequency is necessary to analyse the high-resolution laser data and that the metastable states concerned are so-called Feshbach states and therefore suitable resonance condition is to be taken into account in calculation.

To precisely solve this three-body system, we employ the coupled-rearrangement-channel variational method [6] which is suited for describing the different types of pictures simultaneously and treating the electron-antiproton correlation. The three-body total wave function Ψ_{JM} is described as a sum of amplitudes of three rearrangement channels $c = 1-3$ (Fig. 1).

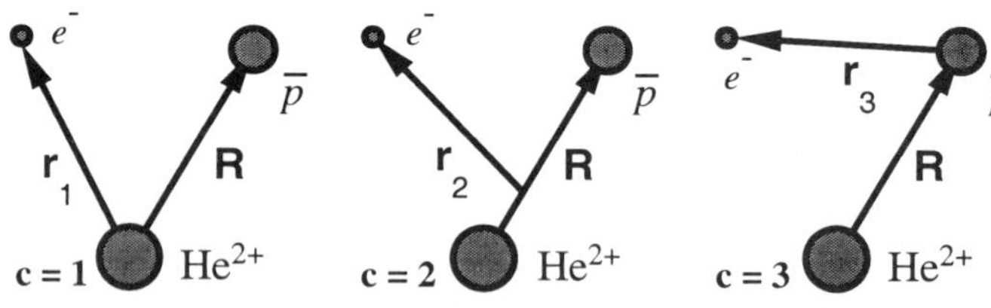

Figure 1. Three rearrangement channels ($c = 1-3$) of $\bar{p}He^+$.

$$\Psi_{JM} = \Phi_{JM}^{(c=1)}(\mathbf{r}_1, \mathbf{R}) + \Phi_{JM}^{(c=2)}(\mathbf{r}_2, \mathbf{R}) + \Phi_{JM}^{(c=3)}(\mathbf{r}_3, \mathbf{R}) \,. \tag{1}$$

Each amplitude is expanded in terms of the Gaussian basis functions of the coordinates $\mathbf{r}_c$ and $\mathbf{R}$:

$$\Phi_{JM}^{(c)}(\mathbf{r}_c, \mathbf{R}) = \sum_{nl,NL} A_{c,nl,NL} \left[\phi_{nl}^{\mathrm{G}}(\mathbf{r}_c)\, \psi_{NL}^{\mathrm{G}}(\mathbf{R})\right]_{JM} \quad (c = 1-3) \,, \tag{2}$$

$$\phi^{G}_{nlm}(\mathbf{r}) = r^l\, e^{-\nu_n r^2}\, Y_{lm}(\widehat{\mathbf{r}}) \qquad (n = 1 - n_{\max})\,, \tag{3}$$

$$\psi^{G}_{NLM}(\mathbf{R}) = R^L\, e^{-\lambda_N R^2}\, Y_{LM}(\widehat{\mathbf{R}}) \qquad (N = 1 - N_{\max}). \tag{4}$$

The Gaussian ranges are chosen to lie in a geometrical progression [6]. This type of choice of Gaussian ranges is very suited for describing both the short-range correlations and long-range tail behaviour.

Channel $c = 1$ is suited for describing the atomic picture mentioned above. Channel $c = 2$ is for describing the molecular picture. Channel $c = 3$ is for describing the electron-antiproton correlation which plays an essential role in long lifetime of the atoms by reducing the Auger transition.

Energies are calculated with the complex coordinate rotation method to take a resonant boundary condition of metastable states into account. Using the first order perturbation theory, the relativistic and QED corrections are considered [3]. We carefully evaluate the accuracy of our calculations and conclude that we have achieved an eight-digits accuracy in the transition frequency.

First, we assume $m_{\bar{p}} = m_p$, $e_{\bar{p}} = -e_p$ and calculate the transition frequencies for $(J_i, v_i) - (J_f, v_f)$ and compare the results with the latest experimental values [7] in Table 1. The calculated values are in good agreement with the experimental values. The best agreement is the transition (33,1)-(32,1). The worst agreement is the transition (34,1)-(33,3); the discrepancy is due to a unique character of the daughter state which we do not discuss here.

Table 1. Transition frequencies between metastable states of $\bar{p}He^+$; theoretical values ($\nu_{\rm th}$), experimental values ($\nu_{\rm exp}$)[7], discrepancies between $\nu_{\rm th}$ and $\nu_{\rm exp}$, transition frequency shifts due to the scaling of antiproton mass and charge ($\nu_{\rm th}/\frac{d\nu_{\rm th}}{dx}$) and uncertainty of antiproton mass ($\frac{\Delta m_{\bar{p}}}{m_p} = (\nu_{\rm th} - \nu_{\rm exp})/\frac{d\nu_{\rm th}}{dx}$) .

$(J_i, v_i) - (J_f, v_f)$	$\nu_{\rm th}$ (GHz)	$\nu_{\rm exp}$ (GHz)	$\frac{\nu_{\rm th}-\nu_{\rm exp}}{\nu_{\rm th}}$ (ppm)	$\nu_{\rm th}/\frac{d\nu_{\rm th}}{dx}$	$\frac{\Delta m_{\bar{p}}}{m_p}$ (ppm)
(33,1)-(32,1)	804 633.127(5)	804 633.11(11)	0.021	2.56	0.054
(32,0)-(31,0)	1 012 445.559	1 012 445.52(17)	0.039	2.45	0.094
(35,3)-(34,3)	501 948.828(8)	501 949.01(13)	-0.36	4.44	-1.6
(35,1)-(34,3)	412 885.131(8)	412 885.18(12)	-0.12	6.05	-0.72
(34,2)-(33,4)	420121.53(1)	420121.9(10)	-0.88	2.73	-2.4
(34,1)-(33,3)	486 104.43(7)	486 102.6(7)	3.8	3.40	13

Second, we examine uncertainty of the antiproton mass by slightly changing the antiproton mass when the calculated values are in good agreement with the experimental values. We scale the antiproton mass by a small fraction x as well as the charge: $m_{\bar{p}} = (1 + x)m_p$, $e_{\bar{p}} = (1 + x)e_p$ which keeps the measured charge-to-mass ratio of antiproton. We recalculate transition frequencies as a function of the scaling parameter x. An example for the transition (33,1)-(32,1) is shown in Fig. 2. The uncertainty of antiproton mass (or equivalently charge) is given by $\Delta m_{\bar{p}}/m_p(\Delta e_{\bar{p}}/e_p) = \Delta x = (\nu_{\rm th} - \nu_{\rm exp})/\frac{d\nu_{\rm th}}{dx}$ where $\frac{d\nu_{\rm th}}{dx}$ is obtained from the slope of Fig. 2.

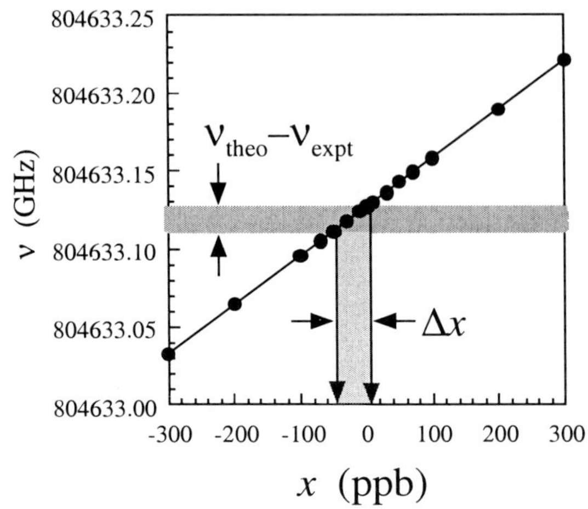

Figure 2. The calculated frequencies of the transition (33,1)-(32,1) as a function of the $\bar{p}$ mass shift parameter x. Closed circles show calculated values and solid line is a guide for eyes. $\Delta m_{\bar{p}}/m_p(\Delta e_{\bar{p}}/e_p) = \Delta x = (\nu_{\rm th} - \nu_{\rm exp})/\frac{d\nu_{\rm th}}{dx}$

In conclusion, we obtained the best limit of the antiproton mass with 5.4×10^{-8} uncertainty. In the experimental paper [7], based on this value but taking some more statistics, $\Delta m_{\bar{p}}/m_p < 6 \times 10^{-8}$ was reported and this value was cited in Particle Listings 2002 [8]. The uncertainty is ten times smaller than the previously estimated value, in Particle Listings 2000 [1], based on our calculation [2, 3, 4] and the experimental data [5].

References

1. D.E. Groom et al. (Particle Data Group): Eur. Phys. J. **C15**, 1 (2000)
2. Y. Kino, M. Kamimura, H. Kudo: Nucl. Phys. **A31**, 649 (1998)
3. Y. Kino, M. Kamimura, H. Kudo: Hyperfine Interact. **119**, 201 (1999)
4. Y. Kino, M. Kamimura, H. Kudo: Few-Body Syst. Suppl. **12**, 40 (2000)
5. H.A. Torii et al.: Phys. Rev. **A59**, 223 (1999)
6. M. Kamimura: Phys. Rev. **A38**, 621 (1988)
7. M. Hori et al.: Phys. Rev. Lett. **87**, 093401 (2001)
8. K. Hagiwara et al. (Particle Data Group): Phys. Rev. **D66**, 010001 (2002)

Few-Body Systems Suppl. 14, 167–170 (2003)

Few-Body Systems

Interaction of Heavy Charged Particles with Atomic Hydrogen

R. Lazauskas*, J. Carbonell†

Institut des Sciences Nucléaires, 53 av. des Martyrs, 38026 Grenoble, France

Abstract. Faddeev equations in configuration space are employed to study low energy scattering of heavy positively charged particles on hydrogen atoms. Special interest is devoted to pH system, which was found to have a resonant value of scattering length, indicating the existence of a neartreshold H_2^+ bound state. Its binding energy B=1.13×10^{-9} a.u. below the H ground state, the smallest bound ever predicted.

The system of three charges constitutes a genuine 3-body problem studied from the birth of quantum mechanics. Since, much progress was done in solving bound state problem with explicit studies of molecular structures as H_2^+, $p^+\mu^-p^+$, etc. On the other hand the rigorous solutions of three-body Coulomb scattering problem are still limited to the simplest cinematical cases as e^--H and e^+-H, where one particle is much heavier than others.

This contribution is devoted to extend these studies by considering elastic scattering of heavy positively charged particles, with mass m in the range $m_e \leq m \leq m_p$, on atomic hydrogen at energies bellow inelastic thresholds. We will use all along the paper electronic atomic units ($m_e = e^2 = \hbar = 1$).

The main difficulty of solving three-body Coulomb problem is related to long-range character of the potential. The primary equations of three body problem, the Faddeev equations, suppose free asymptotic behavior of the particles, in case of long range interaction become ill behaved due to noncompactness of their kernel. Anyway, they can still provide satisfactory solution for bound state problem when asymptotic conditions are implemented by vanishing the total wave function. They are however completely unapplicable for scattering states. The technical reason is that their right hand sides do not decrease fast enough to ensure the decoupling of Faddeev amplitudes in the asymptotic region and to allow unambiguous implementation of boundary conditions. In order to circumvent this problem, Merkuriev [1] proposed to split

* *E-mail address* lazauskas@isn.in2p3.fr
† *E-mail address:* carbonel@isn.in2p3.fr

the Coulomb potential V into two parts (short and long range), $V = V^s + V^l$, by means of some arbitrary cut-off function χ:

$$V^s(x,y) = V(x)\chi(x,y) \qquad V^l(x,y) = V(x)[1-\chi(x,y)]\,. \tag{1}$$

One is then left with a system of equivalent equations

$$(E - H_0 - W_\alpha - V_\alpha^s)\Psi_\alpha = V_\alpha^s \sum_{\alpha\neq\beta} \Psi_\beta \qquad W_\alpha = V_\alpha^l + V_\beta^l + V_\gamma^l \tag{2}$$

with r.h.s. containing only short range contributions (V_s) and with some 3-body potential (W_α) which incorporates the long range parts. This approach was found very efficient in calculating the e^+Ps and e^+H cross sections [3, 4].

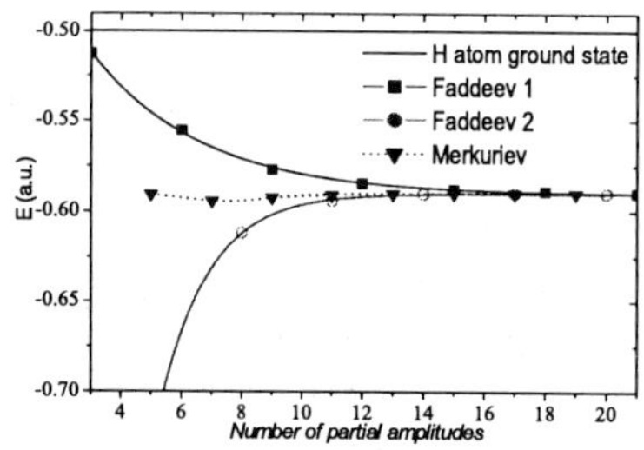

Figure 1. Convergence of Merkuriev and Faddeev equations in μ^+H ground state calculations.

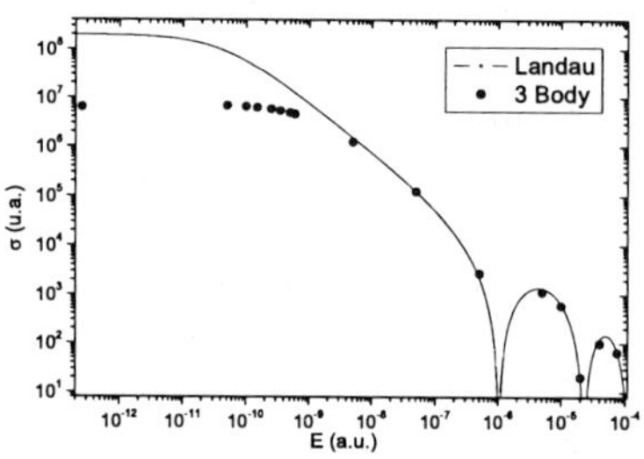

Figure 2. Low energy pH scattering in the pp triplet state, compared to the results of effective 2-body Landau potential.

Equations (2) were solved by expanding Ψ_i in the bipolar harmonics basis and their components $\varphi_{i\alpha_i}$ in terms of two-dimensional splines. Its worth mentioning that the Merkuriev equations when applied to bound state problem are advantageous with respect to standard ones, because corresponding components are smoother functions and needs smaller bipolar harmonic basis to be described Fig. 1. In this figure, two different convergence schemes of Faddeev equations are presented: Faddeev 1 indicates a scheme where the number of the partial amplitudes is equal in all three components, whereas in Faddeev 2 the number of the partial amplitudes in repulsive components is by 1 smaller than in attractive ones.

When dealing with a three-body Coulomb scattering problem some additional care should be taken when introducing boundary conditions – 2-body state is never free due to long range interaction – as well as when extracting the scattering observables from the asymptotic solution at finite distance. The long range polarization force makes the convergence of the phase shifts as a function of the X^+-H distance very slow and requires an additional analysis [2].

Some interesting features of the 3-body Coulomb system can be learned from Fig. 3, where the zero energy X^+H cross section as a function of the projectile mass m_X is displayed. Each peak corresponds to the appearance of a new S-wave bound state. The critical mass values m_i at which they occur, enable to generalize the ground state stability triangle [5] to higher excitations. The calculated scattering lengths for physical projectiles μ^+ and π^+ are $a_{\mu^+H} = 69.1$ and $a_{\pi^+H} = 24.4$ respectively. Other interesting feature about the $X^+ - H$ system can be learned from studying the zero energy scattering wave functions. All the nodal structure of wave function is situated along the line connecting X^+ and H atoms center of mass. This corresponds to the "vibrational" bound states and contains information about their number.

Scattering of protons on H atoms presents special interest. The pH 3-body wave function is antisymmetric with respect to the protons exchange. This can be realized in two different ways following the proton spin coupling. When two protons spins are antiparallel (singlet) the spatial part of the wave function is symmetric, while for the parallel case (triplet) it is antisymmetric. In Born-Oppenheimer approach, these two cases give rise to completely different 2-body effective potentials. The singlet case has a broad attractive well which supports a great number of bound states. They have been calculated since the first days of Quantum Mechanics and presently are known with a very high precision (see e.g. [6, 7] and reference therein). Our 3-body calculations cannot reach this kind of accuracy for bound states but are in 5-6 digits agreement for the lower excitations. Our calculations provide the first result for the pH scattering length $a_s = -29.3$. We notice that the zero energy scattering wave function shows 20 nodes in X^+H-direction, indicating the existence of 20 L=0 σ_g energy levels for H_2^+.

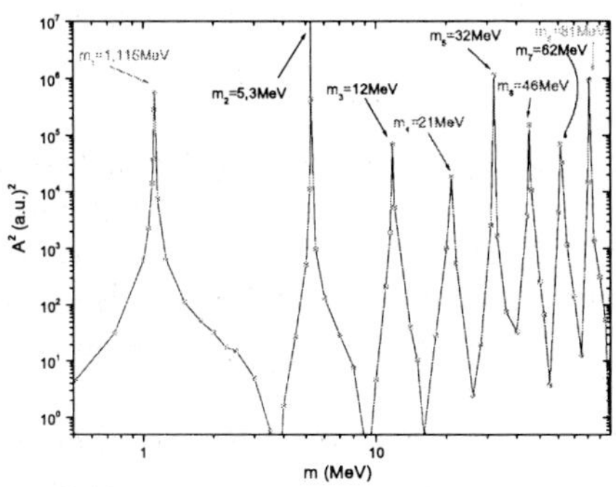

Figure 3. 3-body zero energy X^+H cross section as a function of the projectile mass.

The triplet case – for which the 2-body potential was successfully modelled by Landau [9] – is dominated by the Pauli repulsion between the two protons, overbalanced at $r \sim 10$ by the attractive polarization forces. Our 3-body calculations give a scattering length of $a_t = 750 \pm 5$. The nodal structure of the Faddeev amplitudes indicates that such a big value is due to the existence of a first excited L=0 state with extremely small binding energy. By using the modified effective range theory [10] we are able to determine its value B=$(1.135 \pm 0.035) \times 10^{-9}$ below the H ground state. This result was recently confirmed by direct bound state calculation [11]. To our knowledge, this is the weakest bond ever predicted, three times smaller than ^{4}He atomic dimer [8].

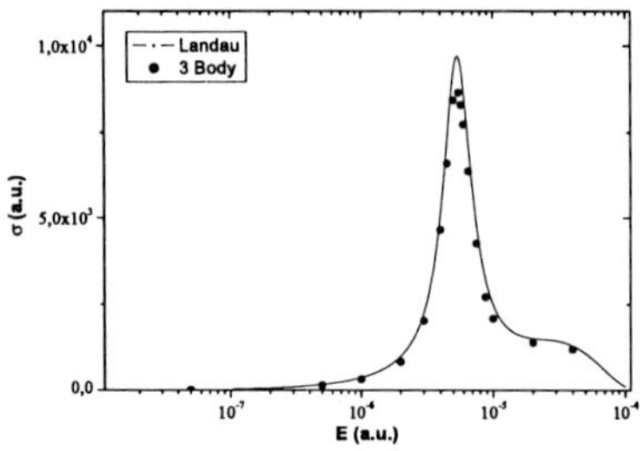

Figure 4. pH elastic cross section for L=3 in the pp spin triplet state.

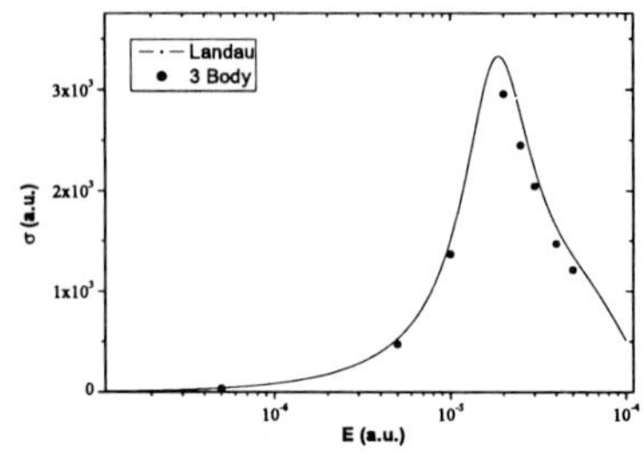

Figure 5. pH elastic cross section for L=4 in the pp spin triplet state.

The pH cross sections for higher partial waves have also been calculated. They exhibit narrow resonances in several partial waves. Figs. 4-5 show the elastic cross sections for the L=3 and L=4 states. The position and width of these resonances were estimated to be $E=(5.13\text{-}1.61i)\times 10^{-6}$ for L=3 and $E=(1.56\text{-}0.94i)\times 10^{-5}$ for L=4.

Acknowledgement. We are grateful to C. Gignoux for useful discussions. Numerical calculations were performed at CGCV (CEA Grenoble) and IDRIS (CNRS). We thank the staff members of these organizations for their support.

References

1. S.P. Merkuriev: Ann. Phys. **130**, 395 (1980)
2. R. Lazauskas, J. Carbonell: Few-Body Systems **31**, 125 (2002) and references therein
3. A.A. Kvitsinsky, J. Carbonell, C. Gignoux: Phys Rev. **A46** (1992) 1310
4. A.A. Kvitsinsky, J. Carbonell, C. Gignoux: Phys. Rev. **A51** (1995) 2997
5. A. Martin, J.-M. Richard and T.T. Wu, Phys. Rev. **A46** (1992) 3697
6. J.M. Taylor, Z. Yan, A. Dalgarno, J.F. Babb: Molec. Phys. **97**, 25 (1999)
7. L. Hilico, N. Billy, B. Gremaud, D. Delande: Eur. Phys. J. **D12**, 449 (2000)
8. F. Luo, G. McBane, G. Kim, C. Giese, W. Gentry: J. Chem. Phys. **98** 3564 (1993)
9. L. Landau, E. Lifshits: *Mecanique Quantique*, Moscou: Ed. Mir (1975)
10. T.F. O'Malley, L. Spruch, L. Rosenberg: Phys. Rev. **125** 491 (1961)
11. J. Carbonell, R. Lazauskas, D. Delande, L. Hilico, S. Kilic: physics/0207007 (2002)

Few-Body Systems Suppl. 14, 171–174 (2003)

Few-Body Systems

Configuration Space Calculations for the $^{16}O_3$ Molecule

M.L. Lekala[1,*], S.A. Sofianos[1], V. Roudnev[2], M. Braun[1]

[1] Department of Physics, University of South Africa, Pretoria, South Africa

[2] Department of Mathematics & Computational Physics, Institute for Physics St-Petersburg State University 198504 Ul'yanovskaya 1, Petrodvorets, St. Petersburg, Russia

Abstract. The low-lying states of the ozone molecule are calculated within the framework of the Faddeev equations in configuration space. The Faddeev equations describing the system are solved as three-dimensional equations in the internal coordinates. The eigenvalues for the various states are obtained via the restarted Arnoldi algorithm in combination with a minimization technique.

1 Introduction

The ozone molecule system, $^{16}O_3$, has a rich vibrational spectrum of excited states which are within a few eV from the ground state [1]. These excited states could result from various possible orbital configurations which give rise to problems in constructing the various basis functions needed for such calculations.

In the present work we perform configuration space calculations for the ground and excited states using Faddeev equations within the framework of the total angular momentum representation and without resorting to partial wave decomposition. In this way we can address, for example the following three questions: Firstly, how good is a Hamiltonian with intermolecular forces acting in a pairwise manner in describing complex systems such as that of $^{16}O_3$? Secondly, how important are the three-body forces in these systems? And thirdly, how many states can be reliably extracted using Faddeev equations?

Employing the orthogonal collocation method results in three-dimensional matrices of very large order, which cannot be handled by standard linear algebra methods.

* *E-mail address:* lekala@science.unisa.ac.za

2 Formalism

We assume the Hamiltonian of the three-particle system that contains both the two- and three-body forces, namely,

$$H = H_0 + \sum_{i=1}^{3} V_{i,2B}(\mathbf{x}_i) + V_{3B}\,, \tag{1}$$

where H_0 is the three free particles Hamiltonian, $V_{i,2B}$ is the two-body interaction, and V_{3B} is the three-body force. The wave function Ψ of the system is obtained in the usual way as a sum of three Faddeev components Φ_i satisfying the Faddeev equations

$$\begin{aligned}\left[-\nabla_x^2 - \nabla_y^2 + V_{i,2B}(\mathbf{x}_i) + V_{3B}(\mathbf{x}_i, \mathbf{y}_i) - E\right] \Phi_i(\mathbf{x}_i, \mathbf{y}_i) &= \\ -V_{i,2B}(\mathbf{x}_i) \sum_{j\neq i} \Phi_j(\mathbf{x}_j, \mathbf{y}_j)\,. \end{aligned} \tag{2}$$

Here E is the energy of the system. For three identical bosons the projection of Eq. (2) in a subspace with zero total angular momentum leads to a system of equations [2]

$$(H_0^0 + V_{2B} + V_{3B} - E)\phi^0(x, y, z) = V_{2B}(x)P\phi^0(x, y, z)\,; \tag{3}$$

where the, now redundant, subscript i is omitted and the intrinsic coordinates are defined by

$$x = |\mathbf{x}|\,, \quad y = |\mathbf{y}|\,, \quad z = \frac{(\mathbf{x}, \mathbf{y})}{xy}\,, \quad x, y \in [0, \infty)\,, \quad z \in (-1, 1) \tag{4}$$

The Faddeev components ϕ^0 are known to have the asymptotic behaviour [2]

$$\phi^0(x, y, z) \sim \varphi_2(x)\mathrm{e}^{-k_y y} + A(x/y, z)\frac{\mathrm{e}^{-k_3(x^2+y^2)^{1/2}}}{(x^2 + y^2)^{1/4}}\,, \tag{5}$$

where $\varphi_2(x)$ is the two-body bound state wave function. The first term in Eq. (5) correspond to the virtual decay into a particle and a two-body bound state while the second term corresponds to the virtual decay into three single particles. In our calculations this virtual decay term is neglected. In Eq. (3) H_0^0 and ϕ^0 are, respectively, the projections of the free Hamiltonian H_0 and the Faddeev component Φ. P is the operator that permutes the coordinates.

3 Numerical Methods

Equation (3) is solved as a three-dimensional differential equation without resorting to partial wave decomposition according to the method described in [2, 3]. In other words, following the standard procedure, a regular solution of

Eq. (3) fulfilling the boundary conditions (5) is approximated by an expansion in terms of the basis functions as

$$\phi^0(x,y,z) = \sum_i f_i B_i(x,y,z) , \tag{6}$$

where now i stands for the multi-index $\{i_x, i_y, i_z\}$ and $B_i(x,y,z) = B_{i_x}(x)B_{i_y}(y)B_{i_z}(z)$. After substitution of Eq. (6) into Eq. (3) followed by the orthogonal collocation procedure, the original equation, Eq. (3), is transformed into an eigenvalue problem

$$-(\hat{H}_1 - E\hat{I})^{-1}\hat{H}_2\mathbf{f} = \lambda(E)\mathbf{f} , \tag{7}$$

where $\hat{H}_1$ is an operator such that $(\hat{H}_i - E\hat{I}_i)^{-1}$ can be explicitly constructed. $\hat{H}_2$ can be considered as the remainder. The physical solutions correspond to $\lambda(E) = 1$. The eigenvalue problem (7) is treated using the restarted Arnoldi Method incorporating convergence acceleration using Chebyshev polynomials. The spectrum of the Eq. (7) can be calculated by a direct application of the Arnoldi algorithm as described in [2]. However, because of the large number of the states for the weakly bound molecular systems, such as the ozone molecule, great care is required in calculating the low-lying excited states, especially when they lie very close to one another. To this end, we employ the Arnoldi method in conjunction with a minimization procedure called MERLIN [4]. In the MERLIN procedure Eq. (7) is transformed into a constrained minimization problem of the form

$$\hat{L}\mathbf{f} = \lambda\mathbf{f} . \tag{8}$$

The operator $\hat{L}$ contains all the operators appearing in Eq. (7). If $\lambda(E_0)$ is the eigenvalue corresponding to the physical solution, the problem of solving Eq. (8) for λ corresponds to a minimization problem of the function

$$min(E(p)) = \sum_i [\lambda(\mathbf{p}, E_i) - \lambda(E_0)]^2 , \tag{9}$$

where $\mathbf{p}$ represent the set of parameters on which λ depends. MERLIN, unlike many minimization methods, uses many algorithms to obtain the minimum. Applying this approach we obtain the ground state and the low-lying states excited states of the ozone molecule. It should be noted that this method would be very useful when one fixes the binding energies and searches via minimization for the corrected 3B-parameters that reproduce these states.

4 Results

In order to investigate the importance of three-body forces we calculate few low-lying spectrum of $^{16}O_3$. We employed the potential of Varandas et al. [5]. The results are shown in Table 1 where one immediately sees that the neglect of the three-body forces gives binding energies of lower magnitudes. Throughout this study we used $\hbar^2/M = 2.614 \times 10^{-4}$ eV Å^2. A comparison of the relative energies with the results of Jeffrey and co-workers [6] shows the Faddeev equation method can be reliably used for studying the spectrum of molecules such as ozone.

Table 1. The low-lying spectrum of the ozone $^{16}O_3$ molecule. ΔE's refer to the vertical energies relative the energy of the ground state. In column 4 ΔE is with a 3B force. Energies are in eV.

State	Without 3B force	With 3B force	ΔE	ΔE [6]
1	−10.051	−13.423	—	—
2	−9.9014	−12.095	1.328	1.11
3	−8.0902	−11.432	1.993	1.79
4	−8.5100	−11.286	2.139	1.96
5	−8.0012	−10.523	2.900	2.02

5 Conclusion

We presented the three-body bound state calculations results for a few low-lying vibrational states of the ozone molecule using the Faddeev equations method. Usually in the chemical physics theoretical calculations, the spectra of ozone molecule is calculated using the *ab initio* or variational approaches. To our knowledge, this is the first study for the ozone molecule using the Faddeev equations approach. We mention also that the investigation was a test case in that only one potential model was used. Further investigations using different potentials such as the more realistic single-valued and multi-valued potentials are underway.

References

1. P.J. Hay, T.H. Dunning, W.A. Goddard III: Chem. Phys. **62**, 3912 (1975)
2. V.A. Roudnev, S.L. Yakovlev, S.A. Sofianos: physics/0204025
3. M.L. Lekala, S.A. Sofianos, M. Braun: contribution in these proceedings
4. D.G. Papageorgiou, I.N. Demetropoulos, I.E. Lagaris: Comput. Phys. Commun. **109**, 109 (1998)
5. A.J.C. Varandas, J.N. Murrell: Chem. Phys. Lett. **88**, 1 (1982)
6. P. Jeffrey Hay, T.H. Dunning Jr: J. of Chem. Phys. **67**, 2290 (1977)

Few-Body Systems Suppl. 14, 175–178 (2003)

Few-Body Systems

Correlation Effects in Fluorescence Decay of Doubly Excited States in Helium

M. Žitnik, K. Bučar

Jožef Stefan Institute, Jamova 39, Ljubljana, SI-1000, Slovenia

Abstract. According to the recent calculations the fluorescence lifetimes of low-lying doubly excited states in helium below the second ionization threshold converge to series specific value generally larger than the fluorescence lifetime of $2p \rightarrow 1s$ transition in He^+. To study this phenomena we generated various probability densities from the correlated doubly excited state wavefunctions. The different fluorescence rates may be clearly associated to the behavior of the unconditional probability density of a single electron close to the nucleus. The degeneracy is destroyed by the Stark effect occurring in the body-frame when the outer electron polarizes the inner one into specific linear combination of $2s$ and $2p$ orbitals.

It is known that electron-electron correlations in doubly excited states of helium lead to strong rearrangement of intensities in the autoionization channel [1]. The intensity of the $(0,1)^+$ series[1] below N=2 threshold is observed to be about 100 times larger than intensity of the $(1,0)^-$, and the weakest of the three $^1P^o$ series, $(-1,0)^0$, was firstly seen in photoionization 10 years ago [2].

The fluorescence from these states was recently a subject of many experimental and theoretical studies [3]. For low n the relative intensity of the radiative decay channel with respect to the autoionization is usually small but increases toward higher members of the series. The fluorescence decay rates of doubly excited states below $N=2$ are generally smaller than 10 ns^{-1}, the fluorescence rate of $2p \rightarrow 1s$ transition in He^+, and converge to certain series specific value for moderate n. This work investigates how the presence of outer electron quenches the fluorescence transition happening in the core. Similar to [4] we study the behavior of reduced probability density obtained from the correlated wavefunction of doubly excited state.

Our wavefunctions are obtained by truncated diagonalization of the nonrelativistic Hamiltonian [5]. The lowest 44 discrete eigenstates of either $^1P_1^o$, $^3P_1^o$ or $^3D_1^o$ symmetry were calculated above the first ionization threshold. The states

[1]The series of states are labeled $(\mathrm{K,T})^A$, notation proposed by [6].

were expanded in the basis set of two-electron configurations $n_1 l_1 n_2 (l_1+1)$ of the matching symmetry,

$$| \, i \; ^{2S+1}LM\rangle = \sum_{n_1 n_2 l_1} c^i_{n_1 n_2 l_1} |n_1 l_1 n_2 (l_1+1) \; ^{2S+1}LM\rangle, \tag{1}$$

with n_1, n_2 going up to 12 and l_1 going up to 3 [3].

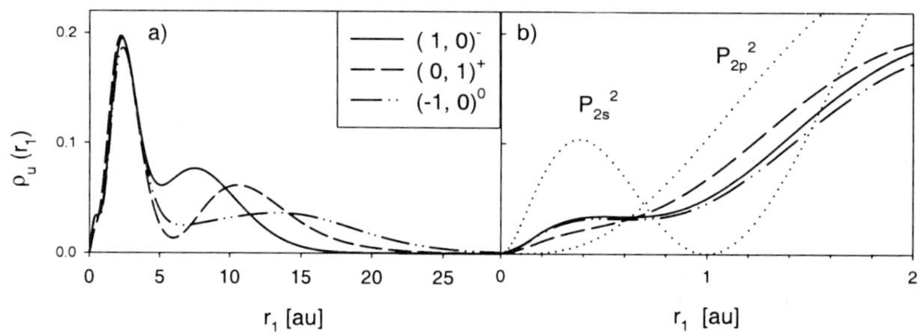

Figure 1. a) Unconditional probability density of the three $^1P^o$ doubly excited states with n=3, N=2 in He. b) Blow-up of a) with squares of 2s and 2p orbital.

The reduced probability density of a given state equals two-electron probability density integrated over the three Euler angles which determine orientation of the electron-nucleus-electron system. If we choose r_1, r_2, θ_{12} for internal coordinates we can write in general

$$\rho^i(r_1, r_2, \theta_{12}) = \sum_K F^i_K(r_1, r_2) P_K(\cos\theta_{12}), \tag{2}$$

where P_K is Legendre polynomial of order K and F_k are linear combinations of products of hydrogen-like helium radial orbitals and expansion coefficients from (1). To visualize the correlations we calculate the conditional probability density of one electron to be found at given r_2, θ_{12} when the other electron is at the distance r_1 from the nucleus:

$$\rho^i_c(r_1|r_2, \theta_{12}) = \frac{\rho^i(r_1, r_2, \theta_{12})}{\rho^i_u(r_1)} r_1^2 r_2^2.$$

The unconditional probability density ρ^i_u represents the probability for one electron to be at r_1 in the i-th resonance state, regardless of the other electron:

$$\rho^i_u(r_1) = r_1^2 \int_0^\infty r_2^2 \, dr_2 \int_0^\pi \sin\theta_{12} d\theta_{12} \; \rho^i(r_1, r_2, \theta_{12}). \tag{3}$$

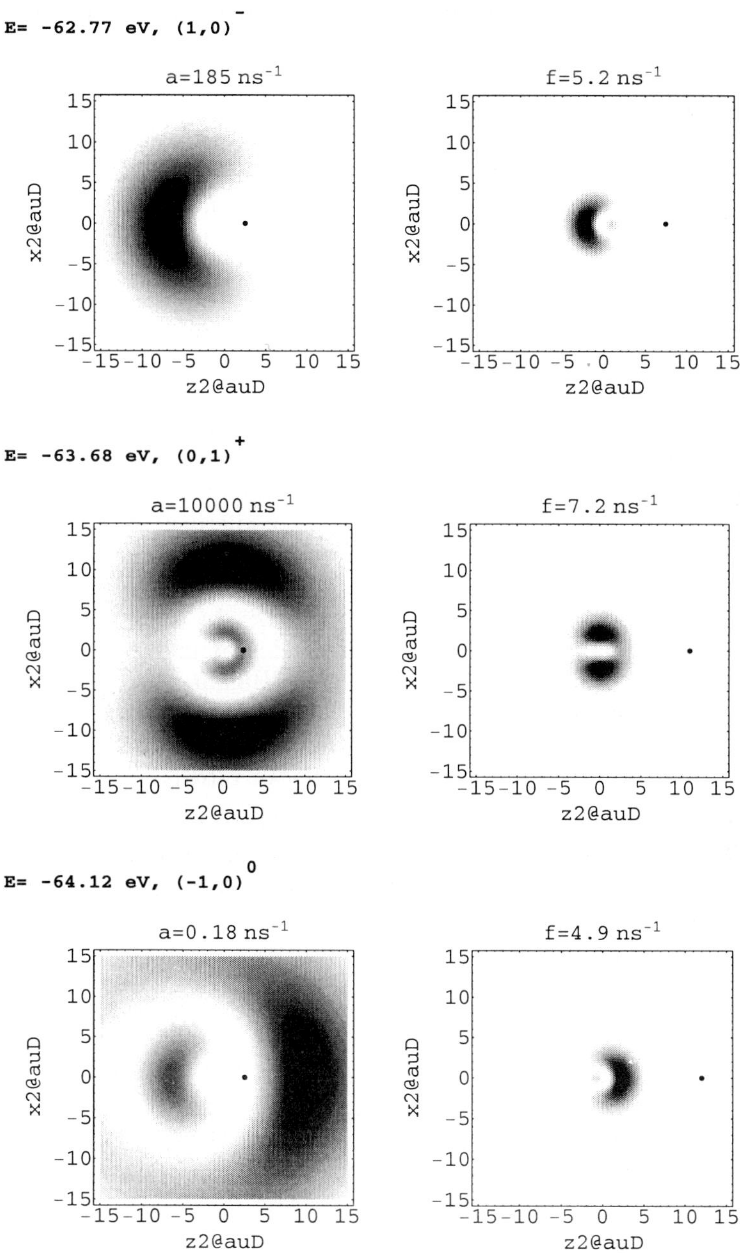

Figure 2. Conditional probability densities of the three $^1P_1^o$ doubly excited states with n=3, N=2 in He. The nucleus is in the center of the plot. The position of the first electron is marked by the black circle. For each $(K,T)^A$ state we denote the calculated energy E, the autoionization decay rate a and the fluorescence decay rate f.

Besides reflecting the shell structure the unconditional densities in Fig. 1 display a wiggle at $r_1 \approx 0.5$ au. Its shape is a fingerprint of 2p/2s mixing. It is less pronounced when the weight of the 2p orbital is larger which corresponds to a larger fluorescence decay rate.

To illustrate the connection of $(\mathrm{K,T})^A$ numbers with the autoionization and fluorescence decay rate we present an array of density plots in Fig. 2. For each of the three states we fix one electron at radial distance corresponding to the two maxima of the unconditional probability density in Fig. 1a. That way we present the conditional probability density of an outer (inner) electron with an inner (outer) electron at its most probable distance from the nucleus. For each group of states with given (N, n, L, S, π) there is a full frame of quasi hydrogen-like Stark states as seen in the right part of Fig. 2 for the three $^1P^o$ states with $n=3$. When an outer electron approaches the nucleus a particular short range correlation pattern evolves with the antinode (node) at $\boldsymbol{r}_1 = \boldsymbol{r}_2$ for +(-) state (left of Fig. 2). The Stark mixing of hydrogen-like helium in the weak electric field leads to the equal energy separation of the three states and to the relative fluorescence rates ratio of 1:2:1. Even if these predictions are not followed in detail the energy ordering of $^1P^o$ states does match the simple model and to some extend also the relative fluorescence rates. In fact, the ability of (K,T) numbers to label the states was realized previously by introduction of DESB wavefunctions which incorporate intra shell Coulomb mixing; they were shown to approximately diagonalize the nonrelativistic Hamiltonian [6].

We conclude that the fluorescence decay rate of doubly excited states most strongly depends on the long-range dipole interaction classified by (K,T). This is different from autoionization where the decay rate is determined mainly by the short-range electron correlations described by quantum number A. We believe that prediction power of the simple Stark model about the relative fluorescence rates can be still improved by accounting for the inter shell mixing in the hydrogen-like helium due to the relatively strong electric field induced by outer electron.

References

1. J.W. Cooper, U. Fano, F. Prats: Phys. Rev. Lett. **10**, 518 (1963)
2. M. Domke, G. Remmers, G. Kaindl: Phys. Rev. Lett. **69**, 1171 (1992)
3. M. Žitnik, K. Bučar, M. Štuhec, F. Penent, R. I. Hall, P. Lablanquie: Phys. Rev. **A65**, 32520 (2002)
4. G.S. Ezra, R.S. Berry: Phys. Rev. **A28**, 1974 (1983)
5. L. Lipsky, R. Anania, M.J. Conneely: At. Data & Nucl. Data Tables **20**, 127 (1977)
6. C.D. Lin: Phys. Rev. **A25**, 76 (1982)

Few-Body Systems Suppl. 14, 179–184 (2003)

Few-Body Systems

News in Charged-Composite Particle Scattering: Theory and Applications

E.O. Alt

Institut für Physik, Universität Mainz, D-55099 Mainz, Germany

Abstract. A brief summary is given of recent developments in the theory of charged-composite particle scattering, and of practical applications.

1 Introduction

Scattering of charged particles represents one of the remaining challenges in few-body theory which is why it continues to attract widespread interest. Although remarkable results have been obtained both in the formulation of the theory and in application to physical problems, still several aspects have not yet been explored in sufficient detail. This talk aims at summarizing recent theoretical advances and at reporting on interesting and novel applications.

2 Theoretical Developments

2.1 Momentum Space Integral Equations

2.1.1 Status of the Theory

Until recently the status could be summarized as follows. If only two particles are charged and one is neutral, integral equations for (suitably modified) *three-body operators* are available for all energies, with kernels which become compact after a suitable number of iterations [1]. If, however, all three particles are charged, the modified Faddeev-Veselova equations for amplitudes for binary reactions could be shown to possess compact kernels only for energies below the three-body threshold [2]. In fact, it has even been argued that the kernels of these or equivalent equations should loose their compactness property for energies above that threshold, due to severe singularities arising from the long-range Coulomb forces. This conjecture has for a long time discouraged attempts to tackle the problem. However, recently the solution has come up as will be discussed below.

But is should be stressed that the "screening and renormalisation approach" provides an independent method for calculating the amplitudes for all binary (if all three particles are charged) and also for breakup processes (only two of the particles are charged). For a recent review see [3]. For this reason, it has had widespread applications to proton-deuteron elastic scattering ([4] and references therein) and deuteron breakup by proton impact [5].

2.1.2 Compact Kernel Equations for Binary Amplitudes for $E > 0$

For three charged particles interacting via short-range and purely repulsive Coulomb potentials the Hamiltonian is $H = H_0+\sum_{\nu=1}^{3}(V_\nu^S+V_\nu^C)$. The reaction amplitude corresponding to the transition from channel α to channel β satisfies a LS-type equation ($z = E+i0$, $E = \bar{q}_\alpha^2/2M_\alpha - B_\alpha$, with $B_\alpha > 0$ denoting the binding energy of the pair (β,γ), $\bar{q}_\alpha$ the on-shell momentum in channel α, and $M_\alpha^{-1} = m_\alpha^{-1} + (m_\beta + m_\gamma)^{-1}$), which in operator form reads as

$$\mathcal{T}_{\beta\alpha}(z) = \mathcal{V}_{\beta\alpha}(z) + \sum_{\nu=1}^{3} \mathcal{V}_{\beta\nu}(z)\mathcal{G}_{0;\nu}(z)\mathcal{T}_{\nu\alpha}(z). \tag{1}$$

If, for simplicity, the short-range potentials are assumed to be of separable form, $V_\alpha^S = | \chi_\alpha\rangle\lambda_\alpha\langle\chi_\alpha |$, $\alpha = 1,2,3$, the effective potentials are defined as

$$\mathcal{V}_{\beta\alpha}(\boldsymbol{q}'_\beta, \boldsymbol{q}_\alpha; z) := \langle \boldsymbol{q}'_\beta, \chi_\beta \mid G^C(z) - \delta_{\beta\alpha}G_\alpha^C(z) \mid \chi_\alpha, \boldsymbol{q}_\alpha\rangle, \tag{2}$$

where $G^C(z) = (z - H_0 - \sum_{\nu=1}^{3} V_\nu^C)^{-1}$ is the resolvent of the three-particle Coulomb Hamiltonian, and $G_\alpha^C(z) = (z - H_0 - V_\alpha^C)^{-1}$.

The question of applicability of Eq. (1) depends on the singularity structure of its kernel. The effective propagator $\mathcal{G}_{0;\alpha}(z)$ for a bound state of (β,γ) and the free particle α possesses only the well-known pole:

$$\mathcal{G}_{0;\alpha}(\boldsymbol{q}_\alpha; z) \sim [z - q_\alpha^2/2M_\alpha + B_\alpha]^{-1} \sim [(\bar{q}_\alpha^2 - q_\alpha^2)/2M_\alpha + i0]^{-1}. \tag{3}$$

As shown in [6], for $\beta \neq \alpha$ the leading singularity of the effective potential (2) is of the form ($\mu_\alpha^{-1} = m_\beta^{-1} + m_\gamma^{-1}$, $k_\alpha = |\boldsymbol{q}'_\beta + \mu_\alpha \boldsymbol{q}_\alpha/m_\gamma|$)

$$\mathcal{V}_{\beta\alpha}(\boldsymbol{q}'_\beta, \boldsymbol{q}_\alpha; z) \sim \left[z - q_\alpha^2/2M_\alpha - k_\alpha^2/2\mu_\alpha\right]^{-1+i(\hat{\eta}_\alpha+\hat{\eta}_\beta)}, \tag{4}$$

where, e.g., the Coulomb parameter of the pair (β,γ) is defined as $\hat{\eta}_\alpha = e_\beta e_\gamma \mu_\alpha/\sqrt{2\mu_\alpha(z - q_\alpha^2/2M_\alpha)}$. It is obvious that this singularity can never coincide with the propagator singularity (3) and, thus, is harmless.

The leading singularity of the diagonal part of (2) is shown in [7] to be

$$\mathcal{V}_{\alpha\alpha}(\boldsymbol{q}'_\alpha, \boldsymbol{q}_\alpha; z) = \frac{\tilde{I}(\boldsymbol{q}'_\alpha, \boldsymbol{q}_\alpha; z)}{(\boldsymbol{q}'_\alpha - \boldsymbol{q}_\alpha)^2} + O\left(\frac{1}{|\boldsymbol{q}'_\alpha - \boldsymbol{q}_\alpha|}\right). \tag{5}$$

The function $\tilde{I}$ is non-singular for repulsive Coulomb potentials (but develops a severe singularity for the attractive case), and reduces on the energy shell to $\tilde{I} \to 4\pi e_\alpha(e_\beta + e_\gamma)$. Hence, $\mathcal{V}_{\alpha\alpha}$ can be decomposed as

$$\mathcal{V}_{\alpha\alpha}(\boldsymbol{q}'_\alpha, \boldsymbol{q}_\alpha; z) = v_{\alpha,\mathrm{cm}}^C(\boldsymbol{q}'_\alpha, \boldsymbol{q}_\alpha) + \mathcal{V}_{\alpha\alpha}^{SC}(\boldsymbol{q}'_\alpha, \boldsymbol{q}_\alpha; z), \tag{6}$$

where $v^C_{\alpha,\mathrm{cm}}(\boldsymbol{q}'_\alpha, \boldsymbol{q}_\alpha) = 4\pi e_\alpha(e_\beta + e_\gamma)/(\boldsymbol{q}'_\alpha - \boldsymbol{q}_\alpha)^2$ is the well-known *center-of-mass Coulomb potential* which acts between the total charges of the two colliding fragments, and $\mathcal{V}^{SC}_{\alpha\alpha}$ is less singular. Obviously, the first term in (6) leads together with (3) to a non-integrable singularity of the kernel. But since it is of two-body type it can be inverted explicitly, yielding the on-shell relation

$$\mathcal{T}_{\beta\alpha}(\bar{\boldsymbol{q}}'_\beta, \bar{\boldsymbol{q}}_\alpha) := \delta_{\beta\alpha} t^C_{\alpha,\mathrm{cm}}(\bar{\boldsymbol{q}}'_\alpha, \bar{\boldsymbol{q}}_\alpha) + \langle \psi^{C(-)}_{\bar{\boldsymbol{q}}'_\beta,\mathrm{cm}} | \mathcal{T}^{SC}_{\beta\alpha}(E + i0) | \psi^{C(+)}_{\bar{\boldsymbol{q}}_\alpha,\mathrm{cm}} \rangle. \tag{7}$$

Here, $t^C_{\alpha,\mathrm{cm}}(\bar{\boldsymbol{q}}'_\alpha, \bar{\boldsymbol{q}}_\alpha)$ is the amplitude, and $|\psi^{C(+)}_{\bar{\boldsymbol{q}}_\alpha,\mathrm{cm}}\rangle$ the scattering state, resulting from a potential $v^C_{\alpha,\mathrm{cm}}$ at an energy $\bar{q}^2_\alpha/2M_\alpha$, both of which are known analytically. The amplitude $\mathcal{T}^{SC}_{\beta\alpha}(z)$ is solution of an integral equation similar to (1), but with a kernel which contains only the less singular potential part $\mathcal{V}^{SC}_{\beta\alpha}(z) := \mathcal{V}_{\beta\alpha}(z) - \delta_{\beta\alpha} v^C_{\alpha,\mathrm{cm}}$ and thus becomes compact after a suitable number of iterations. Therefore, this equation can be used for practical calculations.

2.2 Coulomb-Fourier Transformation of the Three-Body Hamiltonian

A novel, alternative approach to calculate charged-composite particle scattering observables is based on constructing a unitary transformation of the Hamiltonian such that the interaction parts of longest range are explicitly eliminated.

2.2.1 Illustration by Means of two Charged Particle Scattering

Consider the Hamiltonian $H = H^C + V^S$, $H^C = H_0 + V^C$, $H^C|\psi^C_{\boldsymbol{k}}\rangle = k^2/2\mu|\psi^C_{\boldsymbol{k}}\rangle$. If, for simplicity, we assume the Coulomb potential to be repulsive the states $\{|\psi^C_{\boldsymbol{k}}\rangle\}$ form a basis. Thus, any operator O is represented by matrix elements between basis states $O \Longrightarrow \mathcal{O}(\boldsymbol{k}', \boldsymbol{k}) := \langle \psi^C_{\boldsymbol{k}'}|O|\psi^C_{\boldsymbol{k}}\rangle$, which are called *Coulomb transform* ($\mathcal{CT}$) of O. For instance, the $\mathcal{CT}$ of H is

$$\mathcal{H}(\boldsymbol{k}', \boldsymbol{k}) = \delta(\boldsymbol{k}' - \boldsymbol{k})\frac{k^2}{2\mu} + \langle \psi^C_{\boldsymbol{k}'}|V^S|\psi^C_{\boldsymbol{k}}\rangle =: \mathcal{H}_0(\boldsymbol{k}', \boldsymbol{k}) + \mathcal{V}^{SC}(\boldsymbol{k}', \boldsymbol{k}), \tag{8}$$

and is seen to contain no explicit Coulomb potential. Thus, any equation valid for short-range potentials applies also in the $\mathcal{CT}$ theory as, e.g., the LS equation

$$\mathcal{T}(\boldsymbol{k}', \boldsymbol{k}; z) = \langle \psi^C_{\boldsymbol{k}'}|V^S|\psi^C_{\boldsymbol{k}}\rangle + \int d^3k'' \langle \psi^C_{\boldsymbol{k}'}|V^S|\psi^C_{\boldsymbol{k}''}\rangle \frac{1}{z - k''^2/2\mu} \mathcal{T}(\boldsymbol{k}'', \boldsymbol{k}; z), \tag{9}$$

the solution of which is the familiar Coulomb-modified short-range amplitude $\mathcal{T}(\boldsymbol{k}', \boldsymbol{k}; z) \equiv T^{SC}(\boldsymbol{k}', \boldsymbol{k}; z)$. And from it the on-shell amplitude corresponding to the original Hamiltonian H can be recovered simply as

$$T(\bar{\boldsymbol{k}}', \bar{\boldsymbol{k}}) = T^C(\bar{\boldsymbol{k}}', \bar{\boldsymbol{k}}) + T^{SC}(\bar{\boldsymbol{k}}', \bar{\boldsymbol{k}}; \bar{k}^2/2\mu + i0). \tag{10}$$

2.2.2 Three-Body System with two Charged and one Neutral Particle

The Hamiltonian is $H = H^C_1 + \boldsymbol{Q}^2_1/2M_1 + \sum_{\nu=1}^3 V^S_\nu$, with $H^C_1 = \boldsymbol{K}^2_1/2\mu_1 + V^C_1$, $H^C_1|\psi^C_{\boldsymbol{k}_1}\rangle = \boldsymbol{k}^2_1/2\mu_1|\psi^C_{\boldsymbol{k}_1}\rangle$. If the Coulomb potential is assumed to be repulsive, the states $\{|\psi^C_{\boldsymbol{k}_1}, \boldsymbol{q}_1\rangle\}$ form a basis. Hence, any operator O is represented by matrix elements between these basis states $O \Longrightarrow \mathcal{O}(\boldsymbol{k}'_1, \boldsymbol{q}'_1; \boldsymbol{k}_1, \boldsymbol{q}_1) :=$

$\langle \psi^C_{\boldsymbol{k}'_1}, \boldsymbol{q}'_1 | O | \psi^C_{\boldsymbol{k}_1}, \boldsymbol{q}_1 \rangle$. They will be called *Coulomb-Fourier transform* ($\mathcal{CFT}$) of O. For instance, the $\mathcal{CFT}$ of $\{H^C_1 + \boldsymbol{Q}^2_1/2M_1\}$ is the free Hamiltonian

$$\mathcal{H}_0(\boldsymbol{k}'_1, \boldsymbol{q}'_1; \boldsymbol{k}_1, \boldsymbol{q}_1) := \delta(\boldsymbol{k}'_1 - \boldsymbol{k}_1)\delta(\boldsymbol{q}'_1 - \boldsymbol{q}_1)\left\{\frac{\boldsymbol{k}^2_1}{2\mu_1} + \frac{\boldsymbol{q}^2_1}{2M_1}\right\}. \tag{11}$$

Thus, the $\mathcal{CFT}$ of H is (for local potentials)

$$\mathcal{H}(\boldsymbol{k}'_1, \boldsymbol{q}'_1; \boldsymbol{k}_1, \boldsymbol{q}_1) = \mathcal{H}_0(\boldsymbol{k}'_1, \boldsymbol{q}'_1; \boldsymbol{k}_1, \boldsymbol{q}_1) + \sum_{\nu=1}^{3} \mathcal{V}^{SC}_\nu(\boldsymbol{k}'_1, \boldsymbol{q}'_1; \boldsymbol{k}_1, \boldsymbol{q}_1), \tag{12}$$

with ($\boldsymbol{\Delta}_1 := \epsilon_{\alpha 1}(\boldsymbol{q}_1 - \boldsymbol{q}'_1)$, $\epsilon_{\alpha 1}$ being a sign factor and λ a mass ratio)

$$\mathcal{V}^{SC}_1(\boldsymbol{k}'_1, \boldsymbol{q}'_1; \boldsymbol{k}_1, \boldsymbol{q}_1) = \delta(\boldsymbol{q}_1 - \boldsymbol{q}'_1)\, \langle \psi^C_{\boldsymbol{k}'_1} | V^S_1 | \psi^C_{\boldsymbol{k}_1} \rangle, \tag{13}$$

$$\mathcal{V}^{SC}_\alpha(\boldsymbol{k}'_1, \boldsymbol{q}'_1; \boldsymbol{k}_1, \boldsymbol{q}_1) = \hat{V}^S_\alpha(\boldsymbol{\Delta}_1)\, \mathcal{L}^C_\alpha(\boldsymbol{k}'_1, \boldsymbol{q}'_1; \boldsymbol{k}_1, \boldsymbol{q}_1),\ \alpha = 2, 3. \tag{14}$$

$$\mathcal{L}^C_\alpha(\boldsymbol{k}'_1, \boldsymbol{q}'_1; \boldsymbol{k}_1, \boldsymbol{q}_1) = \int d\boldsymbol{x}_1 e^{i\lambda \boldsymbol{x}_1 \cdot \boldsymbol{\Delta}_1} \psi^{C*}_{\boldsymbol{k}'_1}(\boldsymbol{x}'_1)\psi^C_{\boldsymbol{k}_1}(\boldsymbol{x}_1),\ \alpha = 2, 3, \tag{15}$$

is a universal function as it is independent of short-range potentials and, therefore, needs to be calculated only once and for all. For details we refer to [8].

Thus, the $\mathcal{CFT}$ of the three-body Hamiltonian for a system of two charged and one neutral particles, Eq. (12), contains no explicit long-range Coulomb potential. Consequently, standard short-range (Faddeev, AGS, etc) techniques are applicable. And the corresponding results for the original Hamiltonian H can be obtained from those for $\mathcal{H}$ by inverse $\mathcal{CF}$ transformation.

Let us add a final comment. As follows from Eq. (14), those original two-body potentials are converted into three-body quantities which describe the short-range interaction between the neutral and one of the charged particles *in the presence of the other charged particle.* This feature accounts for the long-range Coulombic three-body correlations which are known to exist in three-body systems with (at least) two charged particles [9]. This fact does, however, not present any difficulty as (genuine or effective) three-body forces can be incorporated into three-body theory in the standard manner.

3 Practical Applications

3.1 Coulomb Corrections to Bose-Einstein n-Particle Correlations

One of the most important tasks of high energy heavy-ion studies is to prove the existence of the elusive quark-gluon plasma and study the properties of this new state of matter. Intensity or Hanbury-Brown Twiss interferometry of identical particles has become an important tool as it can be used to measure the evolving space-time geometry of the interaction region which is expected to be sensitive to the formation of QGP stage. For a recent review see, e.g., [10].

For this purpose, n-particle Bose-Einstein correlation functions for identical bosons with similar 3-momenta are being investigated. They are defined as

$$C_n(\boldsymbol{k}_1, \cdots, \boldsymbol{k}_n) = P_n(\boldsymbol{k}_1, \cdots, \boldsymbol{k}_n)/\{P_1(\boldsymbol{k}_1) \cdots P_1(\boldsymbol{k}_n)\}$$

where $P_n(\boldsymbol{k}_1, \cdots, \boldsymbol{k}_n)$ is the probability to observe n particles with momenta $\boldsymbol{k}_1, \cdots, \boldsymbol{k}_n$. In a wave function picture one has (for incoherent emission)

$$P_n(\boldsymbol{k}_1, \cdots, \boldsymbol{k}_n) = \int \prod_{i=1}^{n} d^3x_i \rho(\boldsymbol{x}_1) \cdots \rho(\boldsymbol{x}_n) \left| \Psi^{(+)\mathcal{S}}_{\boldsymbol{k}_1 \cdots \boldsymbol{k}_n}(\boldsymbol{x}_1, \cdots, \boldsymbol{x}_n) \right|^2 . \quad (16)$$

Here, $\rho(\boldsymbol{x}_i)$ is the emission probability density of the source for particle i, conventionally chosen as a Gaussian of width R, $\rho(\boldsymbol{x}_i) \sim \exp\left\{-\boldsymbol{x}_i^2/2R^2\right\}$. $\Psi^{(+)\mathcal{S}}_{\boldsymbol{k}_1 \cdots \boldsymbol{k}_n}(\boldsymbol{x}_1, \cdot, \boldsymbol{x}_n)$ denotes the properly symmetrized wave function of the n emitted particles. ¿From this follows that measurement of momentum correlations (C_n) yields information about the spatial structure ($\rho(\boldsymbol{x})$) of the source.

The quantitative interpretation depends, however, critically on the understanding of the rôle of the Coulomb interaction between the detected particles. To study its effect one defines the Coulomb correction to correlation function

$$K_{Coulomb} = \frac{\int d^3x_1 \cdots d^3x_n \rho(\boldsymbol{x}_1) \cdots \rho(\boldsymbol{x}_n) \left| \Psi^{(+)\mathcal{S}}_{\boldsymbol{k}_1 \cdots \boldsymbol{k}_n}(\boldsymbol{x}_1, \cdots, \boldsymbol{x}_n) \right|^2}{\int d^3x_1 \cdots d^3x_n \rho(\boldsymbol{x}_1) \cdots \rho(\boldsymbol{x}_n) \left| \Psi^{(0)\mathcal{S}}_{\boldsymbol{k}_1 \cdots \boldsymbol{k}_n}(\boldsymbol{x}_1, \cdots, \boldsymbol{x}_n) \right|^2} . \quad (17)$$

Here, $\Psi^{(0)}_{\boldsymbol{k}_1 \cdots \boldsymbol{k}_n}(\boldsymbol{x}_1, \cdots, \boldsymbol{x}_n) = \prod_{j=1}^{n} e^{i\boldsymbol{k}_j \cdot \boldsymbol{x}_j}$ is a plane wave describing n free particles, while $\Psi^{(+)}_{\boldsymbol{k}_1 \cdots \boldsymbol{k}_n}(\boldsymbol{x}_1, \cdots, \boldsymbol{x}_n)$ includes fsi effects. A simple model which has been conjectured to at least qualitatively account for Coulomb fsi, is provided by choosing $\Psi^{(+)(G)}_{\boldsymbol{k}_1 \cdots \boldsymbol{k}_n}(\boldsymbol{x}_1, \cdots, \boldsymbol{x}_n) = \Psi^{(0)} \prod_{i<j=1}^{n} N_{ij}$ where $N_{ij} = e^{-\pi\eta_{ij}/2}\,\Gamma(1 + i\eta_{ij})$ is the Gamov factor, $\eta_{ij} = e_i e_j \mu_{ij}/k_{ij}$ the Coulomb parameter, $\boldsymbol{k}_{ij} = \boldsymbol{k}_i - \boldsymbol{k}_j$ the relative momentum, and $\mu_{ij} = m_i m_j/(m_i + m_j)$ the reduced mass of the pair ij. This yields

$$K^{(G)}_{Coulomb} = \prod_{i<j=1}^{n} |N_{ij}|^2 : \text{standard “Gamov” approximation.} \quad (18)$$

However, a much better substantiated ansatz is provided by the so-called Redmond-Merkuriev wave function $\Psi^{(+)(RM)}_{\boldsymbol{k}_1 \cdots \boldsymbol{k}_n}(\boldsymbol{x}_1, \cdots, \boldsymbol{x}_n) = \Psi^{(0)} \prod_{i<j=1}^{n} N_{ij}$ $F[-i\eta_{ij}, 1; i(k_{ij} r_{ij} - \boldsymbol{k}_{ij} \cdot \boldsymbol{r}_{ij})]$ since it is exact solution of the n-charged particle Schrödinger equation in a large region of (asymptotic) configuration space. Here, $\boldsymbol{r}_{ij} = \boldsymbol{x}_i - \boldsymbol{x}_j$, and $F[a, b; x]$ is the confluent hypergeometric function.

Model calculations [11] show that the “Gamov” approximation underestimates Coulomb fsi effects in the three-body correlation factor up to $\sim 10\%$, and in the five-body correlation factor by up to $\sim 40\%$, if the source parameter is 5 - 10 fm which is the range of interest in heavy ion collisions. Hence, in order to enhance the reliability of the parameters (e. g., the width R) used to characterize the source, as extracted from momentum correlation measurements, the better substantiated wave function approximation should be used.

3.2 Proton-Deuteron Elastic Scattering with Realistic Interactions

Proton-deuteron (*pd*) scattering is the prototype reaction for applying charged-composite particle scattering theory. Its practical importance derives from the fact that it allows for more stringent tests of the fundamental nuclear interaction, of three-nucleon force models, etc than the nd reaction since experimental data are much more abundant and of much higher precision. Although sophisticated calculations with coordinate space methods exist, it is indispensable to perform calculations also with the alternative momentum space integral equations as each approach has its specific strengths and weaknesses.

We use the screening and renormalisation approach. Of the advantages over differential equations we mention only two: no problems arise either from lack of the completeness of the boundary condition, or from lack of convergence of Coulomb partial wave series. Recent results for *pd* elastic scattering with the Paris potential (PEST1) have been published in [4].

Finally, I remind of older calculations of *pd* breakup observables in [5] which till now are the only ones which correctly take into account the Coulomb repulsion between the protons. Although only simple separable potentials could be used at that time, nevertheless qualitative to semiquantitative agreement with experimental data has been achieved.

References

1. E.O. Alt, W. Sandhas, and H. Ziegelmann: Phys. Rev. **C17**, 1981 (1978)
2. A.M. Veselova: Theor. Math. Phys. **3**, 542 (1970); **13**, 369 (1972)
3. E.O. Alt and W. Sandhas: in: *Coulomb Interactions in Nuclear and Atomic Few-Body Collisions*, p. 1. New York: Plenum 1996 (F.S. Levin and D. Micha, eds.)
4. E.O. Alt, A.M. Mukhamedzhanov, M.M. Nishonov, and A.I. Sattarov: Phys. Rev. Lett. **81**, 4820 (1998); Phys. Rev. **C65**, 064613 (2002)
5. E.O. Alt and M. Rauh: Phys. Rev. **C49**, R2285 (1994); Few-Body Syst. **17**, 121 (1994)
6. A.M. Mukhamedzhanov, E.O. Alt, and G.V. Avakov: Phys. Rev. **C61**, 064006 (2000)
7. A.M. Mukhamedzhanov, E.O. Alt, and G.V. Avakov: Phys. Rev. **C63**, 044005 (2001)
8. S.B. Levin, E.O. Alt, and S.L. Yakovlev, in preparation
9. E.O. Alt and A.M. Mukhamedzhanov, Phys. Rev. **A47**, 2004 (1993)
10. R.M. Weiner, Phys. Rep. **327**, 249 (2000)
11. E.O. Alt, T. Csörgő, B.L. Lörstad, and J. Schmidt-Sørensen, Phys. Lett. **B 458**, 407 (1999); Eur. Phys. J. **C 13**, 663 (2000); in *Proc. XXXth Int. Symp. Multiparticle Dynamics*, p. 546. Singapore: World Scientific 2001 (T. Csorgo, S. Hegyi, and W. Kittel, eds.)

Few-Body Systems Suppl. 14, 185–190 (2003)

Few-Body Systems

Quasilinearization Method and its Application to Physical Problems

V.B. Mandelzweig*

Racah Institute of Physics, Hebrew University, Jerusalem 91904, Israel

Abstract. The general properties of quasilinearization method (QLM), particularly its fast quadratic convergence, monotonicity and numerical stability are analyzed and illustrated on different physical problems. The method, whose mathematical basis in physics was discussed recently by the present author approaches the solution of a nonlinear differential equation by approximating the nonlinear terms by a sequence of the linear ones, and is not based on the existence of some kind of a small parameter, which enables the summation of many orders of the perturbation theory as well of the $1/N$ expansion series.

In a series of recent papers [1, 2, 3] the possibility of applying a very powerful approximation technique called the quasilinearization method (QLM) to physical problems has been discussed.

The QLM is designed to confront the nonlinear aspects of physical processes. The method, whose iterations are constructed to yield rapid convergence and often monotonicity, was introduced years ago by Bellman and Kalaba [4]. Modern developments and applications of the QLM to different fields are given in a monograph [5].

However, the QLM was never systematically studied or extensively applied in physics. The reason for the sparse use of the QLM in physics is that the convergence of the method has been proven only under rather restrictive conditions [4], which generally are not fulfilled in physical applications. Recently, though, by the present author it was shown [1] that a different proof of the convergence can be provided so that the applicability of the method is extended to incorporate realistic physical conditions of forces defined on infinite intervals with possible singularities at certain points.

The quasilinearization approach is applicable to a general nonlinear ordinary or partial n-th order differential equation in N-dimensional space or to their systems. Here, however, we will restrict ourselves to the case of nonlinear

* *E-mail*: victor@phys.huji.ac.il

ordinary differential equations in one variable since the purpose of this contribution is to show on simple examples how this method works in physics. The further examples could be found in refs. [1, 2, 3].

Let us start for the purpose of transparency from the first order ordinary nonlinear differential equation. Physically this covers the quantum mechanics of one particle in a central field since in this case the Schrödinger equation for a wave function could be rewritten as the Ricatti equation for the inverse logarithmic derivative. $\phi(r) = \kappa\chi(r)/\chi'(r)$ of the wave function. Indeed, defining a new function $a(r)$ which has the dimension of length with the help of the relation $\phi(r) = \kappa(r + a(r))$ and substituting it in the Schrödinger equation one obtains the equation

$$\frac{da(r)}{dr} = f(a(r), r), \quad a(0) = 0 \tag{1}$$

where $f(a(r), r) = (E - V(r))(r + a(r))^2$ and the units $2m = 1$ are used. For zero energy this equation reduces to the Calogero equation [6] where $a(r)$ has the meaning of the variable s-wave scattering length.

The aim of QLM is to obtain the solution of Eq. (1) as a limit of a sequence of linear differential equations. The sequence should be constructed in such a way as to obtain quadratic convergence and, if possible, monotonicity. Making an identical transformation and subtracting from both parts of Eq. (1) a term $f_a(a(r), r)a(r)$ where $f_a(a(r), r) = \delta f(a(r), r)/\delta a(r) = 2(E - V(r))(r + a(r))$ one obtains the first order differential equation, whose integral form, satisfying the boundary condition $a(0) = 0$, could be written as

$$a(r) = \int_0^r ds[f(a(s), s) - f_a(a(s), s)\, a(s)] \exp \int_s^r dt f_a(a(t), t). \tag{2}$$

We can now set up an iteration scheme:

$$a_{n+1}(r) = \int_0^r ds[f(a_n(s), s) - f_a(a_n(s), s)\, a_n(s)] \exp \int_s^r dt f_a(a_n(t), t). \tag{3}$$

Differentiation of this equation gives a differential form of the iteration scheme:

$$a'_{n+1}(r) = f(a_n(r), r) + \delta a_{n+1}(r)\, f_a(a_n(r), r), \quad \delta a_{n+1}(r) = a_{n+1}(r) - a_n(r) \tag{4}$$

Subtraction of the same expression with n changed to $n - 1$ and use of the mean value theorem for functionals

$$f(a_n(r), r) - f(a_{n-1}(r), r) = f_a(a_{n-1}(r), r)\, \delta a_n(r) + \frac{1}{2} f_{aa}(\overline{a}_n(r), r)\, \delta a_n^2(r) \tag{5}$$

where $\overline{a}_n(r)$ lies between $a_n(r)$ and $a_{n-1}(r)$, gives

$$\delta a'_{n+1}(r) - f(a_n(r), r)\, \delta a_{n+1}(r) = \frac{1}{2} \delta a_n^2(r)\, f_{aa}(\overline{a}_n(r), r), \tag{6}$$

or

$$\delta a_{n+1}(r) = \frac{1}{2} \int_0^r ds f_{aa}(\overline{a}_n(s), s)\, \delta a_n^2(s) \exp \int_s^r dt f_a(a_n(t), t) \tag{7}$$

We see, that the difference δa_{n+1} between two subsequent iterations is proportional to an integral, containing a square of the difference δa_n of the two previous subsequent iterations. In addition, if the sign of the second functional derivative $f_{aa}(\overline{a}_n(s), s)$ is definitive, δa_{n+1} has the same sign. The convergence of our iteration sequence is thus quadratic, and, in case of the definitive sign of the second functional derivative of $f(a(r), r)$ also monotonic. The convergence, naturally, depends on a zero iteration $a_0(r)$, which has to be chosen from physical and mathematical considerations.

As our first example we consider the QLM solution [3] of the Thomas-Fermi equation

$$\sqrt{x}\; u''(x) = u^{3/2}(x), \quad u(0) = 1, \; u(\infty) = 0, \tag{8}$$

which is an equation for the electron density around the nucleus of the atom.

As the initial guess we choose $u_0(x) \equiv 1$, which satisfies the boundary condition at zero. The results of QLM calculations are presented in Fig. 1 which displays the exact solution together with the first four QLM iterations.

As our second example let us consider the Lane-Emden equation

$$y''(r) + \frac{2}{r}\, y'(r) + y^n(r) = 0, \quad y(0) = 1, \; y'(0) = 0 \tag{9}$$

which is a nonlinear second order differential equation which arises in the study of stellar structure. Setting $y = u/r$ transforms the equation to a more conve-

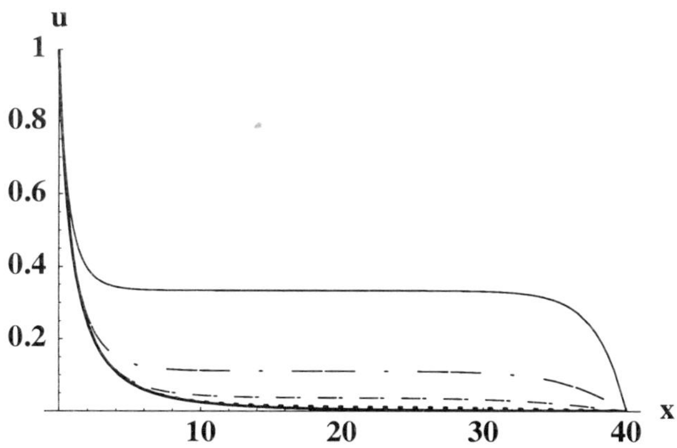

Figure 1. Convergence of QLM iterations for the Thomas-Fermi equation and comparison with the numerically obtained exact solution. Thin solid, dot-dashed, short-dashed and dotted curves correspond to the first, second, third and fourth QLM iteration, respectively, while the thick solid curve displays the exact solution. The difference between the exact solution and the eighth QLM iteration for all x in the figure is less than 10^{-7}.

nient form without a first derivative:

$$u''(r) + \frac{u^n(r)}{r^{n-1}} = 0, \; u(0) = 0, \; u'(0) = 1. \qquad (10)$$

The physically interesting and analytically nonsolvable case corresponds to $n = 4$. The initial guess, satisfying the boundary conditions is $u_0(r) = r$. Comparison of the quasilinear solutions [3] of Eq. 10 corresponding to the first five iterations with the numerically computed exact solution are given in Fig. 2.

Consider now the relationship between QLM and expansions in small parameters. Let us denote by Δa_n a difference between n-th iteration and the exact solution. One can prove [1] that in QLM Δa_{n+1} is proportional to $(\Delta a_n)^2$. Therefore, if initial guess a_0 is correct in zero order of a small parameter λ, that is $\Delta a_0 \sim \lambda$, then $\Delta a_1 \sim \lambda^2$, $\Delta a_2 \sim \lambda^4$ etc., so that $\Delta a_{n+1} \sim \lambda^{2^n}$. Thus n-th iteration of QLM reproduces exactly at least 2^n terms of the expansion of the solution in the powers of the small parameter. As an example, consider the Calogero equation for the inverse square potential

$$V(r) = \frac{g}{4r^2}\Theta(R-r), \qquad (11)$$

where λ is the dimensionless coupling constant. The exact scattering length in units of R is $A = -1 - 2/g(1 - \sqrt{1+g})$. The solution has a singularity, namely a branch point, at $g = -1$, which marks the beginning of interval $-\infty < g < -1$ where a fall to the center takes place. The QLM

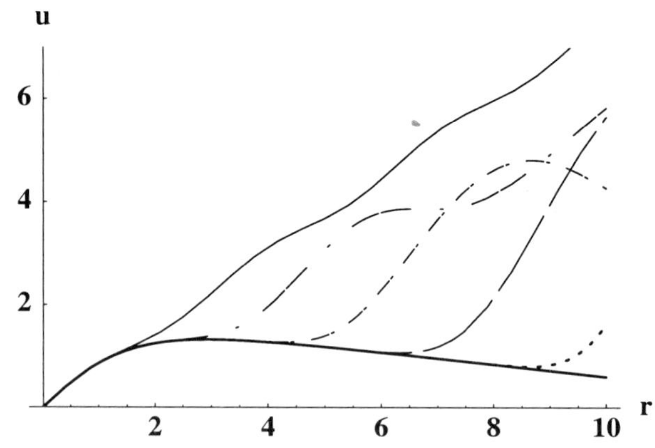

Figure 2. Convergence of QLM iterations for the Lane-Emden equation and comparison with the numerically obtained exact solution. Thin solid, dot-dashed, short-dashed, long-dashed and dotted curves correspond to the first, second, third, fourth and fifth QLM iteration, respectively, while the thick solid curve displays the exact solution. The difference between the exact solution and the eighth QLM iteration for all r in the figure is less than 10^{-11}.

iteration sequence for this potential [1] reduces to the algebraic expression $A_{n+1} = -g1 - A_n^2/4 + 2g(1 + A_n)$ which enables easily to calculate higher iterations. For example, the expansion of the difference between exact scattering length and the sixth QLM iteration in powers of g has the form $A - A_6 = c_1 g^{127} + c_2 g^{128} + c_3 g^{129} + \ldots$, which is even better than one expects from the above formulated rule. In addition, one can see that the next terms of the perturbation series are also reproduced extremely well, their difference with the precise terms being infinitesimally small. Namely, the coefficient c_1 is $3.45 \cdot 10^{-76}$, c_2 is $2.19 \cdot 10^{-74}$, c_3 is $7.04 \cdot 10^{-74}$ and so on.

Let us compare now the QLM and the $1/N$ expansion method. The latter is used extensively extensively for calculating energy eigenvalues of complicated systems in chemistry and atomic physics. Recently the behavior of this method was investigated at high orders [7] and it was shown that the $1/N$ expansion is asymptotic, that is convergent at low orders but divergent at higher ones. To check the power of QLM and to verify that higher QLM iterations contain no divergencies we display in Table 1 taken from work [2] the binding energies

Table 1. QLM and exact binding energies E for different potentials $V(r)$. E are taken from citations in Ref. [7]. E_p and N_p are the energies by the $1/N$ perturbation method of Ref. [7] and the corresponding ranges of orders of the perturbation where the $1/N$ expansion converges; a finite range means that the expansion diverges for larger N; stable digits are given only. n is the principal quantum number of the state, m is the reduced mass.

$V(r)$	m	n	QLM	E	E_p	N_p
$2^{7/2}r$	1	1	9.352429642	9.35243	9.352	10–20
		2	16.35179778	16.3518	16.352	10–15
		3	22.08223931	22.08224	22	10–28
$r^{3/2}$	1/2	1	2.708092416	2.70809	2.71	14–15
$\ln r$	1/2	1	1.044332	1.0443	1.04	113–14
$-\frac{1}{r}$	1	1	0.499999999	0.5	1 ± 10^{-10}	29–∞
		2	0.125000001	0.125	0.25 ± 10^{-8}	29–∞
r^5	1/2	1	4.089159315	4.08916	4.	6–7
$r^2 + r^4$	1/2	1	4.648812183	4.64881	4.6	10–11
$\frac{(r^2-16)^2}{128}$	1	1	0.4830541244	0.483053433 0.483053390	0.48302	12–13

for all potentials for which the $1/N$ expansion is divergent at higher orders. In the caption of the table $V(r)$ is the potential and n denotes the number of the excited state; m is the reduced mass of the particle and is given different values for different potentials in order to enable comparison of the QLM bound state energies with those obtained by the $1/N$ expansion method in Ref. [7] where m takes on values $m = 1$ or $1/2$ depending on the interaction. ¿From Table 1 one can conclude that QLM is extremely precise. Energies and the wave function for both ground and excited states typically converge to the order of 10 significant digits after about $10 - 20$ iterations. Our calculations confirm numerically the conclusion following from the proof in Ref. [3] that once the quasilinear iteration sequence starts to converge, it will continue to do so, unlike the perturbation expansions in powers of the coupling constant or in powers of $1/N$, which are often given by the asymptotic series and therefore converge only up to a certain order and diverge thereafter.

Fast convergence of the QLM iterations to the exact solution confirms numerically the uniform and quadratic law of convergence proved in Refs. [1] for realistic physical interactions defined on infinite intervals with possible singularities at certain points of the intervals. For convergence it is enough that an initial guess for the zeroth iteration is sufficiently good. In all the examples considered in the paper the simplest initial guess satisfying one of the boundary conditions was usually enough to produce a rapid convergence.

Acknowledgement. The collaborators in this project were Drs. R. Krivec and F. Tabakin. The numerous discussions with them are gratefully acknowledged. The research was supported by the Israel Science Foundation grant 131/00.

References

1. V.B. Mandelzweig: J. Math. Phys. **40**, 6266 (1999)
2. R. Krivec and V.B. Mandelzweig: Computer Physics Comm. **138**, 69 (2001); ibid., 2002, in press
3. V.B. Mandelzweig and F. Tabakin: Computer Physics Comm. **141**, 268 (2001)
4. R. Kalaba: J. Math. Mech. **8**, 519 (1959); R. E. Bellman and R.E. Kalaba: *Quasilinearization and Nonlinear Boundary-Value Problems.* New York: Elsevier Publishing Company 1965
5. V. Lakshmikantham and A. S. Vatsala: *Generalized Quasilinearization for Nonlinear Problems*, Mathematics and its Applications, Vol. 440. Dordrecht: Kluwer Academic Publishers 1998
6. F. Calogero: *Variable Phase Approach to Potential Scattering.* New York: Academic Press 1965
7. N. Bjerrum-Bohr: J. Math. Phys. **41**, 2516 (2000)

Few-Body Systems Suppl. 14, 191–196 (2003)

Atoms in Strong Magnetic Field: A Stochastic Variational Study

K. Varga*

Solid State Division, Oak Ridge National Laboratory, Oak Ridge, TN-37831, USA

Abstract. A special deformed correlated Gaussian variational basis is introduced to solve few-electron problems in strong magnetic field. The bound state energies of the helium and lithium atoms and other few-electron systems are calculated by the stochastic variational method. The calculation significantly improves the accuracy of the energies of these systems.

1 Introduction

The theoretical interest [1, 2, 3, 4, 5, 6, 7, 8] in the studies of atomic and molecular systems in magnetic field is motivated by a number of experimental observations. One of these is the discovery of the strong magnetic field on (magnetic) white dwarfs [9] ($\approx 10^7$ G) and on the surface of neutron stars [10] ($\approx 10^{12}$ G). (The strongest available laboratory magnetic field strength is about 10^6 G.) On the other hand there are a lots of experiments on excitons, excitonic complexes and quantum dots in semiconductors. The small effective masses and large dielectric constant results in very large effective field strength in these systems.

Beyond the perturbation regime, the Coulomb interaction is rivaled by the external magnetic field and the calculation of the binding energies of multielectron-systems becomes very complicated. In these systems the spherical symmetry of the Coulomb interaction is strongly broken by the magnetic field. The intricate geometry of the highly anisotropic charge distribution is very challenging for conventional theoretical approaches and only the hydrogen and helium atom has been solved with accuracy required by the spectroscopical needs. Precise calculation of the spectra of light elements in magnetic field

* *On leave from:* Institute for Nuclear Research of the Hungarian Academy of Sciences (ATOMKI), PO Box 51, 4001 Debrecen, Hungary. Supported by OTKA grants (Hungary) T029003 and T037991.

would greatly help the interpretation of the optical absorption spectra and therefore the understanding of the composition of the atmosphere of stars.

The Correlated Gaussians (CG) are very successfully applied in atomic physics and quantum chemistry [11]. The weak points of the CG's (they do not satisfy the cusp condition and their convergence is relatively slow) are far outscored by their advantages: (1) their matrix elements are analytically available for a general N-particle system (2) they are flexible enough to approximate rapidly changing functions (3) the permutation symmetry can be easily imposed and (4) one can make a simple transformation between single particle and relative coordinate systems. The conventional form of the CG's are spherically symmetric, so for example for a two-particle system they are assumed in the simple $\exp\{-\alpha_1\mathbf{r}_1^2-\alpha_2\mathbf{r}_2^2-\beta\mathbf{r}_1\mathbf{r}_2\}$ form. In the case of a strong magnetic field the particles are in the field of an isotropic harmonic oscillator in the x-y plane. The spherical CG basis does not provide sufficient flexibility to describe this situation because its parameters in the x-y and z direction are equal. The simplest extension of the CG's to tackle this problem is to choose different nonlinear parameters in the x-y plane and in the z direction: $\exp\{-\alpha_1(x_1^2+y_1^2)-\alpha_{1z}z_1^2-\alpha_2(x_2^2+y_2^2)-\alpha_{2z}z_2^2-\beta(x_1x_2+y_1y_2)-\beta_x z_1z_2\}$. By choosing equal parameters in the x-y plane and z-direction the deformed CG (DCG) basis trivially reduces to the spherical one, but in the case of magnetic field it has the required flexibility to be able to describe of the harmonic oscillator wave function in the x-y plane and motion governed by the Coulomb interaction in the z direction. At the same time, the basis allows for the correlation between the electrons. The basis parameters can be efficiently chosen by the stochastic variational method [11].

2 The Formalism

The Hamiltonian:

$$H=\sum_{i=1}^{N}\frac{1}{2m_i}(\mathbf{p}_i+\frac{e_i}{c}\mathbf{A}_i)^2+\sum_{i<j}\frac{e_ie_j}{\kappa r_{ij}}\,, \tag{1}$$

where $\mathbf{A}_i=\frac{1}{2}\mathbf{r}_i\times\mathbf{B}$ and the magnetic field is $\mathbf{B}=(0,0,B)$, m_i, e_i are the masses and charges. The Hamiltonian, in detailed form:

$$H=\sum_{i=1}^{N}\frac{1}{2m_i}\left(-\hbar^2\Delta_i+\frac{e_i^2B^2}{4c^2}(x_i^2+y_i^2)+\frac{e_i\hbar B}{c}l_{zi}\right)+\sum_{i<j}\frac{e_ie_j}{\kappa r_{ij}}\,, \tag{2}$$

where l_{zi} is the z component of the orbital momentum of the ith particle.

The deformed geometry, described by the Hamiltonian requires a special deformed variational basis. The Correlated Gaussians [11] are proved to be a very flexible basis in nuclear and atomic physics. The correlated Gaussians are defined as

$$\exp\{-\frac{1}{2}\mathbf{r}A\mathbf{r}\}=\exp\{-\frac{1}{2}\sum_{i,j=1}^{N}A_{ij}\mathbf{r}_i\cdot\mathbf{r}_j\}\,, \tag{3}$$

where $\mathbf{r} = (\mathbf{r}_1, \ldots, \mathbf{r}_N)$ are the coordinates of the particles. This basis is spherical, to extend it to a deformed nonspherical systems one usually multiplies it by product of spherical harmonics coupled to a total orbital momentum "L". This form, however, is not ideally suited for systems in magnetic field, because the basis functions belonging to different orbital momenta are coupled by the Hamiltonian. This coupling would require an infinite series of orbital momentum states which is obviously out of question. To avoid this, we choose a deformed form of the correlated Gaussians (DCG):

$$\exp\{-\frac{1}{2}\sum_{i,j=1}^{N} A_{ij}\rho_i \cdot \rho_j - \frac{1}{2}\sum_{i,j=1}^{N} B_{ij}z_i \cdot z_j\} , \tag{4}$$

where the nonlinear parameters are different (and independent) in the "xy" and "z" directions ($\rho_i = (x_i, y_i)$). This extension brings a great deal of flexibility by allowing a separate description on the "xy" plane and along the "z" axis. The Hamiltonian does not commute with L^2 but it has common eigenfunctions with L_z. The above form of the DCG belongs to $M = 0$. To allow for $M \neq 0$ states we multiply the basis by

$$\prod_{i=1}^{N} \xi_{m_i}(\rho_i) , \tag{5}$$

where

$$\xi_m(\rho) = (x + iy)^m. \tag{6}$$

Thus our variational trial function reads as

$$\Phi_M(\mathbf{r}) = \mathcal{A}\{\left(\prod_{i=1}^{N} \xi_{m_i}(\rho_i)\right) \exp\{-\frac{1}{2}\sum_{i,j=1}^{N} A_{ij}\rho_i \cdot \rho_j - \frac{1}{2}\sum_{i,j=1}^{N} B_{ij}z_i \cdot z_j\}\}, \tag{7}$$

where $M = m_1 + m_2 + ... + m_N$, and m_i are integers.

3 Results

To test this formalism we have calculated the binding energies of the ground and excited states of the Helium atom and compared them to results of other methods in the literature. There are various approches [1, 2, 3, 4, 5, 6, 7, 8] used to calculate the energy of the He atom in strong magnetic field. To test the accuracy of our approach we have compared it to the results of (1) Hylleraas-like explicitly correlated (HEC) functions [2]; (2) B-spline basis [6]; (3) Gaussian expansion (GE) [5]. The results are presented in Tables I., II and III. Our results are converged in all digits shown in the Tables. The agreement with other approaches is very good. We believe that we have significantly improved the accuracy of the calculated energies of the Helium atom in strong fields. The formalism is general and it is readily applicable for larger atoms/systems.

Table 1. Energies of singlet states of helium with $M = 0$.

β	SVM	HEC [2]
0.0	−2.90372337	
0.1	−2.89583445	−2.89583
0.2	−2.87287076	−2.87287
0.3	−2.83646731	−2.83646
0.4	−2.78842446	−2.78842
0.5	−2.73000000	−2.73038

Table 2. Energies of singlet states of helium with $M = -1$.

β	SVM	HEC [2]	B-splines [6]	GE [5]
0.0	−2.12384089			−2.123774
0.1	−2.14527502	−2.14527	−2.1448	
0.2	−2.10615755	−2.10618	−2.1055	−2.0105987
0.3	−2.04474860	−2.04486	−2.0441	
0.4	−1.96985631	−1.97011	−1.9692	−1.969560

The excitonic complexes are another example of applications of the method. The excitonic complexes are confined systems of electrons and holes described in the framework of the effective mass approximation. In the effective mass approximation the holes are positively charged quasi-particles with masses slightly larger (2-5 times) than the mass of the electron. The confinement is modeled by square-well potential in the z-direction and the particles interact by Coulombic force. Several such system has been the subject of our recent works [12].

To show an example for even larger system we present the density distribution of a two-dimensional harmically confined 16 electron system in magnetic field. In the case of weak magnetic field, the Coulomb repulsion dominates the dynamics and the electrons form a crystal-like (Wigner-crystal) structure. Figure 1 shows that the electrons are ordered into rings, one electron in the middle, 5 electrons in the inner and 10 electrons in the outer rings. Similar systems has been studied in [13].

In summary we presented a formalism to describe the correlated motion of charged particles in strong magnetic fields. Test examples show that the approach is very accurate and powerful and it is readily applicable for $N = 2-8$ particle problems.

References

1. D.M. Larsen: Phys. Rev. **B20** 20 (1979)
2. A. Scrinzi: Phys. Rev. **A58**, 3879 (1998)
3. M. Braun, W. Schweizer and H. Elster: Phys. Rev. **A57** 3739 (1998)

Table 3. Energies of singlet states of helium with $M = -2$.

β	SVM	HEC [2]
0.0	−2.05561573	
0.1	−2.07937321	−2.07949
0.2	−2.03241589	−2.03308
0.3	−1.96555463	−1.96735
0.4	−1.88652446	−1.89021

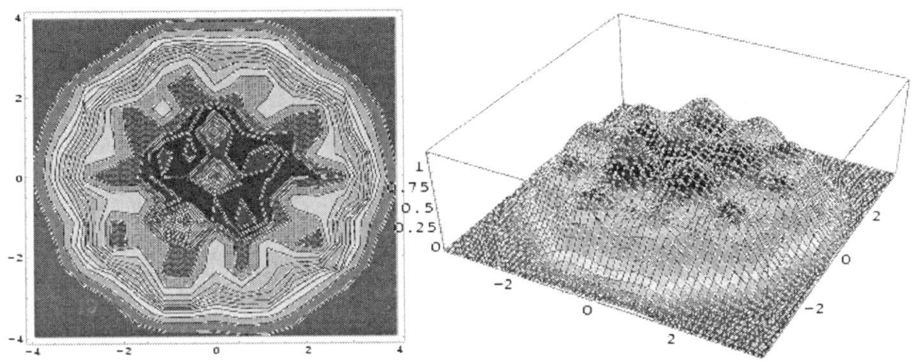

Figure 1. Wigner crystal of 16 electrons confined by a two-dimensional harmonic oscillator potential.

4. J.S. Heyl and L. Hernquist: Phys. Rev. **A58** 3567 (1998)
5. W. Becken and P. Schmelcher: J. Phys. **B33** 545 (2000)
6. H. Qiao and B. Li: Phys. Rev. **A60** 3134 (1999)
7. M. Vincke and D. Baye: J. Phys. **B22** 2089 (1989)
8. M.D. Jones, G. Ortiz, and D. Ceperley: Phys. Rev. **E55** 6202 (1997)
9. J.P. Ostriker and F.D.A. Hartwick: Astrophys. J. **173**, 797 (1968)
10. J. Trümper, W. Pietsch, C. Reppin, W. Voges, R. Stauben, and E. Kendziorra, Astrophys. J. **219**, L105 (1978)
11. Y. Suzuki, K. Varga: *Stochastic variational approach to quantum mechanical few-body problems.* Berlin, Heidelberg: Springer-Verlag 1998
12. C. Riva, F.M. Peeters, and K. Varga: Phys. Rev. bf B64, 5301 (2001); Phys. Rev. **B63**, 5302 (2001)
13. K. Varga, P. Navratil, J. Usukura, and Y. Suzuki: Phys. Rev. **B63**, 205308 (2001)

Few-Body Systems Suppl. 14, 197–202 (2003)

Few-Body Systems

Solution of Nd Breakup Scattering Problem in Configuration Space*

V.M. Suslov[1,†], B. Vlahovic[2,3,**]

[1] Dept. of Mathematical and Computational Physics, St. Petersburg State University, 198504 Ul'yanovskaya 1, Petrodvorets, St. Petersburg, Russia

[2] Department of Physics, North Carolina Central University, 1801 Fayetteville Street, Durham, NC 27707, USA

[3] Jefferson Lab, 12000 Jefferson Avenue, Newport News, VA 23606, USA

Abstract. The configuration-space Faddeev equations for the breakup scattering problem are solved using a new method of partial inversion. Unlike other computations this one keeps the incoming wave in the asymptotic conditions. Using the standard spline-decomposition for the angle variable and Numerov's method for the hyperradius optimizes considerably the inversion due to the sparse block structure of the matrix. We report on calculations of the inelasticities and phase shifts, as well as the breakup amplitudes for nucleon-deuteron scattering for laboratory energies 14.1 and 42.0 MeV. The results are compared with calculations of other authors.

1 Introduction

Considerable progress has been achieved by many groups [1, 2, 3, 4] in calculating the $n-d$ scattering states in three-nucleon systems. The results obtained using different approaches are in good agreement. Recently, accurate computations on $p-d$ breakup scattering have been performed by the Pisa group [5]. Despite much progress some questions remain that require further explanation, for example the A_y puzzle.

In this article we present results of our nucleon-deuteron breakup scattering calculations in the framework of the Faddeev differential equations in configuration space. Here we briefly describe the method and compare our initial results using the MT I-III potential with those of other authors. The method can be extended straightforwardly to realistic potentials.

*This work was partially sponsored by the RFFI under Grant No. 02-02-16562

† *E-mail address:* suslov@mph.phys.spbu.ru
** *E-mail address:* branko@jlab.org

2 Faddeev Equations in Configuration Space

In an s-wave approach the system of integro-differential Faddeev equations [6] for the spin-doublet case, in the polar coordinates has the form

$$
\begin{aligned}
&\{-\frac{\partial^2}{\partial\rho^2}-\frac{1}{\rho^2}\frac{\partial^2}{\partial\theta^2}+V_c^t(\rho,\theta)+V^t(\rho,\theta)-\frac{1}{4\rho^2}-E\}U^t(\rho,\theta)=-\frac{1}{4}V^t(\rho,\theta)\\
&\times\int_{-1}^{+1}du\frac{\sin\theta\cos\theta}{\sin\theta'\cos\theta'}(U^t(\rho,\theta')-3U^s(\rho,\theta'))\,,\\
&\{-\frac{\partial^2}{\partial\rho^2}-\frac{1}{\rho^2}\frac{\partial^2}{\partial\theta^2}+V_c^s(\rho,\theta)+V^s(\rho,\theta)-\frac{1}{4\rho^2}-E\}U^s(\rho,\theta)=-\frac{1}{4}V^s(\rho,\theta)\\
&\times\int_{-1}^{+1}du\frac{\sin\theta\cos\theta}{\sin\theta'\cos\theta'}(-3U^t(\rho,\theta')+U^s(\rho,\theta'))\,,
\end{aligned}
\tag{1}
$$

where we eliminate the first derivative in the radius. In Eq. (1) the $s-wave$ Coulomb potential $V_c^{(t,s)}(\rho,\theta)$ is given by

$$
\begin{aligned}
V_c^t(\rho,\theta)&=\frac{n\mu^t(\theta)}{\rho},\quad \mu^t(\theta)=\begin{cases}2/\sqrt{3}\sin\theta, & \theta>30^\circ\\ 2/\cos\theta, & \theta\le 30^\circ\end{cases}\\
V_c^s(\rho,\theta)&=\frac{n\mu^s(\theta)}{\rho},\quad \mu^s(\theta)=\frac{1}{3}\left(\frac{2}{\cos\theta}+\mu^t(\theta)\right).
\end{aligned}
\tag{2}
$$

The Malfliet-Tjon MT I-III potential (ref. [3]) was chosen as the nuclear potential $V^{t,s}(\rho,\theta)$. The set of partial differential equations (Eq. (1)) must be solved for functions satisfying the regularity conditions

$$
U^{t,s}(0,\theta)=U^{t,s}(\rho,0)=U^{t,s}(\rho,\pi/2)=0\,, \tag{3}
$$

and the following asymptotic conditions [6, 7]:

$$
\begin{aligned}
&U^t\mid_{\rho\to\infty}\sim\sqrt{\rho}\varphi_d(\rho,\cos\theta)\{F_o(\gamma,p\frac{\sqrt{3}}{2}\rho\sin\theta)+a(p)[G_o(\gamma,p\frac{\sqrt{3}}{2}\rho\sin\theta)\\
&+iF_o(\gamma,p\frac{\sqrt{3}}{2}\rho\sin\theta)]\}+\mathcal{A}^t(\theta)\exp[i\sqrt{E}\rho-i\frac{n\mu^t(\theta)}{2\sqrt{E}}\ln(2\sqrt{E}\rho)]\,,\\
&U^s\mid_{\rho\to\infty}\sim\mathcal{A}^s(\theta)\exp[i\sqrt{E}\rho-i\frac{n\mu^s(\theta)}{2\sqrt{E}}\ln(2\sqrt{E}\rho)]\,.
\end{aligned}
\tag{4}
$$

Here, $\varphi_d(\rho,\cos(\theta))$ is the deuteron wave function, F_o and G_o are the regular and irregular Coulomb functions, $\gamma=2n/3p$ is the Coulomb parameter with p the momentum in the centre-of-mass system and $\mu^{t,s}(\theta)$ is defined in Eqs. (2). The unknown functions $a(p)$ and $\mathcal{A}^{t,s}$ are the elastic and breakup scattering amplitudes,

$$
a(p)=\frac{\eta\exp(i2\delta)-1}{2i}\,, \tag{5}
$$

and η and δ are the inelasticity and phase shift, respectively.

3 Method of Partial Inversion

To simplify the numerical solution of the problem (1)-(4) we expand the Faddeev component $U^{t,s}$ as a function of the angular variable in the basis of the cubic Hermitian splines. To approximate the radial derivative, we used Numerov's method. As a result, we have the matrix problem of the form:

$$(D * U)_i = -\delta_{in} D^+ U_{n+1}, \tag{6}$$

where D^+ is the part of the finite difference operator for the second partial derivatives in ρ, which is just a multiplication of the component U_{n+1} by a matrix of the dimensionality $2 \cdot N_\theta \times 2 \cdot N_\theta$, and N_θ is the number of collocation points in the angular variable θ. Index $n+1$ stands for hyperradius $\rho_{n+1} = R_{max}$, where R_{max} is the cutoff radius at which the asymptotic conditions Eq.(4) are implemented. By formal inversion of the matrix D in Eq. (6) we can write the solution of the problem:

$$U_j = -D^{-1}_{jn} D^+ U_{n+1}, \qquad j = 1, 2....N_\rho.$$

In this equation we consider the last two components of vector U:

$$U_{n-1} = -D^{-1}_{n-1n} D^+ U_{n+1}\ , \qquad U_n = -D^{-1}_{nn} D^+ U_{n+1}. \tag{7}$$

Provided R_{max} is large enough we may substitute for U_{n-1}, U_n on the left side of Eqs. (7) the corresponding vectors obtained by evaluating Eqs. (4) at the radii $\rho = \rho_{n-1}$ and $\rho = \rho_n$, and obtain a set of linear equations for unknown amplitudes a and $\mathcal{A}$:

$$v_{n-1} \cdot a + m_{n-1} \cdot \mathcal{A} = \mathcal{F}_{n-1}\ , \qquad v_n \cdot a + m_n \cdot \mathcal{A} = \mathcal{F}_n. \tag{8}$$

As $R_{max} \to \infty$, this set of equations has a constant a as a solution. At finite R_{max} its solution is a vector a with generally different components corresponding to different angels. We follow the method of S.P. Merkuriev et al. [1], which consists in sellecting the components of a in the region of the maximum of the deuteron wave function. Where a turns out to be independent of the angle.

4 Numerov's Method

The idea of Numerov's method consists in using the initial differential equation to calculate higher derivatives in the expansion of unknown function in Taylor's series. According to Numerov, one has to use the following finite-difference approximation of the second radial derivative:

$$\left.\frac{\partial^2 U(\rho,\theta)}{\partial \rho^2}\right|_{\rho_i} = \frac{U(\rho_{i+1},\theta) - 2U(\rho_i,\theta) + U(\rho_{i-1},\theta)}{\Delta\rho^2} - \frac{\Delta\rho^2}{12} U^{IV}_\rho(\rho_i,\theta) + O(\Delta\rho^4). \tag{9}$$

The fourth radial derivative of Faddev component has to be found by differentiating the corresponding Faddeev equation. Substituting the fourth radial derivative evaluated into Eq. (9), we obtain the finite-difference approximation for the second derivative up to $4th$ order in $\Delta\rho$. Finally, replacing the second radial derivative in the Faddeev equations by the expression obtained, we get the analog of Numerov's method for the spin-quartet Faddeev equation as follows:

$$-\Big[\frac{U(\rho_{i+1},\theta)-2U(\rho_i,\theta)+U(\rho_{i-1},\theta)}{\Delta\rho^2}-\Big(1+\frac{\Delta\rho^2}{12}\frac{\partial^2}{\partial\rho^2}\Big)\Big\{-\frac{1}{\rho^2}\frac{\partial^2}{\partial\theta^2}$$
$$+V_c(\rho,\theta)+V(\rho,\theta)-\frac{1}{4\rho^2}-E\Big\}U(\rho,\theta)\Big]_{\rho_i,\theta_j} \tag{10}$$
$$=\frac{2}{\sqrt{3}}\Big(1+\frac{\Delta\rho^2}{12}\frac{\partial^2}{\partial\rho^2}\Big)\Big[V(\rho,\theta)\int_{\theta-}^{\theta+}d\theta' U(\rho,\theta')\Big]_{\rho_i,\theta_j}.$$

5 Results of the Calculations

The elastic amplitude a, and breakup amplitude $\mathcal{A}$ for $n-d$ and $p-d$ scattering are computed at E_{lab}= 14.1 and 42.0 MeV. The following values are used for the parameters of the calculation: $N_\rho\sim 10000$, $N_\theta\sim 600$ and values of the hyperradius R_{max} as large as 800 fm.

Table 1. The nd and pd elastic-scattering phase shifts and inelasticities.

	LA/Iowa ref. [3]	Bochum ref. [3]	Pisa ref. [5]	present work	Pisa ref. [5]	present work
14.1 MeV		doublet	n-d		p-d	
Re(δ)	105.48	105.50	105.48	105.47	108.44	108.06
η	0.4648	0.4649	0.4649	0.4649	0.4984	0.4929
		quartet	n-d		p-d	
Re(δ)	68.95	68.96	68.952	68.93	72.604	73.64
η	0.9782	0.9782	0.9782	0.9782	0.9795	0.9202
42 MeV		doublet	n-d		p-d	
Re(δ)	41.34	41.37	41.341	41.34	43.667	43.47
η	0.5024	0.5022	0.5022	0.5022	0.5056	0.5071
		quartet	n-d		p-d	
Re(δ)	37.71	37.71	37.722	37.70	39.947	39.19
η	0.9035	0.9033	0.9033	0.9034	0.9046	0.866

As one can see from Table 1, our results for the elastic phase shifts and inelasticities for $n-d$ breakup are in a very good agreement with calculations of other groups. For $p-d$ breakup they differ from those of the Pisa group [5].

It is important to note that in our s-wave approach we have not included higher partial waves of the Coulomb potential. Nonetheless the accuracy of this approach is about 1% for energies exceeding 1 MeV, as was already pointed out by S.P. Merkuriev et al. in ref. [2].

Table 2. The nd and pd spin-quartet reduced breakup amplitude presented in the format x.xx ≡ x.xx 10^{-1}, E_{lab} = 14.1 MeV.

θ (deg)	0	10	20	30	40	50	60	70	80
				present work, **nd** results, $R_{max} \to \infty$					
Re(^{3}S$_1$)	-1.91	-1.93	-1.94	-1.89	-1.75	-1.58	-1.47	-1.51	-1.78
Im(^{3}S$_1$)	3.65	3.67	3.70	3.72	3.73	3.81	4.00	4.32	4.62
				LA/Iowa, **nd** results, $R_{max} \to \infty$, ref.[4]					
Re(^{3}S$_1$)	-1.92	-1.93	-1.94	-1.89	-1.75	-1.58	-1.47	-1.51	-1.78
Im(^{3}S$_1$)	3.65	3.67	3.70	3.72	3.73	3.81	4.00	4.31	4.62
				present work, **pd** results, $R_{max} \to \infty$					
Re(^{3}S$_1$)	-2.21	-2.24	-2.31	-2.44	-2.07	-1.82	-1.68	-1.71	-1.96
Im(^{3}S$_1$)	3.21	3.19	3.14	2.99	3.34	3.57	3.83	4.14	4.39

Agreement between our results and those of Los-Alamos and Bochum groups [4] are excellent as one can see from Table 2. It should be noted that one cannot calculate the reduced breakup amplitude straight for the angle $\theta = 90°$ solving the Faddeev equation. Comparing the breakup amplitudes for the **nd** and **pd** in Table 2, it appears that the Coulomb interaction effects are noticeable even for such a large energy. Obviously, this effect should be important for calculation of the spin depended observables with more realistic nucleon-nucleon potentials taken into account.

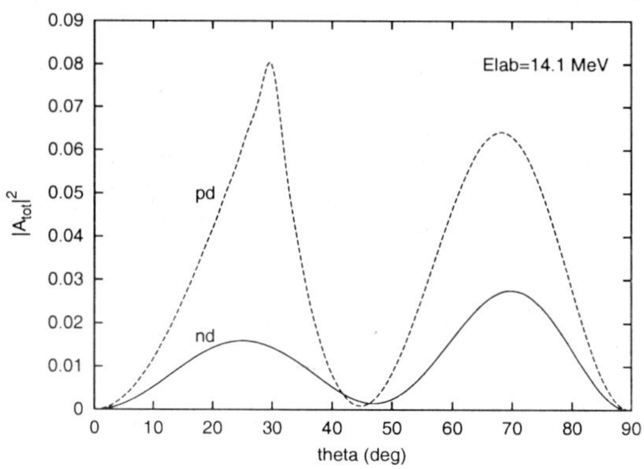

Figure 1. Squares of the moduli of the physical quartet breakup amplitudes.

From Fig. 1 it follows that the Coulomb interaction brings a large influence on the total quartet $p - d$ breakup amplitude. To verify our results, we used the optical theorem. For spin-quartet case in s-wave approach [1], it reads

$$Im(a(p)) = |\, a(p)|^2 + \frac{K}{p}\int_0^{\pi/2} d\theta A^*_{tot}\mathcal{A}\,, \quad A_{tot}(\theta) = \mathcal{A}(\theta) - \frac{2}{\sqrt{3}}\int_{\theta^-}^{\theta^+} d\theta' \mathcal{A}(\theta')\,.$$

Table 3. Optical theorem results.

	nd–quartet		pd–quartet	
Elab (MeV)	l.h.s	r.h.s	l.h.s	r.h.s
14.1	0.8626	0.8626	0.8871	0.8776
42.0	0.3860	0.3860	0.4127	0.4086

Table 3 clearly confirms the accuracy of our results, and the estimation of S.P. Merkuriev et al. [2] on the contribution of higher coulomb partial waves under an s-wave approach, which is less than 1%.

6 Conclusion

We have shown that by using Numerov's method very accurate calculation can be performed with minimal computation resources (PC). By retaining the incident wave in the asymptotics for the Faddeev components, unnecessary additional computations are eliminated casting the problem in a form that enables partial inversion. Finally, the stability of our solutions for relatively large values of R_{max} illustrates its advantages for the investigation of the asymptotic behavior of solutions. Numerov's method allows us to compute the breakup amplitudes with high accuracy for R_{max}=800 fm and more.

Acknowledgement. Authors wish to thank M.A. Braun for useful discussion and J.L. Friar for clear explanation on the breakup amplitude extrapolation. V.M.S. would like to acknowledge financial support from Jefferson Lab and North Carolina Central University.

References

1. S.P. Merkuriev, C. Gignoux, A. Laverne: Ann. Phys. **99**, 30 (1976)
2. A.A. Kvitsinsky, Yu.A. Kuperin, S.P. Merkuriev, A.K. Motovilov, S.L. Yakovlev: Fiz. Elem. Chastis At. Yadra **17**, 267 (1986)
3. J.L. Friar, B.F. Gibson, G. Berthold, J. Haidenbauer, Y. Koike, G.L. Payne, J.A. Tjon, W.M. Kloet: Phys. Rev. **C42**, 1838 (1990)
4. J.L. Friar, G.L. Payne, W. Glöckle, D. Hüber, H. Witała: Phys. Rev. **C51**, 2356 (1995)
5. A. Kievsky, M. Viviani, S. Rosati: Phys. Rev. **C64**, 024002 (2001)
6. S.P. Merkuriev, L.D. Faddee: *Quantum Scattering Theory for Few-Particle Systems* (in Russian). Moscow: Nauka 1985
7. S.P. Merkuriev: Ann. Phys. **130**, 3975 (1980); S.P. Merkuriev: Acta Physica Austriaca Suppl. **XXIII**, 65 (1981)

Few-Body Systems Suppl. 14, 203–206 (2003)

Few-Body Systems

Improved Effective Interaction for the Hyperspherical Formalism

N. Barnea[1,2,*], W. Leidemann[3,4], G. Orlandini[3,4]

[1] Racah Institute of Physics, Hebrew University, 91904 Jerusalem, Israel

[2] European Center for Theoretical Studies in Nuclear Physics and Related Areas, Strada delle Tabarelle 286, I-38050 Villazzano (Trento), Italy

[3] Dipartimento di Fisica, Università di Trento, I-38050 Povo, Italy

[4] Istituto Nazionale di Fisica Nucleare, Gruppo Collegato di Trento

Abstract. We present an improvement of the recently developed effective interaction method for the hyperspherical formalism (EIHH). The method is extended to include contributions of three-body effective forces and the non adiabatic kinetic energy term. It is shown that the rate of convergence is improved considerably, opening up the way for application of the method to realistic 6-body calculations.

1 Introduction

In the last decade quite a progress was made in microscopic calculations of few-body nuclei. In fact several theoretical approaches were developed and a few of them like the Green Function Monte Carlo (GFMC), the Stochastic Variational Method (SVM), the No Core Shell Model Method (NCSM) and the Effective Interaction in the Hyperspherical Harmonics basis (EIHH) [2] were even applied to systems with more than four nucleons. Very recently various of these methods were tested in a benchmark calculation for the α-particle ground state with a realistic NN force [4]. A very good agreement among the results of the different calculations was found. This shows the high standard of present day few-body physics calculations and opens up the possibility to tackle even more complicated microscopic ab initio calculations in the near future.

In particular the effective interaction approach is a very promising new tool in few-body physics. In this field it has been first introduced by Navratil and Barrett [1]. In the EIHH approach [2, 3] the division of the total HH

* *E-mail address:* nir@wiener.phys.huji.ac.il

space in P and Q spaces is realized via the HH quantum number K (P (Q) space: $K \leq (>) K_{max}$). The total Hamiltonian is written in hyperspherical coordinates,

$$H = T_\rho + T_K(\rho) + V(\Omega,\rho) = -\frac{\Delta_\rho}{2m} + \frac{\hat{K}^2}{2m\rho^2} + \sum_{i<j} v_{ij} \,, \tag{1}$$

where ρ is the hyperradius and Δ_ρ contains derivatives with respect to ρ only. The grand-angular momentum operator $\hat{K}^2$ is a function of the variables of particles A and $(A-1)$ and of $\hat{K}_{A-2}$, the grand angular momentum operator of the $(A-2)$ residual system. Then from the total Hamiltonian one can extract an adiabatic "two-body" Hamiltonian of particles A and $(A-1)$

$$\mathcal{H}^{[2]}(\rho) = T_K(\rho) + v_{A,(A-1)} \,, \tag{2}$$

which, however, contains the hyperspherical part of the total kinetic energy. Since the HH functions of the $(A-2)$ system are eigenfunctions of $\hat{K}^2_{A-2}$ one has an explicit dependence of $\mathcal{H}^{[2]}$ on the quantum number K_{A-2} of the residual system, i.e. $\mathcal{H}^{[2]} \to \mathcal{H}^{[2]}_{K_{A-2}}$. Applying the Lee-Suzuki similarity transformation [5] to $\mathcal{H}^{[2]}$, the effective interaction is obtained from

$$v^{[2]\,eff}_{K_{A-2}}(\rho) = U^\dagger(\rho)\mathcal{H}^{[2]}U(\rho) - T_K(\rho) = \mathcal{H}^{[2]\,eff}_{K_{A-2}}(\rho) - T_K(\rho) \,. \tag{3}$$

This $v^{[2]\,eff}(i,j)$ replaces v_{ij} in Eq. (1) when we project the solution on the P-space. This effective potential has the following property: $v^{[2]\,eff}(i,j) \to v_{ij}$ for $P \to 1$. Due to the "effectiveness" of the operator the solution of the Schrödinger equation converges faster to the true one. The HH formulation leads to various advantages: (i) $v^{[2]\,eff}$ itself is ρ dependent, therefore it contains some information on the "medium", (ii) because of the above mentioned K_{A-2} dependence the $(A-2)$ residual system is not a pure spectator, and (iii) an additional confining potential is not needed, since the presence of ρ in Eq. (2) automatically confines the two-body system to the range $0 \leq r_{A,(A-1)} < \sqrt{2}\rho$.

The aim of this paper is to present possible extensions of the original formulation of the EIHH method outlined above, consisting of: (i) an inclusion of the hyperradial kinetic energy and (ii) the incorporation of a three-body effective interaction.

2 Beyond the Adiabatic Approximation

In the EIHH approach outlined above the Lee-Suzuki similarity transformation is applied to the adiabatic *quasi two-body* hamiltonian of Eq. (2). The effective potential is obtained by subtraction of the hypercentrifugal term from the effective hamiltonian, Eq.(3). One can go beyond this adiabatic approach by applying $U(\rho)$ to the non adiabatic *quasi two-body* hamiltonian $\tilde{\mathcal{H}}^{[2]}(\rho) = T_\rho + \mathcal{H}^{[2]}(\rho)$. This leads to a non adiabatic effective *quasi two-body* hamiltonian

$$\tilde{\mathcal{H}}^{[2]\,eff} = U^\dagger(\rho)\,T_\rho\,U(\rho) + \mathcal{H}^{[2]\,eff}(\rho) = T_\rho + \Delta T^{eff}_\rho + \mathcal{H}^{[2]\,eff}(\rho) \,. \tag{4}$$

A new effective interaction is then obtained by subtracting the full kinetic energy

$$\tilde{v}^{[2]\,eff}_{A,A-1} = \tilde{\mathcal{H}}^{[2]\,eff} - T_\rho - T_K(\rho) = v^{[2]\,eff}_{A,A-1} + \Delta T^{eff}_\rho \ . \tag{5}$$

The new effective potential $\tilde{v}^{[2]\,eff}_{A,A-1}$ differs from the previous $v^{[2]\,eff}_{A,A-1}$ by the term $\Delta T^{eff}_\rho \approx \frac{\partial U^\dagger}{\partial\rho}\frac{\partial U}{\partial\rho}$, which can also be interpreted as an *effective kinetic energy.*

3 The Three-Body Effective Interaction

The natural way to construct the three-body effective interaction is the generalization of the procedure outlined above, namely: replace the adiabatic *quasi two-body* Hamiltonian of Eq. (2) with

$$\begin{aligned}\mathcal{H}^{[3]}(\rho) &= T_K(\rho) + v_{A,(A-1)} + v_{(A-1),(A-2)} + v_{(A-2),A} \\ &= T_K(\rho) + v^{[3]}(A, A-1, A-2)\,;\end{aligned} \tag{6}$$

diagonalize it and apply the Lee-Suzuki similarity transformation to obtain $v^{[3]\,eff}$. Then, solve the A-body problem with

$$H^{[A]\,eff} = T_\rho + T_K + \frac{1}{A-2}\sum_{i<j<k} v^{[3]\,eff}(i,j,k)\ . \tag{7}$$

The actual calculation of the three-body effective interaction is quite complicated and technically rather challenging. The difficulty has its origin in the separation of the A-body system into two subsystems (the interacting three- and the residual $(A-3)$-body systems). Expressing the A-body matrix elements of the interaction in terms of the matrix elements for the 3-body subsystems one has to pay the price of evaluating the overlap between the anti-symmetric A-body basis functions and basis functions constructed in the $(3, A-3)$ scheme. Another difficulty is the numerical task involved in the solution of the *quasi three-body* Hamiltonian Eq. (6). This task can be reduced if we replace the bare two-body interactions in Eq. (6) by effective ones.

4 Results and Discussion

At this point let us demonstrate the merits of applying the ideas presented in the previous sections to few-body calculations. In Fig. 1 we study the combined effect of the non-adiabatic correction and the 3-body effective interaction on the convergence rate for the ground state energy of ^{4}He with the Argon AV6 NN force. It can be seen that contrary to the 2-body effective interaction where the binding energy converges from below, the new method ($v^{[3]\,eff} + \Delta T_\rho$) oscillates around the correct value. This effect is mainly attributed to the non-adiabatic correction ΔT_ρ. Moreover, with these corrections one can estimate the ground state energy with accuracy of 100-200 keV even at very low values of K_{max}. These results open up the way for applying the EIHH method for p-shell nuclei with realistic interactions.

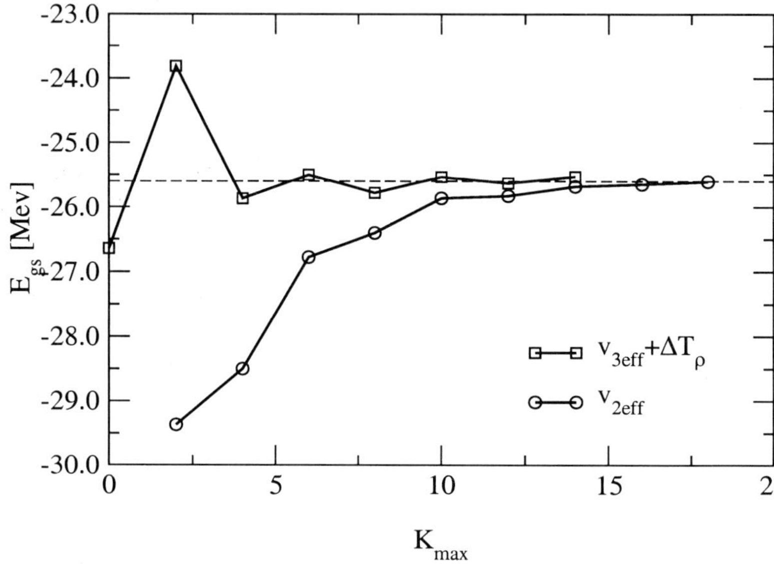

Figure 1. The convergence of the ground state energy for ^{4}He with the AV6 NN potential. The squares stand for non-adiabatic calculation with effective 3-body interaction, and the circles for adiabatic calculation with effective 2-body interaction. The dotted line stands for the best result obtained with the effective 2-body interaction.

References

1. P. Navrátil and B.R. Barrett: Phys. Rev. **C59**, 1906 (1999)
2. N. Barnea, W. Leidemann, and G. Orlandini: Phys. Rev. **C61**, 054001 (2000)
3. N. Barnea, W. Leidemann and G. Orlandini: Nucl. Phys. **A693**, 565-578 (2001)
4. H. Kamada et al.: Phys. Rev. **C64**, 044001 (2001)
5. K. Suzuki and S.Y. Lee: Progr. Theor. Phys. **64**, 2091 (1980)

Few-Body Systems Suppl. 14, 207–210 (2003)

Few-Body Systems

Rotational States of the Helium Trimer in the Symmetry-Adapted Hyperradial-Adiabatic Approach

A.V. Matveenko[1], E.O. Alt[2]

[1] Bogoliubov Laboratory of Theoretical Physics, Joint Institute for Nuclear Research, 141980 Dubna, Moscow, Russia

[2] Institut für Physik, Universität Mainz, D-55099 Mainz, Germany

Abstract. We have searched for bound rotationally excited states of the helium trimer using the symmetry-adapted hyperradial adiabatic approach. Since the calculated $J^p = 2^+$ and $J^p = 1^-$ potential curves are both completely repulsive, we infer that there are no bound rotational states of $^4He_3$. A recent adiabatic calculation [1] based on the direct solution of the Coriolis-coupled Schrödinger equation agrees with this conclusion.

1 Introduction

In the frequently used adiabatic approximation, the vibrational motion is treated so as to produce an effective potential that governs the hyperradial motion. The rotational degrees of freedom are then taken into account by adding the centrifugal term $\sim J(J+1)$ to the channel potentials and introducing rotational (Coriolis) couplings into conventional close-coupling calculations. Of course, a perturbative treatment of the rotational motion is justified only as long as the rotational energy is sufficiently small. If, however, the rotation is as energetic as the vibrational motion, both should be treated on the same footing in order to produce physically relevant adiabatic potentials.

Recently, the symmetry-adapted hyperradial-adiabatic (SAHA) approach has been developed as a new, efficient tool to investigate rotationally excited states of arbitrary three-body systems [2, 3]. As an application we investigate here rotational bound states of the $^4He_3$ trimer. Preliminary results were reported at the International RIKEN Conference "Muon Catalyzed Fusion and Related Exotic Atoms" (April 22-26, 2001, Shimoda, Japan) [4]. Later, a paper by Lee et al. [1] appeared in which the same problem was studied, in a similar manner and reaching the same conclusion. These combined results allow us to formulate as final conclusion that $^4He_3$ trimers do not possess bound rotationally excited states. This result is mainly due to quantum statistics.

It is worthwhile to recall some of the interesting properties of small ^{4}He clusters: the $^4He_2$ dimer is bound by only $E_d = 1.1$ mK [5], the $^4He_3$ trimer by $E_t^{(0)} \approx 130$ mK. In addition, at least one more bound state of total angular momentum $J = 0$, of binding energy $E_t^{(1)} \approx 2$ mK, has been predicted by several calculations using different methods (see [6] and references therein). This is of particular interest because the smallness of the dimer energy suggests that this excited trimer state may represent a first physical example of an Efimov state. Moreover, rotationally excited states of $^4He_3$, if they exist at all, should have binding energies of less than 1 mK [7].

2 Hyperspherical Approach

In hyperspherical coordinates, the symmetry-projected adiabatic Hamiltonian of three spinless particles reads, for states with fixed total angular momentum J and parity p, [2, 3]

$$\mathbf{H}_{\hat{\omega}}^{Jp}(R,\xi,\eta) = -\frac{1}{2M}\frac{1}{R^5}\frac{\partial}{\partial R}R^5\frac{\partial}{\partial R} + \mathbf{h}_{\hat{\omega}}^{Jp}(R;\xi,\eta), \tag{1}$$

$$\mathbf{h}_{\hat{\omega}}^{Jp}(R;\xi,\eta) = \frac{[\Lambda^2(\xi,\eta)]_{\hat{\omega}}^{Jp}}{2MR^2} + V. \tag{2}$$

It depends on the hyperradius R and two internal hyperspherical angles ξ, η, and is, for states with parity $p = (-)^J$, a $(J+1)\times(J+1)$ matrix operator. As the matrix structure of (1) (Coriolis couplings) arises only from the hyperangular momentum operator $[\Lambda^2(\xi,\eta)]_{\hat{\omega}}^{Jp}$, it can be resolved in a natural manner if variational primitives are chosen in the vector-column form

$$||\psi_l^{(i)}(R,\xi,\eta)||_{\hat{\omega}}^{Jp} = \mathcal{R}_l^{(i)}(R,\sin\alpha_i)||y_l^{(i)}||_{\hat{\omega}}^{Jp}; \quad i = a,b,c; \quad l = 0,\cdots J. \tag{3}$$

The functions $\mathcal{R}_l^{(i)}$ can be chosen rather arbitrarily. For a given pair angular momentum l, the intrinsic hyperspherical harmonics (IHH) are defined as

$$||y_l^{(i)}||_{\hat{\omega}}^{Jp} = \sin\alpha_i^L \cos\alpha_i^l \, \mathbf{d}^{Jp}(\omega_i) \begin{pmatrix} p_l^0(\theta_i) \\ \vdots \\ p_l^m(\theta_i) \\ \vdots \\ p_l^l(\theta_i) \\ 0 \\ \vdots \\ 0 \end{pmatrix}, \quad L = J - l, \; m = 0, \ldots J. \tag{4}$$

They are solutions of the matrix differential equation in two variables

$$\left([\Lambda^2(\xi,\eta)]_{\hat{\omega}}^{Jp} - J(J+4)\right) ||y_l^{(i)}||_{\hat{\omega}}^{Jp} = 0. \tag{5}$$

Here, for each channel i we have introduced Jacobi vectors $\mathbf{x}_i, \mathbf{y}_i$, with reduced masses $M_i^{-1} = m_j^{-1} + m_k^{-1}$, $\mu_i^{-1} = m_i^{-1} + (m_j + m_k)^{-1}$, $\{i,j,k\} = \{a,b,c\}$. Moreover, $\cos\omega_i = \hat{\boldsymbol{x}}_i \cdot \hat{\boldsymbol{\omega}}$, $\cos\theta_i = \hat{\boldsymbol{x}}_i \cdot \hat{\boldsymbol{y}}_i$, $\alpha_i = \arctan(\sqrt{M_i}x_i/\sqrt{\mu_i}y_i)$; $p_l^m(\theta)$ are specially normalized associated Legendre polynomials [2, 3], and $\mathbf{d}^{Jp}(\omega_i)$ is the parity preserving combination of the Wigner d-matrices. The subscript $\hat{\boldsymbol{\omega}}$ on the IHH (4) serves as a reminder that for any possible choice of channel index i we are in the same body-fixed coordinate frame, with the direction of the z'-axis being specified by the unit vector $\hat{\boldsymbol{\omega}}$. Finally, the hyperradius R with $M = M_c$ and $\mu = \mu_c$ is defined by $MR^2 = Mx^2 + \mu y^2$.

3 Results and Discussion

For the wave function of the $^4He_2$ bound state we employ basis functions of the form [8] $f_d(r_{ij}) = \exp\left[-(a/r_{ij}^5 + b/r_{ij}^2 + cr_{ij})\right]$ and fix eight sets of parameters such that, for the Aziz and Slaman LM2M2 potential [9], the energy value $E_d = 1.3$mK of the $^4He_2$ bound state is reproduced. With these parameters we perform our trimer calculations.

Using functions of type (3) the eigenvalues of the SAHA Hamiltonian (2) can easily be calculated. Clearly, as we deal with three identical bosons, the physical wave function for given J^p should be

$$||\Psi(R,\xi,\eta)||_{\hat{\omega}}^{Jp} = \sum_{i=a,b,c;\ l=0,\cdots J} c_l^{(i)} ||\psi_l^{(i)}(R,\xi,\eta)||_{\hat{\omega}}^{Jp}. \tag{6}$$

As can be seen from Fig. 1, we obtain completely repulsive adiabatic poten-

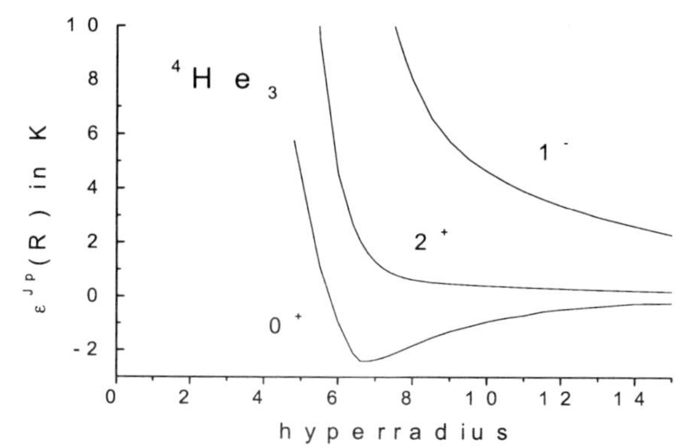

Figure 1. SAHA potential curves for the $^4He_3$ trimer with R in Å. Neglecting boson identity in the 1^- case we find three ground states of different "exchange symmetry", the third being the physical one shown above (see below).

tials for 1^- and 2^+ states. Interestingly enough, as a consequence of quantum statistics the $J^p = 2^+$ potential curve is even less repulsive than that for 1^-.

At this stage our results are in a complete agreement with those of [1]. In order to get deeper insight into the differences between these two, seemingly nearly identical adiabatic methods, we have performed a trial calculation with the same Hamiltonian (2) but without imposing the symmetrization condition (6). As a typical example we present the coefficients $c_l^{(i)}$ of the adiabatic expansion with the ansatz (3) for a hyperradius $R = 10$ Å:

$n = 1$	$n = 2$	$n = 3$	Jacobi channel
.02	4.62	1.07	a
−.02	4.62	1.07	b
.00	−9.25	1.07	c
−1.15	−0.66	−1.98	a
1.15	−0.66	−1.98	b
.00	1.33	−1.98	c
−4.14	−2.38	.08	a
4.14	−2.38	.08	b
.00	4.77	.08	c

Inspection shows immediately that only the third state ($n = 3$) is physical. In contrast, in [1] a forth solution has been identified as the one having the proper symmetry. This is due to different mathematics used in the two approaches (see [2] for details): we have summed out the Coriolis couplings and, thus, the magnetic quantum number disappeared from the analysis, while Lee et al. could not get rid of this approximate quantum number.

Acknowledgement. This work has been supported by the Heisenberg-Landau Program of the joint German-JINR collaboration.

References

1. T.G. Lee et al.: J. Phys. **B34**, L203 (2001)
2. A.V. Matveenko, E.O. Alt and H. Fukuda: Few-Body Syst. Suppl. **13**, 140 (2001)
3. A.V. Matveenko: Few-Body Syst. **30**, 259 (2001)
4. A.V. Matveenko and E.O. Alt: Preprint MZ-TH 01/17, to appear
5. R.E. Grisenti et al.: Phys. Rev. Lett. **85**, 2284 (2000)
6. A.K. Motovilov et al.: Eur. Phys. J. **D13**, 33 (2001)
7. L.W. Bruch: J. Chem. Phys. **110**, 2410 (1999)
8. M.V. Rama Krishna and K.B. Whaley: J. Chem. Phys. **93**, 6738 (1990)
9. R.A. Aziz and M.J. Slaman: J. Chem. Phys. **94**, 8047 (1991)

Few-Body Systems Suppl. 14, 211–214 (2003)

Few-Body Systems

Efficient Technique for Solving Few-Body Scattering Problems by the Wave-Packet Continuum Discretisation

O.A. Rubtsova*,†, V.I. Kukulin**

Institute of Nuclear Physics, Moscow State University, 119992, Leninskie gory, Moscow, Russia

Abstract. Brief description of the new general approach to solving few-body scattering problems on the L_2 basis is given. The formalism is based on the wave-packet continuum discretisation method which leads to purely algebraic scattering equations with non-singular kernels. Practical solving few-body collision problems by the method requires much less numerical cost as compared with a solution of an appropriate Faddeev–Yakubovsky integral equations.

1 Introduction

Despite of the impressive progress in recent years, it is well known that practical solution of few-body scattering equations meets very serious problems in many cases, e.g. for $N > 3$ (or even for $N = 3$, when the Coulomb interaction is fully included) or if the pairwise input interaction is non-hermitian etc. On the other hand, the treatment of bound state problems is quite feasible still for $N \sim 5 \div 11$ or even higher if one applies a Monte-Carlo technique. The reason of this is very complicated boundary conditions which should be satisfied in solution of many-body scattering problems. There are however a few techniques which reduce solving the scattering problems to solution of bound-state like problems, still incorporating the proper boundary conditions. One of the most efficient such technique is the wave-packet continuum discretisation (WPCD), which has been developed recently by the present authors [1,2]. The main advantage of this new method is its evident simplicity, universality and much wider scope of applicability then the conventional few-body techniques.

*This work was supported by the Russian Foundation for Basic Research (Grants NN. 01-02-16621, 02-02-06671).

† *E-mail address:* rubtsova@nucl-th.sinp.msu.ru

** *E-mail address:* kukulin@nucl-th.sinp.msu.ru

2 The Basic Formulation of the WPCD Method

Let us consider two-body scattering with central interaction potential V. The WPCD method consists of few steps: (i) Discretisation of the continuous spectrum of the free Hamiltonian H_0 (or of the total Hamiltonian H) by decomposing the full continuum onto finite (infinite in the limit) number of energy bins ($0 = \mathcal{E}_0 < \ldots < \mathcal{E}_i < \ldots < \mathcal{E}_N$) with widths $\Delta_i = \mathcal{E}_i - \mathcal{E}_{i-1}$. (ii) Construction of the stationary wave packets from the exact continuum eigenfunctions of the free, $|\Psi_0(E,l)\rangle$, and the total Hamiltonians, $|\Psi(E,l)\rangle$, in each bin:

$$|X_i\rangle = \frac{1}{\sqrt{\Delta_i}} \int_{\mathcal{E}_{i-1}}^{\mathcal{E}_i} |\Psi_0(E,l)\rangle dE, \quad |Z_i\rangle = \frac{1}{\sqrt{\Delta_i}} \int_{\mathcal{E}_{i-1}}^{\mathcal{E}_i} |\Psi(E,l)\rangle dE. \tag{1}$$

The 'free' packets (FP) $\{|X_i\rangle\}_{i=1}^N$ can be easily shown [2] to form an orthonormalized system while the 'interaction' packets (IP) $\{|Z_i\rangle\}_{i=1}^N$ form a orthonormalized set jointly with respective bound-state vectors $\{\Psi_{nl}^B\}_{n=1}^K$ of the Hamiltonian H. Thus one can construct the projection operators onto FB- and IP-subspaces:

$$P_F = \sum_{i=1}^{N} |X_i\rangle\langle X_i|; \quad P_I = \sum_{n=1}^{K} |\Psi_{nl}^B\rangle\langle\Psi_{nl}^B| + \sum_{i=1}^{N} |Z_i\rangle\langle Z_i|. \tag{2}$$

By their definitions the vectors $|X_i\rangle$ and $|Z_i\rangle$ represent eigenstates of the projected Hamiltonians $P_F H_0 P_F$ and $P_I H P_I$ respectively.[1] It is very important that the free Green operator $G_0^(\pm)(E) = [E + i0 - H_0]^{-1}$ is diagonal in this representation. We have derived for the projection of this resolvent $G_{0P}^{(\pm)}(E) \equiv P_F \hat{G}_0^{(\pm)}(E) P_F$ an accurate analytical formula [1,2]:

$$G_{0P}^{(\pm)}(E) = \sum_{i=1}^{N} \frac{|X_i\rangle\langle X_i|}{\Delta_i} \ln\left|\frac{\mathcal{E}_{i-1} - E}{\mathcal{E}_i - E}\right| \mp i\pi \frac{|X_k\rangle\langle X_k|}{\Delta_k}, \quad E \in (\mathcal{E}_{k-1}, \mathcal{E}_k). \tag{3}$$

It should be stressed that the (matrix) operator G_{0P} has a purely discrete spectrum, though its imaginary part has a proper discontinuity across positive real axis as function of energy. Such a convenient finite-dimensional representation for the resolvent operator is a crucial point of the WPCD method. Using further this representation and Lippmann–Schwinger equation for the t-operator it is easy to get a finite-dimensional representation for t-matrix both off- and on- the energy shell and also to derive discrete (matrix) counterparts for any operators in scattering theory like the Möller wave operators, S-matrix etc. (see the details in [2]). It should be stressed the discretised S-matrix satisfies the unitarity conditions because it can be considered formally as the exact one for the finite rank potential $V_P = P_F V P_F$. Some important practical realizations of the WPCD approach (as applied to non-local or non-hermitian complex valued interactions) have been published in ref. [2].[2]

---

[1] These projected Hamiltonians have discrete spectra only

[2] The developed formalism makes it possible to replace, in real calculations, the IP basis states $|X_i\rangle$ with eigenvectors of the H_0 matrix taken on some L_2 basis. This results in

2.1 WPCD Approach in the Three-Body Scattering

By generalizing the two-body wave-packet representation one can introduce the three-body wave packets for each channel Hamiltonian $H_\alpha = H_0 + V_\alpha$ (we treat here the three-body system with total Hamiltonian $H = H_0 + V_a + V_b + V_c$, where $V_a \equiv V_{bc}$ are the pairwise potentials). In particular, one can define WP in the channel 'a' as a superposition of two-body IP (in subsystem $\{bc\}$) and FP along the relative motion coordinate between $\{bc\}$ and the third particle a:

$$|S_a, \Lambda M\rangle \equiv |Z_i^{l_\alpha}, X_j^{L_\alpha}, \Lambda M\rangle, \quad S_a = (i, l_a, j, L_a).$$

Here l_a and L_a are orbital momenta in the subsystems $\{bc\}$ and $\{a - \{bc\}\}$ respectively, and $\boldsymbol{\Lambda} = \boldsymbol{l}_a + \boldsymbol{L}_a$ are the total three-body orbital momentum. It can be proven the properties of the above three-body WP are quite similar to those of IP and FP in two-body case. In particular, $|S_a, \Lambda M\rangle$ form the orthonormalized system. The projector P_a onto WP-subspace in a-channel is a superposition of two-body projectors $P_F^{L_a}$ and $P_I^{l_a}$. An extraordinary convenience of the three-body basis chosen is the fact that the exact formulas for 'mixed' resolvent projections $(G_a^{(+)}(E))_{ab} \equiv P_a G_a^{(+)}(E) P_b$ for the channel resolvent operators $G_a^{(+)}(E) = [(E + i0) - H_a]^{-1}$ can be found in fully analytical form:

$$(G_a^{(+)}(E))_{ab} \approx \sum_{S_a, S_b} |S_a, \Lambda M\rangle\langle S_b, \Lambda M|[G_a^{(+)}(E)]_{S_a}\langle S_b, \Lambda M|S_a, \Lambda M\rangle, \tag{4}$$

where $[G_a^{(+)}(E)]_{S_a} = \langle S_a, \Lambda M|G_a^{(+)}(E)|S_a, \Lambda M\rangle$ — are eigenvalues of the matrix projection of the channel resolvent, which is *diagonal* in its own channel WP basis. In particular, $G_a^{(+)}(E)$ can be taken as a convolution of two-particle subresolvent g_a (for $\{bc\}$ subsystem) and g_0 (the free resolvent for the relative motion $a - \{bc\}$), and thus can be decomposed in two parts: the bound-continuum (BC) and continuum-continuum (CC) ones). The decomposition corresponds to the bound state and continuum parts in g_a. We display here, for illustration, the analytical formula for the BC part of the channel resolvent:

$$[G_a^{(+)}]_{S_a}^{BC} = \frac{1}{\Delta_j} \ln \left| \frac{\mathcal{E}_{j-1}^{L_a} + E_n^{l_a} - E}{\mathcal{E}_j^{L_a} + E_n^{l_a} - E} \right| - \begin{cases} \frac{i\pi}{\Delta_j}, & E \in (\mathcal{E}_{j-1}^{L_a} + E_n^{l_a}, \mathcal{E}_j^{L_a} + E_n^{l_a}) \\ 0, & E \notin (\mathcal{E}_{j-1}^{L_a} + E_n^{l_a}, \mathcal{E}_j^{L_a} + E_n^{l_a}). \end{cases}$$

Here $E_n^{l_a}$ are the bound state energies in $\{bc\}$ subsystem, and $\mathcal{E}_j^{L_a}$ are points of the discretised continuum in the subsystem $a - \{bc\}$. The matrix elements for CC part of $G_a(E)$ can be also expressed in a fully analytical form. We emphasize here that despite of the fact the projections $(G_a^{(+)})_{ab}$ represent completely 'discretised' (i.e. matrix-like) operators their imaginary parts (which are responsible for the satisfaction of correct three-body unitarity conditions) have quite correct logarithmic energy dependence and correct discontinuity across the logarithmic cuts.

---

finite-dimensional representation for basic operators in scattering theory in *any* complete basis.

2.2 *Practical Solution of the Three-Body Scattering Problem*

The general three-body amplitude for the rearrangement process $a \to b$ has a form:

$$R^{\Lambda M}_{bm,an}(E) = \langle \Psi^{\Lambda M}_{bm}(E)|\bar{V}_b|\Psi^{\Lambda M}_{an}(E)\rangle + \langle \Psi^{\Lambda M}_{bm}(E)|\bar{V}_b G \bar{V}_a|\Psi^{\Lambda M}_{an}(E)\rangle, \quad (5)$$

in which the channel asymptotic wave functions ('in' and 'out') are taken as a superposition of bound states and plane waves (in the channel), and $\bar{V}_\alpha = H - H_\alpha$ — is an external interaction in the channel α. It is important here that the total three-body resolvent is truncated effectively in the coordinate space due to short-range potentials and bound-state wave functions in the above asymptotic wave functions. Hence, it is quite sufficient to find only the 'mixed' projection of the above total resolvent $G_{ba} \equiv P_b G P_a$. As a result of this substitution, one has no need to solve the exact Faddeev or Yakubovsky equations in our approach. Instead of this, we have derived (using various resolvent identities) a symmetric matrix equations for the different channel projections of the total few-body resolvent:

$$G_{\beta a}(E) = (G_a(E))_{\beta a} + \sum_{\gamma \neq \beta} (G_a(E))_{\beta\gamma} \hat{V}_\gamma G_{\gamma a}(E), \quad \beta = a, b, c. \quad (6)$$

Here the Born terms represent the wave packet projections of the channel resolvents. $\hat{V}_\alpha \equiv P^{\Lambda M}_\alpha V_\alpha P^{\Lambda M}_\alpha$ are projections of respective interaction operators (no matter they are local or non-local in our approach). Thus the practical solving the three-body scattering problem $a \to b$ includes: (i) calculation of the potential projections: $\hat{V}_a$, $\hat{V}_b$ and $\hat{V}_c$; (ii) finding the channel resolvent projections $(G_a)_{aa}$, $(G_b)_{ba}$ and $(G_c)_{ca}$ from (4); (iii) solving the algebraic system (6) for the total resolvent projection G_{ab} and (iv) straightforward calculation of the process amplitude by explicit formula:

$$R^{\Lambda M}_{bm,an}(E) = \frac{\langle S^0_b|\hat{U}_b + \hat{U}_b G_{ba}(E)\hat{U}_a|S^0_a\rangle}{\sqrt{\Delta^{\lambda_a}_{j_a}\Delta^{\lambda_b}_{j_b}}}, \text{ here } \hat{U}_\alpha = P^{\Lambda M}_\alpha \bar{V}_\alpha P^{\Lambda M}_\alpha.$$

3 Summary

We have developed practically efficient new approach for treatment of few-body scattering processes in continuum. The technique employs simple matrix equations for the channel projections of the total resolvent. This approach can be straightforwardly generalizated to numerous four- and many-body scattering problems still keeping its main advantages.

References

1. V.I. Kukulin, O.A. Rubtsova: Theor. Math. Phys. **130**, 54 (2002)

2. V.I. Kukulin, O.A.Rubtsova: Theor. Math. Phys. (in print).

Few-Body Systems Suppl. 14, 215–216 (2003)

Few-Body Systems

Microscopic Calculation of Inelastic Reaction of Six-Body Nuclear Systems

S. Bacca[1], M.A. Marchisio[1], N. Barnea[2], W. Leidemann[1], G. Orlandini[1]

[1] Dipartimento di Fisica, Università di Trento and INFN (Gruppo Collegato di Trento), I-38050 Povo (Trento), Italy

[2] The Racah Institute of Physics, The Hebrew University, 91904 Jerusalem, Israel

The inelastic response of a system is a fundamental physical quantity, since it contains important information about the dynamics of the system. A deeper understanding can be obtained studying in a microscopic way the responses of the system to different external excitations. We investigate the six-body nuclear systems, i.e. the ^{6}Li and ^{6}He nuclei, with respect to electromagnetic E1, C0 ad C2 excitations. To this aim we make use of the Lorentz Integral Transform (LIT) method [1] and of the recently developed Effective Interaction approach in the Hyperspherical Harmonics formalism (EIHH) [2].

In an inclusive process the response function is given by

$$R(\omega) = \int d\Psi_{\rm f} \left| \langle \Psi_{\rm f} | \, \hat{O} \, | \Psi_0 \rangle \right|^2 \delta(E_{\rm f} - E_0 - \omega), \tag{1}$$

where $|\Psi_{0/{\rm f}}\rangle$ and $E_{0/{\rm f}}$ denote wave function and energy of ground and final state, respectively, while $\hat{O}$ is the excitation operator. The extremely complicated microscopic calculation of the final state wave function can be avoided via the use of the LIT method, where the full final state interaction is properly taken into account. To this end, one firstly calculates the LIT of $R(\omega)$

$$L(\sigma_{\rm R}, \sigma_{\rm I}) = \int d\omega \frac{R(\omega)}{(\omega - \sigma_{\rm R})^2 + \sigma_{\rm I}^2} = \left\langle \widetilde{\Psi} | \widetilde{\Psi} \right\rangle, \tag{2}$$

where $|\widetilde{\Psi}\rangle$ is found as unique solution of the "Schrödinger-like" equation

$$(H - E_0 - \sigma_{\rm R} + {\rm i}\sigma_{\rm I})|\widetilde{\Psi}\rangle = \hat{O}|\Psi_0\rangle. \tag{3}$$

Since the right-hand side of the above equation is localized and the energy has an imaginary part $\sigma_{\rm I}$, one can apply bound-state techniques for its solution.

Once the Lorentz transform is obtained, one gets $R(\omega)$ by inversion of the transform. In solving Eq. (3) we use a wave function expansion in terms of Hyperspherical Harmonics. In order to speed up the convergence of this expansion we use an effective interaction, EIHH [2]. For the LIT we calculate directly the quantity $\langle\widetilde{\Psi}|\widetilde{\Psi}\rangle$ via the Lanczos algorithm [4].

In the following we discuss results for the response of the six-body nuclei to the dipole operator E1. We have performed our calculations using simple semirealistic interaction [5, 6]. Results for inclusive photoabsorption cross section of ^{6}Li and ^{6}He are shown in Fig. 1. While ^{6}Li has a single broad giant resonance, there are two well separated peaks for ^{6}He corresponding to the breakup of the neutron halo and of the α core, respectively. For ^{6}Li a similar structure is not found probably because the break-up into two three-body nuclei, ^{3}He and ^3H, fills the gap between the two peaks. In case of ^{6}He the corresponding reaction, $^6\text{He} \rightarrow {}^3\text{H}+{}^3\text{H}$, is not induced by the dipole operator (no dipole moment between two identical particles). The responses of these systems to the isoscalar monopole C0 and quadrupole C2 have also been studied.

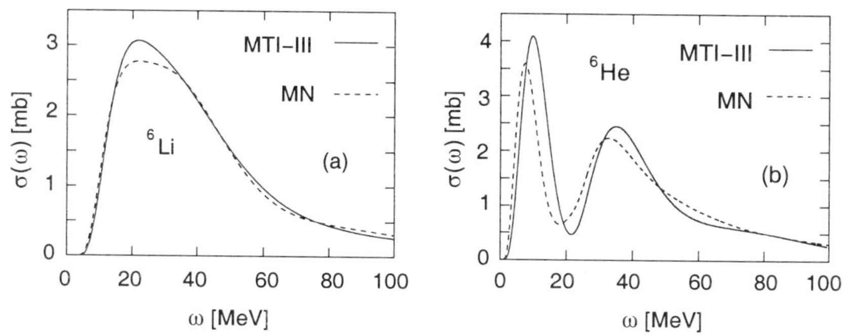

Figure 1. Six-body photoabsorption cross sections with MTI-III [5] and MN potentials [6]: ^{6}Li (a) and ^{6}He (b).

References

1. V.D. Efros, W. Leidemann, G. Orlandini: Phys. Lett. **B338**, 130 (1994)

2. N. Barnea, W. Leidemann, G. Orlandini: Phys. Rev. **C61**, 054001 (2000); Nucl. Phys. **A693**, 565 (2001)

3. S. Bacca, M.A. Marchisio, N. Barnea, W. Leidemann, G. Orlandini: Phys. Rev. Lett. **89**, 052502 (2002)

4. M.A. Marchisio, N. Barnea, W. Leidemann, G. Orlandini: nucl-th/0202009

5. R.A. Malfliet, J. Tjon: Nucl. Phys. **A127**, 61 (1969)

6. D.R. Thomson, M. LeMere, Y.C. Tang: Nucl. Phys. **A286**, 53 (1977); I. Reichstein, Y.C. Tang: Nucl. Phys. **A158**, 529 (1970)

Few-Body Systems Suppl. 14, 217–218 (2003)

Few-Body Systems

Variational Wave Equations for Fermions Interacting via Scalar and Vector Fields

J.W. Darewych[1], A. Duviryak[2]

[1] York University, Toronto, ON M3J 1P3, Canada

[2] Inst. of Condensed Matter Phys., Lviv, UA-79011, Ukraine

We consider a method for deriving relativistic two-body wave equations for fermions with scalar, pseudoscalar, vector, pseudovector and tensor coupling. The Lagrangian of the theory is reformulated by partially eliminating the mediating field by means of covariant Green's functions. The reformulated Lagrangian $\mathcal{L} = \sum_{a=1}^{2} \mathcal{L}_a^{\text{free}} + \mathcal{L}^{\text{int}}$ contains local free terms

$$\mathcal{L}_a^{\text{free}} = \bar{\psi}_a(x) \left\{ \mathrm{i}\, \gamma^\mu \partial_\mu - m_a \right\} \psi_a(x)$$

and nonlocal interaction terms of the type

$$\mathcal{L}^{\text{int}} = -\tfrac{1}{2} \int d^4x' J(x) D(x - x') J(x')$$

in which the mediating-field propagators $D(x - x')$ are the symmetric Green's functions, sandwiched between the fermionic particle currents $J(x) = \sum_a g_a \bar{\psi}_a(x) \Gamma \psi_a(x)$; here m_a are masses of fermions, g_a are coupling constants, and the matrices Γ depend on the tensor structure of the interaction (1 for scalar, γ_5 for pseudoscalar, γ_μ for vector, $\gamma_\mu\gamma_5$ for pseudovector and $\sigma_{\mu\nu}$ for tensor interactions).

The time-nonlocal term $\mathcal{L}^{\text{int}}$ is reduced approximately to the local expression by using the expansion series in a time deviation and keeping two lowest terms:

$$\mathcal{L}^{\text{int}} \approx -\tfrac{1}{2} \int d^3x' J(\boldsymbol{x}) G(r) J(\boldsymbol{x}') + \tfrac{1}{4} \int d^3x' \dot{J}(\boldsymbol{x}) Q(r) \dot{J}(\boldsymbol{x}');$$

here $G(r) = \int_{-\infty}^{\infty} d\lambda D(\lambda^2 - r^2)$, $Q(r) = \int_{-\infty}^{\infty} d\lambda D(\lambda^2 - r^2)\lambda^2$, $\dot{J} = \partial J/\partial t$ and $r = |\boldsymbol{r}| = |\boldsymbol{x} - \boldsymbol{x}'|$. Using the double-zero method [1], we eliminate the time derivatives of fields from $\mathcal{L}^{\text{int}}$. Then, performing an appropriate change of variables we reduce the Lagrangian $\mathcal{L}$ to the canonical form and thus arrive at the equal-time Hamiltonian formalism. Finally we perform the canonical quantization of the theory.

Two-fermion variational states

$$|2\rangle = \int d^3x\, d^3y\, F_{\alpha\beta}(\boldsymbol{x},\boldsymbol{y})\psi^\dagger_{1\alpha}(\boldsymbol{x})\psi^\dagger_{2\beta}(\boldsymbol{y})|\tilde{0}\rangle,$$

(summation on $\alpha, \beta = 1, 2, 3, 4$ implied) are constructed by employing an unconventional "empty" vacuum state $\psi_{a\alpha}(\boldsymbol{x})|\tilde{0}\rangle = 0$. Relativistic two-fermion wave equations are then derived by means of the variational principle

$$\delta\langle 2|\hat{H} - E|2\rangle = 0,$$

where $\hat{H}$ is the Hamiltonian operator of the QFT. The resulting two-fermion eigenvalue equation has the structure,

$$\{h_1(\boldsymbol{x}) + h_2(\boldsymbol{y}) + U(\boldsymbol{x},\boldsymbol{y}) + V(\boldsymbol{x},\boldsymbol{y}) - E\}\, F(\boldsymbol{x},\boldsymbol{y}) = 0,$$

where $h_a(\boldsymbol{x}) = -\mathrm{i}\,\boldsymbol{\alpha}_a \cdot \boldsymbol{\nabla}_x + m_a\beta_a$, $U(\boldsymbol{x},\boldsymbol{y}) = g_1g_2G(r)\beta_1\Gamma_1 * \Gamma_2\beta_2$ ($*$ denotes the convolution over tensor indexes of Γ), and $V(\boldsymbol{x},\boldsymbol{y})$ has rather cumbersome form. In the case of scalar interaction:

$$\begin{aligned} V_{\mathrm{s}}(\boldsymbol{x},\boldsymbol{y}) &= \tfrac{1}{2}g_1g_2\beta_1\beta_2\,\{Q_{\mathrm{s}}(r)h_1(\boldsymbol{x})h_2(\boldsymbol{y}) + h_1(\boldsymbol{x})h_2(\boldsymbol{y})Q_{\mathrm{s}}(r) \\ &\quad + h_1(\boldsymbol{x})Q_{\mathrm{s}}(r)h_2(\boldsymbol{y}) + h_2(\boldsymbol{y})Q_{\mathrm{s}}(r)h_1(\boldsymbol{x})\}\,; \end{aligned}$$

here the action of the operators β_a and Γ_a ($a = 1, 2$) is defined as follows: $\beta_1F = \beta F$, $\beta_2F = F\beta^T$ etc. In the case of the electromagnetic (vector) interaction we obtain the Breit equation. The superposition of scalar, pseudoscalar, vector, pseudovector and tensor interactions is described by the sum of corresponding potentials U and V. Details will be presented in a forthcoming article.

The application of this approach to the Wick-Cutkosky and Yukawa models and to QED is discussed in [2, 3, 4] and [5], respectively. The generalization to three-fermion systems is straightforward. The use of the more realistic Dirac vacuum, rather than the "empty" vacuum, is under study.

References

1. B.M. Barker, R.F. O'Connell: Phys. Lett. **A78**, 231 (1980)

2. M. Barham, J. Darewych: J. Phys. **A31**, 3481 (1998)

3. J. Darewych: Condensed Matt. Phys. **1**, 593 (1998)

4. V. Shpytko, J. Darewych: Phys. Rev. **D64**, 045012 (2001)

5. J. Darewych, A. Duviryak: Phys. Rev. **A66**, 032102 (2002)

Few-Body Systems Suppl. 14, 219–220 (2003)

Few-Body Systems

Three-Body Bound State Calculations Without Explicit Partial Wave Decomposition

M.L. Lekala*, M. Braun, S.A. Sofianos

Department of Physics, University of South Africa, Pretoria, South Africa

1 Introduction

The problem of calculating three-body bound state properties has been tackled by several authors for a considerable time. In the usual approach the partial wave decomposition of the solution is sought, resulting in the reduction of the Faddeev equations to a set of two-dimensional coupled equations, which are easily amenable to numerical solution. This approach yields results which are in good agreement with experiment if only a few partial waves are necessary to describe the system, as for example in nuclear systems. However, for weakly bound systems interacting via a force having a very strong repulsive core it would be necessary to include lots of partial waves resulting in an intractable numerical problem.

The method of solving the three-dimensional Faddeev equations in configuration space directly without resorting to partial wave decomposition of the solutions, was first proposed by Kostrykin *et. al.* [1]. In this work we present an algorithm based on the framework of this method to calculate, as a test, the binding energy for triton using the Malfliet-Tjon two-body potential.

2 Theoretical Tools and Numerical Methods

The $\mathbf{L} = 0$ three-particle configuration space Faddeev equations for bound states reads

$$(-\nabla_x^2 - \nabla_y^2 + V_i(\mathbf{x}_i) - E)\psi_i(\mathbf{x}_i, \mathbf{y}_i) = -V_i(\mathbf{x}_i)\sum_{j\neq i}\psi_j(\mathbf{x}_j, \mathbf{y}_j) \ (i,j = 1,2,3) \quad (1)$$

The detailed structure of these equations depends on the nature of the system considered. For example, for three identical particles, it can be shown that the

* *E-mail address:* lekala@science.unisa.ac.za

projected Faddeev equations for the amplitudes take the form

$$(H_0 + V(x) - E)\psi_1^0(x, y, z) = -V(x)P\psi_1^0(x, y, z) \tag{2}$$

where $P = P^+ + P^-$ is the permutation operator, $z = \cos\theta$ is the cosine of the angle between the Jacobi vectors $\mathbf{x}$ and $\mathbf{y}$. Expanding $\psi \equiv \psi_1^0$ in a spline basis in all three variables leads to

$$\psi(x, y, z) = \sum_{l=1}^{L} \sum_{m=1}^{M} \sum_{n=1}^{N} \mathbf{a}_{lmn} S_l(x) S_m(y) S_n(z) \,. \tag{3}$$

Substitution of Eq. (3) into Eq. (2) and use of the orthogonal collocation method of ref. [3] yield an algebraic equation for the expansion coefficients

$$\sum_{lmn} (H_{ijk,lmn} - E I_{ijk,lmn}) a_{lmn} = \sum_{lmn} C_{ijk,lmn} a_{lmn} \,. \tag{4}$$

The matrices have a tensor structure, and thus the tensor trick [2] can be used to handle them.

3 Results and Conclusion

Table 3.1 summarizes the results using the MTI-III, and its variants. The results in brackets are those of [2]. Convergence was reached with typical grid size of $16 \times 16 \times 6$ in x, y and z respectively.

Table 3.1. Ground state binding energy for ^{3}H (in MeV) using different potentials.

MTI-III	Zabolitzky	Friar
-8.53160	-8.2706	-7.7804
(-8.53578)	(-8.25399)	(-7.7221)

The calculated results agree with the literature values. We mention also that the method produces results in good agreement with literature for the cluster nuclear system ^{6}Li and ozone molecule $^{16}O_3$. The results for $^{16}O_3$ are reported elsewhere in these proceedings; and for ^{6}Li in a forthcoming paper.

References

1. V.V. Kostrykin, A.A. Kvitsinsky, S.P. Merkuriev: Few-Body Syst. **6**, 97 (1989)

2. N.W. Schellingerhout: *Factorizability in the Numerical Few-Body Problem*, PhD thesis, University of Gröningen, (1995) [and references therein]

3. G.L. Payne: *Lecture Notes in Physics* **273**, 64 (1987)

Few-Body Systems Suppl. 14, 221–222 (2003)

Few-Body Systems

Coulomb Fourier Transformation: Application to a Three-Body Hamiltonian with One Attractive Coulomb Interaction

S.B. Levin[1,2,*], E.O. Alt[3], S.L. Yakovlev[2]

[1] SCFAB, Stockholm University, 10691 Stockholm, Sweden

[2] V.A. Fock Institute for Physics, St. Petersburg University, 198904 St. Petersburg, Russia

[3] Institut für Physik, Universität Mainz, D-55099 Mainz, Germany

Consider a three-body system consisting of one neutral particle 1 and two charged particles characterized by the indices 2 and 3 with charges of opposite sign, i.e., $e_2e_3 < 0$. We use the following notation: $(\mathbf{x}_\nu, \mathbf{y}_\nu)$, $\nu = 1, 2, 3$, denotes the (mass-renormalized) coordinate vector within the pair ν, and between the center of mass of the pair ν and particle ν, respectively. The corresponding canonically conjugate momenta are $(\mathbf{k}_\nu, \mathbf{p}_\nu)$.

The Hamiltonian of the three-body system is $H = H_0 + V_1^C + \sum_{\nu=1}^{3} V_\nu^S$. In the coordinate space representation the free Hamiltonian is $H_0 = -\Delta_{\mathbf{x}_\nu} - \Delta_{\mathbf{y}_\nu}$, $\nu = 1, 2, 3$. Moreover, $V_\nu^S(\mathbf{x}_\nu)$ denotes the short-range two-body potential within the pair ν which for simplicity we assume to be local, and $V_1^C(\mathbf{x}_1) = e_2e_3/x_1$ the Coulomb potential between the charged particles 2 and 3.

Denote by $\{\psi_d^C(\mathbf{x}_1)\} \equiv \{\psi_{n,l,m}^C(\mathbf{x}_1)\}$ the complete set of eigenstates belonging to the discrete spectrum of $H_1^C = -\Delta_{\mathbf{x}_1} + V_1^C(x_1)$ where n, l, and m are the main, the orbital and the magnetic quantum number, respectively, and by $\{\psi_c^C(\mathbf{x}_1)\} \equiv \{\psi_{\mathbf{k}_1}^C(\mathbf{x}_1)\}$ those belonging to the continuous spectrum, i.e.,

$$H_1^C\psi_{n,l,m}^C(\mathbf{x}_1) = E_n\psi_{n,l,m}^C(\mathbf{x}_1),\ H_1^C\psi_{\mathbf{k}_1}^C(\mathbf{x}_1) = \mathbf{k}_1^2\psi_{\mathbf{k}_1}^C(\mathbf{x}_1),\ E_n = -e_2e_3/4n^2. \tag{1}$$

Moreover, let $\varphi_{\mathbf{p}_1}(\mathbf{y}_1) = (2\pi)^{-3/2}e^{i\mathbf{y}_1\cdot\mathbf{p}_1}$ represent the plane wave describing the free motion of particle 1 relative to the center of mass of the pair $(2, 3)$.

The set $\{\psi_d^C, \varphi_{\mathbf{p}_1}\} \cup \{\psi_{\mathbf{k}_1}^C, \varphi_{\mathbf{p}_1}\}$ represents a basis in three-particle space which is complete, orthogonal, and normalized (partially to δ-functions). Thus,

*Supported by Deutscher Akademischer Austauschdienst and a US DOE Grant of the Harvard-Smithsonian Center for Astrophysics

operators are represented by matrix elements, to be called "Coulomb-Fourier" ($\mathcal{CF}$) transforms, which are conveniently arranged in matrix form. Specifically, the $\mathcal{CF}$ transform of the three-body Hamiltonian H appears as

$$\underline{\mathcal{H}} \equiv \begin{pmatrix} \mathcal{H}^{(d'd)} & \mathcal{H}^{(d'c)} \\ \mathcal{H}^{(c'd)} & \mathcal{H}^{(c'c)} \end{pmatrix}.$$

The elements $\mathcal{H}^{(d'd)}$ between the discrete states (1) are easily seen to be of the form $\mathcal{H}^{(d'd)} := \langle \psi_{d'}^C \varphi_{\mathbf{p}_1'} | H | \psi_d^C \varphi_{\mathbf{p}_1} \rangle = \delta_{d'd} \delta(\mathbf{p}_1' - \mathbf{p}_1) \mathcal{E}_d + \sum_{\nu=1}^3 \mathcal{V}_\nu^{(d'd)}$, while those between discrete-continuum and continuum-continuum states are $\mathcal{H}^{(d'c)} := \langle \psi_{d'}^C \varphi_{\mathbf{p}_1'} | H | \psi_{\mathbf{k}_1}^C \varphi_{\mathbf{p}_1} \rangle = \sum_{\nu=1}^3 \mathcal{V}_\nu^{(d'c)} = \mathcal{H}^{(cd')*}$, and $\mathcal{H}^{(c'c)} := \langle \psi_{\mathbf{k}_1'}^C \varphi_{\mathbf{p}_1'} | H | \psi_{\mathbf{k}_1}^C \varphi_{\mathbf{p}_1} \rangle = \delta(\mathbf{k}_1' - \mathbf{k}_1) \delta(\mathbf{p}_1' - \mathbf{p}_1) \mathcal{E}_c + \sum_{\nu=1}^3 \mathcal{V}_\nu^{(c'c)}$. Here, we have introduced the notation $\mathcal{E}_d = E_n + \mathbf{p}_1^2$ with $d \equiv (nlm)$, and $\mathcal{E}_c = \mathbf{k}_1^2 + \mathbf{p}_1^2$.

The following observations can be made:

1. Any explicit appearance of the long-range Coulomb potential has disappeared in $\underline{\mathcal{H}}$. Hence, $\underline{\mathcal{H}}$ is an ordinary (matrix) Hamiltonian to which three-body short-range scattering approaches can be applied.

2. All Coulomb effects resides in the $\mathcal{CF}$-transformed short-range potential matrices $\underline{\mathcal{V}}_\nu$. They are found to be of two types. The one which acts between the charged particles (i.e., $\underline{\mathcal{V}}_1$) simply consists of the short-range potential V_1 in the "Coulomb representation", i.e., taken between the wave functions (1), multiplied by $\delta(\mathbf{p}_1' - \mathbf{p}_1)$. The other two which act between one of the charged and the neutral particle ($\underline{\mathcal{V}}_\nu$, $\nu = 2, 3,$) are given as the Fourier-transformed short-range potentials V_ν multiplied by a function matrix $\underline{\mathcal{L}}$ which generally is of tree-body type. It, thus, represents the well-known long-range three-body correlations resulting from the action of the other charged particle on the considered pair ν [2].

3. The matrix $\underline{\mathcal{L}}$ which is of the same structure as $\underline{\mathcal{H}}$, has been analyzed in detail. The most important property is its "universality", i.e., all elements are independent of the short-range potentials and, thus, only need to be calculated once and for all. This has been accomplished. The element $\mathcal{L}^{(c'c)}$ which is the most singular one, and which is in fact the only one if the Coulomb potential is repulsive, has already been described in [1]. From the analytical expressions for the elements $\mathcal{L}^{(d'd)}$ and $\mathcal{L}^{(cd)} = \mathcal{L}^{(dc)*}$ interesting results can be derived as, e.g., their behavior as function of the principal quantum numbers n and n'. In particular, $\mathcal{L}^{(d'd)}$ has been found to decrease along the diagonal ($n = n'$) as n^{-3}, and along the sides of the matrix ($n \gg n'$) as $n^{-3/2}$.

References

1. S.B. Levin, E.O. Alt, and S.L. Yakovlev: to be published

2. E.O. Alt and A.M. Mukhamedzhanov: JETP Lett. **56**, 435 (1992); Phys. Rev. **A47**, 2004 (1993)

Few-Body Systems Suppl. 14, 223–228 (2003)

Few-Body Systems

Small Clusters Made of Helium Atoms

R. Guardiola[1,*], J. Navarro[2]

[1] Departamento de Física Atómica, Molecular y Nuclear, Universidad de Valencia, Avda. Dr. Moliner, 50, 46100-Burjassot, Spain

[2] IFIC, CSIC-Universidad de Valencia, Apdo. 22085, E-46.071-Valencia, Spain

Abstract. Helium atoms interact very weakly through a van der Waals potential. Nevertheless, they are able to form aggregates or drops with a small number of atoms. This work analyzes the stability of clusters made of ^{4}He atoms, of bosonic nature, clusters made of ^{3}He atoms, of fermionic nature and also mixed aggregates with both kinds of constituents. Some of these drops are predicted to be unstable.

1 Introduction

Microscopic studies of the ground state properties of systems made of helium atoms began in 1965 with the pioneering work of McMillan [1] describing properties of liquid ^{4}He. The study was variational, based in a trial wave function of Jastrow type, and actual calculations were carried out by means of the Monte Carlo method. This work used a Lennard-Jones form for the He-He interaction, with parameters adjusted by deBoer and Michiels [2] to properties of the gas above 60 K, and the most remarkable result was that the obtained energy per particle was very close to the experimental value of 7.20 K. It was soonly recognized the need of improving the He-He interaction, and for many years the preferred form was the ab initio potential of Aziz and coworkers [3]. Later on, this potential was improved [4], giving rise to the so-called *new Aziz* or HFD-B(HE) potential, as well as other new forms of the interaction potential [5, 6].

Almost twenty years later [7] came the study of ^{4}He droplets, i.e., aggregates of a finite and rather small number of ^{4}He atoms. The main interest of this study was to establish a *mass formula*, analogous to the Weizsäcker mass formula of nuclei, with the typical terms of *volume*, *surface* and *curvature*. The conclusion

*E-mail address: Rafael.Guardiola@uv.es

found in this work is that all clusters made of ^{4}He atoms are stable, the energy per particle growing steadily with the number of atoms. This study was made by using the Green Function Monte Carlo method, mainly developed by M. Kalos and coworkers [8], which is one method of integrating stochastically the many-body Schrödinger equation. Modern studies of ^{4}He drops may be found in [9, 10, 11].

It is worth mentioning that the ^{4}He dimer is a slightly bound state, its binding energy being around $1.3 - 1.6$ mK. Very recently the dimer has been experimentally detected [14], and its binding energy has been estimated to be close to the theoretical predictions.

Systems made of ^{3}He are more difficult to study theoretically, the main reason being the need of properly describe their fermionic nature. For the extended system (the liquid), there exist highly reliable calculations of its properties (see, e.g. [12]) in full agreement with the experimental determination [13] of the binding energy per particle and other properties. The binding energy is much smaller that the binding energy of liquid ^{4}He, as a result of the combined effects of a smaller mass (and the corresponding increase of kinetic energy) and the Pauli exclusion principle. With regard to finite systems there are only few microscopic calculations [15, 16, 17] as well as some analysis made with the use of density functional theories [18, 19, 20, 21]. For our purposes the main result of these analysis is the requirement of a minimum number of ^{3}He atoms to have a bound state, this number being between 34 [16] and 29 [21]. Such small fermionic drops have not yet been detected.

2 Mixed Drops

As it was mentioned above, the binding energy of the bosonic dimer is very small. Just by solving the two-body Schrödinger equation for the ^{4}He–^{3}He system one realizes that the most elementary mixed drop is not bound. The reason is the 25% mass difference between ^{3}He and ^{4}He, and the corresponding growing of the kinetic energy of the mixed dimer with respect to the pure ^{4}He dimer. Of course, the pure ^{3}He dimer is again not bound. One may ask oneself what happens after adding another ^{4}He atom, and another, and so on: knowing the rapid increase of the binding energy with the number of bosons one expects to have bound states made of a single ^{3}He atom and a number N_B of ^{4}He atoms.

These systems have been studied by several microscopic methods [22, 23, 24] and it has been established theoretically that with $N_B \geq 2$ and one single atom of ^{3}He all systems are bound. Experimental diffractive studies by Schöllkopf and Toennies [25] have established the existence of such drops with a small number of bosons and a single ^{3}He atom.

Calculations carried out within the framework of density functional theories [26] predict the existence of mixed drops with large values of ^{4}He and ^{3}He atoms, with an interesting geometrical structure: basically the bosons form a core whereas the fermions are distributed in the surface of the drop. It has been shown that in correspondence what happens in the liquid, ^{3}He atoms form Andreev states in the surface of the drop.

Considering together these facts and the peculiarities of pure fermionic drops one should naturally ask the question of what happens when a small bosonic drop, i.e., a drop with two, three ... atoms, is doped with an increasing number of ^{3}He atoms. One may expect to obtain bound states while the number of fermions is small, but afterwards the system could become unstable. However, if a sufficiently large number of ^{3}He atoms is added there should form again stable systems. This is the question addressed in this paper.

3 A Family of Variational Wave Functions

We will describe the system of N_B bosons and N_F fermions by means of a Jastrow correlated configuration interaction wave functions. This kind of trial functions were successfully applied to the description of pure bosonic drops [11], producing values for the ground state energies close to those obtained with the (statistically) exact Diffusion Monte Carlo calculation. These functions must be adapted to the mixture of atoms.

The trial wave function has three pieces. One piece is the *model* or *reference* wave function, whose objective is to incorporate the boson/fermion statistics as well as the confining properties. In our case we have used the form

$$\Psi_{\text{Ref}} = \prod_{i<j}^{N_B} e^{-\alpha_B r_{ij}} \prod_{m<n}^{N_F} e^{-\alpha_F r_{mn}} \prod_{i=1}^{N_B} \prod_{m=1}^{N_F} e^{-\alpha_M r_{im}} \times D_\uparrow \times D_\downarrow \, , \tag{1}$$

where indices $i, j \ldots$ refer to bosons, and $m, n, \ldots$ to fermions. The labels B, F and M refer to bosons, fermions and mixed. The parameters α_X fix the inverse sizes of the three parts of the wave functions and should be adjusted variationally. In addition there are the Slater determinants $D_\uparrow$ and $D_\downarrow$ corresponding to the ^{3}He atoms with spin up and down, respectively. These Slater determinants are constructed with the polynomial part of the harmonic oscillator potential and are translationally invariant. Harmonic oscillator orbitals are filled up within each shell by requiring the maximum allowable spin (the atomic Hund's rule). Moreover, the parts associated to fermions are corrected so as to include the Feynman-Cohen backflow [27] by means of the coordinate transformation introduced by Schmidt and coworkers [28].

The second piece is a Jastrow factor, for which we have assumed the simple McMillan form,

$$\Psi_J = \prod_{i<j}^{N_B} \exp\left(-\frac{b_B^\nu}{r_{ij}^\nu}\right) \prod_{m<n}^{N_F} \exp\left(-\frac{b_F^\nu}{r_{mn}^\nu}\right) \prod_{i=1}^{N_B} \prod_{m=1}^{N_F} \exp\left(-\frac{b_M^\nu}{r_{im}^\nu}\right) \, , \tag{2}$$

with $\nu = 5.2$ and the three parameters b_X are again variational parameters.

Finally there is the configuration interaction (CI) part, which has been taken in coordinate space including up to $2p - 2h$ excitations in an explicit translationally invariant form

$$\Psi_{CI} = 1 + \sum_{i<j}^{N_B+N_F} f(r_{ij}) \,. \tag{3}$$

As in our previous work the CI correlation function f has been expanded in a set of exponentials

$$1 + \sum_{i<j} f(r_{ij}) = \sum_{p=1}^{N_\beta} C_p E_p, \tag{4}$$

with

$$E_p = \sum_{i<j}^{N_B+N_F} \mathrm{e}^{-\beta_p r_{ij}} \,. \tag{5}$$

It is convenient to fix one of the parameters $\{\beta_p\}$, say β_1, to zero, in order to explicitly consider the term without CI correlations in the left hand side of Eq. (4). Our previous experience with nuclear systems as well as with ^{4}He and ^{3}He drops indicates that the selection of the set of parameters $\{\beta_p\}$ is not very critical, as far as a sufficiently large interval of length ranges $(1/\beta_p)$ is included. In the calculations presented below we have used the set $\{\beta_p\} = \{0, 1, 2, 4, 6, 8\} \times N_{\mathrm{cons}} \times \min(\alpha_B, \alpha_F, \alpha_M)$.

The configuration interaction part depends of a set of amplitudes C_p which are determined, for fixed values of the parameters α_X and b_X, by solving a generalized eigenvalue problem with hamiltonian and norm matrices defined by

$$\begin{aligned} \mathcal{H}_{pq} &= \langle \Psi_{\mathrm{Ref}} \Psi_J | E_p H E_q | \Psi_J \Psi_{\mathrm{Ref}} \rangle, \\ \mathcal{N}_{pq} &= \langle \Psi_{\mathrm{Ref}} \Psi_J | E_p H E_q | \Psi_J \Psi_{\mathrm{Ref}} \rangle. \end{aligned} \tag{6}$$

Finally, these matrices are obtained by means of a Monte Carlo random walk controlled by the probability function $|\Psi_{\mathrm{Ref}} \Psi_J|^2$.

4 Results

The calculated energies for systems with 2, 3, 4 and 8 bosons plus up to 20 fermions are plotted in Fig. 1. Data represented by diamonds correspond to truly bound states, whereas those plotted with an open circle are metastable states, with total energy higher than the system which has less fermions. Even if their energy is negative, they do not correspond to truly bound states and, in practice, they are spurious results of the variational method.

We briefly resume the main characteristics of our study:

1. Systems with two bosons may bind up to two fermions. Afterwards the system is unstable until at least 18 fermions are added.

2. There is a clear *shell effect* in the cases of few bosons (2, 3 and 4), resulting in instabilities after filling the 1s shell and the 1p shell.

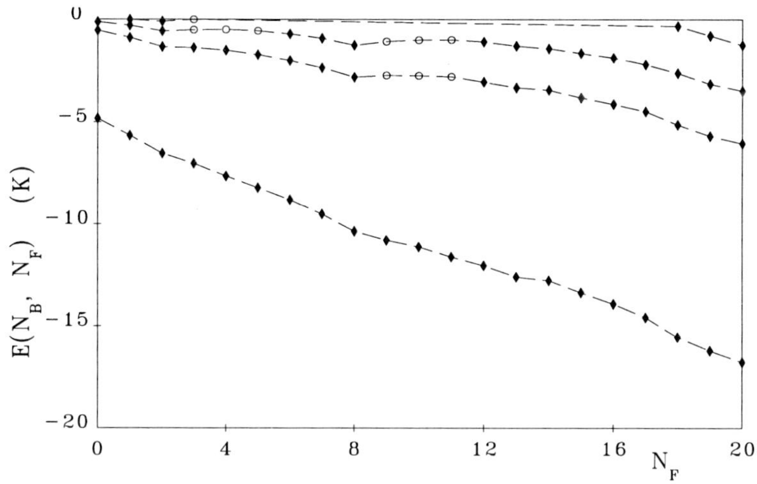

Figure 1. Binding energies of mixed clusters as a function of N_F. From top to bottom, N_B= 2, 3, 4 and 8. Diamonds indicate truly bound states, and open circles correspond to states which energy is negative, but higher than the system with a fermion less. Dashed lines serve just to connect data with the same values of N_B.

3. One may observe small kinks around the middle of the shells, corresponding to $N_F = 6$ and $N_F = 14$, related to the end of putting all spins parallel within the active ($1p$ and $2s - 1d$, respectively) shells (Hund's rule).

In conclusion, there are regions corresponding to few bosons (up to four) and 3, 4 and 5 fermions, as well as 9, 10 and 11 fermions which appear to be unstable. The results are however variational and could change when using a more flexible form for the fermionic wave function or when using other parameterizations of the He-He interaction. Thus, we encourage further theoretical work and, certainly, we hope that these results will stimulate additional experimental work.

This work has been supported by MCyT/FEDER (Spain), grant number BFM2001-0262, and Generalitat Valenciana, grant number GV01-216. One of us (RG) acknowledges the organizers of the *XVIIIth European Conference in Few-Body Problems in Physics* for financial support.

References

1. W.L. McMillan: Phys. Rev. **138**, A442 (1965)
2. J. deBoer and A. Michiels: Physica **5**, 945 (1938) and **6**, 97 (1939)
3. R.A. Aziz, V.P.S. Nain, J.S. Carley, W.L. Taylor and G.T. McConville: J. Cehm. Phys. **70**, 4330 (1979)
4. R.A. Aziz, F.R. McCourt and C.C.K. Wong: Mol. Phys. **61**, 1487 (1987)

5. R.A. Aziz and M.J. Slaman: J. Chem. Phys. **94**, 8047 (1991)
6. K.T. Tang, J.P. Toennies and C.L. Yiu: Phys. Rev. Lett. **74**, 4571 (1995)
7. V.R. Pandharipande, J.G. Zaboliztky, S.C. Pieper, R.B. Wiringa and U. Helmbrecht: Phys. Rev. Lett. **50**, 1676 (1983)
8. D.M. Ceperley and M.H. Kalos: In: *Monte Carlo Methods in Statistical Physics*, K. Binder (ed.). Berlin: Springer, 1979
9. K.B. Waley: Int. Rev. Phys. Chem. **13**, 41 (1994)
10. M. Lewerenz: J. Chem. Phys. **106**, 4596 (1997)
11. R. Guardiola, M. Portesi and J. Navarro: Phys. Rev. **B60**, 6288 (1999)
12. J. Casulleras and J. Boronat: Phys. Rev. Lett. **84**, 3121 (2000)
13. R. De Bruyn Ouboter and C.N. Yang: Physica **B44**, 127 (1987)
14. R.E. Grisenti, W. Schöllkopf, J.P. Toennies, G.C. Hegerfeldt, T. Köhler, and M. Stoll: Phys. Rev. Lett. **85**, 2284 (2000)
15. V.R. Pandharipande, S.C. Pieper and R.B. Wiringa: Phys. Rev. **B34**, 4571 (1986)
16. R. Guardiola and J. Navarro: Phys. Rev. Lett. **84**, 1144 (2000)
17. R. Guardiola: Phys. Rev. **B62**, 3416 (2000)
18. S. Stringari and J. Treiner: J. Chem. Phys. **87**, 5021 (1987)
19. S. Weisberger and P.-G. Reinhard: Z. Phys. **D23**, 275 (1992)
20. M. Barranco, D.M. Jezek, S.E. Hernández, J. Navarro and Ll. Serra: Z. Phys. **D28**, 257 (1993)
21. M. Barranco, J. Navarro and A. Poves: Phys. Rev. Lett. **78**, 4729 (1997)
22. A. Belić, F. Dalfovo, S. Fantoni, and S. Stringari: Phys. Rev. **B49**, 15253 (1994)
23. D. Bressanini, M. Zavaglia, M. Mella, and G. Morosi: J. Chem. Phys. **112**, 717 (2000)
24. E. Krotscheck and R. Zillich: J. Chem. Phys. **115**, 10161 (2001)
25. W. Schöllkopf and J.P. Toennies: Private Communication.
26. M. Barranco, M. Pi, S.M. Gatica, E.S. Hernández, and J. Navarro: Phys. Rev. **B56**, 8997 (1997)
27. R.P. Feynman and M. Cohen: Phys. Rev. **102**, 1189 (1956)
28. K.L. Schmidt, M.A. Lee, M.H. Kalos, and G.V. Chester: Phys. Rev. Lett. **47**, 807 (1981)

Few-Body Systems Suppl. 14, 229–232 (2003)

Few-Body Systems

Model Study of Bremsstrahlung in Alpha Decay

W. van Dijk[1,2,*,†], Y. Nogami[2,†]

[1] Redeemer University College, Ancaster, Ontario, Canada L9K 1J4

[2] Department of Physics and Astronomy, McMaster University, Hamilton, Ontario, Canada L8S 4M1

Abstract. Recent experiments have triggered an interest in bremsstrahlung associated with α decay. We propose a fully time-dependent quantum mechanical calculation, treating the electromagnetic interaction by first-order perturbation theory, of the bremsstrahlung probability coincident with α decay. Preliminary results indicate that model calculations of this type are feasible.

1 Introduction

During the past decade two experiments measuring coincident bremsstrahlung and α decay have been reported [1]. The interpretation of the measurements were discussed subsequently [2] and further experiments are indicated. The measurements were the impetus of a number of theoretical studies. The simple classical model of an α particle accelerating in the Coulomb field overestimates the bremsstrahlung yields by an order of magnitude [3]. It suggests a non-negligible contribution to the radiation due to the α particle tunneling through the barrier. In a quasi-classical analysis [4] the photon emission amplitude is obtained by treating the tunneling as a classical motion in complex time.

In a fully quantum mechanical calculation, Papenbrock and Bertsch [5] used Fermi's golden rule regarding the electromagnetic interaction. They assumed appropriate wave functions of the α particle in the initial and final states but the time-dependent aspect of the α decay process was not taken fully into account. They found only a small contribution to the bremsstrahlung due to tunneling. A more extensive study along these lines [6] suggests that the bremsstrahlung spectrum is a result of interference of a number of effects, such as, tunneling, Coulomb acceleration, as well as others. An alternative approach [7] involves the time-dependent wave function of an emitted α particle which tunnels through a repulsive barrier to calculate the expectation value of the radial momentum

* *E-mail address:* vandijk@physics.mcmaster.ca

† Supported by the Natural Sciences and Engineering Research Council of Canada.

of the α particle as a function of time. This expectation value, in an ad hoc manner, is used in the equation for classical bremsstrahlung. The resulting bremsstrahlung spectrum is suppressed. It shows peaks due to the interference of different resonance components of the α wave function.

The method we propose in this paper deals in full with the time-dependent quantum aspects of the α decay and the accompanying radiation process. The electromagnetic interaction, however, is treated by first-order perturbation theory. As far as we know no such calculation has been done to date.

2 Description of the Model

The Hamiltonian for an α decaying system and the radiation field is $H = H_0 + H_1$ with

$$H_0 = \frac{\mathbf{p}^2}{2M} + V(r) + \sum_{\lambda=1,2} \int d^3k\, \omega a_\lambda^\dagger(\mathbf{k}) a_\lambda(\mathbf{k}), \tag{1}$$

where M is the mass, $\mathbf{p} = -i\nabla$ the momentum, and $\mathbf{r}$ the coordinate of the α particle. The potential $V(r)$ includes a repulsive barrier and supports one or more unstable bound states of the α particle within the barrier. The $a_\lambda^\dagger(\mathbf{k})$ creates a photon of momentum $\mathbf{k}$, energy $\omega = k = |\mathbf{k}|$, and polarization $\boldsymbol{\epsilon}_\lambda$. We use natural units, $c = \hbar = 1$, and the radiation gauge for the photon field. The interaction is given by $H_1 = -(2e/M)\mathbf{p} \cdot \mathbf{A}(\mathbf{r}, t)$, where

$$\mathbf{A}(\mathbf{r}, t) = \sum_\lambda \int d^3k f(k) \boldsymbol{\epsilon}_\lambda [a_\lambda^\dagger(\mathbf{k}) e^{i(\mathbf{k}\cdot\mathbf{r}-\omega t)} + a_\lambda(\mathbf{k}) e^{-i(\mathbf{k}\cdot\mathbf{r}-\omega t)}]. \tag{2}$$

The $2e$ is the charge of the α particle and $f(k) = [(2\pi)^3(2\omega)]^{-\frac{1}{2}}$.

The state vector of the entire system can be written as $\Psi = \Psi_0 + \Psi_1$ with $\Psi_0 = \psi_0(r, t)|0\rangle$ and

$$\Psi_1 = \frac{2e}{M} \sum_\lambda \int d^3k f(k) \psi_1(\mathbf{k}, \lambda, \mathbf{r}, t) a_\lambda^\dagger(\mathbf{k}) e^{i(\mathbf{k}\cdot\mathbf{r}-\omega t)} |0\rangle, \tag{3}$$

where $|0\rangle$ is the photon vacuum. The Schrödinger equation $(i\partial/\partial t - H)\Psi = 0$ can be reduced to

$$\left[i\frac{\partial}{\partial t} - \frac{\mathbf{p}^2}{2M} - V(r) \right] \psi_0(r, t) = 0, \tag{4}$$

where we consider only an S state for the α wave function $\psi_0(r, t)$, and

$$\left[i\frac{\partial}{\partial t} - \frac{\mathbf{p}^2}{2M} - V(r) - \omega \right] \psi_1(\mathbf{k}, \lambda, \mathbf{r}, t) e^{i(\mathbf{k}\cdot\mathbf{r}-\omega t)} = -(\mathbf{p} \cdot \boldsymbol{\epsilon}_\lambda) \psi_0(r, t) e^{i(\mathbf{k}\cdot\mathbf{r}-\omega t)}. \tag{5}$$

We ignore terms of order e^2. In Eq. (5), $\partial/\partial t$ and $\mathbf{p}$ operate on $e^{i(\mathbf{k}\cdot\mathbf{r}-\omega t)}$ also. The $\psi_1(\mathbf{k}, \lambda, \mathbf{r}, t)$ depends on angles through $(\mathbf{k} \cdot \mathbf{r})$ and $(\boldsymbol{\epsilon}_\lambda \cdot \mathbf{r})/r$.

We choose $V(r)$ such that Eq. (4) can be solved analytically for any value of r and t [8]. We assume an initial wave function $\psi_0(r,0)$ which is confined within the potential barrier. The $\psi_0(r,t)$ slowly leaks out tunneling through the potential barrier. We solve the inhomogeneous equation Eq. (5) for $\psi_1(\mathbf{k},\lambda,\mathbf{r},t)$ for various values of $\mathbf{k}$ and $\lambda = 1$ and 2. The initial condition is $\psi_1(\mathbf{k},\lambda,\mathbf{r},0) = 0$. We still need to solve this inhomogeneous wave equation numerically.

The probability for having one photon at time t is given by

$$P(t) = \sum_{\lambda} \int d^3k \; P(\mathbf{k},\lambda,t), \tag{6}$$

$$P(\mathbf{k},\lambda,t) = (2e/M)^2 f^2(k) \int d^3r |\psi_1(\mathbf{k},\lambda,\mathbf{r},t)|^2. \tag{7}$$

Because of the symmetry of the initial nucleus and the subsequent α-particle wave function $\psi_0(r,t)$, the function $\sum_\lambda P(\mathbf{k},\lambda,t)$ is spherically symmetric in $\mathbf{k}$ space. We make the partial wave expansion

$$\psi_1(\mathbf{k},\lambda,\mathbf{r},t)e^{i\mathbf{k}\cdot\mathbf{r}} = \frac{1}{r}\sum_{l,m} u^{\lambda}_{lm}(k,r,t)Y_{lm}(\theta,\phi), \tag{8}$$

where the z axis has been chosen along the $\mathbf{k}$ direction. Making the similar known expansion for $e^{i\mathbf{k}\cdot\mathbf{r}} = e^{ikz}$ in Eq. (5), we obtain differential equations for the u^{λ}_{lm}. For the sake of simplicity we will consider the first few terms of the expansion. The lowest order equation involves $u^{\lambda}_{00}(k,r,t)$. Since $\psi_1(\mathbf{k},\lambda,\mathbf{r},0) = 0$, $u^{\lambda}_{lm}(k,r,0) = 0$ for all l and m. Since the inhomogeneous term of the differential equation for u^{λ}_{00} is zero, $u^{\lambda}_{00}(k,r,t) = 0$. For $l = 1$, $u^{\lambda}_{10}(k,r,t) = 0$ since it also satisfies a homogeneous differential equation, but $u^{\lambda}_{1,\pm1}(k,r,t)$ is a solution of the inhomogeneous differential equation

$$\left[i\frac{\partial}{\partial t} + \frac{1}{2M}\frac{\partial^2}{\partial r^2} - \frac{1}{Mr^2} - V\right] u^{\lambda}_{1,\pm1} = is\sqrt{\frac{2\pi}{3}}\, r\frac{\partial\psi_0(r,t)}{\partial r}[j_0(kr) + j_2(kr)], \tag{9}$$

where $s = \mp1$ for $\lambda = 1$ and $s = -1$ for $\lambda = 2$, and $u^{\lambda}_{1,\pm1}(k,r,0) = 0$. Since s is a phase factor $|u^{\lambda}_{1,\pm1}|$ is independent of λ. Thus we need to solve Eq. (9) only once. The probability to have a photon emitted in coincidence with an α particle per unit photon energy by time t is given by

$$\frac{dP_{l=1}(k,t)}{d\omega} = \frac{1}{\pi^2}\left(\frac{2e}{M}\right)^2 \omega \int_0^{\infty} dr |u^1_{11}(k,r,t)|^2. \tag{10}$$

3 Results and Discussion

As a model potential we use the square-barrier like the one (with reduced width) used for ^{210}Po [9], $V(r) = V_1\theta(R_1 - r) + V_2\theta(R_2 - r)(r - R_1)$ with $R_1 = 8.992$ fm, $R_2 = 9.891$ fm, $V_1 = 4.82$ MeV and $V_2 = 9.55$ MeV. With this potential Eq. (9) was integrated numerically for $k = 100$ keV to obtain

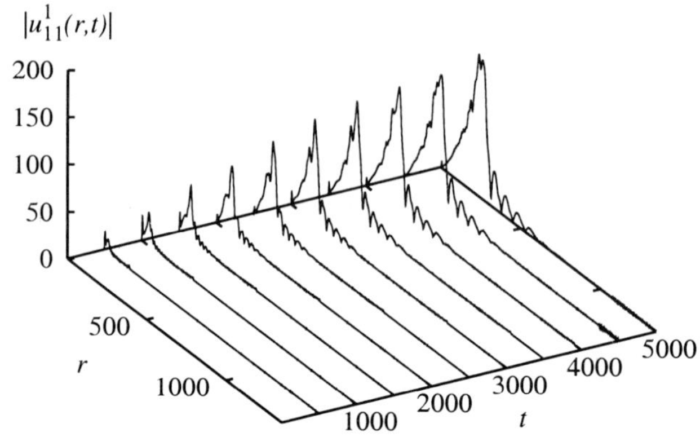

Figure 1. The bremsstrahlung amplitude for 100 keV at different times.

the bremsstrahlung amplitude as shown in Fig. 1. The probability of photon emission per unit energy, Eq. (10), is expected to level off asymptotically in time. The probability over the time period shown in Fig. 1, however, continues to increase in time. This preliminary result is surprising and we are investigating it further.

On the basis of these results we conclude that the approach is viable as a fully time-dependent quantum mechanical model analysis. Provided that the half-life is sufficiently short, we will be able to compare the results of this calculation with those of the other approaches. For calculations on realistic systems, e.g., ^{210}Po, the distance and time scales make the calculation numerically prohibitive and one needs to employ approximations, or less likely, exact solutions of the type obtained for the decaying α particle wave function [8].

References

1. A. D'Arrigo et al.: Phys. Lett. **B332**, 25 (1994); J. Kasagi et al.: Phys. Rev. Lett. **79**, 371 (1997)
2. N. Eremin et al.: Phys. Rev. Lett. **85**, 3061 (2000); J. Kasagi et al.: Phys. Rev. Lett. **85**, 3062 (2000)
3. J. Kasagi et al.: J. Phys. G: Nucl. Part. Phys. **23**, 1451 (1997)
4. M.I. Dyakonov and I. V. Gornyi: Phys. Rev. Lett. **76**, 3542 (1996)
5. T. Papenbrock and G. F. Bertsch: Phys. Rev. Lett. **80**, 4141 (1998)
6. N. Takigawa et al.: Phys. Rev. **C59**, R593 (1999)
7. C.A. Bertulani et al.: Phys. Rev. **C60**, 031602 (1999)
8. W. van Dijk and Y. Nogami: Phys. Rev. Lett. **83**, 2867 (1999); Phys. Rev. **C65**, 024608 (2002)
9. E.V. Tkalya: Phys. Rev. **C60**, 054612 (1999)

Few-Body Systems Suppl. 14, 233–236 (2003)

Few-Body Systems

Three-Body Dynamics in One Dimension

T. Melde[1], L. Canton[1,2], J.P. Svenne[3]

[1] Dipartimento di Fisica, Università di Padova, Via F. Marzolo 8, Padova, Italy, I-35131

[2] Istituto Nazionale di Fisica Nucleare, Sezione di Padova, Via F. Marzolo 8, Padova, Italy, I-35131

[3] Winnipeg Institute for Theoretical Physics and Department of Physics and Astronomy, University of Manitoba, Winnipeg, MB, Canada R3T 2N2

Abstract. We discuss the general three-particle quantum scattering problem, for motion restricted to the full line. Specifically, we formulate the three-body problem in one dimension in terms of the (Faddeev-type) integral equation approach. As a first application, we develop a spinless, one-dimensional (1-D) model that mimics three-nucleon dynamics in one dimension. In addition, we investigate the effects of pionic 3NF diagrams.

1 Introduction

The quantum mechanics of one-dimensional systems has been extensively investigated for a variety of purposes [1]. Quantum scattering in one dimension has also been considered because of its value in pedagogical works [2], since it retains sufficient complexity to embody many of the physical concepts and behaviours that occur in the more complex three-dimensional processes, but without the many technical aspects required for dealing with quantum scattering in higher dimensions. Here, we are interested in the formulation of the quantum three-body theory on the full, $(-\infty, +\infty)$ line, based on a system of coupled integral equations for the three-body transition operators [3] (AGS equations [4]). We illustrate in this paper an approach that harmonizes the typical two-channel structure of the "tunneling" problems, due to the presence of the transmission and reflection processes, with the cluster structure of the three-body problem, thus leading to an overall reduced S-matrix for the two-fragment processes which is 6×6 [5]. The formulation presented herein, with straightforward modifications, is suited also to describe the tunneling (or potential-barrier) effects due to the scattering of a two-body one-dimensional cluster impinging on an external barrier potential.

2 Theory

The Schrödinger equation for a one-dimensional system on the whole line,

$$-\frac{d^2\Psi_k(x)}{dx^2} + U(x)\Psi_k(x) = k^2\Psi_k(x) \tag{1}$$

defines the transmission-reflection problem for positive energies and has two independent solutions with incident waves from the left and right, respectively. The asymptotic form for the wave function, with incidence from the left is

$$\Psi_k^L(x) \to \begin{cases} e^{ikx} + R_L(E)e^{-ikx}, & \text{for } x \to -\infty \\ T_L(E)e^{ikx}, & \text{for } x \to +\infty \end{cases} \tag{2}$$

where T_L, R_L are the transmission and reflection coefficients, with a similar expression for incidence from the right. The S-matrix is given according to the definition

$$S(E) = \begin{pmatrix} T_L(E) & R_R(E) \\ R_L(E) & T_R(E) \end{pmatrix} \tag{3}$$

which reduces to the identity matrix when the potential goes to zero. Off the energy-shell the t-matrix is introduced according to

$$t(E)|\Phi_q\rangle = V|\Psi_q\rangle \tag{4}$$

where $|\Phi_q\rangle$ is the free (plane-wave) solution. The S-matrix in respect to the on-shell t-matrix is

$$S(E) = \begin{pmatrix} 1 & 0 \\ 0 & 1 \end{pmatrix} - \frac{im}{\hbar^2 k}\begin{pmatrix} t(k,k;E) & t(k,-k;E) \\ t(-k,k;E) & t(-k,-k;E) \end{pmatrix} \tag{5}$$

where $t(k,k')$ are the on-energy-shell matrix elements of the t-matrix. In the one-dimensional system, on the energy shell for a given energy E, there exist exactly two possible momenta $k = k_\pm = \pm\frac{\sqrt{2mE}}{\hbar}$. In the three-body system it is possible to define a S-matrix for two-fragment scattering in a similar way.

Generally, a system of three particles is described by the Hamiltonian

$$H = H_0 + \sum_a V_a \tag{6}$$

and it is well established to introduce so called channel states, which are eigenstates of the channel Hamiltonians $H_a = H_0 + V_a$ with $a = 1,2,3$. In the one-dimensional system the channel states can be described by the following wave functions

$$\Phi_{\pm a} = \Phi_{\pm q_a}(x_a, y_a) = e^{\pm i q_a y_a}\phi_a(x_a) \tag{7}$$

where $\phi_a(x_a)$ denotes the properly normalized bound-state wave function of the two-particle fragment and $e^{\pm i q_a y_a}$ denotes the relative motion of the particle a with respect to the c.m. of the two-particle fragment (bc). The asymptotic

1-D plane wave incoming from the left can have three different transmitted two-cluster waves, which leads to the asymptotic form

$$\Psi_a^L(x,y) \to \begin{cases} \Phi_{+b}\delta_{ab} + R_{ab}^L(E)\,\Phi_{-b}, & \text{for } y_b \to -\infty \\ T_{ab}^L(E)\,\Phi_{+b}, & \text{for } y_b \to +\infty \end{cases} \tag{8}$$

and the reduced S-matrix for two-fragment scattering is

$$\tilde{S}_{ab}(E) = \begin{pmatrix} T_{ab}^L(E) & R_{ab}^R(E) \\ R_{ab}^L(E) & T_{ab}^R(E) \end{pmatrix} \tag{9}$$

This definition produces a 6×6 scattering matrix connecting all six possible "two-fragment asymptotic channels". We have shown [5], that it is possible to define the two-fragment S-matrix in one dimension in the following way

$$\begin{aligned} \tilde{S}_{ab}(E) &= \langle \phi_{\pm a} | \, \tilde{S}_{ab} \, | \phi_{\pm b} \rangle \\ &= \delta_{ab} \begin{pmatrix} 1 & 0 \\ 0 & 1 \end{pmatrix} - 2\pi \frac{iM}{\hbar^2 |q_b|} \begin{pmatrix} t_{ab}(+|q_a|, +|q_b|; E) & t_{ab}(+|q_a|, -|q_b|; E) \\ t_{ab}(-|q_a|, +|q_b|; E) & t_{ab}(-|q_a|, -|q_b|; E) \end{pmatrix} \end{aligned} \tag{10}$$

where t_{ab} are related to the AGS channel-transition operators, U_{ab}. It easily seen that the structure of this equation is very similar to the corresponding one in the two-body system, with the difference that here the matrix space is expanded to account for the 2+1 partitions.

3 Results

As first application, we develop a one-dimensional spinless model that mimics three-nucleon dynamics in one dimension. In particular, we investigate numerically the effects of irreducible "pionic" corrections [6, 7] in the one-dimensional three-body bound state by solving the homogeneous version of the AGS equation. Three two-body potentials that give the same 2N binding at −2.225MeV but with different "stiffness" of the repulsion terms have been considered. In

	2NF	2NF+OPE3	2NF+TPE3	2NF+OPE3+TPE3
V_5	-7.722	-7.708	-8.733	-8.718
V_{10}	-7.684	-7.662	-8.670	-8.650
V_{20}	-7.668	-7.642	-8.631	-8.608

Table 1. 'Triton' energies in MeV for the potentials with different "stiffness", including different 3NF corrections.

table 1 we show the effect of one-pion (OPE3) and two-pion exchange (TPE3) three-nucleon force (3NF) type corrections. For the details of the explicit form of these correction terms we refer to the literature [7, 5]. It is seen that unlike the TPE, the OPE corrections have only a small effect on the 'triton' binding energies. In Fig. 1 we show the effect of an effective parameter required in the OPE 3NF [8, 9] (set equal to one in table 1). It is seen that for values not equal

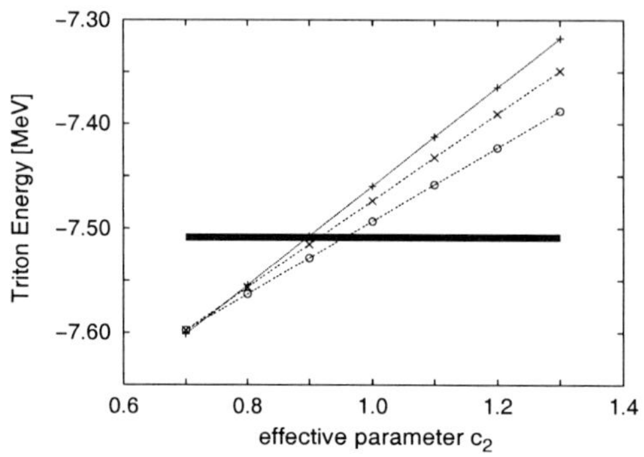

Figure 1. Dependence of the binding energy on the parameter c_2 for all three potentials including the diagonal correction terms. The thick line denotes the triton energy without any correction terms for all three potentials.

to one, the effects on the 'triton' binding energy can be non-negligible and the size of the effect depends directly on the 'stiffness' of the potential.

In conclusion we have developed an integral formulation of the one-dimensional scattering problem on the full line. As first application we tested the effects of different 3NF-type correction terms on the triton binding energy. It would be interesting to extend the investigations of the 2+1 fragment 1D-system to the scattering regimes.

Acknowledgement. We acknowledge support from the Italian MURST-PRIN Project "Fisica Teorica del Nucleo e dei Sistemi a Più Corpi". J.P.S. acknowledges support from NSERC, Canada.

References

1. D. Bolle, F. Gesztesy, M. Klaus: J. Math. Analysis and Appl. **122**, 496 (1987); D. Bolle, F. Gesztesy, F.J. Wilk: J. Operator Theory **13**, 3 (1985)
2. V. Barlette, M. Leite, S. K. Adhikari: Am. J. Phys. **69**, 1010 (2001)
3. T. Melde: Ph.D. Thesis. Univ. of Manitoba 2001 (unpublished)
4. E. O. Alt, P. Grassberger, W. Sandhas: Nucl. Phys. **B2**, 167 (1967)
5. T. Melde, L. Canton, J.P. Svenne, Few-Body Systems, in press (2002), quant-ph/0203021
6. L. Canton: Phys. Rev. **C58**, 3121 (1998)
7. L. Canton, T. Melde, J.P. Svenne: Phys. Rev. **C63**, 034004 (2001)
8. L. Canton, W. Schadow: Phys. Rev. **C64**, 031001(R) (2001)
9. L. Canton, W. Schadow, J. Haidenbauer: Eur. J. Phys. **A14**, 225 (2002)

Few-Body Systems Suppl. 14, 237–240 (2003)

Few-Body Systems

Borromean Molecules

J.-M. Richard*

Institut des Sciences Nucléaires, Université Joseph Fourier – CNRS-IN2P3, 53, avenue des Martyrs, F-38026 Grenoble Cedex, France

Abstract. For some values of the constituent masses m_i, the hydrogen-like molecule $(m_1^+, m_2^+, m_3^-, m_4^-)$ is stable with respect to spontaneous dissociation into two neutral atoms, while none of its three-body subsystems such as (m_1^+, m_2^+, m_3^-) is stable. This occurs in particular in the neighbourhood of the (M^+, m^+, M^-, m^-) configuration with $M/m = 2$, as for instance for the molecule $(p, d, \bar{p}, \bar{d})$ involving an antiproton and an antideuteron.

The word "Borromean" is often used to identify bound states whose subsystems are unbound [1]. It comes after the Borromean rings, which are interlaced in such a subtle topological way, that if any one of them is removed, the two others become unlocked. For instance, the ^{6}He isomer of ordinary Helium is stable, while ^{5}He is not. In a three-body picture, this means that the (α, n, n) system is bound, while neither (α, n) nor (n, n) are stable against dissociation.

For short-range forces, Borromean binding occurs explicitly in simple quantum-mechanical models. Let g be the strength of the pair potential V acting between N identical particles and containing attractive parts. Binding two bosons requires a certain minimum strength, say $g > g_2$. For three bosons, this is $g > g_3$. The remarkable observation is that $g_3 < g_2$. Typically, $g_3 \simeq 0.8\, g_2$ for simple monotonic potentials such as Yukawa, Gaussian, exponential, etc. See, e.g., [1], and references therein. It can be proved that $g_3/g_2 \geq 2/3$ [1]. This lower limit is approached for potentials with an external barrier, whereas $g_3/g_2 \to 1$ for those with a strong repulsive core [1]. Similarly $g_4 < g_3$, etc. A 4-body system is Borromean for $g_4 < g < g_2$, if one adopts the following definition: *A bound state is Borromean if there is no path to build it via a series of stable states by adding the constituents one by one.*

The phenomenon of Borromean binding is intimately linked to the Thomas collapse [2] and the Efimov effect [3]. Let E_N be the ground-state energy of H_N. For $g \to g_2+$, $E_2/E_3 \to 0$, thus if E_2 is kept fixed by rescaling, $E_3 \to -\infty$. The occurrence of many excited states near $g = g_2$ implies that the ground state already exists at this point.

* *E-mail address:* jean-marc.richard@isn.in2p3.fr

Borromean binding is hardly conceivable for long-range attractive potentials. For the gravitational $V(r) \propto -1/r$ potential, the critical strength is $g_N = 0$, $\forall N$, by scaling considerations. The situation is less obvious for a superposition of attractive and repulsive terms, and different masses. Still the existence of a bound state survives any overall rescaling of the strength factors or masses, but depends rather crucially on the balance between the repulsive and attractive terms, and on the mass ratios.

The systems of three unit charges have been studied extensively [4]. While $\mathrm{He}^+(e^-, p, p)$, $\mathrm{Ps}^-(e^+, e^-, e^-)$, and $\mathrm{H}^-(p, e^-, e^-)$, for instance, are stable against dissociation, configurations such as (e^-, e^+, p) or $(p, \bar{p}, e^-)$ are unstable. The stability frontier is shown in Fig. 1. Thanks to scaling, any state is represented by its normalised inverse masses $\alpha_i = m_i^{-1}/\sum_k m_k^{-1}$. The domain $\alpha_1 + \alpha_2 + \alpha_3 = 1$, $\alpha_i > 0$, is limited by an equilateral triangle, see Fig. 1. There is a stability band around the symmetry axis, in between two instability domains which are convex [4].

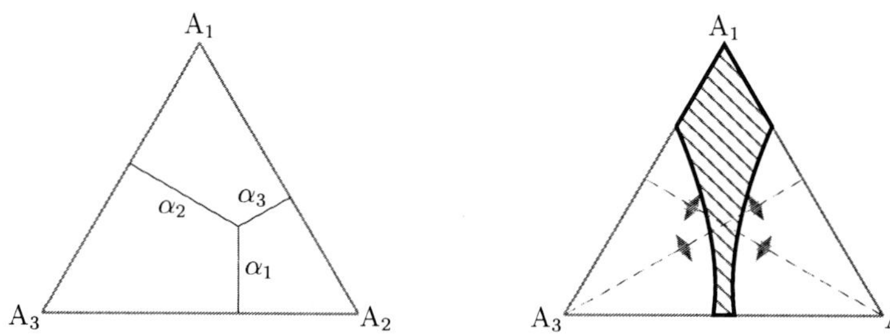

Figure 1. Geometrical representation of the stability domain for 3-body systems (m_1^+, m_2^-, m_3^-) represented by their normalised inverse masses α_i. The arrows indicate the range explored by the 3-body subsystems of $[(A \pm \epsilon)^+, a^+, (A \mp \epsilon)^-, a^-]$ and $[A^+(a \pm \eta)^+, A^-(a \mp \eta)^-]$, for inverse masses in ratio $a/A = 2$. Here $\epsilon = 0.1$ and $\eta = 0.25$, but the stability border is drawn without accurate computation.

The $(m^\mp, M^\pm, m^\pm)$ configurations have been studied in some detail. They correspond to the oblique symmetry axes in Fig. 1. For $m = M$, one gets the stable Ps^-. For $M \gg m$, (e^-, p, e^+) is unstable. For $M \ll m$, $(p, e^-, \bar{p})$ is also unstable. The stability range has been estimated by Mitroy [5],

$$0.70 \lesssim M/m \lesssim 1.64 \ , \tag{1}$$

using a powerful and reliable stochastic variational method [6].

There are also some rigorous or empirical results on $(m_1^+, m_2^+, m_3^-, m_4^-)$ systems with charges $\{q_i\} = \{\pm 1, \pm 1, \mp 1, \mp 1\}$ [4]. The configuration $(p, e^+, \bar{p}, e^-)$ is unstable, as it cannot compete with the deeply bound protonium associated with the lowest threshold $(p, \bar{p}) + (e^+, e^-)$. The hydrogen molecule $\mathrm{H}_2(p, p, e^-, e^-)$, the positronium molecule $\mathrm{Ps}_2(e^+, e^+, e^-, e^-)$, the positronium hydride $\mathrm{PsH}(p, e^+, e^-, e^-)$ are known to be stable. Any configuration with two identical particles, e.g., $m_3 = m_4$, is stable against spontaneous dissociation. In particular, any hydrogen-like state (M^+, M^+, m^-, m^-) is stable with respect to its lowest threshold $2(M^+, m^-)$, with stability improved if M/m is increased or decreased, starting from the Ps_2 case $M/m = 1$ [4]

The above examples share a common property: Ps_2 contains the stable $(e^\pm, e^\mp, e^\mp)$ subsystems; for H_2, both (e^-, p, p) and (p, e^-, e^-) are stable; for PsH, (e^-, e^+, p) is unstable, but (p, e^-, e^-) and (e^+, e^-, e^-) are stable; any (m_1^+, m_2^+, m^-, m^-) state is stable, with at least (m_1^+, m^-, m^-) and (m_2^+, m^-, m^-) having a stable bound state. It is tempting at this point to guess that any stable molecule has at least one stable 3-body subsystem.

This conjecture is wrong, as can be seen by examining the (M^+, m^+, M^-, m^-) configuration. For $M \gg m$ or $M \ll m$, it corresponds to the unstable $(p, e^+, \bar{p}, e^-)$, while for $M = m$, it leads to a rescaled Ps_2 which is stable. The stability domain has been estimated by accurate methods [7]

$$\frac{1}{2.2} \lesssim \frac{M}{m} \lesssim 2.2 \; . \tag{2}$$

Comparing the results (1) and (2) indicates a window for Borromean binding. In particular, for $M/m = 2$, which is the deuteron to proton mass ratio, the molecule (M^+, m^+, M^-, m^-) is bound, but neither $(M^\pm, m^\mp, M^\pm)$ nor $(m^\pm, m^\mp, M^\pm)$ are stable. The molecule experiences a new type of binding mechanism, which is more fragile.

One could also resume the numerical investigations of Ref. [7] and determine the position of stability border of $(m_1^+, m_2^+, m_3^-, m_4^-)$ in the vicinity of the $(m_1 = m_3,\ m_2 = m_4)$ line. Already, some minimal extension of the domain of Borromean binding can be made without any new 4-body calculation, provided the stability border is accurately known in the 3-body case. Consider for instance the configuration given by the *inverse* masses

$$[(A \pm \epsilon)^+, a^+, (A \mp \epsilon)^-, a^-] \; , \; \text{or} \;\; [A^+ (a \pm \eta)^+, A^- (a \mp \eta)^-] \; , \tag{3}$$

where $A = M^{-1}$ and $a = m^{-1}$ links to the previous notation, for $\epsilon = \eta = 0$. If $|\epsilon| < |a - A|$, the lowest threshold remains given by $[(A \pm \epsilon)^+, (A \mp \epsilon)^-] + [a^+, a^-]$, which has the same energy as for $\epsilon = 0$, since the inverse masses average out when building the reduced mass for each atom. On the other hand, the four-body Hamiltonian for the first configurations can be written

$$H(\epsilon) = H(0) \pm \frac{\epsilon}{2} (\mathbf{p}_1^2 - \mathbf{p}_3^2) \; . \tag{4}$$

The second term breaks charge-conjugation symmetry and lowers the ground state. This can be seen by applying the variational principle with the symmetric ground-state wave-function of $H(0)$ as a trial wave-function for $H(\epsilon)$. Hence, the 4-body molecule has an improved stability for $\epsilon \neq 0$ in (4). As ϵ departs from 0, the 3-body subsystems follow transverse and longitudinal paths starting from the $\alpha_1 = \alpha_3$ or $\alpha_1 = \alpha_2$ symmetry axis of the triangle, as pictured schematically in Fig. 1. It should be not too difficult to determine up to which values of ϵ the 3-body systems remain unstable. The same reasoning can be applied to the second configurations.

For example, if the mass ratio of the protonium molecule $(p, p, \bar{p}, \bar{p})$ is modified by replacing a proton and an antiproton by a deuteron and an antideuteron, the resulting molecule $(p, d, \bar{p}, \bar{d})$ becomes Borromean, without any bound 3-body subsystem. This can be compared with the observation by Bertini et al. that if one screens the Coulomb potential in the hydrogen molecule, that is to say, $r_{ij}^{-1} \to \exp(-\mu r_{ij})/r_{ij}$ for all pairs, there is a range of values of μ for which H_2 remains bound while neither H_2^+ nor H are stable [8].

This investigation grew out of discussions with Dario Bressanini and André Martin. Comments by E.A.G. Armour, A.S. Jensen and K. Varga are also gratefully acknowledged. I benefited from the stimulating atmosphere of the XVIII European Few-Body Conference at Bled, where this work was presented and discussed with my colleagues.

References

1. M.V. Zhukov, B.V. Danilin, D.V. Fedorov, J.M. Bang, I.S. Thompson, J.S. Vaagen: Phys. Rep. **231**, 151 (1993); J.-M. Richard, S. Fleck: Phys. Rev. Lett. **73** 1464 (1994); S. Moszkowski, S. Fleck, A. Krikeb, L. Theußl, J.-M. Richard, K. Varga: Phys. Rev. **A62**, 032504 (2000)

2. L.H. Thomas: Phys. Rev. **47**, 903 (1935); A. Delfino, K. Adhikari, L. Tomio, T. Frederico: Phys. Rev. **C46**, 471 (1992)

3. V. Efimov: Phys. Lett. **B33**, 560 (1970); Sov. J. Nucl. Phys. **12**, 581 (1971); R.N. Amado, R.V. Noble: Phys. Lett. **B35**, 25 (1971); Phys. Rev. **D5**, 1992 (1972); S. Albeverio, R. Høegh-Krohn, T.T. Wu: Phys. Lett. **A83**, 101 (1981); D.V. Fedorov, A.S. Jensen: Phys. Rev. Lett. **71**, 4103 (1993)

4. For a review and further references, see, e.g., E.A.G. Armour, W. Byers Brown: Accounts of Chemical Research **26**, 168 (1993); E.A.G. Armour, J.-M. Richard, K. Varga, "*Stability of Few-Charge Systems*", in preparation

5. J. Mitroy: J. Phys. B: At. Mol. Opt. Phys. **33**, 5307 (2000)

6. K. Varga, Y. Suzuki: Phys. Rev. **C52**, 2885 (1995)

7. D. Bressanini, M. Mella, G. Morosi: Phys. Rev. **A55**, 200 (1997); K. Varga: In: B. Desplanques et al. (eds.): Proc. 16th European Conference on Few-Body Problems in Physics, June 1998, Autrans, France; Few-Body Systems Suppl. **10**, 11 (1999)

8. L. Bertini, M. Mella, D. Bressanini, G. Morosi: "*A non-adiabatic quantum Monte-Carlo study of the critical stability of H_2^+ and H_2 with Yukawa potential*", Preprint (2002)

Few-Body Systems Suppl. 14, 241–244 (2003)

Few-Body Systems

Nuclear Clustering Quest in Relativistic Multifragmentation

V. Bradnova[1], M.M. Chernyavsky[2], L. Just[3], M. Haiduc[4], S.P. Kharlamov[2], A.D. Kovalenko[1], V.A. Krasnov[1], V.G. Larionova[2], A.I. Malakhov[1], G.I. Orlova[2], N.G. Peresadko[2], N.G. Polukhina[2], P.A. Rukoyatkin[1], V.V. Rusakova[1], N.A. Salmanova[2], S. Vokál[5], P.I. Zarubin[1]

[1] Joint Institute for Nuclear Research, Dubna, Russia

[2] P. N. Lebedev Physical Institute, Moscow, Russia

[3] Institute of Experimental Physics, Kosice, Slovakia

[4] Institute of Space Studies, Maguerel-Bucharest, Romania

[5] P. J. Šafarik University, Košice, Slovakia

Abstract. The status of nuclear clustering studies performed by nuclear emulsion irradiations in beams of light relativistic nuclei is briefly reviewed. Thanks to the best spatial resolution and full solid angle acceptance provided by nuclear emulsions, such an approach allows one to obtain a unique and evident observations reflecting cluster-like features in light nuclear structures. Future investigations are suggested in relativistic beams of He, Be, B, C, and N isotopes.

We give a short overview of the results and prospects of nuclear clustering studies on the grounds of the observations of interactions of light nuclei with initial energy above $A \times 1$ GeV in nuclear emulsions. The discussed explorations are provided by beams of the JINR Synchrophasotron and Nuclotron in Dubna.

The track detection is performed in emulsion layers measuring $100 \times 200 \times 0.5$ mm^3. The layers are assembled in few cm thick stacks. The stacks are exposed to a beam in the longitudinal direction. They provide multiple track visualization over the total solid angle with spatial resolution of about 0.5 micron. Their sensitivity extends from slow fragments down to relativistic single charge particles. A mean range of light relativistic nuclei in emulsion is defined by the cross section of their inelastic interaction with emulsion nuclei. It varies from 14 cm for ^{6}Li nuclei to 9 cm for ^{24}Mg in a BR-2 type emulsion.

Nuclear clustering manifests itself in the isotope composition of projectile nucleus fragments. They are collimated mostly in a narrow forward cone limited by an angle $0.2/P_o$, where P_o is the momentum per nucleon of a projectile nucleus in GeV/c. Their tracks stay within a single emulsion layer long enough to be identified. The projectile fragment charge can be measured by ionization density. Tracks of double charge relativistic fragments are recognized most easily. This is an advantage for α clustering studies. Measurements of a track multiple scattering allow one to identify relativistic hydrogen and helium isotopes due to sufficient discreteness of their total momenta accepted in projectile nucleus fragmentation. The isotope composition of projectile nucleus fragmentation is independent of a target nucleus in accordance with the established concept of limiting fragmentation. This circumstance makes the question about the emulsion composition unimportant for the discussed application.

To disclose a clustering picture we select events preserving the total charge of relativistic fragments in the forward cone equal to a projectile nucleus value. This leads to a correlated reduction or even absence of accompanying multiplicity. The excitation scale of a fragmenting system in a projectile nucleus rest frame is of the order of few MeV per fragmenting nucleon. In the considered case of the projectile nucleus fragmentation perfect angular measurements play the determinative role in the estimation of the excitation energy scale while their momentum per nucleon may be taken the same as for an incoming nucleus. Thus, emulsions provide an excellent opportunity of a complete observation and study of light nucleus multifragmentation in flight.

For the first time a coherent multifragmentation was observed and explored in emulsion for ^{12}C nucleus dissociation into three α particles at an initial momentum 4.5·A GeV/c [1]-[3]. The mean free path in emulsions for such events is equal to 5 m, i. e. this study demanded a visual scanning over a long track path to find the statistics of about 100 events. This scale of the statistics is a practical limit of the described approach. Relativistic multifragmentation was also explored for ^{16}O, ^{24}Mg, and ^{28}Si nuclei at an initial momentum 4.5·A GeV/c by means of nuclear emulsion [4]-[7] and hydrogen bubble chamber techniques [8]-[10].

Nowadays we have resumed searches of events for straightforward multifragmentation of a ^{24}Mg nucleus as a natural development of α particle clustering searches. One may expect that this study will clarify a question about relative probabilities of fragmentation into multiple and binary channels and, in particular, the symmetric one [11]. As example, the figure shows one of the multifragmentation events initiated by a quasinucleonic collision. A three dimensional image is reconstructed as a projection by means of a scanning CCD microscope. It can be seen (from top to bottom) how a ^{24}Mg nucleus dissociate into five double and two single charged nuclei. An interaction vertex with a nuclear fragment jet and a single charge relativistic particle (the dotted track going down) directed beyond a fragmentation cone is seen on the top photo. Most probably this track is due to a produced pion. A distict separation of two single charged fragments (tracks going up and down) from a major jet is seen on the middle photo. The bottom photo demonstrates a clear track separation of five double charged fragments.

Exotic nuclei present a bright extention of nuclear clustering. The lightest one among them is a ^{6}Li nucleus. Relativistic fragmentation of 4.5·A GeV/c ^{6}Li

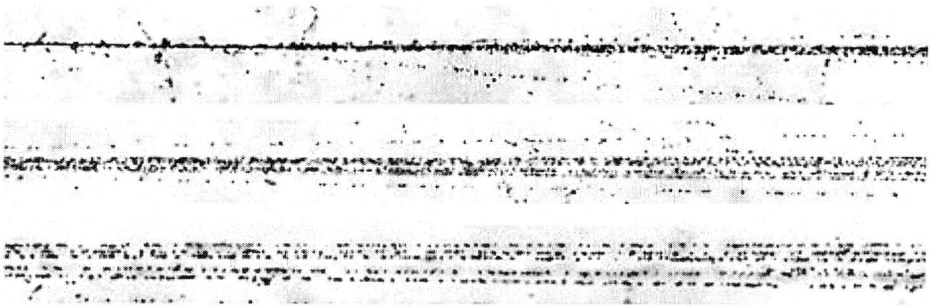

Figure 1. 4.5·A GeV/c ^{24}Mg nucleus dissociation into five double and two single charged fragments in sequential evaluation (top to bottom).

nuclei was explored by means of nuclear emulsion [12]-[16]. It was found that ^{6}Li nuclei have a reduced mean path in emulsion. The obtained value would rather correspond to a nucleus with mass number equal to 11 under the assumption of a uniform nucleon density. The fragmentation channel ^{6}Li$\rightarrow \alpha$ points to a lower value of the mean transverse momentum of α particles, $<p_T>=0.13\pm0.1$ GeV/c when compared with the case ^{12}C$\rightarrow \alpha$ having $<p_T>=0.24\pm0.01$ GeV/c. This also points to an unusually large radius of the nucleon distribution in the ^{6}Li nucleus. Using the geometric overlapping model, its value was estimated to be 2.7±0.1 fm, which is in reasonable agreement with the known data on elastic scattering of protons on a ^{6}Li target. Another distinctive feature of the ^{6}Li nucleus fragmentation is an unusually enhanced yield of relativistic deuterons which is equal to the proton one. In the sample of about 1000 events ^{6}Li nucleus inelastic interactions found in [16] 31 events of ^{6}Li nucleus dissociation aren't accompanied by the target nucleus excitation. Among them 23 events correspond to the dissociation channel α+d, four of them to ^{3}He+t, four to t+d+p and none of them to d+d+d. This topology explains an increased yield of deuterons as a reflection of the structure of weakly bound clusters of the α particle and the deuteron. As extention, we formed a ^{6}He nucleus beam using a charge exchange process ^{6}Li$\rightarrow^6$He. Results can be found in [17].

In the framework of researches with neutron-deficient isotopes, we irradiated emulsions by $A \times 1$A GeV ^{10}B nuclei. The value of the mean free path was found to be in good agreement with the expected one for a nuclei with homogeneous nucleon density. 32 events of ^{10}B dissociation without signals of target excitation or charged meson production have been observed. Their distribution over the final charged states is as follows: 18 events 2+2+1, 9 - 2+1+1+1, 2 - 1+1+1+1+1, 2 - 4+1, and 1 - 2+3. This distribution points to the dominant role of a three-body dissociation. The mass number identification is in a progress now. It will allow one to select events with the complete transfer of a baryon number in a fragmentation cone.

We irradiated emulsions by $A \times 1.2$ GeV ^{7}Be nuclei obtained via a charge exchange process ^{7}Li$\rightarrow^7$Be and a corresponding analysis is in progress now.

It is particularly important to determine the cluster structure of this nucleus for clarification of the problem of the existence of a proton halo in ^{8}B and understand the role of the ^{7}Be as a core. Future irradiations are foreseen with ^{14}N, ^{8}B, ^{10}C, ^{9}C, and nuclei to establish existence and relative intensities of various fragmentation options. The problem of our special interest is a possible role of deuterons and ^{3}He nuclei as structural clusters in these nuclei. We expect that forthcoming studies of multifragmentation of light relativistic nuclei in emulsions will allow one to obtain evident information on nuclear clustering.

References

1. V.V. Belaga et al.: Phys. Atom. Nucl. **58**, 1905 (1995)
2. V.V. Belaga et al.: Phys. Atom. Nucl. **60**, 791 (1997)
3. V.V. Belaga, M.M. Muminov, G.M. Chernov: JETP Letters **62**, 395 (1995)
4. N.P. Andreeva et al.: Phys. Atom. Nucl. **59**, 102 (1996)
5. M.A. Belov et al.: Phys. Atom. Nucl. **65**, 959 (2002)
6. M.L. Allaberdin et al.: e-Print Archive: nucl-ex/0102006
7. A.I. Bondarenko et al.: Sov. J. Nucl. Phys. **55**, 77 (1992)
8. V.V. Glagolev et al.: Eur. Phys. J. **A11**, 285 (2001)
9. V.V. Glagolev et al.: Phys. Atom. Nucl. **63**, 520 (2000)
10. V.V. Glagolev et al.: Phys. Atom. Nucl. **62**, 1388 (1999)
11. V.G. Bogdanov et al.: JETP Lett. **44**, 391 (1986)
12. M.El-Nadi et al.: In: Proceedings of International School of Cosmic Ray Astrophysics, 10th Course, Erice (1996)
13. F.G. Lepekhin, D.M. Seliverstov, B.B. Simonov: JETP Lett. **59**, 332 (1994)
14. F.G. Lepekhin, D.M. Seliverstov, B.B. Simonov: Phys. Atom. Nucl. **58**, 816 (1995)
15. F.G. Lepekhin, D.M. Seliverstov, B.B. Simonov.: Eur. Phys. J. **A1**, 137 (1998).
16. M.I. Adamovich et al.: Phys.Atom Nucl., **62**, 1378 (1999)
17. M.I. Adamovich et al.: Part. and Nucl. Letters. **N1[110]**, 29 (2002); nucl-ex/0206013.

Few-Body Systems Suppl. 14, 245–250 (2003)

Few-Body Systems

Energy Spectra of $^9_\Lambda$Be

E. Cravo[1], A.C. Fonseca[1], Y. Koike[2]

[1] Centro de Física Nuclear, University of Lisbon, Av. Prof. Gama Pinto 2, 1649-003 Lisbon, Portugal

[2] Science Research Center, Hosei University, 2-17-1, Fujimi, Chiyoda-ku, Tokyo 102, Japan

Abstract. The hypernucleus $^9_\Lambda$Be is investigated using the $\alpha + \Lambda + \alpha$ cluster model. The corresponding Faddeev equations are solved for different $\alpha - \alpha$ and $\Lambda - \alpha$ interactions that describe both ^{8}Be and $^5_\Lambda$He energy spectra. For the ground state $J^\pi = \frac{1}{2}^+$ and excited states $\frac{3}{2}^+$ and $\frac{5}{2}^+$ we include the Coulomb repulsion between the α's and calculate, in addition to the energy eigenvalues, the $\alpha-$ and $\Lambda-$particle mean radius, the rms charge radius, the electric quadrupole moment (Q), as well as the magnetic dipole. Structural differences between ^{9}Be and $^9_\Lambda$Be lead to values of Q and μ that have opposite sign. Unlike previous theoretical work we find only two degenerate negative parity resonances $\frac{1}{2}^-(\frac{3}{2}^-)$ but, in addition, we get two degenerate positive parity resonances with $J^\pi = \frac{7}{2}^+(\frac{9}{2}^+)$ at higher energy which, together with the bound states, complete the positive parity rotational band.

1 Introduction

As pointed out by Feshbach [1] a number of years ago, one of the main purposes of hypernucleon physics is to study new many-body spectroscopy through the implant of an "impurity" in the condensed matter sense to unveil dynamical symmetries that are forbidden in ordinary nuclei by the Pauli principle. One nucleus where this phenomenon plays an important role is $^9_\Lambda$Be which, vis-à-vis ^{9}Be, presents, not only a very different spectra of bound states and resonances, but also different dynamical structure, possibly leading to surprising results for observables such as the rms charge radius, the quadrupole moment Q and the magnetic moment μ. This investigation, together with a new search for resonances in the continuum, is the main goal of the present work which is now published in ref. [2]. Our work follows the traditional three-body $\alpha-$cluster approach which has been used before by some of us to study ^{9}Be [3, 4] and also by others [5, 6] for $^9_\Lambda$Be.

Unlike ^{9}Be, which is typically a Boromean nucleus due to neither ^{5}He nor ^{8}Be being stable nuclei, $^{9}_{\Lambda}$Be may be thought as $^{5}_{\Lambda}$He $+\alpha$ due to the existence of $^{5}_{\Lambda}$He as a stable hypernucleus with $B_\Lambda \simeq 3.12$ MeV. Given the small binding energy of $^{5}_{\Lambda}$He relative to the α-particle binding and the large energy gap between Λ+^{4}He and $^{4}_{\Lambda}$He $+ n$, the natural choice is a three-body cluster model based on two alpha particles and the hyperon Λ. The difficulty now lies on the choice of $\alpha - \alpha$ and $\Lambda - \alpha$ interactions that are required to map the low lying spectra of ^{8}Be and $^{5}_{\Lambda}$He and provide, in the case of $\alpha - \alpha$, a reasonable description of Pauli blocking between nucleons in separate clusters.

A number of $\alpha - \alpha$ interactions already exist; some treat Pauli blocking by means of a short range repulsion, others by including Pauli forbidden states which are subsequently removed. Therefore, we use, not only Ali-Bodmer (AB) and Chien-Brown (CB) potentials, but also one derived from resonating group theory (UIM) which is non local due to the exchange nature of the RGM core potential (see ref. [2] for further details).

For the $\Lambda - \alpha$ interaction we use a number of effective potentials that have been developed over the years using different folding techniques or phenomenology. They are the Tang-Herndon (TH) which has a simple Gaussian shape, the Dalitz- (DA) and the Deloff-type (DE) potentials which were derived by folding based on $\Lambda - N$ interactions with a hard-core proposed by Dalitz et al. and Deloff, and the Maeda-Schmid (MS) which uses two Wood-Saxon phenomenological potentials to fit the $\Lambda - \alpha$ interaction also calculated by folding. All potentials reproduce the $^{5}_{\Lambda}$He binding energy reasonably well (see ref. [2] for further details).

For the chosen $\alpha - \alpha$ and $\Lambda - \alpha$ interactions we solve the bound state Faddeev equations [7] including the Coulomb repulsion between the α's. The general form of the equations was first proposed by Lehman et al. [8] and subsequently generalized by Cravo [4] to study the structure of ^{9}Be. The $\alpha - \alpha$ interaction is taken as a local interaction in different $\alpha - \alpha$ partial waves (s-, d-, and g-waves) or as a finite rank operator in the case of the RGM-based potential. The $\Lambda - \alpha$ interaction is always expanded as a sum of separable terms using the method developed by Koike [9]. The convergence of the calculation in terms of rank "r" and number of partial waves in the $\Lambda - \alpha$ interaction is shown in ref. [2]; all s-, p-, and d-waves are included.

2 Structure of $^{9}_{\Lambda}$Be Bound States

In the absence of a $\Lambda - \alpha$ spin-orbit force, two of the three bound states we find are degenerate ($J^\pi = \frac{3}{2}^+$and $\frac{5}{2}^+$). The ground state, $J^\pi = \frac{1}{2}^+$, lies 2.3 - 3.0 MeV below the excited states. Our results are shown in Table 1 for $J^\pi = \frac{1}{2}^+$ and in Table 2 for the excited states $\frac{3}{2}^+(\frac{5}{2}^+)$. The Coulomb interaction has been included up to partial waves L = 6 for convergence. Experimental values are taken from refs. [10] and [11]. The results indicate that the calculated binding energies depend more on the choice of $\Lambda - \alpha$ interaction than on the $\alpha - \alpha$ potential one uses. This is expected given the lack of sufficient experimental information on ${}^{5}_{\Lambda}$He to constrain the $\Lambda - \alpha$ interaction. Although the calculated binding energies for the ground state and excited state span a wide range of energies, the excitation energy between the two states is reasonably independent of the combination of $\alpha - \alpha$ and $\Lambda - \alpha$ potentials one uses, and in close agreement with the experimental excitation energy (3.1 MeV). The best results, compared to data, are obtained with MS $\Lambda-\alpha$ potential in combination with AB (or CB) $\alpha - \alpha$ interaction.

Table 1. $\frac{1}{2}^+$ binding energies (MeV) for different $\alpha - \alpha$ and $\Lambda - \alpha$ interactions. The Coulomb repulsion has been included in all partial waves needed for convergence.

	UIM	AB	CB
TH	−5.96	−5.98	−6.02
MS	−6.50	−6.73	−6.75
DE	−7.08	−7.36	−7.37
DA	−7.74	−8.27	−8.19
Exp. [10]		−6.71	

Table 2. Same as in Table 1 for $\frac{3}{2}^+(\frac{5}{2}^+)$ binding energies (MeV).

	UIM	AB	CB
TH	−3.44	−3.25	−3.17
MS	−4.05	−3.97	−3.80
DE	−4.63	−4.59	−4.41
DA	−5.18	−5.35	−5.05
Exp. [11]		−3.67	

Next we proceed to characterize these two states in terms of $\alpha-$particle and $\Lambda-$particle point rms radius, charge radius, electric quadrupole moment and magnetic dipole. Although most of these observables have not been measured for ${}^{9}_{\Lambda}$Be, they provide a better understanding of the structure of this hypernucleus and allow a comparison with ^{9}Be. All calculations include the Coulomb repulsion between the α's.

Both the $\alpha-$particle and the $\Lambda-$particle r.m.s. radius are calculated in the point-particle approximation. Although both values are slightly bigger for the excited state $\frac{3}{2}^+(\frac{5}{2}^+)$ than for the ground state $\frac{1}{2}^+$, they are comparable and almost independent of the choice of potentials. Therefore we can claim that $\bar{r}_\alpha \simeq 1.83$ fm and $\bar{r}_\Lambda \simeq 2.15$ fm. Comparing with ^{9}Be [4] we find that $\bar{r}_\alpha$ is about the same, but that $\bar{r}_n \simeq 3.1$ fm. Therefore one may conclude that the neutron in ^{9}Be moves well outside the average location of the $\alpha-$particles relative to the center of mass of the system, while the Λ, in ${}^{9}_{\Lambda}$Be, shares its position with the α's. This gives a certain credibility to hypernuclear models that take ${}^{5}_{\Lambda}$He as a cluster.

Next we calculate the values for the charge radius of ${}^{9}_{\Lambda}$Be including the contribution of the alpha particle form factor. The result is also almost independent of the potential choice or state, leading to a value for $r_c \simeq 2.48$ fm. This value compares with $r_c = 2.48$ fm for ^{9}Be calculated with AB potential. This clearly indicates that the size of both nuclei is about the same from the view point of the α subclusters but that the neutron in ^{9}Be sticks much further out than the Λ in ${}^{9}_{\Lambda}$Be. We find no evidence of contraction of the ^{8}Be core due to the substitution of the neutron by the $\Lambda-$ particle as pointed out by others. This may indicate that the Pauli repulsion between the α's is strong enough to block any additional binding effect created by the $\Lambda-$ particle.

Further understanding is obtained by calculating the nonvanishing electric and magnetic moments. The electric quadrupole moment Q is zero for the $\frac{1}{2}^{+}$ state due to Wigner Eckart theorem, but for $\frac{3}{2}^{+}$ and $\frac{5}{2}^{+}$ states the quadrupole moment, as expected, is large but of negative sign: on the average we get $Q \simeq -5.28$ $e\text{fm}^2$ for the $\frac{3}{2}^{+}$ state and $Q \simeq -7.55$ $e\text{fm}^2$ for the $\frac{5}{2}^{+}$ state. Here we observe an important structural difference relative to ^{9}Be that carries a positive quadrupole moment whose experimental value is 5.3 ± 0.3 $e\text{fm}^2$ [12]. This is a direct consequence of the underlying differences in the dominant partial waves that bind ${}^{9}_{\Lambda}$Be and ^{9}Be as three-body cluster nuclei, as explained in ref. [2].

Since the ground state of ${}^{9}_{\Lambda}$Be is $\frac{1}{2}^{+}$ the dominant channels involve $\ell = L = 0$. Therefore the magnetic moment $\mu = \mu_\Lambda$ for such state. Nevertheless the same sign change effect relative to ^{9}Be takes place in the magnetic moment μ of ${}^{9}_{\Lambda}$Be excited states. On the average, we find $\mu \simeq 1.26$ μ_N for the $\frac{3}{2}^{+}$ state and $\mu = 0.38$ μ_N for the $\frac{5}{2}^{+}$ state. Although the $\Lambda-$particle and the neutron have magnetic moments of the same sign, once embedded in ^{8}Be, one gets magnetic moments of opposite sign for reasons similar to the ones explained for Q.

3 Resonances in ${}^{9}_{\Lambda}$Be

In the present work we find two pairs of resonances: one below breakup threshold and one above breakup threshold. The first one is $\frac{1}{2}^{-}(\frac{3}{2}^{-})$ while the second one is $\frac{7}{2}^{+}(\frac{9}{2}^{+})$. The results are shown in Table 3 and 4 respectively. The $\frac{1}{2}^{-}(\frac{3}{2}^{-})$ resonance is located just above the $\Lambda-\alpha$ threshold and has a large decay width. Nevertheless, unlike the work in ref. [5], we find no negative parity states above breakup threshold, but unveil instead a positive parity state $\frac{7}{2}^{+}(\frac{9}{2}^{+})$ that relies on the $\alpha - \alpha$ g-wave interaction. This last resonance has not been found before by other continuum calculations but, together with the $\frac{1}{2}^{+}$ bound state and the $\frac{3}{2}^{+}(\frac{5}{2}^{+})$ excited state, complete the even parity rotational band. On the contrary, the absence of a large number of negative parity states in ${}^{9}_{\Lambda}$Be is two fold: **a)** a direct result of the Pauli principle that allows the Λ to stay predominantly in s-state while keeping the fifth neutron in ^{9}Be in a p-state;

b) proper coupling to the decay channel $\alpha + \Lambda + \alpha$ and correct location of all underlying branch cuts of the $\alpha - \alpha$ system in the complex energy plane. We have tried to search for $\frac{5}{2}^-(\frac{7}{2}^-)$ resonances corresponding to total orbital angular momentum $\mathbf{L} = 3$ but could not find any. Denoting E_x (in MeV) the excitation energy relative to the $\frac{1}{2}^+$ ground state we find that $5.7 < E_x < 8.3$ for the negative parity $\frac{1}{2}^-(\frac{3}{2}^-)$ state, while $8.3 \leq E_x \leq 10.3$ for the positive parity $\frac{7}{2}^+(\frac{9}{2}^+)$ resonance. If we consider, for example, the MS $\Lambda - \alpha$ potential together with CB $\alpha - \alpha$ interaction the $\frac{1}{2}^-(\frac{3}{2}^-)$ resonance corresponds to $\simeq 6.9$ MeV excitation and the $\frac{7}{2}^+(\frac{9}{2}^+)$ to $\simeq 9.7$ MeV.

Table 3. Position and width of the $\frac{1}{2}^-(\frac{3}{2}^-)$ resonance (MeV) for different $\alpha - \alpha$ and $\Lambda - \alpha$ potentials in the absence of the Coulomb repulsion.

	UIM		AB		CB	
	E	$\Gamma/2$	E	$\Gamma/2$	E	$\Gamma/2$
TH	−2.00	0.62	−1.98	0.59	−2.03	0.68
MS	−1.69	0.56	−1.65	0.55	−1.67	0.64
DE	−1.89	0.58	−1.84	0.57	−1.86	0.66
DA	−1.92	0.41	−1.89	0.40	−1.87	0.47

Table 4. Position and width of the $\frac{7}{2}^+(\frac{9}{2}^+)$ resonance (MeV) for different $\alpha - \alpha$ and $\Lambda - \alpha$ potentials in the absence of the Coulomb repulsion.

	UIM		AB		CB	
	E	$\Gamma/2$	E	$\Gamma/2$	E	$\Gamma/2$
TH	1.29	0.14	0.59	0.06	1.66	0.20
MS	0.51	0.06	0.06	0.03	1.09	0.12
DE	−0.06	0.04	−0.51	0.02	0.51	0.09
DA	−0.30	0.04	−0.64	0.02	0.29	0.09

So far the most accepted understanding of the level structure of ${}^9_\Lambda$Be is the work of Yamada et al. [13] where a variational (α + 3N + N) + Λ cluster model that includes $\alpha-$ particle breakup modes is applied to achieve a unified understanding of this hypernucleus in the energy region up to 20 MeV excitation. They find a very rich spectra of rotational bands with different symmetries [14] of which we can only find the lowest one made up of $\frac{1}{2}^+$, $\frac{3}{2}^+(\frac{5}{2}^+)$, and $\frac{7}{2}^+(\frac{9}{2}^+)$ states. The two negative parity bands are absent from our calculation except for the lowest state $\frac{1}{2}^-(\frac{3}{2}^-)$ in the first band which carries total orbital angular momentum $\mathbf{L} = 1$. Two reasons may explain such disagreement: **a)** lack of internal structure of the $\alpha-$particle in our calculation; **b)** lack of appropriate coupling to the continuum in their calculation.

As far as we know, the experimental evidence for the spectra of ${}^{9}_{\Lambda}$Be comes from ref. [15, 16]. In ref. [15] only two peaks are found above the $\alpha + \Lambda + \alpha$ threshold. The second peak, at about 20 MeV above breakup threshold, is clearly out of the domain of applicability of our calculation due to lack of structure in the $\alpha-$ particle. The first peak, also centered at about 5 MeV above breakup, may not correspond to any of the states we have found because of both location and width; the $\frac{1}{2}^-(\frac{3}{2}^-)$ has the right width but is located very close to the breakup threshold once we add about 1.8 MeV for the Coulomb energy, while the $\frac{7}{2}^+(\frac{9}{2}^+)$ has the correct position but is too narrow to correspond to the experimental peak. Instead the experimental work of ref. [16] finds two resonance peaks corresponding to excitation energies of $E_x = 6$ and 10 MeV above the ground state, which may correspond to the states we find. The first one corresponds to the $\frac{1}{2}^-(\frac{3}{2}^-)$ state (1^-), while the second to the $\frac{7}{2}^+(\frac{9}{2}^+)$ state (4^+). In ref. [16] this last state has been denoted as 3^- in accordance with the theoretical predictions of ref. [13] but, as we mentioned above, we find no additional negative parity resonance to justify such assignment.

References

1. H. Feshbach: In: *Proc. of the Summer Study Meeting on Kaon Physics and Facilities*, H. Palevsky (ed.), p. 185 (BNL 18335) 1973
2. E. Cravo, A.C. Fonseca and Y. Koike: Phys. Rev. **C66** 014001 (2002)
3. A.C. Fonseca and M.T. Peña: Nucl. Phys. **A487**, 92 (1988)
4. E. Cravo: Phys. Rev. **C54**, 523 (1996)
5. S. Oryu et al.: Few-Body Systems **28**, 103 (2000)
6. Y. Sunami and H. Narumi: Prog. Theor. Phys. **66**, 355 (1981)
7. L.D. Faddeev: JETP **12**, 1014 (1961).
8. D.R. Lehman et al.: Phys. Rev. **C29**, 1450 (1984)
9. Y. Koike: Phys. Rev. **C42**, R2286 (1990)
10. M. Jurič et al.: Nucl. Phys. **B52**, 1 (1973)
11. H. Akikawa et al.: Phys. Rev. Lett. **88**, 082501 (2002)
12. F. Ajzenberg-Selove: Nucl. Phys. **A490**, 1 (1988)
13. T. Yamada et al.: Phys. Rev. **C38**, 854 (1988) and references therein
14. R.H. Dalitz and A. Gal: Phys. Rev. Lett. **36**, 362 (1976)
15. W. Brüeckner et al.: Phys. Lett. **B55**, 107 (1975); **79B**, 157 (1978)
16. S. Ajimura et al.: Nucl. Phys. **A639**, 93C (1998)

Few-Body Systems Suppl. 14, 251–256 (2003)

Few-Body Systems

The Onset of Binding for $\Lambda\Lambda$ and $\Lambda\Xi$ Hypernuclei

A. Gal*

Racah Institute of Physics, The Hebrew University, Jerusalem 91904, Israel

Abstract. Faddeev calculations suggest that: (i)$B_{\Lambda\Lambda}({}_{\Lambda\Lambda}^{\ 6}\mathrm{He})$ for the recent KEK emulsion event is compatible with fairly weak $\Lambda\Lambda$ potentials $V_{\Lambda\Lambda}$ such as due to the Nijmegen soft-core NSC97 model; (ii) the isodoublet ${}_{\Lambda\Lambda}^{\ 5}\mathrm{H}$ - ${}_{\Lambda\Lambda}^{\ 5}\mathrm{He}$ hypernuclei are particle-stable even in the limit $V_{\Lambda\Lambda} \to 0$; and (iii) ${}_{\Lambda\Lambda}^{\ 4}\mathrm{H}$ considered as a $\Lambda\Lambda d$ system is particle-stable over a wide range of strengths $\bar{V}_{\Lambda\Lambda}$. However, four-body $\Lambda\Lambda pn$ Faddeev-Yakubovsky calculations do *not* produce a bound state for ${}_{\Lambda\Lambda}^{\ 4}\mathrm{H}$ even for $V_{\Lambda\Lambda}$ considerably stronger than required to reproduce $B_{\Lambda\Lambda}({}_{\Lambda\Lambda}^{\ 6}\mathrm{He})$. This is in contrast to the normal situation (e.g. ${}_{\Lambda\Lambda}^{10}\mathrm{Be}$) where a four-body Faddeev-Yakubovsky calculation yields stronger binding than that due to a suitably defined three-body Faddeev calculation.

For stranger systems, Faddeev calculations using $\Lambda\Xi$ interactions which simulate model NSC97 suggest that ${}_{\Lambda\Xi}^{\ 6}\mathrm{He}$ marks the onset of nuclear stability for Ξ hyperons.

1 Introduction and Input

Information on hyperon-hyperon (YY) interactions is not readily available from experiments in free space. It is almost exclusively limited to the study of strangeness $S{=}-2$ hypernuclear systems. This information is crucial for extrapolating into multi-strange hadronic matter, for both finite systems and in bulk, and into neutron stars [1].

The recent unambiguous identification of ${}_{\Lambda\Lambda}^{\ 6}\mathrm{He}$ in the KEK hybrid-emulsion experiment E373 [2] is consistent with a scattering length $a_{\Lambda\Lambda}{\sim}-0.8$ fm [3], indicating a considerably weaker $\Lambda\Lambda$ interaction than that specified by $a_{\Lambda N}{\sim}-2$ fm [4] for the ΛN interaction. With such a relatively weak $\Lambda\Lambda$ interaction, and since the three-body system $\Lambda\Lambda N$ is unbound (comparing it with the unbound Λnn system [5]), the question of whether or not the onset of binding in the $S{=}-2$ hadronic sector occurs at $A{=}4$ becomes highly topical.

Here I review primarily a recent first-ever Faddeev-Yakubovsky calculation for ${}_{\Lambda\Lambda}^{\ 4}\mathrm{H}$ as a four-body $\Lambda\Lambda pn$ system [6]. It takes into account properly *all* the

*Work done in collaboration with I.N. Filikhin, supported by the Israel Science Foundation

rearrangement channels into which the $\Lambda\Lambda pn$ system may be split. I also briefly review calculations for other $\Lambda\Lambda$ hypernuclei and $\Lambda\Xi$ hypernuclei [7].

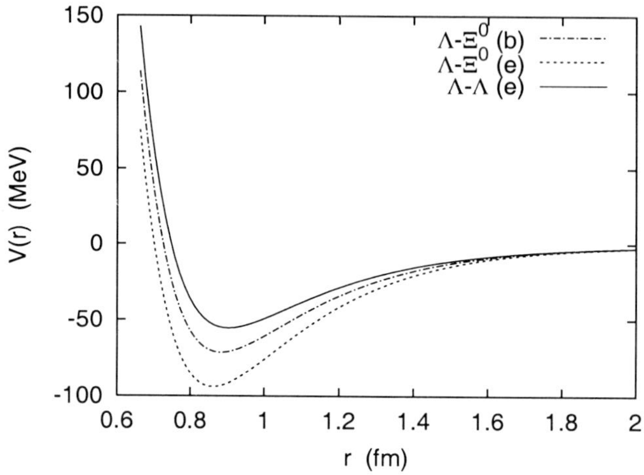

Figure 1. Selected YY potentials simulating NSC97 model interactions.

The YN and YY s-wave interaction input potentials consisted of combinations of three Gaussians with different ranges, such as used by Hiyama et al. [8]. A similar form was used for the pn triplet interaction and the results proved almost identical to those derived using the Malfliet-Tjon potential MT-III [9]. Several YY potentials fitted to reproduce the low-energy parameters of the Nijmegen soft-core NSC97 model [10] are shown in Fig. 1. We note that the $\Lambda\Xi$ interaction is rather strong, considerably stronger within the same version of the model (here e) than the $\Lambda\Lambda$ interaction. The $\Lambda\Lambda$ interaction is fairly weak for all six versions (a)-(f) of model NSC97. Versions e and f provide a reasonable description of single-Λ hypernuclei [4].

2 Results

2.1 ${}^{4}_{\Lambda\Lambda}H$ in a $\Lambda\Lambda d$ Model

$\Lambda\Lambda$ binding-energy values ($B_{\Lambda\Lambda}$) calculated within a $\Lambda\Lambda d$ s-wave Faddeev calculation are shown in Fig. 2 as function of the scattering length $a_{\Lambda\Lambda}$ for two Λd potentials, one of an exponential shape and the other one of an isle shape, both of them fitted to the low-energy parameters of an s-wave Faddeev calculation for Λpn which uses model NSC97f for the underlying ΛN interaction, yielding $B_{\Lambda}({}^{3}_{\Lambda}\text{H}) = 0.19$ MeV.

The roughly linear increase of $B_{\Lambda\Lambda}$ holds generally in three-body $\Lambda\Lambda C$ models (C standing for a cluster) over a wide range of values for the strength parameter $\bar{V}_{\Lambda\Lambda}$ [3]. The onset of particle stability for ${}^{4}_{\Lambda\Lambda}\text{H}(1^+)$ requires a minimum $\bar{V}_{\Lambda\Lambda}$ which is satisfied for the choice of $B_{\Lambda\Lambda}({}^{6}_{\Lambda\Lambda}\text{He})$ [2] as a normalizing datum.

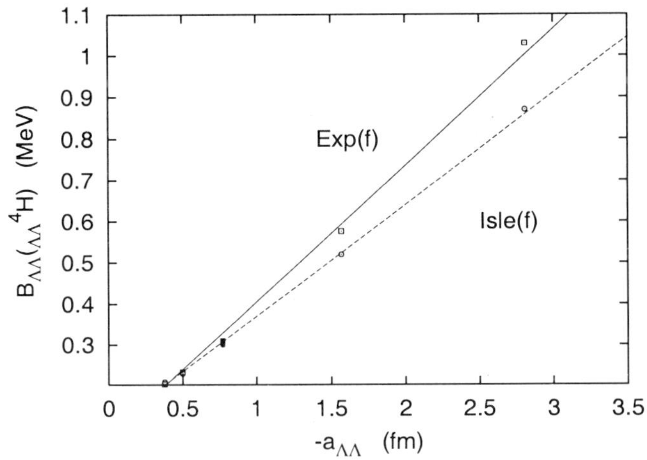

Figure 2. $B_{\Lambda\Lambda}({}^{4}_{\Lambda\Lambda}\mathrm{H})$ calculated in a three-body $\Lambda\Lambda d$ model as function of the scattering length $a_{\Lambda\Lambda}$, for two Λd potentials that reproduce a Faddeev calculation of ${}^{3}_{\Lambda}\mathrm{H}(\frac{1}{2}^{+})$ using NSC97f for the ΛN interaction. The solid squares correspond to the "${}^{6}_{\Lambda\Lambda}\mathrm{He}$" $\Lambda\Lambda$ interaction. The straight lines are drawn only to lead the eye.

Disregarding inessential complications due to spin it can be shown that, for essentially attractive $\Lambda\Lambda$ interactions and for a static nuclear core d, a two-body Λd bound state implies binding for the three-body $\Lambda\Lambda d$ system [11].

2.2 ${}^{4}_{\Lambda\Lambda}H$ in a $\Lambda\Lambda pn$ Model

For two identical hyperons and two essentially identical nucleons (upon introducing isospin) as appropriate to the $\Lambda\Lambda pn$ model calculation of ${}^{4}_{\Lambda\Lambda}\mathrm{H}$, the 18 Faddeev-Yakubovsky components reduce to seven independent components satisfying coupled equations. Six rearrangement channels are involved in our s-wave calculation [6] for ${}^{4}_{\Lambda\Lambda}\mathrm{H}$: $(\Lambda NN)_{S=\frac{1}{2}}+\Lambda$, $(\Lambda NN)_{S=\frac{3}{2}}+\Lambda$, $(\Lambda\Lambda N)_{S=\frac{1}{2}}+N$, for 3+1 breakup clusters; and $(\Lambda\Lambda)_{S=0}+(NN)_{S=1}$, $(\Lambda N)_{S}+(\Lambda N)_{S'}$, with $(S,S')=(0,1)+(1,0)$ and $(1,1)$ for 2+2 breakup clusters.

Using $V_{\Lambda\Lambda}$ which reproduces $B_{\Lambda\Lambda}({}^{6}_{\Lambda\Lambda}\mathrm{He})$, our calculations yield no bound state for the $\Lambda\Lambda pn$ system, as demonstrated in Fig. 3 by the location of the "$\Lambda\Lambda pn$" curve *above* the horizontal straight line marking the "$\Lambda+{}^{3}_{\Lambda}\mathrm{H}$ threshold", where this threshold was in turn obtained as the asymptote of the Λpn s-wave Faddeev calculation.[1] In fact our Faddeev-Yakubovsky calculations exhibit little sensitivity to $\bar{V}_{\Lambda\Lambda}$ over a wide range. Even for considerably stronger $\Lambda\Lambda$ interactions we get a bound ${}^{4}_{\Lambda\Lambda}\mathrm{H}$ only if the ΛN interaction is made considerably stronger, by as much as 40%. With four ΛN pairwise interactions out of a total of six, the strength of the ΛN interaction (about half of that for NN)

[1] Recall that, in agreement with Fig. 2, the asymptote of the curve marked '$\Lambda\Lambda d$' is located *below* this threshold.

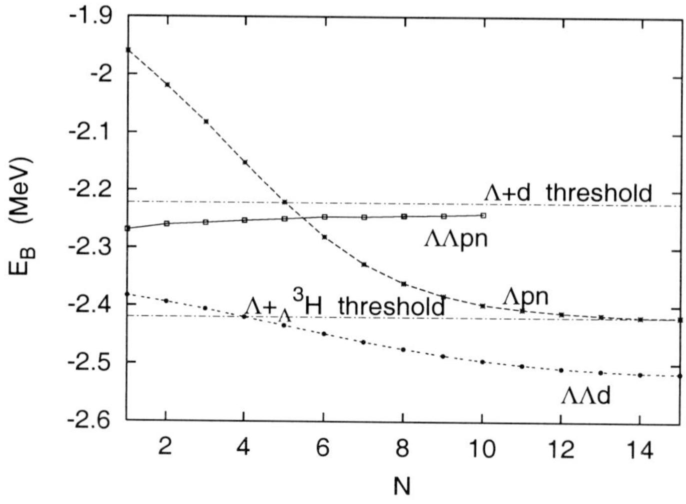

Figure 3. Convergence of Faddeev-Yakubovsky calculations for the binding energy of the Λpn $(\frac{1}{2}^+)$, $\Lambda\Lambda d$ and $\Lambda\Lambda pn$ (1^+) systems with respect to the number N of basis functions. The $\Lambda\Lambda$ interaction was fitted to $B_{\Lambda\Lambda}({}_{\Lambda\Lambda}^{\ 6}\mathrm{He})$.

plays a major role in the four-body $\Lambda\Lambda pn$ problem. Here it appears impossible to prove that, for essentially attractive $\Lambda\Lambda$ interactions and for a *non static* nuclear core d (made out of dynamically interacting proton and neutron), a Λd bound state implies binding for the $\Lambda\Lambda d$ system. It is unlikely that incorporating higher partial waves, and $\Lambda\Lambda - \Xi N$ coupling effects, will change this qualitative feature.

2.3 ${}_{\Lambda\Lambda}^{10}Be$

For heavier $\Lambda\Lambda$ hypernuclei, as confirmed by Faddeev-Yakubovsky calculations [3] for ${}_{\Lambda\Lambda}^{10}\mathrm{Be}$, the relationship between the three-body and four-body models is opposite to that found here for ${}_{\Lambda\Lambda}^{\ 4}\mathrm{H}$: the $\Lambda\Lambda C_1C_2$ calculation provides *higher* binding than a properly defined $\Lambda\Lambda C$ calculation yields (with $C = C_1 + C_2$) due to the attraction induced by the ΛC_1-ΛC_2, $\Lambda\Lambda C_1$-C_2, C_1-$\Lambda\Lambda C_2$ four-body rearrangement channels that include bound states for which there is no room in the three-body $\Lambda\Lambda C$ model. The binding energy calculated within the four-body model increases then "normally" with $\bar{V}_{\Lambda\Lambda}$. This is demonstrated in Fig. 4 for several $\Lambda\Lambda$ interactions, including $V_{\Lambda\Lambda} = 0$ which corresponds to the lowest point on each one of the straight lines. The fairly large value of 1.5 MeV for $\Delta B_{\Lambda\Lambda}({}_{\Lambda\Lambda}^{10}\mathrm{Be})$ in the limit $V_{\Lambda\Lambda} \to 0$ is due to the special $\alpha\alpha$ cluster structure of the ${}^8\mathrm{Be}$ core. Other aspects of the correlation between the ${}_{\Lambda\Lambda}^{10}\mathrm{Be}$ four-body calculation and the ${}_{\Lambda\Lambda}^{\ 6}\mathrm{He}$ three-body calculation, and of the consistency between the various reports on the binding energy of these two species, are discussed in ref. [3] and by Hiyama in these proceedings.

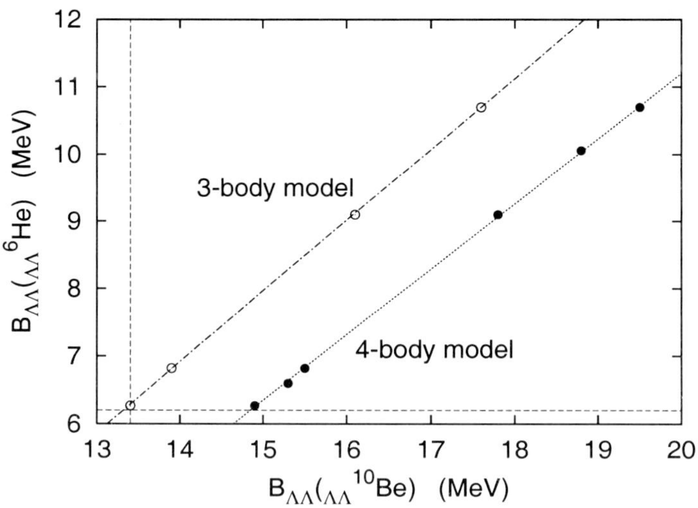

Figure 4. $B_{\Lambda\Lambda}$ calculated for ${}_{\Lambda\Lambda}^{6}$He in a three-body $\alpha\Lambda\Lambda$ model, and for ${}_{\Lambda\Lambda}^{10}$Be in a four-body $\alpha\alpha\Lambda\Lambda$ model (dotted line) *vs* a three-body ^{8}Be$\Lambda\Lambda$ model (dot-dash line). The origin of the dashed axes corresponds to $\Delta B_{\Lambda\Lambda} = 0$, where $\Delta B_{\Lambda\Lambda}({}_{\Lambda\Lambda}^{A}Z) = B_{\Lambda\Lambda}({}_{\Lambda\Lambda}^{A}Z) - 2B_{\Lambda}({}^{(A-1)}_{\Lambda}Z)$.

2.4 The Onset of Ξ Stability

If model NSC97 indeed provides a valid extrapolation from fits to NN and YN data, and recalling the strongly attractive ${}^{1}S_{0}$ $\Lambda\Xi$ potentials due to this model (Fig. 1), it is only natural to search for stability of $A = 6, S = -3$ systems obtained from ${}_{\Lambda\Lambda}^{6}$He replacing one of the Λ's by Ξ. Our Faddeev calculations [7] for the isodoublet hypernuclei ${}_{\Lambda\Xi}^{6}$H and ${}_{\Lambda\Xi}^{6}$He, considered as $\alpha\Lambda\Xi^{-}$ and $\alpha\Lambda\Xi^{0}$ three-body systems respectively, indicate that ${}_{\Lambda\Xi}^{6}$He is particle-stable against Λ emission to ${}_{\Lambda\Lambda}^{5}$He for potentials simulating model NSC97, particularly versions e and f, whereas ${}_{\Lambda\Xi}^{6}$H is unstable since $M(\Xi^{-}) > M(\Xi^{0})$ by 6.5 MeV.[2] However, predicting particle stability for ${}_{\Lambda\Xi}^{6}$He is not independent of the assumptions made on the experimentally unexplored $\Xi\alpha$ interaction which was extrapolated from recent data on ^{12}C [12]; hence this prediction cannot be considered conclusive.

3 Conclusion

We have provided a first ever four-body Faddeev-Yakubovsky calculation for ${}_{\Lambda\Lambda}^{4}$H using NN and ΛN interaction potentials that fit the available data on the relevant subsystems, including the binding energy of ${}_{\Lambda}^{3}$H. No ${}_{\Lambda\Lambda}^{4}$H bound state was obtained for a wide range of $\bar{V}_{\Lambda\Lambda}$, including that corresponding to

[2]The $I = 1/2$ ${}_{\Lambda\Lambda}^{5}$H$-{}_{\Lambda\Lambda}^{5}$He hypernuclei, within a $\Lambda\Lambda C$ Faddeev calculation, are particle stable even in the limit $V_{\Lambda\Lambda} \to 0$

$B_{\Lambda\Lambda}({}_{\Lambda\Lambda}^{6}\mathrm{He})$. This non binding is due to the relatively weak ΛN interaction, in stark contrast to the results of a "reasonable" three-body $\Lambda\Lambda d$ Faddeev calculation. More experimental work is needed to decide whether or not the events reported recently in the AGS experiment E906 [13] correspond to ${}_{\Lambda\Lambda}^{4}\mathrm{H}$.

Accepting the predictive power of model NSC97, our calculations suggest that ${}_{\Lambda\Xi}^{6}\mathrm{He}$ may be the lightest particle-stable $S = -3$ hypernucleus, and the lightest and least strange particle-stable hypernucleus in which a Ξ hyperon is bound (and not ${}_{\Lambda\Lambda\Xi}^{7}\mathrm{He}$ for $S = -4$ as argued in ref. [14]). Unfortunately, the direct production of $\Lambda\Xi$ hypernuclei is beyond present experimental capabilities, requiring the use of Ω^- initiated reactions.

Acknowledgement.

Special thanks are due to Igor Filikhin who has devotedly and patiently introduced me to modern aspects of Few-Body Physics, a research field I have not attended since the pioneering K^-d reactive Faddeev calculation [15].

References

1. J. Schaffner-Bielich and A. Gal: Phys. Rev. **C62**, 034311 (2000); J. Schaffner-Bielich et al.: Phys. Rev. Lett. **89**, 171101 (2002)
2. H. Takahashi et al.: Phys. Rev. Lett. **87**, 212502 (2001)
3. I.N. Filikhin and A. Gal: Nucl. Phys. **A707**, 491 (2002)
4. Th.A. Rijken, V.G.J. Stoks, Y. Yamamoto: Phys. Rev. **C59**, 21 (1999)
5. Y.C. Tang and R.C. Herndon: Phys. Rev. Lett. **14**, 991 (1965)
6. I.N. Filikhin and A. Gal: Phys. Rev. Lett. **89**, 172502 (2002)
7. I.N. Filikhin and A. Gal: Phys. Rev. **C65**, 041001(R) (2002)
8. E. Hiyama et al.: Prog. Theor. Phys. **97**, 881 (1997); Phys. Rev. **C66**, 024007 (2002)
9. R.A. Malfliet and J.A. Tjon: Nucl. Phys. **A127**, 161 (1969)
10. V.G.J. Stoks and Th.A. Rijken: Phys. Rev. **C59**, 3009 (1999)
11. J.-L. Basdevant: private communication by J.-M. Richard (2002)
12. T. Fukuda et al.: Phys. Rev. **C58**, 1306 (1998); P. Khaustov et al.: Phys. Rev. **C61**, 054603 (2000)
13. J.K. Ahn et al.: Phys. Rev. Lett. **87**, 132504 (2001); see, however, I. Kumagai-Fuse and S. Okabe: Phys. Rev. **C66**, 014003 (2002)
14. J. Schaffner et al.: Ann. Phys. [NY] **235**, 35 (1994)
15. G. Toker, A. Gal, J.M. Eisenberg: Phys. Lett. **B88**, 235 (1979); Nucl. Phys. **A362**, 405 (1981)

Few-Body Systems Suppl. 14, 257–262 (2003)

Few-Body Systems

Few-Body Aspects of Hypernuclear Physics

E. Hiyama[1], M. Kamimura[2], T. Motoba[3], T. Yamada[4], Y. Yamamoto[5]

[1] Institute of Particle and Nuclear Studies, High Energy Accelerator Research Organization (KEK), Tsukuba, 305-0801, Japan

[2] Department of Physics, Kyushu Univ., Fukuoka 812-8581, Japan

[3] Osaka Electro-Comm. Univ., Neyagawa 572-8530, Japan

[4] Laboratory of Physics, Kanto Gakuin Univ., Yokohama 236-8501, Japan

[5] Physics Section, Tsuru Univ., Tsuru, Yamanashi 402-8555, Japan

Abstract. On the basis of the three- and four-body structure calculations of ${}^{4}_{\Lambda}\mathrm{H}$ and ${}^{4}_{\Lambda}\mathrm{He}$, ${}^{7}_{\Lambda\Lambda}\mathrm{He}$, ${}^{7}_{\Lambda\Lambda}\mathrm{Li}$, ${}^{8}_{\Lambda\Lambda}\mathrm{Li}$, ${}^{9}_{\Lambda\Lambda}\mathrm{Li}$, ${}^{9}_{\Lambda\Lambda}\mathrm{Be}$ and ${}^{10}_{\Lambda\Lambda}\mathrm{Be}$, it is emphasized that there are many interesting and important few-body problems in hypernuclear physics. By performing four-body calculation with both $3N + \Lambda$ and $3N + \Sigma$ channels with realistic NN and YN interactions, we show that the Σ-channel component plays an important role in making ${}^{4}_{\Lambda}\mathrm{H}$ and ${}^{4}_{\Lambda}\mathrm{He}$. Energy levels of the double-Λ hypernuclei ${}^{7}_{\Lambda\Lambda}\mathrm{He}$, ${}^{7}_{\Lambda\Lambda}\mathrm{Li}$, ${}^{8}_{\Lambda\Lambda}\mathrm{Li}$, ${}^{9}_{\Lambda\Lambda}\mathrm{Li}$, ${}^{9}_{\Lambda\Lambda}\mathrm{Be}$ and ${}^{10}_{\Lambda\Lambda}\mathrm{Be}$ are predicted on the basis of the $\alpha + x + \Lambda + \Lambda$ four-body model with $x = n, p, d, t, {}^{3}\mathrm{He}$ and α, respectively. The *Demachi-Yanagi* event, observed recently for ${}^{10}_{\Lambda\Lambda}\mathrm{Be}$, is interpreted as discovery of its 2^{+} excited state by the present calculation.

1 Introduction

One of the major goals of hypernuclear physics is to study any dynamical change of the nuclear structure by injecting impurity particles, hyperons, into the core nucleus. Another important purpose of the hypernuclear physics is to understand the hyperon-nucleon(YN) and hyperon-hyperon(YY) interactions. Since the YN scattering experiments are extremely limited, so-far proposed YN interactions have a great ambiguity. Therefore, it is very important to obtain useful information on YN and YY interactions from hypernuclear structure study. Especially few-body approach is essentially important. Along this line, in this paper, we discuss ${}^{4}_{\Lambda}\mathrm{H}$, ${}^{4}_{\Lambda}\mathrm{He}$, ${}^{7}_{\Lambda\Lambda}\mathrm{He}$, ${}^{7}_{\Lambda\Lambda}\mathrm{Li}$, ${}^{8}_{\Lambda\Lambda}\mathrm{Li}$, ${}^{9}_{\Lambda\Lambda}\mathrm{Li}$, ${}^{9}_{\Lambda\Lambda}\mathrm{Be}$ and ${}^{10}_{\Lambda\Lambda}\mathrm{Be}$ and emphasize what is interesting from the view points of hypernuclear physics and few-body physics.

2 Calculational Method

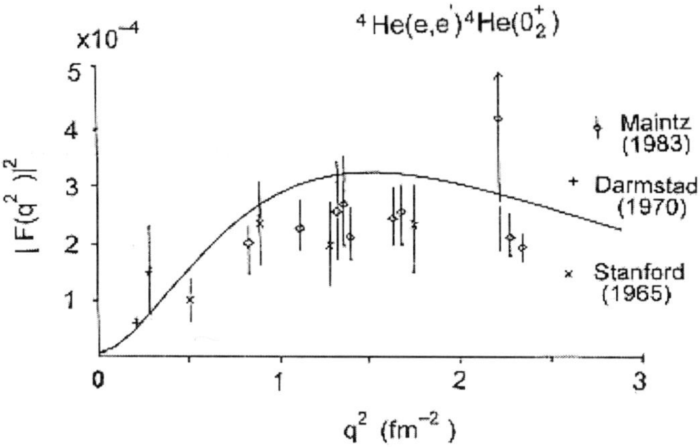

Figure 1. Inelastic electron scattering form factor of ^{4}He with AV8 NN force.

In order to solve three- and four-body problems accurately, we have employed the coupled-rearrangement-channel variational method with the use of Jacobian-coordinate Gaussian basis functions [1, 2]. This method has been successfully applied to the bound states of various three- and four-body systems [1, 2, 3]. The basis functions work excellently in describing both the short-range correlations and the long-range tail behaviour. Further development for solving four-body bound states has been done by the two of the present authors (E. H. and M. K.) by introducing a new representation of the Gaussian basis functions, infinitesimally-shifted Gaussian-lobe basis functions [4], which are mathematically equivalent to the previous ones [1, 2] but are different in expression, so that the calculation of the interaction matrix elements can be performed much more easily. Owing to the use of these new basis functions, applicability of our calculational method is much extended even to four-body systems with complicated interactions. As an example, we made a benchmark test of calculation of four-nucleon bound state (^{4}He) using AV8 NN potential [5]. The results in the binding energy, r.m.s. radius and two-body correlation function are in good agreement with those among seven different calculational methods.

Furthermore, it is interesting to apply the four-body method to the second 0^+ state at 20.1 MeV excitation. So far, Fonseca et al. have studied structure of this state based on the solution of the AGS equations [6].

Recently, we succeeded in performing a four-body calculation of the second 0^+ state using full rank AV8 NN interaction, namely with no truncation of the NN force in the angular momentum space. Since AV8 potential does not reproduce the binding energies of ^{3}H, ^{3}He and ^{4}He, we employed a phenomenological three-body force so as to reproduce these binding energies. As a results, the r.m.s. charge radius of the ground state of ^{4}He is 1.658 fm, which is reproduced precisely the experimental data, 1.671 ± 0.014 fm. It is interesting to see the binding energy and wave-function of the second 0^+ state. The excitation energy is 20.1 MeV which is well reproduced the observed binding energy. We illustrate the form factor of the inelastic electron scattering. As shown in Fig. 1, the result is not inconsistent with the observation. This is the first four-body calculation using realistic NN force.

3 $^4_\Lambda$H and $^4_\Lambda$He

In hypernuclear physics, the study of the role of Σ particle in the Λ hypernuclei is one of the most interesting subjects. In the normal nuclei, the $N \to \Delta$ conversion has been intensively studied by many authors [7, 8]. In hypernuclei, the mass difference between Λ and Σ, which is about 80 MeV, is much smaller than that between N and Δ. Therefore, $\Lambda \to \Sigma$ is expected to play an important role of hypernuclear structure. For example, it is interesting issue to know how important is the $\Lambda \to \Sigma$ conversion in the binding energy and the Σ mixing probability in $^4_\Lambda$H and $^4_\Lambda$He. For this study, it is highly desirable to perform four-body calculation taking both $(3N + \Lambda)$ and $(3N + \Sigma)$ channel explicitly. This calculation is considered to be one of the most difficult four-body calculations. Recently we have successfully performed the extensive four-body calculations without any restriction on the configuration space: Both the $3N+\Lambda$ and $3N+\Sigma$ channels have been incorporated explicitly and all the rearrangement channels of these baryons are taken into account. Jacobian coordinates of $^4_\Lambda$H and $^4_\Lambda$He are illustrated in Fig. 2.

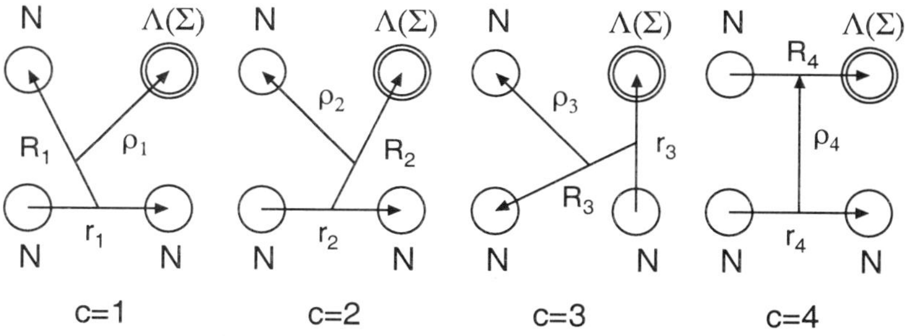

Figure 2. Jacobian coordinates for all the rearrangement channels of $(3N + \Lambda)$ and $(3N + \Sigma)$ of $^4_\Lambda$H and $^4_\Lambda$He. The three nucleons are to be antisymmetrized.

The four-body wave-function are written as

$$\Psi_{JM}({}^4_\Lambda\mathrm{H},{}^4_\Lambda\mathrm{He}) = \sum_{Y=\Lambda,\Sigma}\sum_{c=1}^{4}\sum_{I,j}\sum_{l,L,\lambda}\sum_{s_0,s,S,t_0,t_0}\sum_{n,N,\nu} A^{(c)}_{Y,nlNL\nu\lambda Ijs_0sSt_0t}$$
$$\times \mathcal{A}_N\Big\{\Big[\Big[[\phi^{(c)}_{nl}(\mathbf{r}_c)\psi^{(c)}_{NL}(\mathbf{R}_c)]_I\xi^{(c)}_{\nu\lambda}(\boldsymbol{\rho}_c)\Big]_j$$
$$\times\Big[[[\chi^{(\sigma)}_{1/2}(N_1)\chi^{(\sigma)}_{1/2}(N_2)]_{s_0}\chi^{(\sigma)}_{1/2}(N_3)]_s\chi^{(\sigma)}_{1/2}(Y)\Big]_S\Big]_{JM}$$
$$\times\Big[[[\chi^{(\tau)}_{1/2}(N_1)\chi^{(\tau)}_{1/2}(N_2)]_{t_0}\chi^{(\tau)}_{1/2}(N_3)]_t\chi^{(\tau)}_{t_Y}(Y)\Big]_{T=1/2}\Big\}, \quad (1)$$

where $\mathcal{A}_N$ are the three-nucleon antisymmetrization operator and the isospin $t_Y = 0$ for $Y = \Lambda$ and $t_Y = 1$ for $Y = \Sigma$. Here, n, N and ν denote the size of the Gaussian basis functions for four-body systems. The eigenenergies of Hamiltonian and coefficients C are determined by Rayleigh-Ritz variational method.

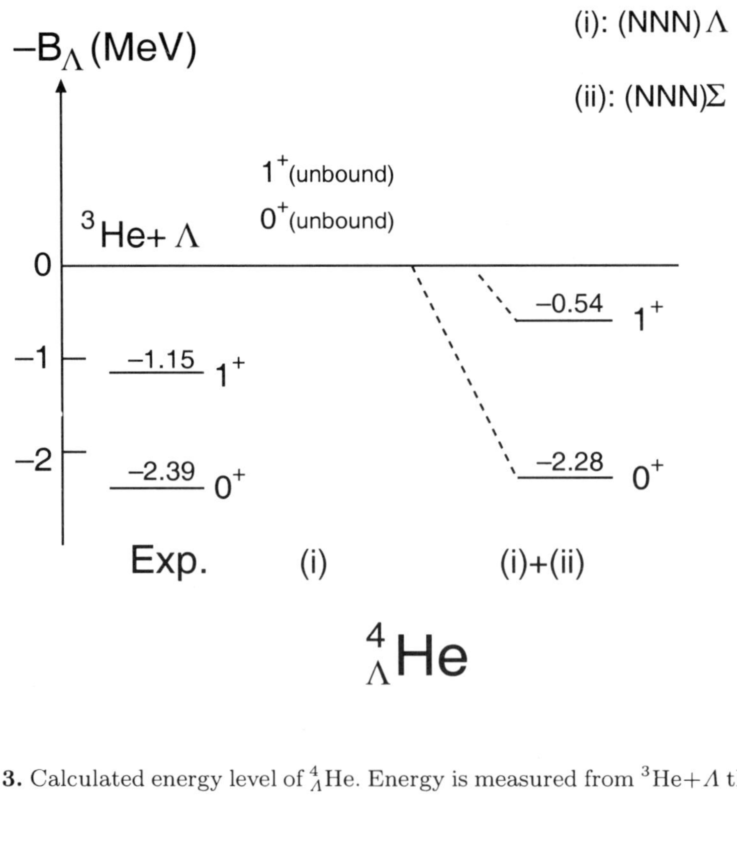

Figure 3. Calculated energy level of ${}^4_\Lambda$He. Energy is measured from ^{3}He+Λ threshold.

As for the YN interaction with $\Lambda N - \Sigma N$ coupling terms, we consider those of Nijmegen soft core '97f, but we simulate them by using a soft-core potential with central, spin-orbit and tensor terms [9] which reproduce the scattering phase shifts obtained by the original interaction. In the following, we refer to this phase-shift equivalent soft-core potential as NSC97f. We employ the AV8 potential as the NN interaction.

All the calculations shown below have been done both for ${}^{4}_{\Lambda}$He and ${}^{4}_{\Lambda}$H, but we present only the case of the former since the results very similar to each other. Calculated energies of the 0^+ ground state and the 1^+ excited state are illustrated in Fig. 3 together with the observed value. In the case of only the $3N + \Lambda$ channel adopted, both of the two states are unbound. When $(3N + \Sigma)$ channel is further included, the 0^+ state is bound and 1^+ state is still unbound. We understand that the Σ-channel component plays a very important role in the energy gain of both the 0^+ and 1^+ states.

4 Double Λ Hypernuclei

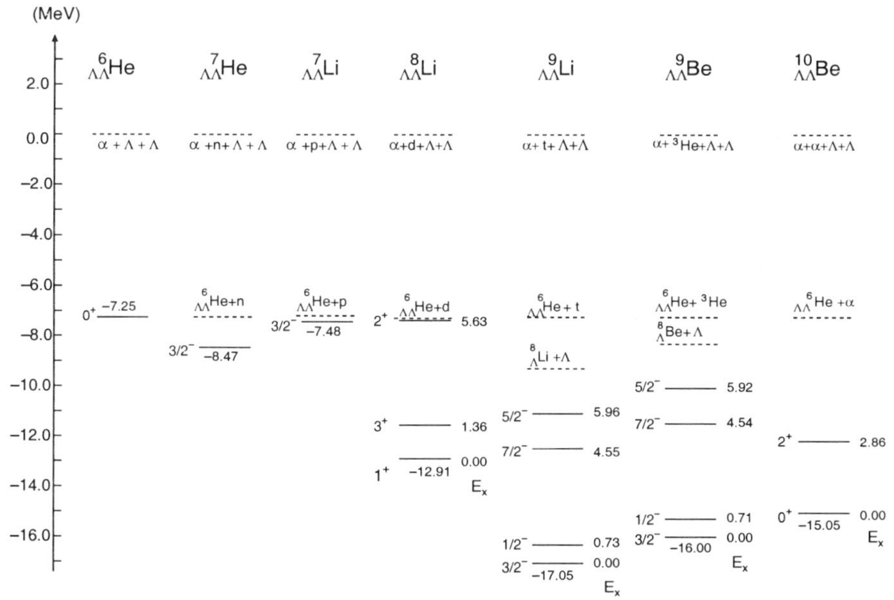

Figure 4. Energy level of double-Λ hypernuclei, ${}^{6}_{\Lambda\Lambda}$He, ${}^{7}_{\Lambda\Lambda}$He, ${}^{7}_{\Lambda\Lambda}$Li, ${}^{8}_{\Lambda\Lambda}$Li, ${}^{9}_{\Lambda\Lambda}$Li, ${}^{9}_{\Lambda\Lambda}$Be and ${}^{10}_{\Lambda\Lambda}$Be.

Last year, ${}^{6}_{\Lambda\Lambda}$He was observed as NAGARA event in KEK-E373 experiment [10]. The obtained $\Lambda\Lambda$-binding energy $B_{\Lambda\Lambda} = 7.25 \pm 0.19^{+0.18}_{-0.11}$ MeV. This observation is giving a great contribution to the study of the structure of double Λ hypernuclei as the entrance to the multi-strangeness world.

The analysis of KEK-E373 experiment is in progress. Furthermore, it will planned to produce many double Λ hypernuclei in the future at BNL and Japan-Hadron-Facility(JHF). Thus, it is required to predict binding energies of ground states and excited states as bound states before measurements. Therefore, hoping to measure much more double Λ hypernuclei at the analysis of E373 experiments, BNL and JHF, we have performed careful theoretical calculations of double-Λ hypernuclei of ${}^{7}_{\Lambda\Lambda}$He, ${}^{7}_{\Lambda\Lambda}$Li, ${}^{8}_{\Lambda\Lambda}$Li, ${}^{9}_{\Lambda\Lambda}$Li, ${}^{9}_{\Lambda\Lambda}$Be and ${}^{10}_{\Lambda\Lambda}$Be.

In these four-body calculations, it is important that all the two-body interactions should be chosen so as to reproduce both the binding energy of any subsystem composed two- and three-consistent particles ad two-body low-energy scattering phase shifts. Among these interactions, the $\Lambda\Lambda$ interaction has been designed within the $\alpha + \Lambda + \Lambda$ three-body model, so as to reproduce the binding energy of ${}^{6}_{\Lambda\Lambda}$He.

The obtained energy spectra of double Λ hypernuclei with $A = 6 \sim 10$ are illustrated in Fig. 4. It is striking that the calculated $B_{\Lambda\Lambda}$ of the 2^+ excited state in ${}^{10}_{\Lambda\Lambda}$Be, 12.28 MeV, agrees with the experimental value $B^{\rm exp}_{\Lambda\Lambda}({}^{10}_{\Lambda\Lambda}{\rm Be}) = 12.33\pm^{+0.35}_{-0.21}$ MeV in the *Demachi-Yanagi* event. We therefore interpret this event as the observation of the 2^+ excited state of ${}^{10}_{\Lambda\Lambda}$Be. It is our intention that these extensive four-body cluster-model calculations should contribute to stimulating comprehensive spectroscopic studies of double Λ hypernuclei.

Acknowledgement. This work was supported in part by Yamada Science Foundation.

References

1. M. Kamimura: Phys. Rev. **A38**, 621 (1988)

2. H. Kameyama, M. Kamimura, Y. Fukushima: Phys. Rev. **C40**, 974 (1989)

3. E. Hiyama, M. Kamimura: Nucl. Phys. **588**, 35c (1995)

4. E. Hiyama: RCNP Physics Report RCNP-P-132, 35 (1994)

5. H. Kamada et al.: Phys. Rev. **C64**, 044001 (2001)

6. A.C. Fonseca, G. Hale, J. Haidenbauer: Few-Body Systems **31**, 139 (2002)

7. B.F. Gibson, A. Goldberg, M.S. Weiss: Phys. Rev. **C6**, 741 (1972)

8. Y. Akaishi, T. Harada, S. Shinmura, Khin Swe Myint: Phys. Rev. Lett. **84**, 3539 (2000)

9. S. Shinmura: Private Communication

10. H. Takahashi et al.: Phys. Rev. Lett. **87**, 212502 (2001)

Few-Body Systems Suppl. 14, 263–268 (2003)

Few-Body Systems

Chiral Dynamics in Systems with Strangeness

E. Oset[1], D. Jido[1], J. Palomar[1], A. Ramos[2], C. Bennhold[3], S. Kamalov[4]

[1] Departamento de Fisica Teorica and IFIC, Universidad de Valencia, Institutos de Investigacion de Paterna, Valencia, Spain

[2] Departament d'Estructura i Constituents de la Materia, Universitat de Barcelona, Diagonal 647, Barcelona, Spain

[3] Center for Nuclear Studies, Department of Physics, The George Washington University, Washington D. C. 20052

[4] Laboratory for Nuclear Physics, JINR Dubna, Russia

Abstract. In this talk a brief review of several problems involving systems with strangeness is made. In the first place one shows how the $\Lambda(1405)$, $\Lambda(1670)$ and $\Sigma(1620)$ states, for $S = -1$, and the $\Xi(1620)$ for $S = -2$ are generated dynamically in the context of unitarized chiral perturbation theory. The results for the $\bar{K}N$ interaction are then used to evaluate the $K^- d$ scattering length. Results obtained for the kaon selfenergy in a nuclear medium within this approach, with application to K^- atoms, are also mentioned. Finally a few words are said about recent developments in the weak decay of Λ hypernuclei and the puzzle of the Γ_n/Γ_p ratio.

1 Spectra of Low Lying Strange Baryons

The combination of unitary techniques together with chiral dynamics of meson and baryon systems already used in [1] and [2] to study low-energy $K^- N$ scattering, has reached a certain maturity, it has been made more systematic and it has also been applied to the study of a large variety of physical processes [3]. The idea behind this work is the use of the unitarity constraints in coupled channels, which provide the imaginary part of the inverse of the scattering T matrix. From there, by means of dispersion relations, or equivalent techniques, one can construct the whole T matrix. In the strangeness $S = -1$ sector the N/D, or dispersion relation method has been applied in [4] with the result that the T matrix in coupled channels can be written as

$$T = [1 - V\,G]^{-1}\,V \tag{1}$$

with V obtained from the lowest order chiral Lagrangian for meson baryon interaction and G a properly regularized loop function of the meson baryon propagators of the intermediate states. The function V is given by

$$V_{ij} = -C_{ij}\frac{1}{4f^2}(2\sqrt{s} - M_{Bi} - M_{Bj})\left(\frac{M_{Bi}+E}{2M_{Bi}}\right)^{1/2}\left(\frac{M_{Bj}+E'}{2M_{Bj}}\right)^{1/2}, \quad (2)$$

with C_{ij} coefficients evaluated in [5] and the G function given by

$$\begin{aligned} G_l &= i2M_l\int\frac{d^4q}{(2\pi)^4}\frac{1}{(P-q)^2 - M_l^2 + i\epsilon}\frac{1}{q^2 - m_l^2 + i\epsilon} \\ &= \frac{2M_l}{16\pi^2}\left\{a_l(\mu) + \ln\frac{M_l^2}{\mu^2} + \frac{m_l^2 - M_l^2 + s}{2s}\ln\frac{m_l^2}{M_l^2}\right. \\ &+\frac{\bar{q}_l}{\sqrt{s}}\left[\ln(s-(M_l^2-m_l^2)+2\bar{q}_l\sqrt{s}) + \ln(s+(M_l^2-m_l^2)+2\bar{q}_l\sqrt{s})\right. \\ &\left.\left. - \ln(-s+(M_l^2-m_l^2)+2\bar{q}_l\sqrt{s}) - \ln(-s-(M_l^2-m_l^2)+2\bar{q}_l\sqrt{s})\right]\right\}, \quad (3) \end{aligned}$$

where μ is the scale of regularization and $a_l(\mu)$ is a subtraction constant which is chosen by fits to the data. Eq. (1) is nothing but the Bethe Salpeter equation where the Kernel V is given by the lowest order chiral amplitude. In [4] it was also found that the subtraction constant is easily related to the cut off employed in [5], and hence the equivalence of the N/D method of [4] and the Bethe Salpeter equation of [5] was established. Yet, the N/D method with the dimensionally regularized G function is preferable to extend the approach to higher energies. The structure of the T matrix in Eq. (1) allows for the existence of poles which correspond to resonances that we call dynamically generated, since they appear thanks to the multiple scattering implicit in the Bethe Salpeter equation, with the lowest order amplitudes playing the role of the potential. In this sense, both in [5] and [4], as well as in [2], one finds very clearly the $\Lambda(1405)$ resonance (see Fig. 1), which has been claimed for long to be a quasibound state of $\bar{K}N$.

More novel have been the findings of [6] where the same approach of [4], with the subtraction constants evaluated from the unique cut off in all channels of [5], has lead to two new resonances, which we identify with the $\Lambda(1670)$ and the $\Sigma(1620)$. In the amplitudes in Fig. 2 one can see how the $\Lambda(1670)$ shows up both in the real and imaginary parts of the amplitudes.

One of the interesting findings concerning the $\Lambda(1670)$ state is that it couples very strongly to the $K\Xi$ states, and quite weakly to the other channels, which allows one to identify the $\Lambda(1670)$ with a quasibound state of $K\Xi$. Even more interesting are the findings of [7], where using also subtraction constants of the natural size found in the study of the $S = -1$ sector, one finds a resonance for $S = -2$ which could be identified in principle with the $\Xi(1620)$ or $\Xi(1690)$ $I = 1/2$ resonances. However, a detailed study of the partial decay widths of the found resonance (which are obtained from the residues of the transition amplitudes at the pole in the second Riemann sheet of the complex plane),

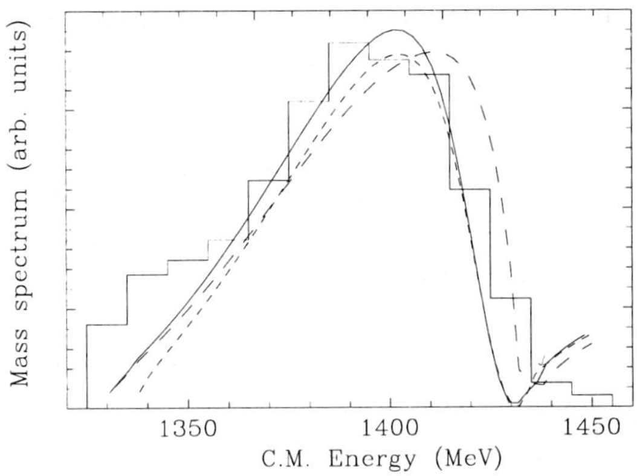

Figure 1. $\Lambda(1405)$ resonance as obtained from the invariant $\pi\Sigma$ mass distribution, with the full basis of physical states (solid line), omitting the η channels (long-dashed line) and with the isospin-basis (short-dashed line).

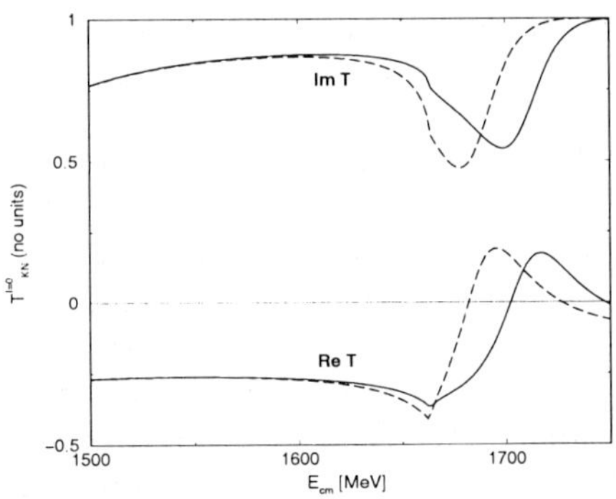

Figure 2. Real and imaginary parts of the $\bar{K}N$ scattering amplitude in the isospin $I = 0$ channel in the region of the $\Lambda(1670)$ resonance.

clearly shows that this resonance is at odds with the properties of the $\Xi(1690)$, and by elimination can only correspond to the $\Xi(1620)$ state, hence allowing one to theoretically determine the spin and parity for this resonance as $1/2^-$, which is unknown in the particle data book. At the same time one makes predictions for the partial decay rates of this resonance which are also unknown so far.

2 K^- Deuteron Scattering Length

One of the results in the study of [5] is the good description of the low energy cross sections of K^-p to different channels. The success of this theory has allowed us to revise the problem of the scattering length of K^-d, which has been approached before [8, 9]. Using the fixed centre approximation to the Faddeev equations, proved to be excellent for this problem in [10], we have studied this scattering length in [11]. The problem is subtle since there is no convergence of the expansion of the multiple scattering series implicit in the Faddeev equations. Second, the results are quite sensitive to the input used for the $\bar{K}N$ amplitudes. The chiral approach has allowed one to re address this latter issue from a new perspective in which special attention is given to questions of unitarity, analyticity, and resonance properties, which make the new amplitudes quite accurate in principle. The results obtained in our approach are $A_{Kd} = -1.61 + i\,1.91$ fm, which differ appreciably from the results quoted as best in Ref. [9] in their multichannel approach of $A_{Kd} = -1.34 + i\,1.04$ fm. This quantity is one of the important quantities to be measured in the near future at Frascati in the DEAR experiment [12] and should put further constraints on the theoretical approaches, and, in particular, a challenge to the chiral unitary theory which makes these novel predictions.

3 K^- Interaction in Nuclei

Another of the applications of this chiral unitary approach in the strange sector is the evaluation of the K^- selfenergy in the nuclear medium. Pauli exclusion effects in the intermediate states were taken into account in [13, 14], which lead to a shift of the $\Lambda(1405)$ resonance to higher energies and a considerable attraction on the kaon. A further selfconsistent treatment of the problem, including the calculated kaon selfenergy in the intermediate meson baryon loop functions, [15], moved the resonance back and produced a moderate attraction. Further work in [16] taking into account the renormalization of the intermediate pions and also the baryon selfenergies, produced new modifications leading to wider K^- spectral functions and still a moderate K^- attraction. This attraction is only of the order of $(40 - i\,50)$ MeV at $\rho = \rho_0$ and threshold, and moderately dependent on the energy. It is surprising that this potential turns out to be so small when it was claimed in the past to be of the order of 200 MeV to be able to interpret the data on K^- atoms. Yet, it was proved in [17] that with this potential one could obtain a good description of K^- atoms, something that has been corroborated later on in [18] and [19]. This potential also leads to very deeply bound K^- states, with about 30-40 MeV binding, but with very large widths of the order of 80-100 MeV. With a potential of this type there are no hopes that one can obtain narrow deeply bound K^- states as claimed in [20]. The reason for the claim in that latter work is that a large binding energy is obtained, of the order of those used in the absence of sefconsistency in previous works, and also that many decay channels, which would be still open at these energies (like Σph), and which are evaluated in [16], are omitted in [20].

Another outcome of these calculations is that kaon condensation in neutron stars is far more unlikely than has been assumed in the past.

4 Weak Decay of Λ Hypernuclei

The weak decay of Λ hypernuclei and particularly, the neutron to proton induced decay ratio, Γ_n/Γ_p, has been a battlefield for years, with simple models based on just one pion exchange providing values around 0.1 for this ratio, while experimentally one had values around unity or bigger, albeit with large errors. A thorough recent study of this process, including kaon exchange and correlated and uncorrelated two pion exchange, was done in [22], where good results for the total rates were found for different nuclei and a ratio Γ_n/Γ_p of the order of 0.54 was found for all nuclei. The contribution of the correlated two pion exchange was done following the lines of the work for σ exchange in the NN interaction of [21], where no σ was put but was simulated by the interaction of two pions in the scalar isoscalar sector. Similar results for Γ_n/Γ_p have also been obtained more recently in [23]. The still puzzling thing is that, in spite of the apparently quite thorough work done in [22], there seems to be still discrepancies with the latest and most accurate analysis from the experimental point of view of [24], which provides values bigger than one for the Γ_n/Γ_p ratio, with errors small enough to make the results incompatible with those of [22]. The puzzle has been solved recently. The proton spectra of [24] were analyzed using the code of [25] that takes into account the final state interaction of the nucleons after the Λ decay. This code contained one fatal error and has been corrected recently [25], leading to quite different spectra than those obtained in the original work. The experimental results of [24] have also been corrected to the light of the new results in [25], with the outcome that the new values for Γ_n/Γ_p are of the order of 0.5-0.6 [26], which are in perfect agreement with the theoretical results in [22].

5 Conclusions

A short overview of different problems related to the field of few body physics in strong and weak interactions have been made, with the common denominator that they could be faced with fresh light from the new perspective provided by the unitary extensions of chiral perturbation theory. The short overview here has served to show the potential of this approach to face different problems at intermediate energies, allowing one to establish connections which could not be done prior to the introduction of these techniques. Unitarized chiral perturbation theory is thus proving to be a a quite useful tool to address hadron and nuclear physics at intermediate energies, providing both a systematic working technique, as well as some novel understanding on the nature of those mesonic and hadronic resonances which could not be accommodated in the traditional quark models.

References

1. N. Kaiser, P.B. Siegel and W. Weise: Nucl. Phys. **A594**, 325 (1995)

2. N. Kaiser, T. Waas and W. Weise: Nucl. Phys. **A612**, 297 (1997)
3. J.A. Oller, E. Oset and A. Ramos: Prog. Part. Nucl. Phys. **45**, 157 (2000)
4. J.A. Oller and U.G. Meissner: Phys. Lett. **B500**, 263 (2001)
5. E. Oset and A. Ramos: Nucl. Phys. **A635**, 99 (1998)
6. E. Oset, A. Ramos and C. Bennhold: Phys. Lett. **B537**, 99 (2002)
7. A. Ramos, E. Oset and C. Bennhold: nucl-th/0204044, Phys. Rev. Lett. (in print)
8. G. Toker, A. Gal and J. M. Eisenberg: Nucl. Phys. **A362**, 405 (1982)
9. M. Torres, R. H. Dalitz and A. Deloff: Phys. Lett. **B174**, 213 (1986)
10. A. Deloff: Phys. Rev. **C61**, 024004 (2000)
11. S.S. Kamalov, E. Oset, A. Ramos: Nucl. Phys. **A690**, 494 (2001)
12. C. Guaraldo et al.: Nucl. Phys. **A623**, (1997) 311c.
13. V. Koch: Phys. Lett. **B337**, 7 (1994)
14. T. Waas, N. Kaiser and W. Weise: Phys. Lett. **B365**, 12 (1996)
15. M. Lutz; Phys. Lett. **B426**, 12 (1998)
16. A. Ramos and E. Oset: Nucl. Phys. **A671**, 481 (2000)
17. S. Hirenzaki, Y. Okumura, H. Toki, E. Oset and A. Ramos: Phys. Rev. **C61**, 055205 (2000)
18. A. Baca, C. Garcia-Recio and J. Nieves: Nucl. Phys. **A673**, 335 (2000)
19. A. Cieply, E. Friedman, A. Gal and J. Mares: Nucl. Phys. **A696**, 173 (2001)
20. Y. Akaishi and T. Yamazaki: Phys. Rev. **C65**, 044005 (2002)
21. E. Oset, H. Toki, M. Mizobe and T.T. Takahashi: Prog. Theor. Phys. **103**, 351 (2000)
22. D. Jido, E. Oset and J.E. Palomar: Nucl. Phys. **A694**, 525 (2001)
23. A. Parreno and A. Ramos: Phys. Rev. **C65**, 015204 (2002)
24. O. Hashimoto et al.: Phys. Rev. Lett. **88**, 042503 (2002)
25. A. Ramos, M.J. Vicente-Vacas and E. Oset: Phys. Rev. **C55**, 735 (1997); Erratum-ibid. **C66**, 039903 (2002)
26. H. Outa: Talk at PANIC2002 Conference, Osaka, October 2002

Few-Body Systems Suppl. 14, 269–272 (2003)

Few-Body Systems

ΛN-ΣN and $\Lambda\Lambda$-ΞN Coupling Effects in Light Hypernuclei

K.S. Myint[1], S. Shinmura[2], Y. Akaishi[3]

[1] Physics Department, Mandalay University, Mandalay, Myanmar
[2] Gifu University, Gifu 501-1112, Japan
[3] Institute of Particle and Nuclear Studies, KEK, Tsukuba 305-0801, Japan

Abstract. The suppression due to ΛN-ΣN coupling solves the overbinding problem in ${}^5_\Lambda$He but it causes a severe underbinding in the four-body systems. This shortage is overcome by the coherent Λ-Σ coupling which is equivalent to the ΛNN three-body force. The Pauli suppression effect due to $\Lambda\Lambda$-ΞN coupling in ${}^6_{\Lambda\Lambda}$He is found to be 0.43 MeV for the coupling strength of the NSC97e potential. This indicates that the free-space $\Lambda\Lambda$ interaction is stronger by about 5° phase shift than that deduced from the empirical data of ${}^6_{\Lambda\Lambda}$He without including the Pauli suppression effect. In ${}^5_{\Lambda\Lambda}$He and ${}^5_{\Lambda\Lambda}$H, an attractive term arising from $\Lambda\Lambda$-ΞN conversion is enhanced by the formation of an alpha particle in intermediate Ξ states.

1 Introduction

The significant role of ΛN-ΣN coupling in s-shell single-Λ hypernuclei and in neutron matter with a Λ has been presented by us in previous papers [1, 2]. It is also our interest to find out the effect of $\Lambda\Lambda$-ΞN coupling in double-Λ hypernuclei, since the mass difference between the Λ and Ξ channels is on the order of 28 MeV, which is smaller than that of the Λ and Σ channels. Recently, in the KEK E373 emulsion counter hybrid experiment a ${}^6_{\Lambda\Lambda}$He, known as Nagara event [3], was observed unambiguously. Filikhin and Gal carried out Faddeev-Yakubovsky calculations to analyze ${}^6_{\Lambda\Lambda}$He and other double-Λ hypernuclei [4]. However, we notice that the Pauli suppression effect has not been included in their calculations, which motivates us to investigate this effect on the ${}^6_{\Lambda\Lambda}$He, ${}^5_{\Lambda\Lambda}$He and ${}^5_{\Lambda\Lambda}$H binding energies. In this paper we reveal an appreciable Pauli suppression effect of $\Lambda\Lambda$-ΞN coupling in ${}^6_{\Lambda\Lambda}$He and a significant enhancement effect in ${}^5_{\Lambda\Lambda}$He.

2 ΛN-ΣN Coupling Effects

We have solved the overbinding problem in ${}^{5}_{\Lambda}$He and fitting Λ separation energies of all the s-shell Λ hypernuclei consistently by introducing coherent Λ-Σ coupling effect [1]. In the process of coherent Λ-Σ coupling, a nucleon remains in its ground state after converting Λ to Σ, giving all the other nucleons an equal chance to interact with the Σ. This coherent Λ-Σ coupling effect can be equivalently represented by three-body ΛNN force of the following form;

$$U_{\Lambda NN} = \sum_{\alpha} W_3^{\alpha}(r_{1\Lambda}, r_{\Lambda 2})[a_\alpha + b_\alpha(\boldsymbol{\sigma}_1\boldsymbol{\sigma}_2) + c_\alpha \frac{1}{2}\boldsymbol{\sigma}_\Lambda(\boldsymbol{\sigma}_1 + \boldsymbol{\sigma}_2)], \quad (1)$$

$$W_3^{ts} \equiv V^t_{\Lambda N,\Sigma N}(r_{1\Lambda})\frac{1}{\Delta M}V^s_{\Sigma N,\Lambda N}(r_{\Lambda 2}) + V^s_{\Lambda N,\Sigma N}(r_{1\Lambda})\frac{1}{\Delta M}V^t_{\Sigma N,\Lambda N}(r_{\Lambda 2}), \quad (2)$$

where $\alpha = tt, ts, ss$ with t and s stand for triplet and singlet, respectively.

We have found that in the case of single-channel YN description, the three-body ΛNN force is repulsive in isospin saturated hypernuclei, and either attractive or repulsive in other hypernuclei. The effect is found to become strongly attractive in the ground states of ${}^{4}_{\Lambda}$He and ${}^{4}_{\Lambda}$H and neutron-rich Λ hypernuclei [2]. Details of the ΛNN force will be given elsewhere [5].

3 $\Lambda\Lambda$-ΞN Coupling Effects

We start with single-channel $\Lambda\Lambda$ interactions obtained phase-shift equivalently from the Nijmegen NSC97e and NHC-F models [6] as

$$V^{\rm sc}_{\Lambda\Lambda} = V_{\Lambda\Lambda,\Lambda\Lambda} - V_{\Lambda\Lambda,\Xi^- p}\frac{1}{\bar{\Delta E}}V_{\Xi^- p,\Lambda\Lambda} - V_{\Lambda\Lambda,\Xi^0 n}\frac{1}{\bar{\Delta E}}V_{\Xi^0 n,\Lambda\Lambda}. \quad (3)$$

We modify the long-range part of $V_{\Lambda\Lambda,\Lambda\Lambda}$ to fit the value $\Delta B_{\Lambda\Lambda} = 1.01$ MeV recommended from the Nagara event. We call this fitted $\Lambda\Lambda$ interaction $V^{\rm e1}_{\Lambda\Lambda}$, where the superscript means that the NSC97e coupling strength is used.

The Pauli suppression effect in ${}^{6}_{\Lambda\Lambda}$He is given by

$$\Delta V_{\rm Pauli} = V_{\Lambda\Lambda,\Xi^- p}\frac{P_\alpha}{\Delta M}V_{\Xi^- p,\Lambda\Lambda} + V_{\Lambda\Lambda,\Xi^0 n}\frac{P_\alpha}{\Delta M}V_{\Xi^0 n,\Lambda\Lambda}, \quad (4)$$

$$\Delta M = \frac{M_{\Xi^0} + M_{\Xi^-}}{2} + \frac{M_p + M_n}{2} - 2M_\Lambda + 2B_\Lambda({}^{5}_{\Lambda}\text{He}) = 32.0 \text{ MeV}, \quad (5)$$

where P_α is the projection operator on the $0s$ nucleon state in α particle. We then define $V^{\rm e2}_{\Lambda\Lambda}$ as

$$V^{\rm e2}_{\Lambda\Lambda} = V^{\rm e1}_{\Lambda\Lambda} - \Delta V^{\rm e}_{\rm Pauli}. \quad (6)$$

This $V^{\rm e2}_{\Lambda\Lambda}$ is the single-channel $\Lambda\Lambda$ interaction in free space which reproduces the Nagara event data. Similarly, $V^{\rm F1}_{\Lambda\Lambda}$ and $V^{\rm F2}_{\Lambda\Lambda}$ are constructed from NHC-F.

The free-space $\Lambda\Lambda$ interactions, $V^{\rm e2}_{\Lambda\Lambda}$ and $V^{\rm F2}_{\Lambda\Lambda}$, give $\Delta B_{\Lambda\Lambda}({}^{6}_{\Lambda\Lambda}\text{He}) = 1.44$ MeV and 1.89 MeV, respectively, and when the Pauli suppression effect is included they reduce to the empirical value, 1.01 MeV. Thus, the Pauli suppression effect due to the $\Lambda\Lambda$-ΞN coupling is 0.43 MeV for NSC97e and 0.88 MeV for NHC-F in $^{6}_{\Lambda\Lambda}$He. The Λ-Λ scattering lengths are $a^{\rm e1}_{\Lambda\Lambda}$ = -0.73 fm, $a^{\rm e2}_{\Lambda\Lambda}$ = -1.03 fm, $a^{\rm F1}_{\Lambda\Lambda}$ = -0.67 fm, and $a^{\rm F2}_{\Lambda\Lambda}$ = -1.32 fm, respectively. Thus, the potential in free space $V^{\rm e2}_{\Lambda\Lambda}$ is much more attractive than $V^{\rm e1}_{\Lambda\Lambda}$. The situation is more pronounced with the NHC-F interaction.

We then investigate the $\Lambda\Lambda$-ΞN coupling effect in $^{5}_{\Lambda\Lambda}$He and $^{5}_{\Lambda\Lambda}$H. The coupling consists of two effects, $\Delta V_{\rm Pauli}$ and $\Delta V_{\rm alpha}$, where the former is the Pauli suppression effect, while the latter is an enhancement effect due to the formation of an alpha particle in the Ξ intermediate state:

$$\Delta V_{\rm Pauli}(^{5}_{\Lambda\Lambda}{\rm He}) = V_{\Lambda\Lambda,\Xi^- p}\frac{P_{\rm h}}{\Delta M_1}V_{\Xi^- p,\Lambda\Lambda} + \frac{1}{2}(V_{\Lambda\Lambda,\Xi^0 n}\frac{P_{\rm h}}{\Delta M_1}V_{\Xi^0 n,\Lambda\Lambda}), \quad (7)$$

$$\Delta M_1 = \frac{M_{\Xi^0}+M_{\Xi^-}}{2} + \frac{M_p+M_n}{2} - 2M_\Lambda + 2B^{\rm av}_\Lambda(^4_\Lambda{\rm He}) = 28.83 \text{ MeV}, \quad (8)$$

and

$$\Delta V_{\rm alpha}(^{5}_{\Lambda\Lambda}{\rm He}) = \frac{1}{2}(V_{\Lambda\Lambda,\Xi^0 n}\frac{P_{\rm h}}{\Delta M_1}V_{\Xi^0 n,\Lambda\Lambda}) - \frac{1}{2}(V_{\Lambda\Lambda,\Xi^0 n}\frac{P_\alpha}{\Delta M_2}V_{\Xi^0 n,\Lambda\Lambda}), \quad (9)$$

$$\Delta M_2 = M_{\Xi^0} + M_\alpha - M_{\rm h} - 2M_\Lambda + 2B^{\rm av}_\Lambda(^4_\Lambda{\rm He}) = 5.69 \text{ MeV}, \quad (10)$$

where $P_{\rm h}$ is the projection operator on the $0s$ state of ^{3}He, and $B^{\rm av}_\Lambda$ is the average of the 0^+ and 1^+ levels of $^4_\Lambda$He as

$$B^{\rm av}_\Lambda = \frac{1}{4}B_\Lambda(0^+) + \frac{3}{4}B_\Lambda(1^+). \quad (11)$$

The values of $\Delta B_{\Lambda\Lambda}$ for the three double-Λ hypernuclei are plotted against the coupling strength in Fig. 1. It is to be noted that $\Delta B_{\Lambda\Lambda}$ for the five-body system is defined by Filikhin and Gal [4] as

$$\Delta B_{\Lambda\Lambda} = B_{\Lambda\Lambda} - 2B^{\rm av}_\Lambda. \quad (12)$$

It is interesting to find that these effects give $\Delta B_{\Lambda\Lambda} = 1.28$ MeV of $^{5}_{\Lambda\Lambda}$He for NSC97e, which is larger than 1.01 MeV of $^{6}_{\Lambda\Lambda}$He. The contributions of $\Delta V^{\rm e}_{\rm Pauli}$ and $\Delta V^{\rm e}_{\rm alpha}$ in this system are 0.23 MeV and -0.41 MeV. Filikhin and Gal predicted the values of $\Delta B_{\Lambda\Lambda}$ for $^{5}_{\Lambda\Lambda}$He and $^{5}_{\Lambda\Lambda}$H, which are smaller than that of $^{6}_{\Lambda\Lambda}$He, and gave a comment that "the heavier is the core nucleus, the larger is the $\Delta B_{\Lambda\Lambda}$." This contradicts our results, where $\Delta B_{\Lambda\Lambda}$ of $^{5}_{\Lambda\Lambda}$He is larger than that of $^{6}_{\Lambda\Lambda}$He.

4 Conclusions

The $\Lambda\Lambda$ interaction $V^{\rm e1}_{\Lambda\Lambda}$ gives a Λ-Λ phase shift of 12° at the maximum, while the free-space $\Lambda\Lambda$ interactions, $V^{\rm e2}_{\Lambda\Lambda}$ and $V^{\rm F2}_{\Lambda\Lambda}$, give 17° and 22° maximum values, respectively. Thus, the free-space $\Lambda\Lambda$ interaction is considerably stronger

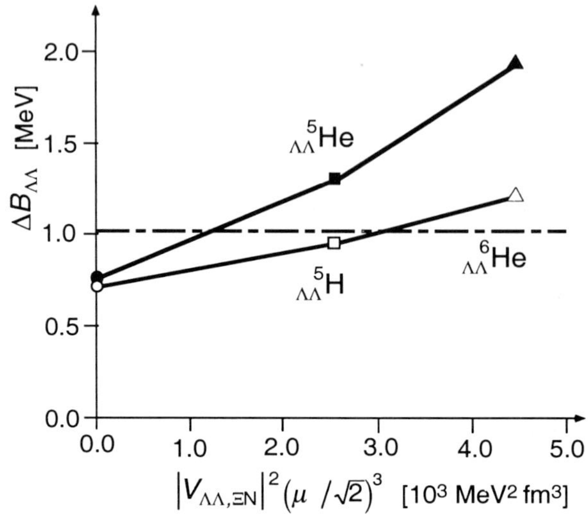

Figure 1. $\Delta B_{\Lambda\Lambda}$ values against the coupling strength. The circles, squares and triangles represent the cases of no, NSC97e and NHC-F coupling strengths, respectively.

than the interaction which is obtained by fitting the Nagara event data while neglecting the Pauli suppression effect.

The coupling effects in the five-body systems consist of the Pauli suppression, ΔV_{Pauli}, and an enhancement, ΔV_{alpha}. In ${}^{5}_{\Lambda\Lambda}$He, the enhancement effect dominates, and the net effect is 0.18 (0.47) MeV attraction for the NSC97e (NHC-F) coupling strength. Observations of ${}^{5}_{\Lambda\Lambda}$He would be critical for determining the $\Lambda\Lambda$-ΞN coupling strength.

References

1. Y. Akaishi, T. Harada, S. Shinmura, K.S. Myint: Phys. Rev. Lett. **84** 3539 (2000)
2. S. Shinmura, K.S. Myint, T. Harada, Y. Akaishi: J. Phys. G: Nucl. Part. Phys. **28** 1 (2002)
3. H. Takahashi et al.: Phys. Rev. Lett. **87** 212502 (2001)
4. I.N. Filikhin, A. Gal: Phys. Rev. **C65**, 041001(R) (2002); nucl-th/0203036
5. K.S. Myint, Y. Akaishi: to appear
6. Th.A. Rijken, V.G.J. Stoks, Y. Yamamoto: Phys. Rev. **C59** 21 (1999); M.M. Nagels, T.A. Rijken, J.J. de Swart: Phys. Rev. **D20**, 1633 (1979)

Few-Body Systems Suppl. 14, 273–276 (2003)

Few-Body Systems

Five-Body Calculation of ${}^5_\Lambda$He with Explicit Σ Admixture

H. Nemura[1], Y. Akaishi[1], Y. Suzuki[2]

[1] Institute of Particle and Nuclear Studies, KEK, Tsukuba, 305-0801, Japan

[2] Department of Physics, Niigata University, Niigata 950-2181, Japan

Abstract. Variational calculations for s-shell hypernuclei are performed by explicitly including Σ degrees of freedom. Two sets of YN interactions (D2 and SC97e(S)) are used. The bound-state solution of ${}^5_\Lambda$He is obtained by using each of YN potentials, and a large energy expectation value of the tensor $\Lambda N - \Sigma N$ transition part is found by using the SC97e(S). The internal energy of the ^{4}He subsystem changes a lot by the presence of a Λ particle with the strong tensor $\Lambda N - \Sigma N$ transition potential.

1 Introduction

^{4}He is one of the most tightly bound nuclei in light mass region. It is well-known that the tensor correlation plays an important role for the binding of the ^{4}He. Considering the fact that the experimental Λ separation energy (B_Λ) of ${}^5_\Lambda$He is only about 3 MeV and that both of ^{4}He and Λ have isospin $I = 0$, one may think that ${}^5_\Lambda$He is a weakly coupled system of Λ and α particle, and effects of the tensor ΛN interaction or of the $\Lambda N - \Sigma N$ coupling would be suppressed. On the other hand, considering the fact that the one-pion-exchange mechanism, which is the origin of the tensor NN interaction in part, induces the $\Lambda N - \Sigma N$ transition for the YN sector, the tensor $\Lambda N - \Sigma N$ transition potential as well as the tensor NN potential could play an important role for the α particle with a Λ particle attached, i.e., ${}^5_\Lambda$He. If this is the case, it means that the structure of the core nucleus, ^{4}He, would be strongly influenced by the presence of a Λ particle. Therefore, a genuine five-body calculation is crucial to explore the exotic structure of ${}^5_\Lambda$He.

If one constructs a phenomenological central ΛN potential, which is consistent with the experimental binding energies for $A = 3, 4$ hypernuclei as well as the Λp total cross section, that kind of potential would overestimate the $B_\Lambda({}^5_\Lambda\text{He})$ value[1, 2]. This is known as an anomalously small binding of ${}^5_\Lambda$He. In view of the aim to pin down a reliable YN interaction, a systematic study for the complete set of the s-shell Λ hypernuclei is desirable.

The purpose of this study is twofold: First is to describe an *ab initio* calculation of ${}^5_\Lambda$He as well as $A = 3, 4$ hypernuclei explicitly including Σ degrees of freedom. Second is to conduct a new view of the ${}^5_\Lambda$He, due to taking account of explicit Σ admixture, beyond $\alpha + \Lambda$ model.

2 Interactions and Method

The hamiltonian of the system comprising $(A-1)$ N and a Y is given by

$$H = (T_c + V_{NN}) + (T_{Y-c} + V_{YN}), \tag{1}$$

where T_c is the kinetic energy of the core nucleus (c) subtracted by the center-of-mass (CM) energy of c, and the first term, $T_c + V_{NN}$, gives the sub-hamiltonian of c. T_{Y-c} is the kinetic energy of the relative motion between Y and the CM of c. The V_{NN} (V_{YN}) is the NN (YN) potential. The V_{NN} and V_{YN} take account of a summation over appropriate particle pairs. For the sake of simplicity, the Coulomb potential is omitted in (1), though the actual calculation includes it. In the present study, we use the G3RS potential[3] for the NN interaction, and two kinds of YN interactions, D2 and SC97e(S)[4], are used. Both YN interactions include the $\Lambda N - \Sigma N$ coupling. The D2 potential consists of only the central component, while the SC97e(S) has tensor and spin-orbit components in addition to the central one.

The trial function is given by a combination of basis functions:

$$\Psi_{JMTM_T} = \sum_{k=1}^{N} c_k \mathcal{A}\Big[\mathrm{e}^{-\frac{1}{2}\mathbf{x}A_k\mathbf{x}} \, v_k^{2K_k+L_k} \Big[Y_{L_k}(\hat{\mathbf{v}}_k) \times \chi_{S_k} \Big]_{JM} \eta_{kTM_T} \Big]. \tag{2}$$

Here, $\mathcal{A}$ is an antisymmetrizer acting on nucleons, and χ_{S_k} (η_{kTM_T}) is the spin (isospin) function. The abbreviation $\mathbf{x} = (\mathbf{x}_1, \cdots, \mathbf{x}_{A-1})$ is a set of relative coordinates. A set of linear variational parameters $(c_1, \cdots, c_N)$ is determined by the Ritz variational principle. The angular momentum part is expressed by the global vector representation, and the global vector, $\mathbf{v}_k$, is given by a linear combination of the relative coordinates:

$$\mathbf{v}_k = \sum_{i=1}^{A-1} (u_k)_i \mathbf{x}_i. \tag{3}$$

The A_k and u_k are sets of nonlinear parameters which characterize the spatial part of the basis function. The variational parameters are optimized by a stochastic procedure. Recently, we performed a five-body calculation of ${}^5_\Lambda$He[5] by using the stochastic variational method[6]. The reader is referred to Refs.[5, 6] for details of the calculation.

3 Results

Table 1 lists the results of the Λ-separation energies. Both D2 and SC97e(S) YN interactions reasonably well reproduce the experimental B_Λ values for

	$B_\Lambda(^3_\Lambda\mathrm{H})$	$B_\Lambda(^4_\Lambda\mathrm{H})$	$B_\Lambda(^4_\Lambda\mathrm{He})$	$B_\Lambda(^4_\Lambda\mathrm{H}^*)$	$B_\Lambda(^4_\Lambda\mathrm{He}^*)$	$B_\Lambda(^5_\Lambda\mathrm{He})$
D2	0.05	2.16	2.11	1.17	1.15	3.39
SC97e(S)	0.10	2.06	2.02	0.92	0.90	2.75
Experiment	0.13(5)	2.04(4)	2.39(3)	1.00(4)	1.24(4)	3.12(2)

Table 1. Λ separation energies, given in units of MeV, of $A = 3 - 5$ Λ-hypernuclei.

the $A = 3, 4$ hypernuclei. Though the calculated $B_\Lambda(^5_\Lambda\mathrm{He})$ is slightly larger or smaller than the experimental value, the discrepancies are rather small in view of the anomalous binding problem. This is a *first ab initio* calculation to produce the bound state of $^5_\Lambda$He with explicit Σ degrees of freedom. The present result implies that the key to resolve the anomalous binding problem is to include explicitly Σ degrees of freedom.

Figure 1 displays the energy expectation values of the kinetic and potential energy terms for $^5_\Lambda$He. The most left-side step is the energy expectation value of the kinetic energy T_c, and the next three steps take the spin-singlet central, the spin-triplet central, and the tensor parts of the NN interaction into account successively. The solid line is the case of no presence of Λ particle, and the horizontal line at $E \approx -25$ MeV gives the energy of $\Lambda + {}^4$He threshold. The energy expectation value obtained with the SC97e(S) (D2) is shown by broken (dotted) line. The presence of a Λ particle reduces the internal energy of the ^{4}He subsystem:

$$\Delta E_c = \Big(\langle T_c\rangle + \langle V_{NN}\rangle\Big)_{^5_\Lambda\mathrm{He}} - \Big(\langle T_c\rangle + \langle V_{NN}\rangle\Big)_{^4\mathrm{He}} \approx \begin{cases} 2.3\mathrm{MeV} & \text{(D2)} \\ 4.7\mathrm{MeV} & \text{(SC97e(S))}. \end{cases} \quad (4)$$

The change of the energy, ΔE_c, for the SC97e(S) is considerably large, despite the fact that the rms radius of N from the CM of ^{4}He for the $^5_\Lambda$He hardly changes from that for ^{4}He. (Both radii are 1.5 fm.) The next step of the energy levels includes the kinetic energy of T_{Y-c}.

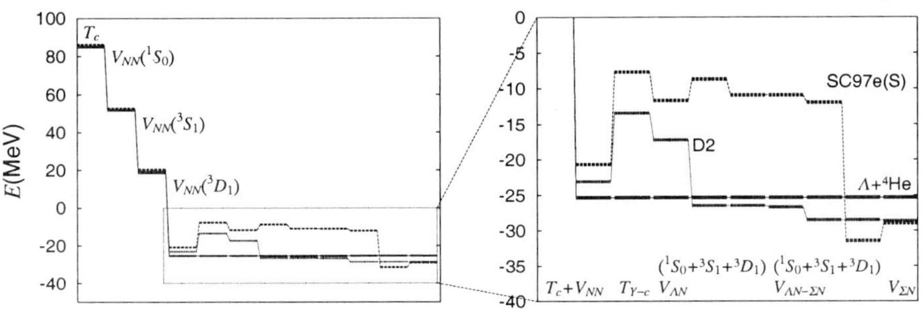

Figure 1. Energy expectation values of the kinetic and potential energy terms for $^5_\Lambda$He and ^{4}He (see the text).

To include the spin-triplet (central) part of the D2 $\Lambda N - \Lambda N$ diagonal potential, the energy level reaches across the $\Lambda + {}^4$He threshold. Therefore, for the ${}^5_\Lambda$He, the main part of the D2 is the (spin triplet) $\Lambda N - \Lambda N$ diagonal component. On the other hand, the $\Lambda N - \Lambda N$ diagonal part of the SC97e(S) is repulsive. The energy levels including $\Lambda N - \Lambda N$ diagonal part of the SC97e(S) are extremely (more than 10 MeV) higher than the $\Lambda + {}^4$He threshold. The tensor $\Lambda N - \Sigma N$ transition part brings a surprisingly large attractive energy (about -20 MeV). This large coupling energy makes ${}^5_\Lambda$He bound for the SC97e(S).

4 Summary

We have made a systematic study of all s-shell hypernuclei based on ab initio calculations using YN interactions with an explicit Σ admixture. Both YN interactions, D2 and SC97e(S), reasonably well reproduce the experimental separation energies. Regarding the anomalous binding problem of ${}^5_\Lambda$He, the present calculation convinces us that one of the keys to resolve the problem is to include the explicit Σ degrees of freedom. A sizable amount of P_Σ was obtained for ${}^5_\Lambda$He by using the SC97e(S). The contribution of the energy from the tensor $\Lambda N - \Sigma N$ coupling is quite large, and this coupling is considerably important to make ${}^5_\Lambda$He bound. The core nucleus, ^{4}He, is no longer rigid in interacting with a Λ particle. A similar situation can occur for strangeness $S = -2$ systems. Investigations into the strength of the $\Lambda\Lambda$ interaction based on the experimental data of the binding energy for double Λ hypernuclei (e.g., ${}^6_{\Lambda\Lambda}$He [7]) should take account of the energy reduction of the core nucleus (ΔE_c for ${}^6_{\Lambda\Lambda}$He is expected to be larger than the present ΔE_c for ${}^5_\Lambda$He).

References

1. R.H. Dalitz, R.C. Herndon, Y.C. Tang: Nucl. Phys. **B47**, 109 (1972)
2. H. Nemura, Y. Suzuki, Y. Fujiwara, C. Nakamoto: Prog. Theor. Phys. **103**, 929 (2000)
3. R. Tamagaki: Prog. Theor. Phys. **39**, 91 (1968)
4. Y. Akaishi, T. Harada, S. Shinmura, Khin Swe Myint: Phys. Rev. Lett. **84**, 3539 (2000)
5. H. Nemura, Y. Akaishi, Y. Suzuki: Phys. Rev. Lett. **89**, 142504 (2002)
6. Y. Suzuki, K. Varga: *Stochastic Variational Approach to Quantum-Mechanical Few-Body Problems* (Lecture Notes in Physics, Vol. m54). Berlin Heidelberg: Springer-Verlag 1998
7. H. Takahashi et al.: Phys. Rev. Lett. **87**, 212502 (2001)

Few-Body Systems Suppl. 14, 277–280 (2003)

Few-Body Systems

Effects of Σ Mixing on Nonmesonic Weak Decay of Λ Hypernuclei

K. Sasaki[1,*], T. Inoue[2], M. Oka[3]

[1] Institute of Particle and Nuclear Studies, High Energy Accelerator Research Organization (KEK), 1-1, Oho, Tsukuba, 305-0801 Ibaraki, Japan

[2] Institut fuer Theoretische Physik, Universität Tübingen Auf der Mogenstelle 14, D-72076 Tübingen, Germany

[3] Department of physics, Tokyo Institute of Technology, 2-12-1, Ookayama, Megro-ku, 152-8551 Tokyo, Japan

Abstract. The effects of virtual Σ component in ${}^4_\Lambda$He and ${}^4_\Lambda$H are estimated. We find that the virtual Σ contribution changes the magnitudes of the non-mesonic decay rate significantly. We point out that observation of final *pp* state from the decay of ${}^4_\Lambda$He will give us a clear evidence of the Σ component in Λ hypernucleus.

1 Introduction

The decay of hypernuclei is brought mainly by a two-body transition, $\Lambda N \to NN$. It is realized that the momentum transfer is large due to the $\Lambda - N$ mass difference, and therefore the short range interaction becomes important. Recently we proposed to treat the short range part using the valence quark picture of the baryon and the effective four quark weak hamiltonian. We found that the direct quark (DQ) process gives significantly large contribution and shows qualitatively different features from the meson exchange contribution, especially in its isospin structure [1, 2].

In our study [3], we found that the direct quark (DQ) process is important to describe the short range physics, and the well-known n/p puzzle is roughly solved by applying the $\pi + K + DQ$ model for $\Lambda N \to NN$ transition.

It is advocated that the mixing of Σ is crucial to describe the binding energy of light hypernuclei [4]. In this paper, we study the non-mesonic weak decay of light hypernuclei with considering the coherent mixing of Σ.

*This work was supported by JSPS Research Fellowships for Young Scientists

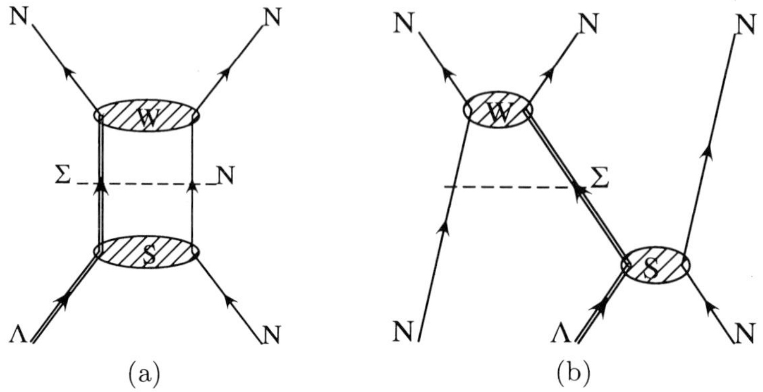

Figure 1. Diagrammatic representation of two and three body type Σ mixing for the $\Lambda N \to NN$ transition.

2 Λ-Σ Mixing Contribution

The strong interaction may produce a virtual Σ from Λ in nuclei through $\Lambda N \to \Sigma N$ process. The virtual ΣN state contributes to the non-mesonic decay as is shown in Fig.1. A previous study [5] considered two-pion exchange process between Λ and N, one of which induces weak transition, (Fig.1(a)). The intermediate ΣN state is restricted to $I = 1/2$. It is, however, possible that the virtual Σ decays with the assistance of a second nucleon, (Fig.1(b)). For $I \neq 0$ hypernuclei, the coherent Σ mixing allows the virtual Σ to interact with all the nucleons equally and therefore the three-body weak process Fig.1(b) becomes important.

Now we consider the Σ mixing effect on the ${}^4_\Lambda$He. Suppose that the Σ+3N state with the same quantum number mixes to the $|{}^4_\Lambda\mathrm{He}\rangle$ state,

$$|{}^4_\Lambda\mathrm{He}\rangle^0_{1/2} = \sqrt{1-\beta^2}|{}^4_\Lambda\mathrm{He}\rangle^0_{1/2} + \beta|{}^4_\Sigma\mathrm{He}\rangle^0_{1/2}\,, \tag{1}$$

where the super- and sub-script stand for the total spin and isospin for hypernuclei. The β is the mixing parameter of Σ and treated as a free parameter. The weak $\Sigma N \to NN$ decay is calculated in the same way as $\Lambda N \to NN$, considering both the meson (π,K) exchange and the direct quark (DQ) process. We perform the similar calculation for the ${}^4_\Lambda$H.

The "three body" type process is significant because it involves the weak interaction of the $\Sigma N(I = 3/2)$ states, which does not contribute in Fig.1(a). Due to the $\Sigma N(I = 3/2)$ states, an interesting observation is a possible $\Sigma^+ p \to pp$ decay, which emits two protons almost back to back. This will be a signal of Σ mixing in hypernuclei as well as the contribution of the $I = 3/2$ state in the non-mesonic decay.

3 Results

It is interesting to see how large the effect of the Σ mixing is. As we discussed in the previous section, we consider virtual Σ-hypernucleus component, whose probability is given by $|\beta|^2$. In Table 1, we show the decay rates of A=4 hypernuclei for two values of β, such that $\beta^2 \sim 1\%$. One sees that the mixing changes the total decay rate by about 20% for $\beta^2 \sim 1\%$. One sees that the results for $\beta = -0.1$ give better agreement with the current experimental data. It has been also found that the negative β is consistent with the study of the magnetic moments due to the Σ mixing in hypernuclei [7]. The negative β is found to reduce the proton induced decay rate, Γ_{pn}, while a positive β will enhance the rates. We find that the mixing does not change the qualitative behaviors of the Γ_{nn}/Γ_{pn} ratio.

A new interesting decay channel is available when we consider the Σ^+ mixing in ${}^4_\Lambda$He. The system consists of a virtual Σ^+, p and two n. When Σ^+ meets the proton, it decays into two p, i.e., $\Sigma^+ p \to pp$ decay. The calculated decay rate for $\Sigma^+ p \to pp$ is

$$\Gamma_{pp} = 0.0003 \Gamma_\Lambda \tag{2}$$

at $|\beta| = 0.1$. Although the branching ratio is tiny, it gives a clean signal as a back to back p-p in the final state. This will be a direct evidence for virtual Σ mixing in Λ hypernuclei.

Table 1. No-mesonic decay rates of ${}^4_\Lambda$He in units of Γ_Λ

${}^4_\Lambda$He	total	Γ_{pn}	Γ_{nn}	Γ_{nn}/Γ_{pn}
π	0.272	0.250	0.022	0.089
π+K	0.155	0.145	0.009	0.064
DQ	0.032	0.021	0.011	0.516
π+K+DQ	0.218	0.214	0.004	0.019
$\beta = -0.1$	0.168	0.165	0.003	0.017
$\beta = +0.1$	0.276	0.270	0.006	0.021
EXP [8]	0.20±0.03	0.16±0.02	0.04±0.02	$0.25^{+0.05}_{-0.13}$
EXP [9]	0.17±0.05	0.16±0.02	$0.01^{+0.04}_{-0.01}$	$0.06^{+0.28}_{-0.06}$

In ${}^4_\Lambda$H case, the $J = 1$ part of Γ_{pn} is absent. As the $\Gamma_{pn}(J = 1)$ is the largest contribution of OPE due mainly to the strong tensor force, the OPE predicts very small total nonmesonic decay rate. The kaon exchange and DQ enhances Γ_{nn}, and the total decay rate reaches up to 0.19. Thus the total nonmesonic decay rate of ${}^4_\Lambda$H is a good indicator of the non-pion contributions to the decay process.

Table 2. Nomesonic decay rates of ${}^{4}_{\Lambda}$H in unit of Γ_{Λ}

${}^{4}_{\Lambda}$H	total	Γ_{pn}	Γ_{nn}	Γ_{nn}/Γ_{pn}
π	0.040	0.011	0.029	2.597
π+K	0.071	0.005	0.067	14.225
DQ	0.040	0.013	0.027	2.028
π+K+DQ	0.187	0.030	0.157	5.318
$\beta = -0.1$	0.201	0.027	0.174	6.519
$\beta = +0.1$	0.171	0.032	0.139	4.317
EXP [9]	0.15±0.13	—	—	—

4 Conclusions

We calculate the nonmesonic decay rates of Λ hypernuclei by using the quark-meson hybrid model and estimate the effect of the virtual Σ excitation which is known to be important when we consider the property of Λ in nuclei. In this paper we add the $\Sigma N \to NN$ amplitude to the $\Lambda N \to NN$ amplitude by using the mixing parameter β. We found that the $\Sigma N \to NN$ amplitude changes the total non-mesonic decay rate by about 20% for $\beta^2 \sim 1\%$, and our estimates agree with the experimental data fairly well for a negative $\beta \simeq -0.1$. We have also pointed out that observation of final p-p state from the decay of ${}^{4}_{\Lambda}$He gives us a chance to show a clear evidence of the Σ component in Λ hypernucleus.

References

1. T. Inoue, S. Takeuchi, M. Oka: Nucl. Phys. **A577**, 281c (1994); Nucl. Phys. **A597**, 563 (1996)
2. T. Inoue, M. Oka, T. Motoba, K. Itonaga: Nucl. Phys. **A633**, 312 (1998)
3. K. Sasaki, T. Inoue, M. Oka: Nucl. Phys. **A707**, 477 (2002)
4. Y. Akaishi, T. Harada, S. Shinmura, Khin Swe Myint: Phys. Rev. Lett. **84**, 3539 (2000)
5. H. Bando, Y. Shono, H. Takaki: Int. J. Mod. Phys. **A3**, 1581 (1988)
6. E. Hiyama: Nucl. Phys. **A670**, 273c (2000)
7. M. Oka: In: *Physics with GeV Electrons and Gamma-Rays* (Edited by T. Tamae et al.). Universal Academic Press 2001
8. V.J. Zeps et al.: Nucl. Phys. **A639**, 261c (1998)
9. H. Outa et al.: Nucl. Phys. **A639**, 251c (1998)

Few-Body Systems Suppl. 14, 281–286 (2003)

Few-Body Systems

Three-Body Analysis of Incoherent Photoproduction of η Mesons on the Deuteron near Threshold*

A. Fix, H. Arenhövel

Institut für Kernphysik, Johannes Gutenberg-Universität, 55099 Mainz, Germany

Abstract. The importance of three-body dynamics in the ηnp system in elastic and inelastic η-deuteron scattering as well as coherent and incoherent η photoproduction on the deuteron in the energy region from threshold up to 30 MeV above has been investigated. It is shown that a restriction to first order rescattering with respect to the NN- and ηN-final state interactions, i.e., restriction to rescattering in the two-body subsystems, does not give a sufficiently accurate approximation to the s-wave reaction amplitude and that higher order terms, as described by the three-body dynamics give very substantial contributions.

1 Introduction

Eta photoproduction on the deuteron is of interest with respect to information on the neutron amplitude, the role of ηN interaction; With respect to the absolute strength of the neutron amplitude, quasi-free production in the incoherent process is favoured; However, with respect to the relative phase between the elementary amplitudes of proton and neutron, the coherent reaction on the deuteron is better suited because the deuteron acts as an isospin filter and the proton and neutron amplitudes interfere coherently.

2 The Elementary η Production Operator

The following pure resonance form with excitation of the $S_{11}(1535)$ is used

$$t_{\gamma\eta}^{(s/v)} = \frac{k_{\gamma N}}{M_{N^*}+M_N}\frac{e\,g_{\gamma NN^*}^{(s/v)}g_{\eta NN^*}}{W_{\gamma N}-M_{N^*}+\frac{i}{2}\Gamma(W_{\gamma N})}\,i\boldsymbol{\sigma}\cdot\boldsymbol{\varepsilon}_\lambda\,.$$

with parameters $M_{N^*} = 1535$ MeV, $\Gamma_{\pi\pi N} = 16$ MeV, $g_{\eta NN^*} = 2.10$, and $g_{\pi NN^*} = 1.19$. The vertex constant $g_{\gamma NN^*}$ is related to the helicity amplitude

*Supported by Deutsche Forschungsgemeinschaft (SFB 443).

$A^N_{1/2}$ by $eg_{\gamma NN^*} = \sqrt{2M_{N^*}(M_{N^*}+M_N)/(M_{N^*}-M_N)}A^N_{1/2}$. This parametrization yields a good description of the total cross section on the proton using $A^p_{1/2} = 0.104$ GeV$^{-1/2}$ as is shown in Fig. 1.

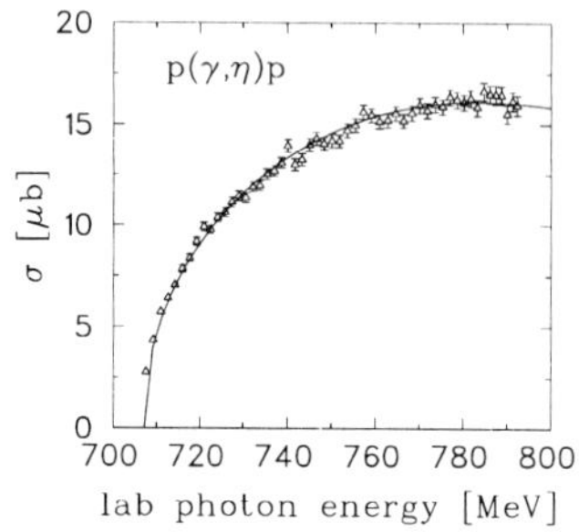

Figure 1. Total cross section for η production on the proton. Data from ref. [1].

3 Inclusion of Final State Interaction in Incoherent η Photoproduction by Rescattering in 2-Body Subsystems

Near threshold the impulse approximation (IA) yields a very small cross section for $d(\gamma,\eta)np$ due to the large momentum mismatch and indeed fails drastically yielding a cross section much too low compared to experiment. Therefore, we first have performed an approximate treatment of final state interaction (FSI) by taking into account only complete rescattering in the two-body ηN and NN subsystems of the final state (see Fig. 2) [2]. In this calculation, deuteron wave

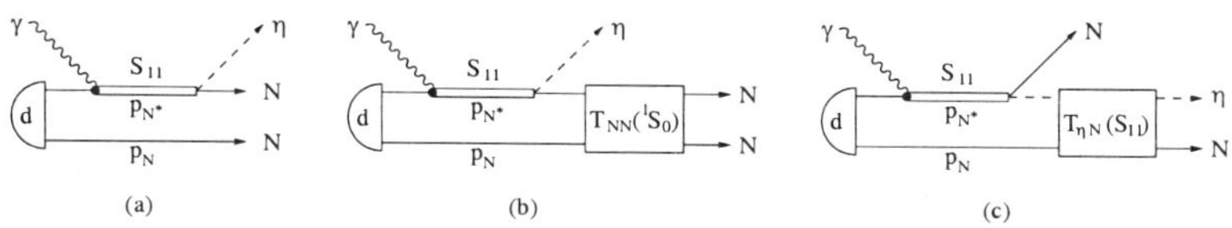

Figure 2. Diagrams for $d(\gamma,\eta)np$: (a) IA, (b) NN rescattering, (c) ηN rescattering.

function and T_{NN} are determined from the Bonn OBEPQ potential while for ηN rescattering the ηN t-matrix is taken in the isobar approach with intermediate excitation of the $S_{11}(1535)$

$$t_{\eta N}(W_{\eta N}) \quad = \quad v^\dagger_{N^*} g_{N^*}(W_{\eta N}) v_{N^*} = g^2_{\eta NN^*}[W_{\eta N} - M_{N^*} + \frac{i}{2}\Gamma(W_{\eta N})]^{-1},$$

In view of the near-threshold region it is sufficient to consider rescattering in s-waves only. This first order rescattering leads to a considerable improvement as is seen in the left panel of Fig. 3. The spectrum of the outgoing η meson shows at low energies a distinct signature of the final state NN rescattering exhibiting the prominant $^1S_0$ peak close to threshold of NN-scattering (see middel and right panels of Fig. 3). Differential cross sections near threshold are shown in Fig. 4 and one notes a considerable improvement over the IA. However, the left panel indicates that first order rescattering still fails to explain quantitatively the enhancement of the experimental data right above threshold. This is corroborated by very recent more precise near-threshold data by Hejny et al. [3].

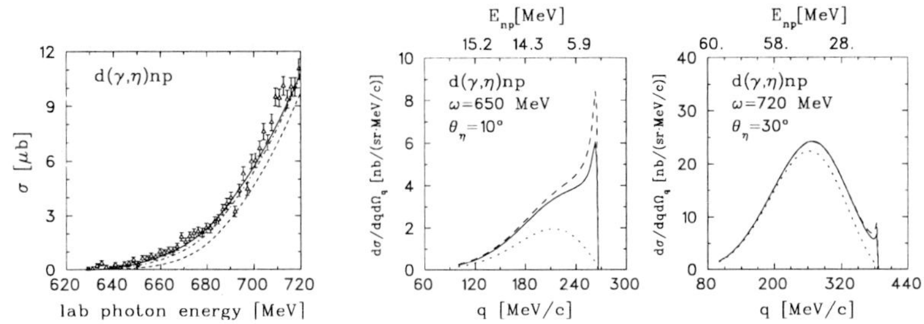

Figure 3. Left panel: Total cross section for $d(\gamma,\eta)np$: dashed: IA, solid: IA + rescattering, dash-dotted: IA + NN rescattering, data: inclusive $\gamma d \rightarrow \eta X$ from from ref. [1]. Middel and right panels: η-meson spectra at forward angles: dotted: IA; solid: IA + rescattering; dashed: without D-wave contribution to NN-rescattering amplitude. Top abscissa indicates final NN-excitation energy E_{np}.

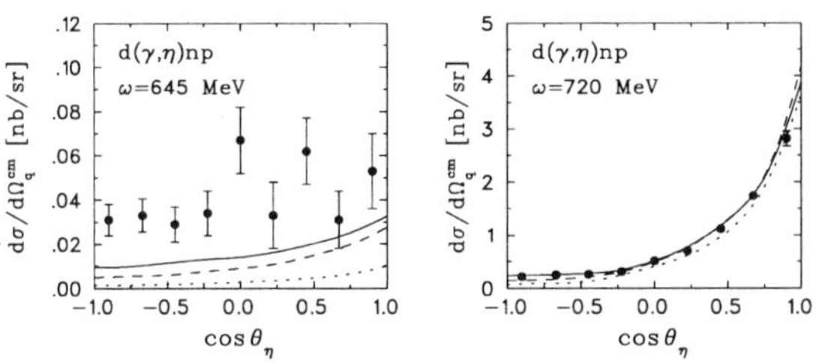

Figure 4. Differential cross section for $d(\gamma,\eta)np$. Dotted: impulse approximation; Dashed: IA plus NN rescattering; Full: complete result. Exp.: inclusive $\gamma d \rightarrow \eta X$ from ref. [1].

4 Three-Body Treatment of Final State Interaction in η-Deuteron Scattering and η Photoproduction

The very strong effect from the hadronic interaction in the final state in first order rescattering suggests that a genuine three-body treatment is required. Defining as channels "N^*" the channel with one spectator nucleon and "d" the channel with the η meson as spectator and taking the interactions in separable form, one obtains from the AGS-3-body equations a set of coupled equations for the channel transition amplitudes (see Fig. 5 for a diagrammatic representation), which reads

$$\begin{aligned} X_{N^*d} &= Z^{(\eta)}_{N^*d} + Z^{(\eta)}_{N^*d}\tau^{(\eta)}_d X^{(\eta)}_d + Z^{(\pi)}_{N^*d}\tau^{(\pi)}_d X^{(\pi)}_d \\ &\quad +(Z^{(\eta)}_{N^*N^*} + Z^{(\pi)}_{N^*N^*})\tau_{N^*}X_{N^*d}\,, \\ X^{(\eta)}_d &= 2Z^{(\eta)}_{dN^*}\tau_{N^*}X_{N^*d}\,, \quad X^{(\pi)}_d = 2Z^{(\pi)}_{dN^*}\tau_{N^*}X_{N^*d}\,, \end{aligned}$$

where the N^* channel comprises two components $N^*(\eta)$ and $N^*(\pi)$ in order to account for the coupling of the ηN-channel to the πN-channel. Because of the near threshold region only s-waves are included. For the NN interaction the tensor force is neglected for reasons of simplicity. In detail, we have

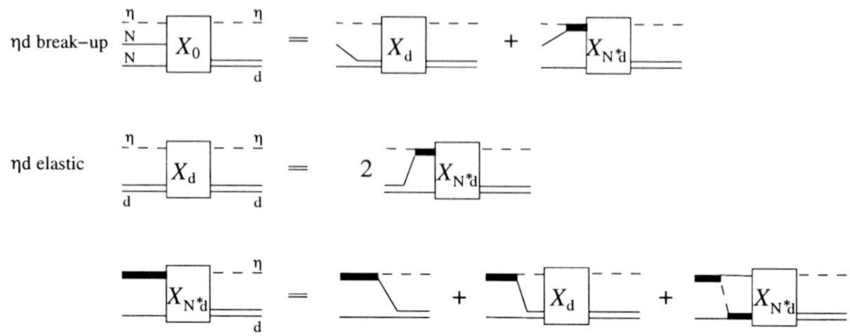

Figure 5. Representation of three-body equations for η-deuteron scattering.

taken for the NN-channel a driving term $V_d(p,p') = g_d(p)\,g_d(p')$ with $g_d(p) = g_d\,\beta_d^2/(p^2+\beta_d^2)$ yielding a t-matrix of the form $t_d(p,p') = g_d(p)\,\tau_d(E)\,g_d(p')$ with

$$\tau_d(E) = -\frac{1}{2M_N}\left[1+\frac{g_d^2\beta_d^2}{16\pi(\sqrt{EM_N}+i\beta_d)}\right]^{-1}.$$

Correspondingly, the t-matrix for the ηN-channel has the form $t_{N^*}^{(ij)}(p,p') = g_{N^*}^{(i)}(p)\,\tau_{N^*}(E)\,g_{N^*}^{(j)}(p')$, where $g_{N^*}^{(i)}(p) = g_{N^*}\,\beta_{N^*}^{(i)\,2}/(p^2+\beta_{N^*}^{(i)\,2})$ and

$$\tau_{N^*}(W) \quad = \quad [W - M_0 - \Sigma_\pi(W) - \Sigma_\eta(W)]^{-1}.$$

For the values of the actual parameters we refer to [4]. This interaction yields a scattering length $a_{\eta N} = 0.75 + i\,0.27$ fm. Results for ηd scattering are displayd in

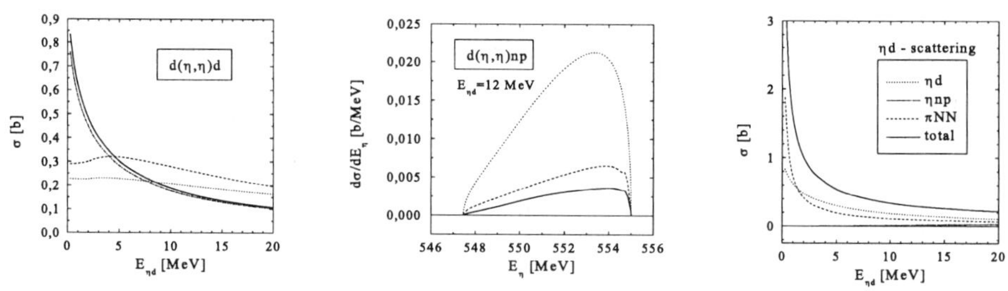

Figure 6. Results for η-deuteron scattering from [4]. Left panel: elastic total cross section, middle panel: spectrum of emitted η mesons in inelastic scattering, notation for left and middel panels: dotted: IA; dashed: first order rescattering; solid: three-body calculation. Right panel: various contributions to total ηd cross section.

Fig. 6. For elastic scattering one notes a rapid increase of the total cross section approaching the threshold (see left panel of Fig. 6), which is explained by the presence of a virtual pole in the ηNN system [5]. This feature is not described by first order rescattering. A similar failure is also seen for the inelastic channel (middel panel of Fig. 6). All contributions to the total cross section are shown in the right panel of Fig. 6.

Turning now to η photoproduction, an analogous representation as in Fig. 5 holds for the coherent and incoherent photoproduction amplitudes, replacing the incoming η by a photon line. Results for total and differential cross sections

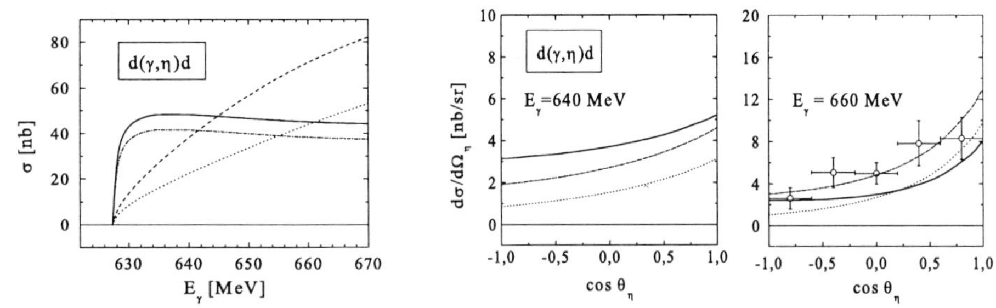

Figure 7. Results for $d(\gamma, \eta)d$ from [4]. Left panel: Total cross section. Middle and right panels: differential cross sections. Dotted: IA; dashed: first order rescattering; solid: complete 3-body model; dash-dot: 3-body model without π-exchange contribution; Exp.: from ref. [7].

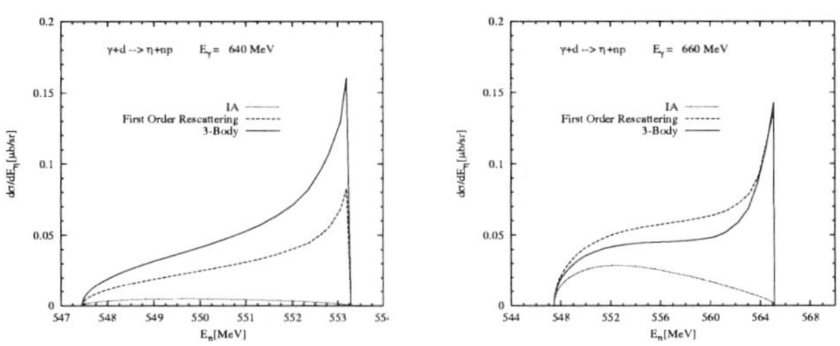

Figure 8. η meson spectrum for $d(\gamma, \eta)np$ close to threshold from [4].

of the coherent photoproduction are shown in Fig. 7. One readily notes a drastic increase of the total cross section over the mere IA right above threshold due to a strong attraction in the $(S = 1, T = 0)$ channel, which the first-order rescattering calculation is not able to reproduce. Also in the differential cross sections in Fig. 7 a large difference between first-order calculation and complete three-body approach is seen. Similar conclusions are reached for the incoherent reaction. The importance of a three-body treatment is demonstrated by the η meson spectrum in Fig. 8. Close to threshold, the first order rescattering underestimates significantly the three-body approach, whereas at higher energies it yields an overestimation. Total and differential cross sections are shown in Fig. 9. The inclusive total cross section data exhibit a distinct enhancement near threshold which is reproduced by the 3-body approach (left panel of Fig. 9). This is also the case for the differential cross sections (middel and right panels of Fig. 9). It remains to be seen, whether a more realistic calculation is also able to describe the data.

5 Conclusions and Outlook

The main conclusions are: (i) Near threshold only a three-body approach gives an adequate description of the ηNN dynamics. The first order rescattering ap-

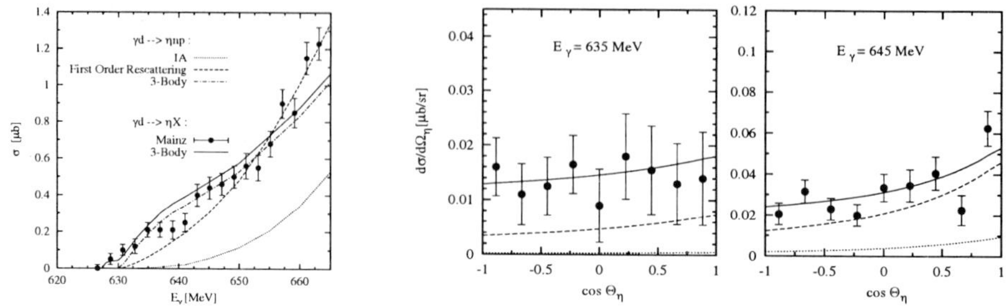

Figure 9. Results for $d(\gamma,\eta)np$ from [4]. Inclusive data from ref. [3]. Left panel: Total cross section: (a) incoherent: dotted: IA; dashed: first order rescattering; dash-dot: complete 3-body model; (b) inclusive: solid. Middel and right panels: Differential cross sections: dotted: IA; dashed: first order rescattering; solid: complete 3-body model.

proximation for the final state interaction fails drastically. (ii) For ηd elastic scattering a very strong enhancement near threshold is found in the three-body calculation which is not born out in first order rescattering. The coupling to the πNN channel is relatively unimportant. (iii) With a Yamaguchi-type separable interaction a satisfactory description of experimental data on η photoproduction on the deuteron is achieved, in particular the enhancement of the total cross section right above the threshold, and the nearly isotropic angular distribution of the outgoing η meson is reproduced in the three-body approach in contrast to the first order rescattering. (iv) The ηN-interaction can be studied in incoherent eta production near threshold. However, a first order rescattering calculation as used, e.g. by [8] is not reliable for that purpose, because right above threshold a three-body approach is mandatory.

As open problems remain (i) the inclusion of additional two-body effects like meson exchange curents as discussed in the coherent process [6], and (ii) the use of realistic NN interactions with inclusion of the tensor force and the deuteron D-wave.

References

1. B. Krusche et al.: Phys. Lett. **B358**, 40 (1995)
2. A. Fix and H. Arenhövel: Z. Phys. **A359** (427) (1997)
3. V. Hejny et al.: Eur. Phys. J **A13**, 493 (2002)
4. A. Fix and H. Arenhövel: Nucl. Phys. **A697**, 277 (2002)
5. A. Fix and H. Arenhövel: Eur. Phys. J **A9**, 119 (2000)
6. F. Ritz and H. Arenhövel: Phys. Rev. **C64**, 034005 (2001)
7. P. Hoffmann-Rothe et al.: Phys. Rev. Lett. **78**, 4697 (1997)
8. A. Sibirtsev et al.: Phys. Rev. **C65**, 044007 (2002)

Few-Body Systems Suppl. 14, 287–292 (2003)

The OOPS Program at Bates

T. Botto*, for the OOPS Collaboration

MIT-Bates, 21 Manning Avenue, 01939 Middleton, MA, USA

Abstract. We report on recent high-precision, exclusive measurements of $\gamma^* \mathrm{N} \to \Delta$ and of virtual Compton scattering from the proton (VCS). These studies are part of the unique program pursued by the OOPS collaboration at the MIT-Bates laboratory. For the first time we made use of high-duty factor, high-energy, extracted electron beams. The full OOPS set-up was commissioned and new data collected in the $\mathrm{H}(e,e'p)\pi^0$, $\mathrm{H}(e,e'\pi^+)n$ and $\mathrm{H}(e,e'p)\gamma$ reactions.

1 Introduction

The Out-of-Plane Spectrometer facility at Bates consists of a unique set of four virtually identical spectrometers. Each OOPS module [1] is a light-weight (16 ton) spectrometer (DQ type, 12° vertical bending angle). The magnetic elements are followed by a compact detection system [2] consisting of tracking devices (HDCs) and 3 layers of plastic scintillators for triggering and PID purposes. Each module provides good angular (1.3 mrad) and momentum ($\Delta p/p = 10^{-2}$) resolutions up to a maximum momentum of 850 MeV/c.

The set-up is completed by the high resolution OHIPS spectrometer, typically used for tagging electrons in coincidence with a p or$\pi^\pm$ in OOPS. OHIPS also provides vertex reconstruction at the target and has the excellent momentum resolution ($\Delta p/p = 10^{-3}$) required for an accurate determination (< 3 mrad) of the direction of the 3-momentum transfer $\boldsymbol{q}$ in the reaction.

In a typical configuration (see Fig. 1) each of the OOPS spectrometers is precisely aligned to given azimuthal angles ϕ_{pq} on the hadron emission cone for constant polar angle θ_{pq} around the direction of the momentum transfer $\boldsymbol{q}$. Owing to its positioning flexibility, the OOPS system offers competitive advantages over more conventional experiments that rely on the detection of "in-plane" particles. In particular one can measure the dependence of the cross section on ϕ_{pq} in A(e,eX)B reactions and simultaneously separate the R_{LT}, R_{TT} interference response functions with high precision and low systematic errors.

* *Email address:* tancredi@lns.mit.edu

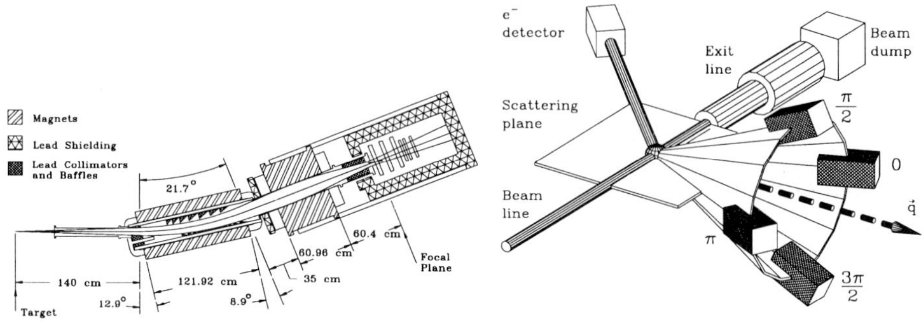

Figure 1. Left: cross-sectional view of an OOPS module. Right: a cluster of four OOPS spectrometers placed around the direction of $\boldsymbol{q}$ in a typical configuration.

With OOPS one can also form asymmetries ratios of events at different ϕ_{pq} angles and over-lapping in phase space. Such "space asymmetries" can be obtained independently from normalization uncertainties. Finally, OOPS provides access to the beam-helicity or fifth structure function ($R_{LT'}$).

In 2000 we commissioned all of the necessary equipment for measurements with a full OOPS set-up, including a fourth module and a new gantry support system. The OHIPS spectrometer momentum acceptance has been extended to $\pm 7\%$. For the first time we used a high-duty factor (unpolarized) beam extracted from the South Hall Ring and the 600 Hz Bates Linac. We used beam energies up to 950 MeV with an extracted current up to 10 μA and a duty factor in excess of 60 %. We took data on a $p(e,e'p)\gamma$ VCS experiment and performed new N-Δ measurements in the $p(e,e'p)\pi^0$ and $p(e,e'\pi^+)n$ channels. These data will provide material for 6 Ph.D. theses.

2 Studies of Virtual Compton Scattering

Compton scattering on the nucleon is well known to give access to its polarizabilities under the stress from the external electric and magnetic fields of the incoming photon. In the virtual reaction (VCS) one can further probe the nucleon polarizability at a varying scale given by the wavelength associated with the momentum of the virtual photon $\boldsymbol{q}$ for a fixed transferred energy (ω).

We studied the $p(e,e'p)\gamma$ reaction below pion threshold at $Q^2 = 0.05$ GeV^2 and for five beam energies corresponding to five values of the outgoing photon energy in the range $E_\gamma = 28 - 115$ MeV [3]. For each setting we measured at a $\phi_{pq} = 90°$ and $180°$, where the contribution from the Bethe-Heitler background is minimized and one retains maximum sensitivity to the electric and magnetic proton polarizabilities. The analysis of these data is well underway (see Fig. 2).

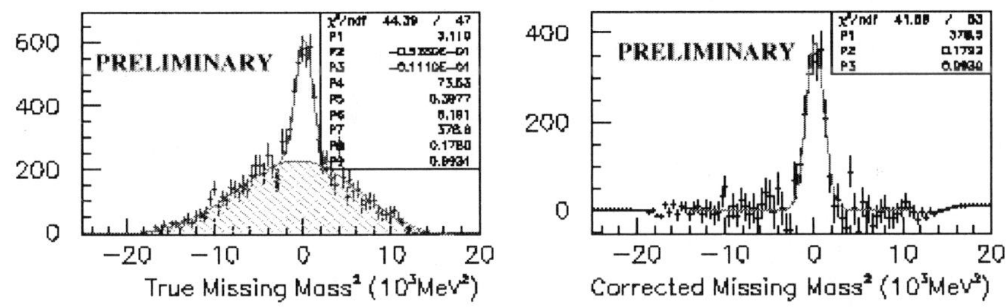

Figure 2. VCS missing mass spectra measured with OOPS [4]. Left: the distribution of true coincidence events as determined by a cut on the coincidence time window. Right: the corrected spectrum after subtraction of random coincidences.

3 Studies of the N-Δ Transition

The $\gamma N \rightarrow \Delta(1232)$ transition connects J=1/2 and J=3/2 baryons. Experimentally it is well known that the reaction is strongly dominated by a magnetic-dipole transition amplitude (M1) which acts by flipping a quark spin in the nucleon. However the transition may also occur via electric (E2) and scalar (C2) resonant quadrupole amplitudes where the incoming photon transfers two units of angular momentum to excite a Δ state. In a CQM picture [5] the quadrupole transitions occur between S and D wave components of the N and Δ wave-functions. In [5] such non-spherical wave-functions are in turn a consequence of a tensor quark interaction required to explain the $N-\Delta$ mass splitting.

Modern experiments have thus focused on the issue of "nucleon deformation" and on the resonant quadrupole amplitudes. Recent experiments from Bates, Mainz and JLab have established that the C2/M1 and E2/M1 ratios (respectively CMR and EMR) are only a few % and that the signal of interest is of the same magnitude of competing non-resonant background amplitudes.

On the other hand, a variety of theoretical approaches have shown that the E2 and C2 amplitudes may in turn be dominated by reaction mechanisms not considered in [5]. In particular, model calculations where $N\pi$ re-scattering effects are taken into account have shown [6, 7] that large contribution to E2 and C2 may arise from the long-range interaction of the pionic cloud surrounding the nucleon, a fact also confirmed in sigma models [8]. One should note that the $\pi - N$ interaction is mostly in p-waves and thus not spherical symmetric [8], a fact that may ultimately be connected with chiral symmetry breaking [9].

Effects from the pion cloud surrounding the nucleon are particularly prominent in our Q^2 range [6]. Early lattice calculation results confirm this, but also point to the fact that the nucleon (and $\Delta(1232)$) may still be intrinsically deformed [11]. On the other hand explicit two-quark interaction currents mediated by meson or gluon exchange (not considered in [5]) may also lead to an apparent signal of nucleon deformation [10].

Clearly the determination of the CMR and EMR ratios is very difficult on

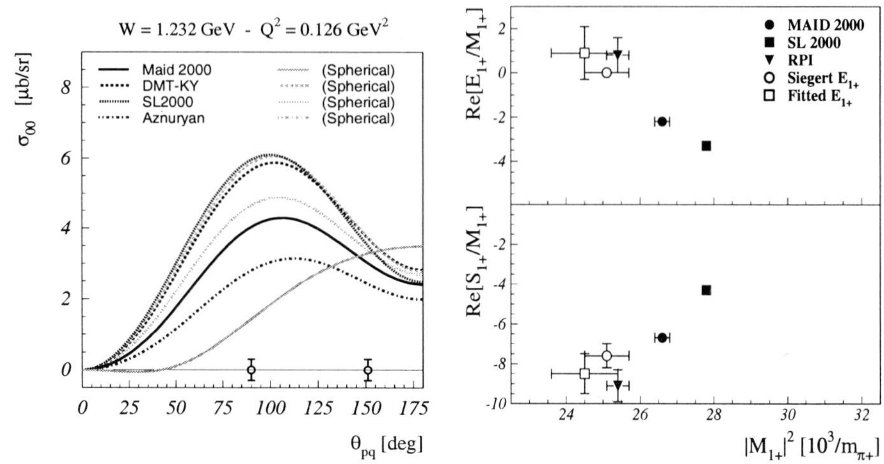

Figure 3. Left panel: R_{00} calculated with MAID [14], the dynamical models of [7, 6] and a calculation from [15]. Empty symbols are projected new OOPS data from 2001, light curves are calculations without deformation. Right panel: correlation between the E_{1+}, S_{1+}, M_{1+} resonance parameters, from fits to published OOPS data [12].

both the experimental and theoretical side, requiring an accurate and extensive data set, and a sophisticated reaction model. To date, the study of $N \to \Delta$ transition is fascinating as it deals with our detailed understanding of the most-prominent reaction in the $\gamma N \pi$ sector and of basic nucleon structure.

In 2001 OOPS collected a significant amount of new data to add to our already robust data-set [12, 13]. A total of 4 data points in the π^0 channel were collected at $Q^2 = 0.127$ GeV^2 and CM energies of $W = 1232$ and 1292 MeV. In addition we took data at $\theta_{\pi q} = 45°$ and $W = 1232$ in the π^+ channel. Our setup was uniquely exploited to study E2 at $Q^2 > 0$ with unprecedented sensitivity. E2 mostly contributes via a $E^*_{1+}M_{1+}$ interference term to nuclear responses dominated by a much larger $|M_{1+}|^2$ term[1]. However, to leading order in M_{1+},

$$R_{00} = R_{TT}(\theta) + R_T(\theta) - R_T(0) = 2(\cos\theta - 1)\mathrm{Re}E^*_{0+}M_{1+} - 12\sin\theta^2\mathrm{Re}E^*_{1+}M_{1+}$$

where θ is the CM angle. To measure R_{00} we combined data along the $\boldsymbol{q}$ direction ($\theta = 0$) and at 3 azimuthal angles for a fixed proton emission angle. Since we did not perform a Rosenbluth separation we approximated $R_{00} \sim R_{TT}(\theta) + R_0\theta - R_0(0)$ where $R_0 = R_T + \epsilon R_L$.

Our expected result for the EMR is shown in Fig. 3. Also shown is the correlation between model-fit results prior to the 2001 run, including two phenomenological fits [12]. The large spread of the results obtained then gives an idea of the size of the theoretical model uncertainty, a fact also due to different parameterizations of the non-resonant background. Most of the data in [12] were sensitive only to the S_{1+}. While it is not surprising that results for S_{1+}

[1]We project the EM multipoles in the $J = 3/2$ channel where C2=S_{1+}, E2=E_{1+}, M1=M_{1+}

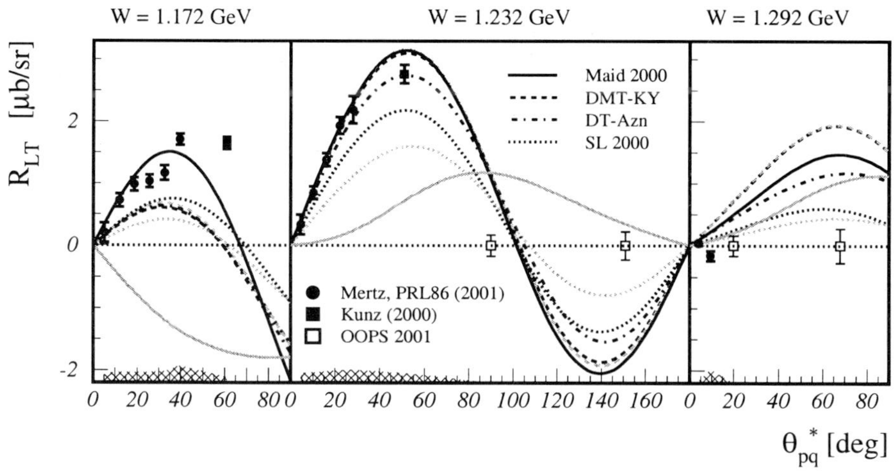

Figure 4. OOPS R_{LT} data at $Q^2 = 0.127$ GeV^2. Full and empty symbols are data from [12, 13] and projected error bars, respectively. The shaded areas are the systematic errors. Curve labels as in Fig. 3

(E_{1+}) and M_{1+} correlate, since they always enter in bilinear combinations, the E_{1+} was found small and poorly constrained. Our new "direct" measurements of the EMR should significantly constrain the extracted resonance parameters.

In Fig 4 we show the R_{LT} response function where the C2 transition amplitude contributes visibly. This sensitivity is unique of electro-production. Our new datum at $\theta = 90°$, where the $S^*_{1+}M_{1+}$ interference vanishes, will provide a measurement of the S-wave non-resonant background. For the first time in 2001, we measured on the low-momentum proton branch, for $\theta_{pq} = 151°$ in the CM. We also added measurements above resonance ($W = 1292$ MeV) where our previous data [12] show a trend in contrast with existing calculations.

4 Out-of-Plane Polarimetry

The general electro-production cross section from $J = 1/2$ targets contains a total of 18 responses. All of these can be uniquely determined when the recoil proton polarization is measured at the OOPS focal plane [16].

Upgrading the OOPS detector package involves installation of additional HDCs (available) for tracking events emerging from a secondary scattering reaction in a movable graphite target. A preliminary design exists [16]. The recoil proton polarization would be measured via the standard $\boldsymbol{p}+^{12}C$ reaction, which is well suited to our energy range.

With such a set-up the physics of interest can be accessed by forming "super-ratios" of focal plane polarization asymmetries measured simultaneously at two spectrometers at symmetric out-of-plane angles. This technique shows great sensitivity, largely cancels systematic uncertainties (such as the polarimeter effective analyzing power) and makes advantageous the use of the OOPS point-like acceptance [16]. Also, it provides a number of important consistency checks.

In contrast with polarization measurements along $\boldsymbol{q}$, and thus in-plane, a total of 14 independent observables can be measured in one setting (equivalent to a large angular scan). In the case of $N \to \Delta$ one can determine all of the 7 E2, C2 sensitive responses. Such experiments would make possible a model-independent multipole analysis at fixed W and Q^2.

5 Summary

The full OOPS facility at Bates has been commissioned to take data with CW extracted beam at luminosities in excess of 10^{36} At/cm^2/s. We have collected exciting new data on VCS and $N \to \Delta$. A study has been initiated to possibly extend the program to out-of-plane measurements of recoil proton polarizations which holds unique promises for a large class of complete experiments.

References

1. S.M. Dolfini et al.: Nucl. Instr. Meth. **A344**, 571 (1994); J. Mandeville et al.: Nucl. Instr. Meth. **A344**, 583 (1994).
2. Z. Zhou et al.: Nucl. Instr. Meth. **A487**, 365 (2002)
3. J.A. Shaw and R. Miskimen (spokespersons): Bates proposal 97-03 (1997)
4. P. Bourgeois: Ph.D. thesis in preparation, UMASS Amherst
5. S.L. Glashow: Physica **96A**, 27 (1979); N. Isgur, G. Karl, R. Koniuk: Phys. Rev. **D25**, 2394 (1982)
6. T. Sato, T.S.H. Lee: Phys. Rev **C63**, 055201 (2001)
7. S.S. Kamalov, S.N. Yang: Phys. Rev. Lett. **83**, 4494 (1999)
8. M. Fiolhais, B. Golli, S. Širca: Phys. Lett. **B373**, 229 (1996)
9. A.M. Bernstein: In: Proceedings of the VII Workshop on Electron-Nucleus Scattering, June 23-28, 2002, Elba, Italy, in preparation
10. A. Buchmann, E. Hernandez, A. Fäßler: Phys. Rev. **C55**, 448 (1997)
11. C. Alexandrou et al.: hep-lat/0209074
12. C. Mertz et al.: Phys. Rev. Lett. **86**, 2963 (2001).
13. C. Kunz: Ph.D.thesis, MIT (2000) (to be submitted for publication)
14. O. Hanstein, D. Drechsel, L. Tiator: Nucl. Phys. **A632**, 561 (1998)
15. I.G. Aznuryan: Phys. Rev **D57**, 2727 (1998)
16. T. Botto (spokesperson): Bates proposal 01-02 (2001), publication to be submitted

Few-Body Systems Suppl. 14, 293–298 (2003)

Few-Body Systems

Pion Production at Threshold as Test of Chiral Perturbation Theory

H. Merkel, for the MAMI A1 Collaboration

Institut für Kernphysik, Universität Mainz, D-55099 Mainz, Germany

Abstract. An overview of recent results in threshold pion production is given. While the photoproduction data are in impressive agreement with calculations in the framework of Chiral Perturbation Theory, electroproduction data at low four-momentum transfer show severe deviations from these calculations.

1 Neutral Pion Photoproduction

Chiral Perturbation Theory is a consistent scheme to utilize the symmetries of QCD to predict observables at the confinement scale. A recent overview of the theoretical status of this field can be found e.g. in ref. [1].

In this talk, the current status of the neutral pion photo- and electroproduction experiments near threshold is discussed. Since the pion is the Goldstone Boson of the chiral symmetry breaking, these experiments are well suited to test the predictions of Heavy Baryon Chiral Perturbation Theory (HBChPT) [2].

First experiments [3] on threshold pion photoproduction were aimed to test the predictions of Low Energy Theorems [4] for the threshold value of the s wave multipole amplitude E_{0+}. The severe disagreement between these theorems and the experiments was resolved in the following years by refined calculations in HBChPT [5], which also gave predictions for the p wave multipole combinations

$$\begin{aligned} P_1 &= 3E_{1+} + M_{1+} - M_{1-} \,, \\ P_2 &= 3E_{1+} - M_{1+} + M_{1-} \,, \\ P_3 &= 2M_{1+} + M_{1-} \,. \end{aligned}$$

The calculations showed, that the s wave amplitude is only slow converging in the chiral expansion, while the p wave combinations P_1 and P_2 are strong predictions in this framework. P_3 is given by a low energy constant of HBChPT and has to be determined by the experiment.

The first experimental access to the multipoles is the measurement of the differential cross section. With the assumption, that only s and p waves contribute at threshold, the angular structure of the cross section is given by

$$\sigma(\theta) = \frac{q}{k}\left(A + B \cdot \cos\theta + C \cdot \cos^2\theta\right)$$

with the phase space factor $\frac{q}{k}$ and three angular coefficients

$$\begin{aligned} A &= E_{0+}^2 + \frac{1}{2}\left({P_2}^2 + {P_3}^2\right) , \\ B &= 2 \cdot Re\left(E_{0+} {P_1}^*\right) , \\ C &= {P_1}^2 - \frac{1}{2}\left({P_2}^2 + {P_3}^2\right) . \end{aligned}$$

Experiments at SAL [6] and MAMI [7, 8] are in good agreement and allow for an extraction of the full angular structure of the differential cross section. But from unpolarized cross section experiments, only the modulus of the s wave multipole $|E_{0+}|$ and the p wave combinations P_1 and $P_{23} = \frac{1}{2}({P_2}^2 + {P_3}^2)$ can be extracted. To further decompose all multipoles, a further observable has to be measured. A convenient choice is the polarized photon asymmetry Σ with the multipole decomposition

$$\Sigma(\theta) \cdot \sigma(\theta) \sim \sin(\theta)\frac{1}{2}\left(P_3^2 - P_2^2\right) .$$

Such an experiment was performed at MAMI [8] at the tagged photon beam of the A2 collaboration with the TAPS detector for the detection of the π^0 decay photons. The polarized photon beam was produced by coherent Bremsstrahlung from a diamond crystal. Figure 1 shows the asymmetry Σ, averaged over the energy range of the experiment.

With this experiment, for the first time a complete separation of the p waves is possible. The results are given in table 1, in comparison with the predictions of Chiral Perturbation Theory (CHPT) [5] and Dispersion Relations(DR) [9]. Within the error bars, the two existing high resolution experiments and the quoted calculations agree. The deviation for P_3 of ChPT can be removed by re-fitting the low energy constants to the new data set.

2 Electroproduction at Low Q^2

Additional information on the pion production mechanism can be extracted from electroproduction experiments. While e.g. the multipole combination P_3 is basically a fit parameter in the description of the photoproduction data, the extention of this quantity to virtual photons is given in ChPT without further degrees of freedom. In addition, the longitudinal s wave amplitude L_{0+} and two further p wave combinations can be extracted.

First experiments at a photon virtuality of $Q^2 = 0.1\,(\mathrm{GeV/c})^2$ aimed to extract the s wave amplitudes at threshold [10, 11]. These experiments were in

Table 1. Experimental multipole amplitudes for photoproduction from MAMI [8] and SAL [6] in comparison with Chiral Perturbation Theory (CHPT)) [5] and Dispersion Relations (DR).

	E_{0+} $(10^{-3/m_\pi})$	P_1 $(qk \cdot 10^{-3}/m_\pi^3)$	P_2 $(qk \cdot 10^{-3}/m_\pi^3)$	P_3 $(qk \cdot 10^{-3}/m_\pi^3)$
MAMI	-1.31 ± 0.08	10.02 ± 0.2	-10.5 ± 0.2	13.1 ± 0.1
SAL	-1.32 ± 0.05	10.26 ± 0.1		
ChPT	-1.16	10.33 ± 0.6	-11.0 ± 0.6	11.7 ± 0.6
DR	-1.22	10.54	-11.4	10.2

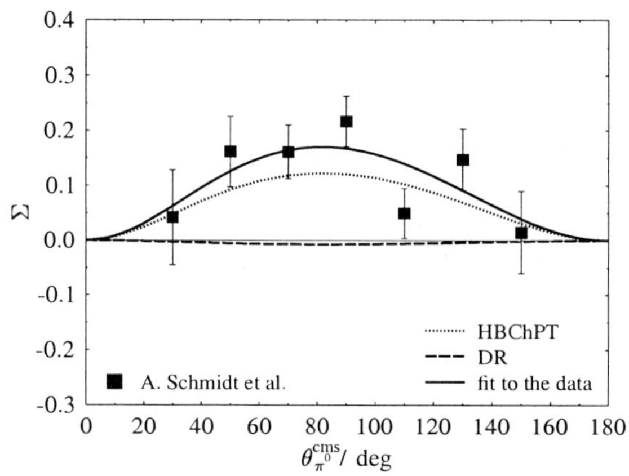

Figure 1. Polarized photon asymmetry measured at MAMI [8] in comparison with calculations in ChPT [5] and in Dispersion Relations formalism [9].

reasonable agreement with calculations [12], but the value of $Q^2 = 0.1\,(\mathrm{GeV/c})^2$ is somewhat to high for the convergence of ChPT.

Therefore, a further experiment at the intermediate value of the photon virtuality $Q^2 = 0.05\,(\mathrm{GeV/c})^2$, half way between photoproduction and the existing data was performed at MAMI [13].

This experiment showed a surprising discrepancy to the calculations. Figure 2 shows the total cross section as function of Q^2. In this observable, the statistical error of all data points is very small and only systematic errors contribute. Thus, the clear discrepancy between data and calculations, but also the increasing discrepancy between ChPT and MAID is visible.

In ref. [13] also an extraction of multipoles in a model independent manner was attempted. While the resolution of the experiments is not yet good enough to perform a complete separation of the multipoles, it seems that the deviation is burried in the multipole combination P_{23}, which is already fixed by photoproduction and can not be adjusted in the calculations to describe the new data set.

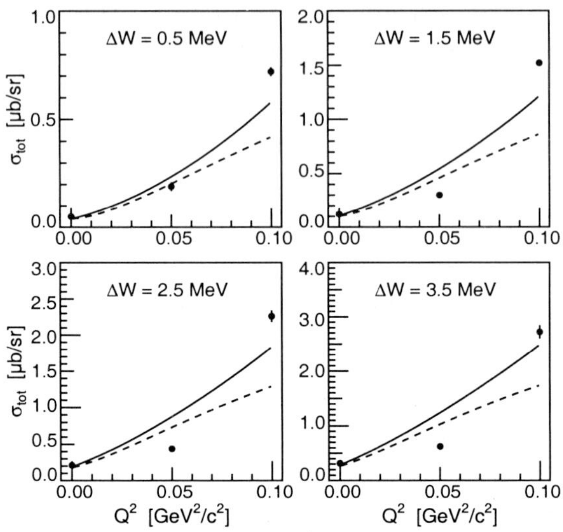

Figure 2. The total cross section σ_{tot} versus Q^2, at a value of $\epsilon = 0.8$. The solid (dashed) line is the prediction of ChPT [12] (MAID [14]), data points at $Q^2 = 0$ and $0.1\,\mathrm{GeV}^2/\mathrm{c}^2$ from [8, 11].

Since the discrepancy is large and surprising, this subject urgently needs further investigation. An experiment at MAMI is planned to cover a continuous range in Q^2, while an independent experiment is planned at JLab [15] with extended kinematical coverage using a large acceptance spectrometer.

3 Electroproduction from the Deuteron

The low energy constants of ChPT were adjusted, as shown above, to describe the existing pion photo- and electroproduction data from the proton. By this, one looses some of the predictive power of CHPT, since in threshold experiments the complete amplitude is already given by only few parameters. On the other hand, from the description of the proton amplitudes one can extract predictions for the pion production from the neutron without introducing further degrees of freedom.

Despite this theoretical advantage, the experimental access to the free neutron amplitude is quite difficult. The most promising method seems to be the coherent pion production from a deuteron target. In impulse approximation, the production amplitude is basically given by the coherent iso-scalar sum of the free proton and free neutron amplitude, corrected by form factors as parameterization of the deuteron structure.

A first measurement of the photoproduction amplitude was performed at SAL [16]. The IGLOO detector was used to detect the decay photons of the π^0 decay in coincidence. By this technique, the missing mass resolution is not sufficient to separate the coherent channel from the deuteron breakup. By calculating this contribution in a simple model the authors were able to extract the threshold value of the s wave amplitude to $E_d = (-1.45 \pm 0.09) \times 10^{-3}/m_\pi$.

This value falls about 20% below the prediction of ChPT [17] of $E_d = (-1.8 \pm 0.6) \times 10^{-3}/m_\pi$ but the agreement seems reasonable within the error bars.

The extension of this experiments to finite Q^2 introduced further experimental difficulties. Due to the background conditions for electroproduction the detection of the pion decay photons had to be replaced by the detection of the recoil deuteron, which suffers at the low energies at threshold from energy loss and multiple scattering. On the other hand, by this technique the coherent channel is clearly separated from the deuteron break up reaction and can be extracted without model assumptions.

A first threshold measurement of $d(e, e'd)\pi^0$ was performed at MAMI [18]. The detection of the deuteron limited this experiment to a four momentum transfer of $Q^2 = 0.1\,(\text{GeV/c})^2$. As for the similar experiments from the proton, the full center of mass angle was covered up to 4 MeV above threshold and a Rosenbluth separation was performed.

The first prediction in ChPT for electroproduction [19] could not explain the extracted s wave amplitudes $|E_d| \approx 0 \cdot 10^{-3}/m_\pi$ and $|L_d| = 0.5 \cdot 10^{-3}/m_\pi$ and missed the experiment by nearly an order of magnitude. The improved cacluation [20] showed that it is necessary to calculate elementary nucleon amplitudes and multi-body currents in a consistent schemes. Figure 3 shows the result of this calculation in comparison with the data. Within the large

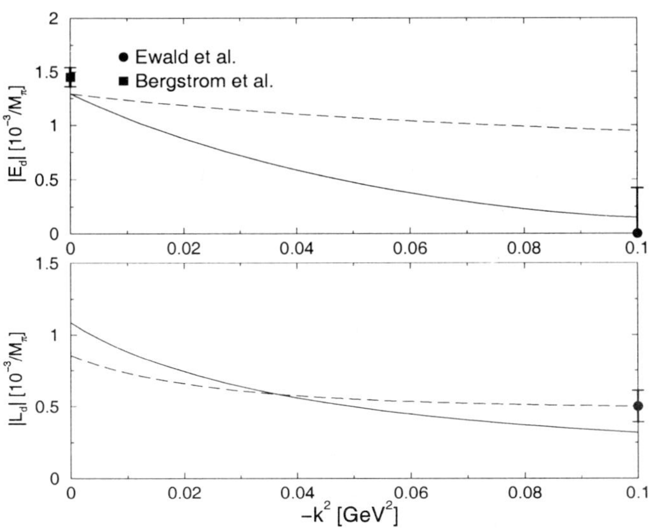

Figure 3. The extracted s wave multipoles from MAMI [18] (circles) in comparison with the prediction of ChPT (from [20]).

systematic error of the data, finally consistency could be achieved. Of course, a measurement at lower Q^2 would be necessary to further test these concepts.

References

1. Proceedings of the Baryons 2002 Conference, World Scientific
2. V. Bernard et al.: Nucl. Phys. **B383**, 442 (1992)
3. E. Mazzucato et al.: Phys. Rev. Lett. **57**, 3144 (1986); R. Beck et al.: Phys. Rev. Lett. **65**, 1841 (1990)
4. P. de Baenst, Nucl. Phys. **B24**, 633 (1970); I. A. Vainshtein and V. I. Żakharov, Nucl. Phys. **B36**, 589 (1972)
5. V. Bernard, N. Kaiser, J. Gasser, and U.-G. Meißner, Phys. Lett. **B268**, 291 (1991); V. Bernard, N. Kaiser, and U.-G. Meißner, Z. Phys. **C70**, 483 (1996)
6. J.C. Bergstrom et al.: Phys. Rev. **C53**, R1052 (1996)
7. M. Fuchs et al.: Phys. Lett. **B368**, 20 (1996); A.M. Bernstein et al.: Phys. Rev. **C55**, 1509 (1997)
8. A. Schmidt et al.: Phys. Rev. Lett. **87**, 232501 (2001)
9. O. Hanstein et al.: Phys. Lett. **B399** (1997) 13
10. H.B. van den Brink et al.: Phys. Rev. Lett. **74**, 3561 (1995); T.P. Welch et al.: Phys. Rev. Lett. **69**, 2761 (1992)
11. M.O. Distler et al.: Phys. Rev. Lett. **80**, 2294 (1998)
12. V. Bernard, N. Kaiser, and U.-G. Meißner, Nucl. Phys. **A607**, 379 (1996); **A633**, 695(E) (1998)
13. H. Merkel et al.: Phys. Rev. Lett. **88**, 012301 (2002)
14. D. Drechsel et al.: Nucl. Phys. **A645**, 145 (1999); S.S. Kamalov et al.: Phys. Lett. **B522**, 27 (2001)
15. R. Lindgren et al.: Experiment E01-014, JLab, 2001
16. J. C. Bergstrom et al.: Phys. Rev. **C57**, 3203 (1998)
17. S.R. Beane et al.: Nucl. Phys. **A618**, 381 (1997)
18. I. Ewald et al.: Phys. Lett. **B499**, 238 (2001)
19. V. Bernard, H. Krebs, U.-G. Meißner, Phys. Rev. **C61**, 58201 (2000)
20. H. Krebs, V. Bernard, Ulf-G. Meißner, nucl-th/0207072

Few-Body Systems Suppl. 14, 299–302 (2003)

Few-Body Systems

On the Relativistic Description of the Multichannel γp Scattering Reactions in the Framework of the Three-Body Field-Theoretical Equations

A.I. Machavariani*,†

LIT JINR, Dubna 141980; Moscow region, Russia

Abstract. A new form of the relativistic three-body equations for the coupled πN and γN scattering reactions with three particle final states $\pi\pi N$ and $\gamma\pi N$ is suggested. These equations are derived in the framework of the time-ordered three dimensional field theory. The solutions of the considered equations satisfy the unitarity condition and are exactly gauge invariant. The form of the suggested three-body equations is not dependent on the choice of the model Lagrangian, and this form is also the same for the formulations with and without quark degrees of freedom. The effective potentials of these equations are defined by the vertex functions with two on-mass shell particles.

The used formulation of the relativistic three-dimensional and three-body equations allows us to overcome a number of approximations which are usually used by practical calculations of the πN and γN scattering reactions.

Nowadays a number of different theoretical models exist which describe with quite a good accuracy the experimental data of the two-body πN and γN reactions independent of three-body $\gamma p \rightarrow \gamma' p'$ and $\pi N \rightarrow \pi' N'$ channels. Moreover, in most papers the γp and πN amplitudes are calculated independently of each other and in addition, the off mass shellness of nucleons and Δ's are neglected. Nevertheless, after different approximations and with a different choice of parameters even for the tree level vertex functions authors have reproduced separately the Compton scattering on the proton, the pion photoproduction and the $\pi N \rightarrow \pi\pi N$, $\pi N \rightarrow \gamma\pi^o N$, $\gamma p \rightarrow \pi\pi^o N$ and $\gamma p \rightarrow \gamma\pi^o N$ reactions with a satisfactory accuracy. Therefore, the next stage of the theoretical investigation of the multichannel πN and γp reactions is a unified description of these reactions with three particle final states.

The general theory of the coupled $\pi N - \gamma N - \pi\pi N - \gamma\pi N$ channels up to the threshold of the creation of the third pi-meson is based on the three-body

* *Alternative address:* High Energy Physics Institute, University of Tbilisi, University str. 9, 380086 Tbilisi, Georgia

† *E-mail address:* machavar@thsun1.jinr.ru, alexm@cv.jinr.ru

field-theoretical equations. Unfortunately, the three-body generalization of the Bethe-Salpeter (BS) equations has a very complicated form and it is too difficult for practical calculations. In particular, there appear two principal difficulties: the ambiguities from the 3D (quasipotential) reductions and determination of three-variable vertex functions which are required for construction of the potential of the BS equation. In practical calculations off-mass variables in these three-variable vertex functions are usually neglected or they are replaced by a product of three one-variable vertex functions.

In my talk I will consider a new form of the three-body field theoretical equations [1, 2, 3] which allow us to overcome a number of approximations by a three-dimensional gauge-invariant formulation of the multichannel πN and γN reactions with the fixed number of intermediate states. The suggested three-dimensional relativistic equation has the following form

$$\mathcal{T}_{\alpha\beta}(E_\beta) = U_{\alpha\beta}(E_\beta) + \sum_{\gamma=1}^{4} U_{\alpha\gamma}(E_\beta)\frac{1}{E_\beta - E_\gamma + i\epsilon}\mathcal{T}_{\gamma\beta}(E_\beta), \tag{1}$$

where $\mathcal{T}_{\alpha\beta}(E_\beta)$ is the $\beta \Rightarrow \alpha$ transition amplitude between the asymptotic $\alpha;\beta = 1,2,3,4 \equiv \pi N, \gamma N, \pi\pi N, \gamma\pi N$ channels and E_α stands for the energy of the asymptotic state α. In the considered field-theoretical formulation $\mathcal{T}_{\alpha\beta}$ is defined through the standard form of the connected part of the field-theoretical $\beta \Rightarrow \alpha$ transition amplitude $< out; \widetilde{\alpha}|j_a(0)|\beta; in >_c$, where a corresponds to one-particle π or γ state, $\alpha = \widetilde{\alpha} + a$ and $j_a(x)$ denotes the source operator of pion $j_\pi(x) = \left(\partial_x^2 + m_\pi^2\right)\Phi_\pi(x)$ or photon $J_\mu(x) = \partial_x^2 A_\mu(x)$. The considered field-theoretical equations are exactly connected with other field-theoretical equations [2], since all of the field-theoretical equations are derived in the framework of the S-matrix reduction theory. Therefore, the results obtained in the framework of these time-ordered 3D equations remain valid in other field-theoretical approaches as well.

A new form of three-body equations [1] differs from standard Faddeev-type equations, since the suggested equations are derived from the S-matrix elements, where the disconnected parts are completely separated from the beginning. Thus, in the considered formulation of the three-body equation the corresponding potential is connected according to definition and therefore double counting problems do not appear there. The resulting three-dimensional relativistic equation for the multichannel πN and γN reactions satisfies all first principles of the quantum field theory and has the following attractive features:

(1.) This formulation is a three dimensional one from the beginning and therefore it is free of the ambiguities which appear due to a three dimensional reduction of the four dimensional Bethe-Salpeter equations.

(2.) The solutions of these equations satisfy the unitarity conditions.

(3.) In the suggested equations for the amplitudes represented in the Coulomb gauge, the gauge invariance condition is exactly satisfied if the requirement of the current conservation for the photon-hadron vertex functions is fulfilled. Therefore, unlike other formulations, here it is not necessary to restrict the number of the calculated diagrams, or to combine some diagrams in the tree approximation, or to make additional assumptions about the Δ propagator in order to ensure the gauge invariant condition.

Equations (1) have the form of the system of the 3D singular integral equations. Analogical form has the BS equations after the 3D (quasipotential) reductions. Major difference between Eq. (1) and other relativistic three-body 3D equations consist in the form of the vertex functions which are required for the construction of the potential $U_{\alpha\gamma}(E)$. The choice of the ingredient input vertex functions for the potentials of these equations is considered as boundary conditions for the solved equations. A scheme of input and output (solutions) of Eq. (1) is given in Fig. 1.

The first kind of the input vertex functions are the vertices with on mass shell hadrons $< \mathbf{p}'_N|J_\mu(0)|\mathbf{p}_N >$ and $< \mathbf{p}'_\pi|J_\mu(0)|\mathbf{p}_\pi >$. These vertices can be obtained from the eN and πe scattering data.

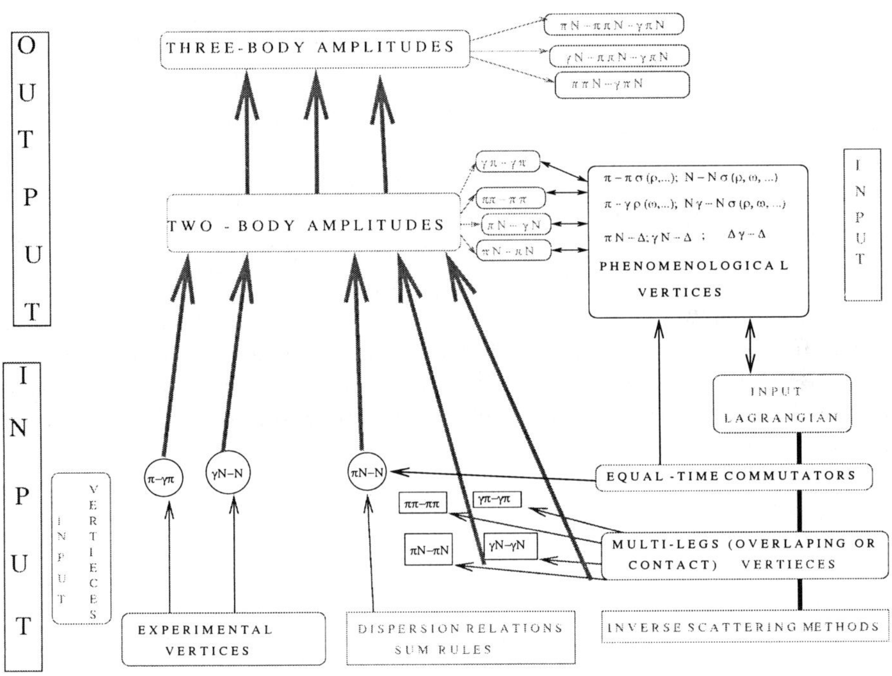

Figure 1. Relations between the solutions of the relativistic 3D field-theoretical equations (1) (output) and the ingredient (input) vertex functions used by construction of potential of these equations. All kind of input vertex functions contains two on-mass shell particle states and one can construct these vertices directly from the experimental data, or using the dispersion relations and other inverse scattering methods.

The πNN vertex functions $< \mathbf{p}'_N|j_\pi(0)|\mathbf{p}_N >$ and $< \mathbf{p}'_N|J_\alpha(0)|\mathbf{p}_\pi >$, where $J_\alpha(x) = \Big((i\nabla_x + m_N)\Psi(x)\Big)_\alpha$ is the nucleon source operator, are included in the second class of the input vertices. They can be determined from the πN phase shifts using the dispersion relations.

The third kind of the input vertex functions are generated by the equal-time commutators which are included in $U_{\alpha\gamma}$ (1) [1, 2]. This part of potential is denoted by $\mathcal{Y}_{\alpha\gamma}$. It consists of t-channel off mass shell particle exchange diagrams and contact (overlapping) terms. $\mathcal{Y}_{\alpha\gamma}$ contains all off mass shell effects including the quark-gluon exchange contributions for the formulations with the quark degrees of freedom [1, 2]. This part of the potential may be determined from the experimental data using the inverse scattering methods [4]. The exact form and the number of overlapping terms of $\mathcal{Y}_{\alpha\gamma}$ are generally determined by the interaction Lagrangian and by the quantization rules. Therefore, this part of the potential may provide insight into the structure of Lagrangians which can be used for a description of experimental data.

The fourth kind of the input vertex functions in Eq. (1) are the phenomenological vertices with Δ-isobar and the heavy meson $\sigma, \rho, \omega, ...$ In these vertex functions only one particle π or γ are off mass shell. The relations between the coupling constants of these vertex functions can be determined according to the chiral symmetry and using the naive $SU(3)$ or $SU(6)$ quark models. Another way to determine of these vertex functions consist in the separable model of the two-body interaction. The separable representation of the P_{33} partial wave of the πN interaction is selfconsistent in the Δ resonance region. But the form of the intermediate Δ propagator is model dependent. In our previous paper [3] we have shown that the experimental data of the γp, $\gamma p - \pi^o p$ and $\gamma p - \gamma \pi^o p$ reactions are strongly dependent on the form of the Δ propagator.

In all other field-theoretical approaches, the effective potentials (or exchange currents) consist also from the above vertex functions including the contact (Schwinger) or overlapping terms. However, in other formulations the effective potential is constructed from more complicated input vertices which are dependent on two or three variables. These vertices are not possible to reproduce directly from experimental data.

References

1. A.I. Machavariani and A. Fäßler: In Preparation

2. A.I. Machavariani: Phys. Part. Nucl. (Sov. J. Part. Nucl.) **24(3)**, 317 (1993)

3. A.I. Machavariani and A. Fäßler: nucl-th/0202060 (to be published)

4. A.I. Machavariani: Phys. Lett. **B540**, 81 (2002)

Few-Body Systems Suppl. 14, 303–306 (2003)

Few-Body Systems

Axial Excitation of the Δ in Chiral Quark Models

S. Širca[1,2], L. Amoreira[3,4], M. Fiolhais[5,4], B. Golli[6,2]

[1] Faculty of Mathematics and Physics, University of Ljubljana, Slovenia
[2] Jožef Stefan Institute, Ljubljana, Slovenia
[3] Department of Physics, University of Beira Interior, Covilhã, Portugal
[4] Centre for Computational Physics, University of Coimbra, Portugal
[5] Department of Physics, University of Coimbra, Portugal
[6] Faculty of Education, University of Ljubljana, Slovenia

Abstract. We discuss electro-magnetic and weak axial N-Δ transition amplitudes in the linear σ-model and the Cloudy-Bag Model as typical representatives of chiral quark models. We show that good qualitative understanding of the transition can be obtained in models which, in addition to the pion cloud, incorporate a fluctuating σ-field inside the baryon.

1 Introduction

The present work was partly motivated by the experience gained in the phenomenological description of the quadrupole electro-excitation of the Δ within the linear σ-model (LSM) and the Chromo-Dielectric Model (CDM) in which the pionic degree of freedom was shown to play a dominant role [1]. In these models, the pion cloud practically saturates the electric and Coulomb quadrupole transition strengths and qualitatively reproduces the Q^2-behaviour of the amplitudes. Furthermore, the results for the ratios of electro-magnetic couplings E2/M1 and C2/M1 at $Q^2 = 0$ were found to be rather insensitive to the details of the models. In the LSM, the absolute values of the transverse helicity amplitudes were well reproduced, although the underprediction of the strength at $Q^2 = 0$ remains an open question. To pin down the pertinent model ingredients, one therefore needs to probe non-zero values of Q^2 where effects of a possibly strong pion cloud in the interiors of the nucleon and the Δ are manifested differently.

We anticipate that the theoretical investigation of axial N-Δ transition amplitudes [2] in chiral quark models may reveal additional information on non-quark degrees of freedom in baryons. Yet due to difficulties in consistent incorporation of the pion field, the model predictions for these amplitudes are very scarce [3], in contrast to the electro-magnetic sector. Experimentally, the structure of the weak axial transition currents is explored by using weak probes [4] or electron scattering [5]. We have a stable world-average for the dominant coupling $C_5^{\mathrm{A}}(0)$ [6], but a very poor knowledge of $C_3^{\mathrm{A}}(0)$, $C_4^{\mathrm{A}}(0)$, and the corresponding form-factors. We calculated the axial amplitudes in the Cloudy-Bag Model (CBM) and in the LSM.

2 Calculation of the Amplitudes

In the linear σ-model and related classes of models involving quarks interacting with chiral fields σ and $\vec{\pi}$ the Hamiltonian can be written as

$$H = H_q^0 + H_\sigma + \int \mathrm{d}\boldsymbol{r} \left\{ \tfrac{1}{2} \left[\vec{P}_\pi^2 + (\nabla^2 + m_\pi^2)\vec{\pi}^2 \right] + U(\sigma, \vec{\pi}) + \vec{j}\vec{\pi} \right\} ,$$

where j_a is the quark source, $\vec{P}_\pi$ is the pion conjugate momentum, H_q^0 and H_σ are the free-quark and the σ-meson terms, and $U(\sigma, \vec{\pi})$ is the meson self-interaction term. In the Cloudy-Bag Model the σ-field and the U-term are absent, while in the linear σ-model all terms are present and U is the well-known Mexican-hat potential. The CDM has an additional scalar-isoscalar field which mimics the glueballs of QCD and dynamically confines the quarks [7].

Evaluating the commutator $[H, \vec{P}_\pi]$ between eigenstates of the Hamiltonian $|\mathrm{N}\rangle$ and $|\Delta\rangle$ (regardless of the model) we obtain a virial constraint of the form

$$(-\Delta + m_\pi^2 - (E_\Delta - E_\mathrm{N})^2)\langle \Delta \,|\, \pi_0(\boldsymbol{r}) \,|\, \mathrm{N}\rangle = -\langle \Delta \,|\, J_0(\boldsymbol{r}) \,|\, \mathrm{N}\rangle \, , \qquad (1)$$

where the source on the RHS of consists of the quark term and the term originating from the meson self-interaction (if present):

$$J_0(\boldsymbol{r}) = j_0(\boldsymbol{r}) + \frac{\partial U(\sigma, \vec{\pi})}{\partial \pi_0(\boldsymbol{r})} \, .$$

In the CBM we assume the usual perturbative form for the pion profiles using the experimental masses for the nucleon and Δ, which fulfills (1). The method we used to impose this constraint in the LSM is described in refs. [8, 9].

In computing the transverse ($\tilde{A}^{\mathrm{A}}$), longitudinal ($\tilde{L}^{\mathrm{A}}$), and scalar ($\tilde{S}^{\mathrm{A}}$) transition helicity amplitudes between states with definite four-momenta, we interpret our localised model states as wave packets of states with good linear momentum. Extending the method explained in [10] we find, to order k^2/M^2:

$$\begin{aligned}
C_6^{\mathrm{A}} &= \frac{M_\mathrm{N}^2}{k^2} \left[-\tilde{A}_{\frac{3}{2}}^{\mathrm{A}} + \sqrt{\frac{3}{2}} \tilde{L}^{\mathrm{A}} \right] \frac{2M_\Delta}{M_\Delta + M_\mathrm{N}} \, , \\
C_5^{\mathrm{A}} &= -\sqrt{\frac{3}{2}} \left(\tilde{L}^{\mathrm{A}} - \frac{k_0}{k} \tilde{S}^{\mathrm{A}} \right) \frac{2M_\Delta}{M_\Delta + M_\mathrm{N}} - \frac{k_0^2 - k^2}{M_\mathrm{N}^2} C_6^{\mathrm{A}} \, , \\
C_4^{\mathrm{A}} &= \frac{M_\mathrm{N}^2}{k M_\Delta} \left[-\sqrt{\frac{3}{2}} \tilde{S}^{\mathrm{A}} + \frac{k_0 k}{M_\mathrm{N}^2} \frac{M_\Delta + M_\mathrm{N}}{2M_\Delta} C_6^{\mathrm{A}} \right] - \frac{M_\mathrm{N}^2}{2M_\Delta^2} C_5^{A} \, .
\end{aligned}$$

For a finite pion mass, the divergence of the axial transition current is given by the PCAC relation $\langle \Delta^+(p') \,|\, \partial^\alpha A_{\alpha\, a} \,|\, \mathrm{N}^+(p)\rangle = -m_\pi^2\, f_\pi \langle \Delta^+(p') \,|\, \pi_a(0) \,|\, \mathrm{N}^+(p)\rangle$, where a is the isospin index and the transition matrix element of the pion field with $a = 0$ is related to the strong form factor by

$$\langle \Delta^+(p') \,|\, \pi_0(0) \,|\, \mathrm{N}^+(p)\rangle = \mathrm{i}\frac{G_{\pi\Delta N}(Q^2)}{2M_\mathrm{N}}\, \frac{\bar{u}_{\Delta\mu}\, q^\mu u_\mathrm{N}}{Q^2 + m_\pi^2} \sqrt{\frac{2}{3}}\,. \tag{2}$$

Assuming that the pion pole dominates the $C_6^\mathrm{A}(Q^2)$ amplitude for $Q^2 \to -m_\pi^2$, the resulting off-diagonal Goldberger-Treiman relation [2, 10, 11] offers an alternative method to compute $C_5^\mathrm{A}(Q^2)$ from the strong $G_{\pi\mathrm{N}\Delta}$ form-factor,

$$C_5^\mathrm{A}(Q^2) = f_\pi\, \frac{G_{\pi\mathrm{N}\Delta}(Q^2)}{2M_\mathrm{N}} \sqrt{\frac{2}{3}} = \frac{2M_\Delta}{M_\Delta + M_N} \sqrt{\frac{2}{3}}\frac{f_\pi}{\mathrm{i}k} \int \mathrm{d}\boldsymbol{r}\, e^{\mathrm{i}\boldsymbol{k}\boldsymbol{r}} \langle \Delta || J_0(\boldsymbol{r}) || \mathrm{N}\rangle\,.$$

3 Results and Discussion

The calculated $C_5^\mathrm{A}(0)$ is 25 % higher than the experimental average, but the Q^2-dependence is reproduced to within a few percent in terms of the dipole cut-off parameter. A better result at $Q^2 = 0$ can be obtained by determining $C_5^\mathrm{A}(Q^2)$ from the calculated strong $\pi\mathrm{N}\Delta$ form-factor through the off-diagonal Goldberger-Treiman relation, yet the Q^2-dependence becomes steeper, with a cut-off of $\approx 0.80\,\mathrm{GeV}$. The disagreement between the two approaches can be attributed to an overestimate of the meson strength, a characteristic feature of LSM where only the meson fields bind the quarks. (Still, the effect of the meson self-interaction is relatively weak in the strong coupling constants.) Essentially the same trend is observed in the nucleon case where we obtain $g_\mathrm{A} = 1.41$. The discrepancy with respect to the experimental value of 1.27 is commensurate with the disagreement in $C_5^\mathrm{A}(0)$. Unfortunately, the overestimate of g_A and $G_\mathrm{A}(Q^2)$ in the LSM seems to persist even if the spurious centre-of-mass motion of the nucleon is removed [12].

In the CBM the picture is reversed. As it has been shown in ref. [8], only the quarks contribute to the C_4^A and C_5^A amplitudes, while C_6^A is almost completely dominated by the pion pole. Hence the calculated values of $C_5^\mathrm{A}(0)$ are too small, ranging from 68 % of the experimental estimate at $R = 0.7\,\mathrm{fm}$ to only 56 % at $R = 1.3\,\mathrm{fm}$. The behaviour of $C_5^\mathrm{A}(Q^2)$ is similar as in the pure MIT Bag Model (to within 10 %), with a fitted dipole cut-off of $\sim 1.2\,\mathrm{GeV\,fm}/R$. The off-diagonal Goldberger-Treiman relation is satisfied in the CBM, but $C_5^\mathrm{A}(Q^2)$ calculated from $G_{\pi\mathrm{N}\Delta}(Q^2)$ has a steeper Q^2-dependence with a cut-off of $\sim 0.8\,\mathrm{GeV\,fm}/R$. The large discrepancy can be partly attributed to the fact that the CBM predicts a too low value for $G_{\pi\mathrm{NN}}$, and consequently for $G_{\pi\mathrm{N}\Delta}$.

The determination of the C_4^A is less reliable because the meson contribution to the scalar amplitude is very sensitive to small variations of the profiles. However, the experimental value is very uncertain as well. Neglecting the non-pole contribution to $\tilde{S}^\mathrm{A}$ and C_6^A, the value of C_4^A is dominated by the term $-(M_\mathrm{N}^2/2M_\Delta^2)\, C_5^\mathrm{A}$, in agreement with the popular value of $C_4^\mathrm{A}(0) = -0.3$.

In accordance with our experience in the electro-magnetic sector, we find that the quark contribution alone strongly underestimates the C_5^{A} amplitude. If only a linear coupling of pions to quarks is added, the situation does not improve since in such a case the pion contribution to C_5^{A} vanishes. On the other hand, the inclusion of meson self-interaction which allows for a substantial deviation of the σ-field from its vacuum value inside the baryon considerably increases C_5^{A}. The LSM overestimates this contribution, but this is not the case in other chiral models which allow for a non-zero fluctuation of the σ-field; in a version of the NJL model with nonlocal regulators [13] the contribution of sea quarks to the nucleon g_{A} is more than a factor of two weaker than the equivalent contribution of chiral mesons in the LSM.

This work was supported by FCT (POCTI/FEDER), Lisbon, and by The Ministry of Science and Education of Slovenia.

References

1. M. Fiolhais, B. Golli, S. Širca: Phys. Lett. **B373**, 229 (1996)
2. C.H. Llewellyn Smith: Phys. Rep. **3C**, 261 (1972)
3. Jùn Líu, N.C. Mukhopadhyay, L. Zhang: Phys. Rev. **C52**, 1630 (1995)
4. S.J. Barish et al.: Phys. Rev. **D19**, 2521 (1979); G.M. Radecky et al.: Phys. Rev. **D25**, 1161 (1982); T. Kitagaki et al.: Phys. Rev. **D42**, 1331 (1990), and references therein
5. A. Liesenfeld et al.: Phys. Lett. **B468**, 20 (1999), and references therein; L. Elouadrhiri, D. Heddle (spokespersons): JLab Proposal E94-005; N. Simićević, S. P. Wells (spokespersons): JLab Proposal PR97-104
6. N.C. Mukhopadhyay et al.: Nucl. Phys. **A633**, 481 (1998)
7. T. Neuber, M. Fiolhais, K. Goeke, J.N. Urbano: Nucl. Phys. **A560**, 909 (1993)
8. B. Golli, S. Širca, L. Amoreira, M. Fiolhais: submitted to Phys. Lett. B
9. B. Golli, L. Amoreira, M. Fiolhais, S. Širca: Bled Workshops in Physics **3**, 55 (2002); S. Širca, L. Amoreira, M. Fiolhais, B. Golli: Bled Workshops in Physics **3**, 67 (2002)
10. T.R. Hemmert, B.R. Holstein, N.C. Mukhopadhyay: Phys. Rev. **D51**, 158 (1995)
11. P.A. Schreiner, F. von Hippel: Nucl. Phys. **B58**, 333 (1973)
12. M. Rosina, M. Fiolhais, B. Golli, S. Širca: In: Proceedings of the Meeting "*Nuclear and Particle Physics with CEBAF at Jefferson Lab*", November 3–10, 1998, Dubrovnik, Croatia, Fizika **B8**, 383 (1999)
13. W. Broniowski, B. Golli, G. Ripka: Nucl. Phys. **A703**, 667 (2002)

Few-Body Systems Suppl. 14, 307–310 (2003)

Few-Body Systems

Final State Interaction Effects in Incoherent Pion Photoproduction on the Deuteron *

E.M. Darwish†, H. Arenhövel, M. Schwamb

Institut für Kernphysik, Johannes Gutenberg-Universität, 55099 Mainz, Germany

In the present work [1] incoherent photoproduction of pions on the deuteron in the $\Delta(1232)$ resonance region is investigated where besides the impulse approximation (IA) complete two-body rescattering in the nucleon-nucleon (NN) and pion-nucleon (πN) final state subsystems is included (see Fig. 1). The elementary $\gamma N \to \pi N$ amplitude including Born and $\Delta(1232)$ resonance contributions is taken from previous work of [2].

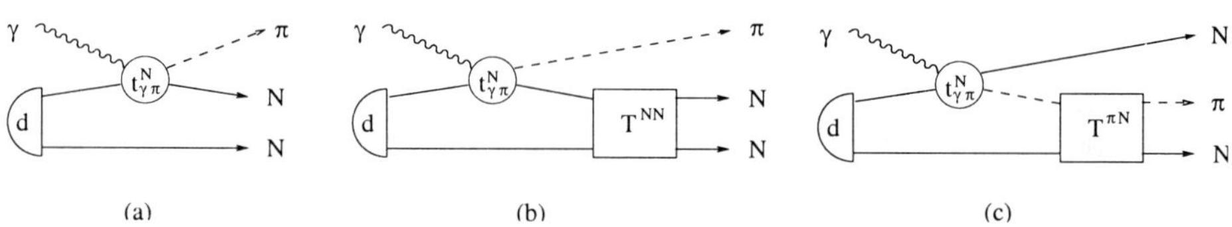

Figure 1. Diagrams for $d(\gamma, \pi)NN$: (a) IA, (b) NN rescattering, (c) πN rescattering.

The t-matrices for NN- and πN-scattering are obtained from the separable interactions of [4], respectively. Final state interaction effects are found to be significant in total (see Fig. 2) and differential cross sections (see Fig. 3), in particular for π^0 production, and lead to a satisfactory agreement with experimental data.

The large effect for π^0 production stems mainly from the fact that in IA quite a fraction of the coherent production is included due to the non-orthogonality of the final plane wave to the bound state. We have also evaluated the contribution to the spin asymmetry of the total cross section which determines the GDH-sum rule. While in IA the spin asymmetry for π^0 production on the deuteron is very similar to the average over neutron and proton

*Supported by Deutsche Forschungsgemeinschaft (SFB 443).

†Supported by Deutscher Akademischer Austauschdienst. *Present address:* Physics Department, Faculty of Science, South Valley University, Sohag, Egypt.

asymmetries this is not the case for charged π production due to the Pauli principle.

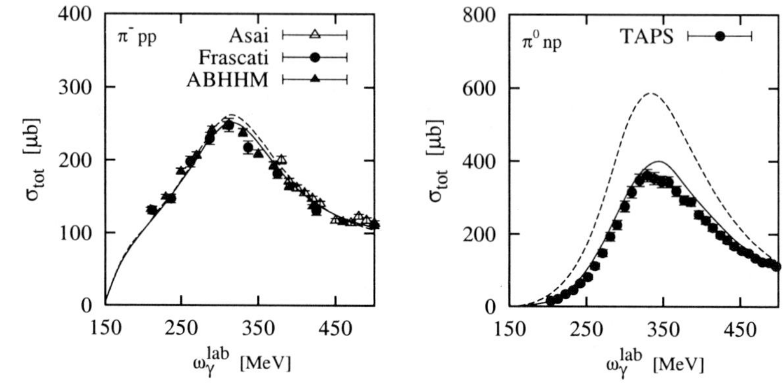

Figure 2. Total cross section for $d(\gamma,\pi)NN$. Dashed: IA; solid: IA+complete rescattering; data from ref. [3].

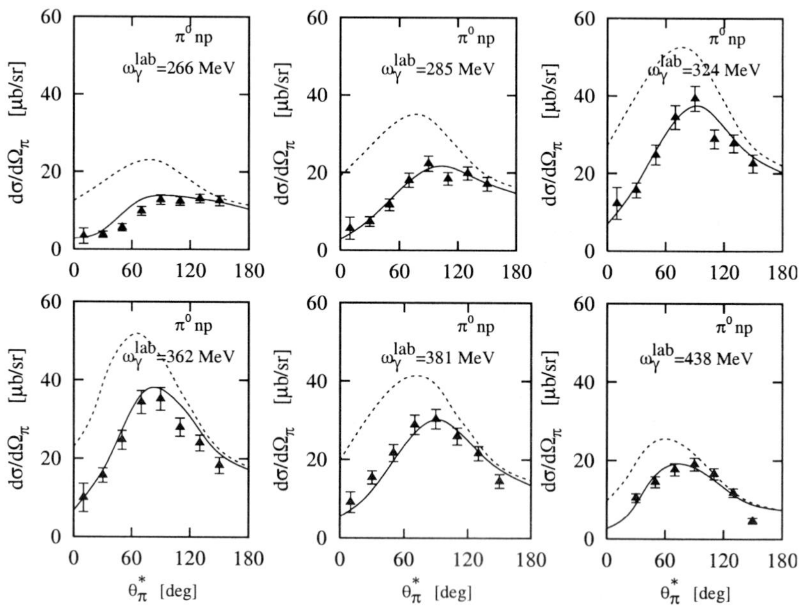

Figure 3. Differential cross sections for $d(\gamma,\pi^0)pn$: Dashed: IA; solid: IA+complete rescattering; Exp. from ref. [5].

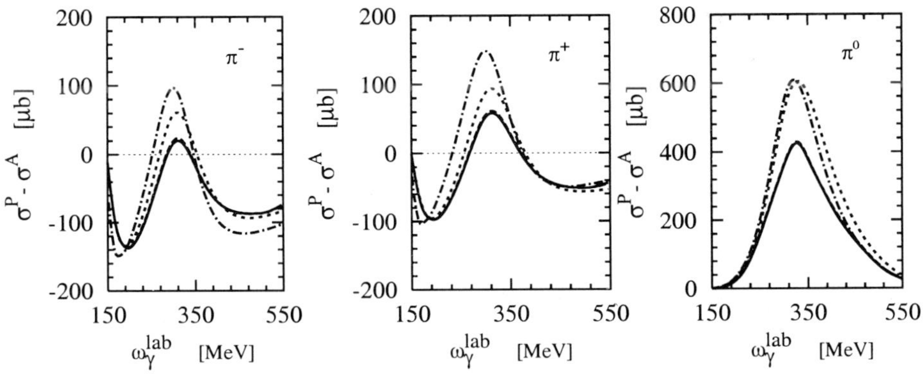

Figure 4. Spin asymmetry of $d(\gamma, \pi^0)pn$: Dotted: IA; dashed: IA+NN rescattering; solid: IA+NN+πN rescattering. Dash-dotted: results for the nucleon, for π^0 average of proton and neutron (right panel).

Moreover, final state rescattering is quite important and cannot be neglected (Fig. 4). For charged π production the effect is much larger in the asymmetry than in the total cross section and comparable to the effect in the neutral π production. For the GDH integral up to 550 MeV, we obtain from $d(\gamma, \pi)NN$ in IA $I^{(IA)}_{d(\gamma,\pi)NN} = 187$ μb which is reduced to $I^{(IA+NN+\pi N)}_{d(\gamma,\pi)NN} = 87$ μb by final state rescattering.

References

1. E.M. Darwish, H. Arenhövel, and M. Schwamb: nucl-th/0208030; Eur. Phys. J. **A** (in print)

2. R. Schmidt et al.: Z. Phys. **A355**, 421 (1996)

3. M. Asai et al.: Phys. Rev. **C42**, 837 (1990); P. Benz et al.: Nucl. Phys. **B65**, 158 (1973); G. Chiefari, E. Drago, M. Napolitano, and C. Sciacca: Lett. Nuovo Cim. **13**, 129 (1975)

4. J. Haidenbauer and W. Plessas: Phys. Rev. **C30**, 1822 (1984); S. Nozawa et al.: Nucl. Phys. **A513**, 459 (1990)

5. B. Krusche et al.: Eur. Phys. J. **A6**, 309 (1999)

Few-Body Systems Suppl. 14, 311–312 (2003)

Few-Body Systems

Eta Production in pp Scattering

S. Ceci*, A. Švarc

Ruđer Bošković Institute, Bijenička Cesta 54, Zagreb, Croatia

1 Introduction

The Zagreb-UCLA-ANL meson-nucleon partial-wave analysis (PWA) [1] has opened the possibility to explain a number of intermediate-energy data. An important result of the PWA are the eta-nucleon elastic T-matrices. The three-body processes can be used to lower the uncertainty of the two-body T-matrices. The idea is to take the two-body T-matrices from existing PWAs and apply them to the three-body processes.

The most convenient process for that purpose is the $pp \rightarrow pp\eta$. It is an isospin-1/2 process, thus enabling us to test half of the amplitudes at a time. Dominant contribution to the production process comes through meson-exchange [2] and decay of nucleon resonances [3, 4, 5].

2 Model

Our model has been described earlier [3]. Because of the large uncertainties in the treatment of the final-state interaction (FSI) of the considered theoretical models [4, 5], we have decided to ignore its influence at the moment. The FSI is significant near threshold, so energies from 100 to 500 MeV would be the most convenient place to test this model. Unfortunately, in this range, experimental data are very hard to obtain. Preliminary results are given in Fig. 1.

3 Discussion

Light and dark lines show constructive and destructive total cross sections, respectively. Enormous discrepancies below few MeV were anticipated because we have completely disregarded FSI. Our preliminary results again do not show a ρ-meson dominance — which would be expected from most of other models

* *E-mail address:* ceci@irb.hr

[5]. To get a better understanding, we plan to calculate further differential cross sections and total cross sections for $pn \to pn\eta$. Complete results will be reported elsewhere.

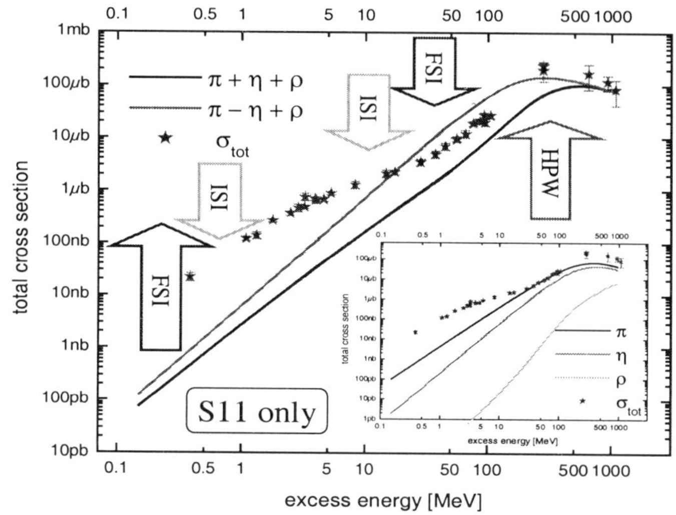

Figure 1. A qualitatively reasonable behavior of the constructive sum of π and η contributions is observed. Added corrections of initial and final interaction will push the theoretical curve in the right direction. The inset shows the contributions for each considered meson exchange. Experimental data are from [6].

References

1. M. Batinić, et al.: Phys. Rev. **C81**, 2310 (1995); M. Batinić, et al.: Phys. Scripta **58**, 15 (1998)
2. R. Machleidt: Adv. Nucl. Phys. **19**, 189 (1989)
3. M. Batinić, A. Švarc, T.-S.H. Lee: Phys. Scripta **56**, 321 (1997)
4. T. Vetter, A. Engel, T. Biró, U. Mosel: Phys. Lett. **B263**, 153 (1991)
5. J.-M. Laget, F. Wellers: Phys. Lett. **B657**, 254 (1991); E. Gedalin, A. Moalem, L. Razdolskaja: Nucl. Phys. **A634**, 368 (1998); A.B. Santra, B.K. Jain: Nucl. Phys. **A634**, 309 (1999); C. Hanhart, K. Nakayama: Phys. Lett. **B454**, 176 (1999)
6. H. Calén et al.: Phys. Lett. **B458**, 190 (1999); P. Moskal et al.: Phys. Lett. **B482**, 356 (2000)

Few-Body Systems Suppl. 14, 313–318 (2003)

Few-Body Systems

Electromagnetic Breakup of Nuclei with $A = 3 \div 6$

W. Leidemann

Dipartimento di Fisica, Università di Trento and INFN (Gruppo collegato di Trento), I-38050 Povo, Italy

Inelastic electromagnetic reactions with nuclei are an important tool to investigate details of the nuclear structure and to study the underlying dynamics. In few-body physics one has the great advantage of not being hampered by many-body approximations. Thus rigorous microscopic calculations can be carried out. On the other hand finding exact solutions for bound and scattering states is a rather difficult task already for the 3N system and a further increasing number of particles leads to even more difficulties.

In this paper I report about recently developed few-body methods making precise few-body calculations more accessible. They enabled us to calculate the following electromagnetic reactions: (i) inclusive reactions with the 3N systems with realistic NN and 3N forces (total photoabsorption cross section, longitudinal (e, e') form factor), (ii) photodisintegration of ^4He, (iii) total photoabsorption cross sections of ^6He and ^6Li. In all those cases ground states are calculated exactly and final state state interaction (FSI) is rigorously taken into account. The inclusion of FSI is achieved by the application of the Lorentz Integral Transform (LIT) method [1], which reduces a continuum state problem to a bound state like problem. For the solution of such bound states we use expansions in hyperspherical harmonics (HH). Correlated HH [2] are employed in case of the above mentioned 3N and 4N reactions, while the very powerful Effective Interaction HH (EIHH) method [3] is used for the six-body calculation.

In order to illustrate the LIT approach let us consider a function $F(\omega)$ which contains information about a specific cross section of a reaction into the continuum. In conventional methods it would be necessary to calculate continuum state wave functions for the evaluation of F, while it is not necessary in the LIT method. In fact one calculates the LIT directly,

$$L(\sigma = \sigma_R + i\sigma_I) = \int d\omega \frac{F(\omega)}{(\sigma_R - \omega)^2 + \sigma_I^2}, \tag{1}$$

without having any knowledge of F. Then in a second step F is obtained by the inversion of the transform.

For inclusive reactions the cross section can be written in terms of response functions $R(\omega)$ of given excitation operators $\hat{O}$ inducing transitions from ground (Ψ_0) to final state wave functions (Ψ_f). The functions R have the general form

$$R(\omega) = \int d\Psi_\mathrm{f} |\langle \Psi_\mathrm{f} | \hat{O} | \Psi_0 \rangle|^2 \delta(\omega - E_\mathrm{f} + E_0) \,, \quad (2)$$

where E_0 and E_f are bound and final state energies, respectively. One finds

$$L(\sigma) = \langle \widetilde{\Psi}_1(\sigma) | \widetilde{\Psi}_1(\sigma) \rangle \,, \quad (H - E_0 - \sigma^*) \widetilde{\Psi}_1(\sigma) = \hat{O} \Psi_0 \,, \quad (3)$$

where H is the Hamiltonian of the system. Note that the above equation for the determination of $\widetilde{\Psi}_1(\sigma)$ has boundary conditions like a bound state due the complex σ and the vanishing of the source term $\hat{O}\Psi_0$ at large distances.

Now we turn to the LIT calculation of the total photoabsorption cross sections of the three-body nuclei. They were carried out with semi-realistic [4] and realistic NN potentials together with realistic 3N forces [5]. The results were obtained in the well know unretarded dipole approximation:

$$\sigma_\mathrm{tot}^{unret}(E_\gamma) = 4\pi^2 \alpha R(E_\gamma) \,, \quad R(E_\gamma) = \int d\Psi_\mathrm{f} |\langle \Psi_\mathrm{f} | \hat{D} | \Psi_0 | \rangle|^2 \delta(E_\gamma - E_\mathrm{f} + E_0) \,, \quad (4)$$

where α and E_γ denote the fine structure constant and the photon energy, respectively, while $\hat{D}$ is the unretarded dipole operator. It was found in those calculations that the 3N force affects the cross section considerably. At low energies, i.e in the peak region, there is a decrease of about 10 % and at higher E_γ one has an increase of similar size. In comparison with experimental data one finds a rather good agreement, but the comparison is not too conclusive, since experimental errors are quite large. These studies also show that at low energies semi-realistic s-wave forces lead to rather similar results as realistic NN potentials. For example for the ^3He polarizability, which can be evaluated as an integral of $\sigma_\mathrm{tot}^{unret}(E_\gamma)/E_\gamma^2$, one obtains: 0.151 fm^3 (MT-I/III potential [6]), 0.159 fm^3 (AV18 potential [7]), 0.145 fm^3 (AV18 + URBIX 3N force [8]).

The LIT results for the 3N photodisintegration were confirmed in a recent benchmark calculation [9], where LIT and Faddeev results for $\sigma_\mathrm{tot}^{unret}$ are compared (see also the contribution of G. Orlandini in this proceedings). In addition the Faddeev results show that effects beyond the dipole approximation (retardation, additional multipoles) are tiny at low energies (less than 1 and 5 % at $E_\gamma = 30$ and 60 MeV, respectively), but become somewhat more important at higher energies (about 15 and 30 % at $E_\gamma = 100$ and 140 MeV, respectively).

The work of [5] is presently being extended to the inclusive (e, e') cross sections of the three-body nuclei. First results for the longitudinal response function R_L have been obtained. Note that for the LIT calculation one has to separate isoscalar and isovector parts of R_L. This allows to introduce isoscalar and isovector nucleon form factors in a proper way after the inversion of the LIT. In the present results only final states up to a total angular momentum $J_\mathrm{f} = 9/2$ are taken into account. Up to now six momentum transfers q are considered, namely 250, 300, 350, 400, 450, and 500 MeV/c. With the restriction of $J_f \leq 9/2$ one has the following exhaustions of the ^3H Coulomb sum rule (ordering according increasing q): 99.6, 98.5, 97.2, 96.4, 92.8, and 89.9 %. It illustrates that convergence is almost reached for the two lower q, while for the higher q still strength is missing from final states with $J_f > 9/2$. In Fig. 1 we show the effect of the 3N force. There is a 5-10 % reduction of R_L in the region of the quasielastic peak due to the 3N force. At higher energies 3N forces have the opposite effect and increase the longitudinal cross section up to about 15 %. In Fig. 2 we show our preliminary results together with experimental data for the two lower q values. One finds a rather good agreement of theory and experiment, but one cannot draw a definite conclusion, since the experimental errors are not small enough to confirm the effect of the 3N force.

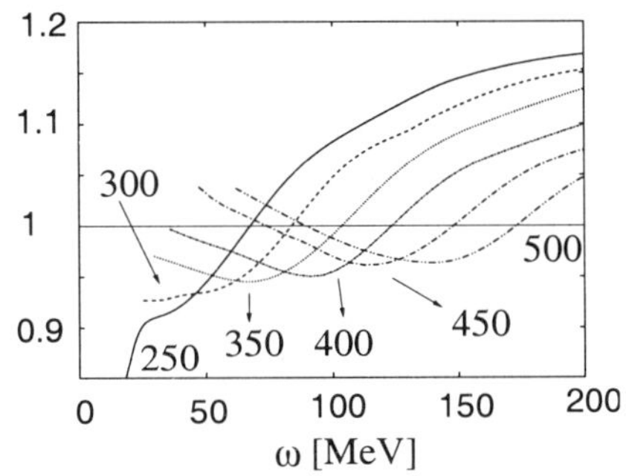

Figure 1. Ratios of F_L(with 3N force)/F_L(without 3N force) for triton at various momentum transfers (q-values indicated in the figure in units of MeV/c).

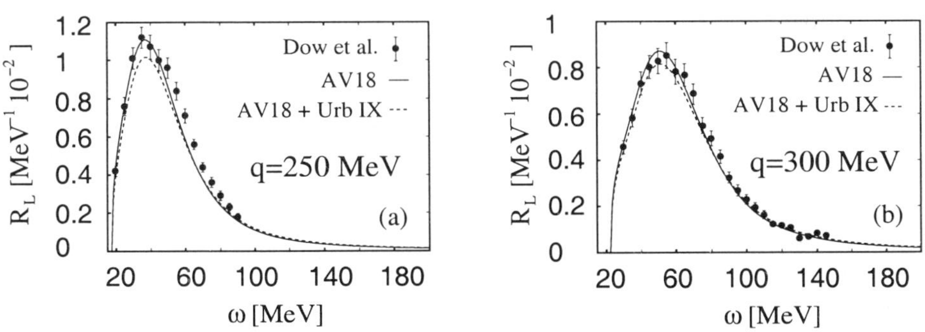

Figure 2. F_L of triton with and without 3N forces at q = 250 MeV/c (a) and 300 MeV/c (b). Experimental data from [10].

Next we discuss electromagnetic processes with ^4He. Also in this case inclusive reactions were calculated with the LIT method, but only semi-realistic NN forces were considered up to now. Nevertheless a very good agreement with experimental data was found for the longitudinal (e, e') responses R_L at q= 300, 400, 500, and 600 MeV/c [11]. Very different is the situation for the total ^4He photoabsorption cross section. At low energies the theoretical $\sigma_{\rm tot}^{unret}$ [12] is considerably higher than the sum of the cross sections of the two open channels ($^4He(\gamma,n)^3He$, ($^4He(\gamma,p)^3H$). Considering on the one hand the results with the semi-realistic potential models for $R_{\rm L}$ of ^4He and on the other hand the 3N total photoabsorption cross sections it does not seem very likely that a more realistic interaction could change the theoretical result for the total ^4He photoabsorption cross section very much. Thus also here the obtained ^4He polarizability of 0.076 fm^3 (MT-I/III potential) should be rather realistic.

For a better comparison with experiment we have started to calculate the measured two-body break up channel $^4He(\gamma,n)^3He$. I would like to emphasize that also such an exclusive reaction can be calculated applying the integral transform method [13]. The reliability of such exclusive LIT results was tested for the $d(e,e'N)N$ reaction [14]. In comparison with a conventional calculation with explicit final state wave functions a very high precision of the LIT results was obtained. First results with the LIT method for the $^4He(\gamma,n)^3He$ reaction can be found in the contribution of S. Quaglioni in this proceedings. It also contains a more detailed discussion on the LIT method for exclusive reactions.

The calculation of $\sigma_{\rm tot}^{unret}$ of the six-body nuclei was made possible by the EIHH method [3]. This recently introduced method enabled us to calculate the necessary bound and LIT bound state like states. Thus I give here a brief outline of the new approach. If one considers an A-nucleon system and neglects 3N forces one has the usual A-body Hamiltonian consisting of kinetic energy and two-body potential $V_{\rm ij}$. One can define an effective interaction in a model space P with projection operator P

$$H_{\rm P} = P\left[\sum_{\rm i=1}^{A} \frac{p_{\rm i}^2}{2m}\right]P + P\left[\sum_{\rm i<j} V_{\rm ij}\right]_{\rm eff} P\,, \tag{5}$$

where $p_{\rm i}$ is the momentum of the i-th particle and m the nucleon mass. The effective interaction $[V_{\rm ij}]_{\rm eff}$ leads to the same solution as the original full space problem, but its construction is as difficult as the solution of the original problem. However, a few years ago it was pointed out by Navrátil and Barrett that using the Lee-Suzuki method [15] one can introduce an effective two-body interaction instead of the full A-body $[V_{\rm ij}]_{\rm eff}$. Important for the whole approach is a continuous increase of the P space up to the point that convergence is observed for the observable under investigation. Note that a steady increase of the P space makes $[V_{\rm ij}]_{\rm eff}$ converge to the original $V_{\rm ij}$ and thus a converged result for an observable should correspond to the true result. The actual calculation of Navrátil and Barrett was carried out in the No-Core-Shell-Model and impressive results for few-body bound states were obtained [16].

The EIHH method follows a similar strategy but uses HH functions as basis set instead of harmonic oscillator functions. The steady increase of the P space is realized via the increase of the HH quantum number K. In the EIHH method the two-body effective interaction is constructed from a pseudo two-body Hamiltonian consisting of the total hyperspherical kinetic energy of the

A-body system and the two-body potential of one NN pair:

$$H_2 = \frac{1}{2m}\frac{\hat{K}^2_{\mathrm{A}-1}}{\rho^2} + V_{12}\,. \tag{6}$$

Here $\hat{K}_{\mathrm{A}-1}$ denotes the HH grand angular momentum operator and ρ is the hyper radius. Using such an H_2 leads to a ρ-dependent two-body $[V_{\mathrm{ij}}]_{\mathrm{eff}}$. The ρ-dependence reflects the fact that due to the total hyperspherical kinetic energy also effects due to the residual (A-2)-system are taken into account.

The EIHH method leads to a rapid convergence of the HH expansion [3]. There is also a recent benchmark calculation for the ^4He ground state with a realistic NN potential [17], where seven different theoretical methods, among them the EIHH method, were applied by seven different theoretical groups. A very high degree of agreement of the calculations was found for various ground state properties. Recent improvements of the EIHH method can be found in the contribution of N. Barnea in this proceedings.

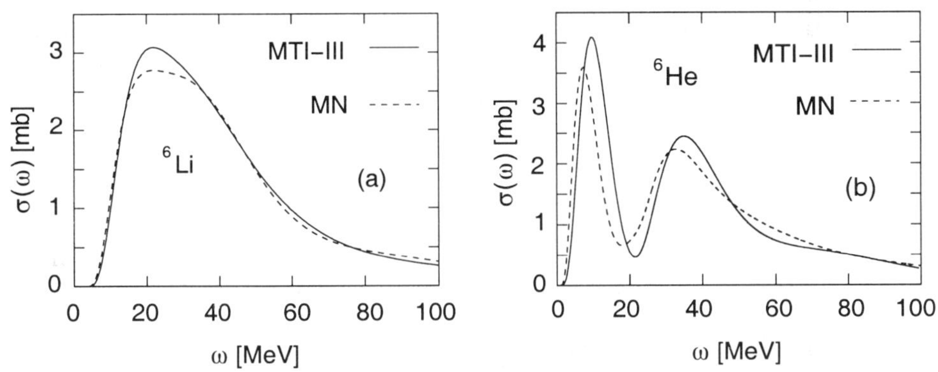

Figure 3. Six-body photoabsorption cross sections with MTI-III [6] and MN potentials [19]: ^6Li (a) and ^6He (b).

Using LIT and EIHH methods together has enabled us to calculate the six-nucleon total photoabsorption cross sections [18]. Again $\sigma_{\mathrm{tot}}^{\mathrm{unret}}$ is considered and semi-realistic NN potentials [6, 19] are used. The results of the calculation are illustrated in Fig. 3. While one finds two cross section peaks for ^6He, there is a single peak in case of ^6Li. Thus in the ^6He photodisintegration one succeeds to separate a halo response of the two neutrons from an α-core response. A similar separation of deuteron and α-core responses is not seen for ^6Li, probably because the break up into two three-body nuclei (^3H, ^3He) has filled up the gap (break up of ^6He into two ^3H nuclei is not induced by the dipole operator). A more detailed discussion of the six-nucleon photodisintegration is given in the contribution of S. Bacca in this proceedings.

Acknowledgement. I would like to thank S. Bacca, N. Barnea, V.D. Efros, A. La Piana, M. Marchisio, G. Orlandini, S. Quaglioni, and E.L. Tomusiak for their important contributions of the work presented here.

References

1. V.D. Efros, W. Leidemann, G. Orlandini: Phys. Lett. **B338**, 130 (1994)
2. Yu.I. Fenin, V.D. Efros: Sov. J. Nucl. Phys. **15**, 497 (1972)
3. N. Barnea, W. Leidemann, G. Orlandini: Phys. Rev. **C61**, 054001 (2000)
4. V.D. Efros, W. Leidemann, G. Orlandini: Phys. Lett. **B408**, 1 (1997)
5. V.D. Efros, W. Leidemann, G. Orlandini, E.L. Tomusiak: Phys. Lett. **B484**, 223 (2000); Nucl. Phys. **A689**, 421 (2001)
6. R.A. Malfliet, J. Tjon: Nucl. Phys. **A127**, 161 (1969)
7. R.B. Wiringa, V.G.J. Stoks, R. Schiavilla: Phys. Rev. **C51**, 38 (1995)
8. R.B. Wiringa: Phys. Rev. **C43**, 1585 (1991); B.S. Pudliner, V.R. Pandharipande, J. Carlson, C.S. Pieper, R.B. Wiringa: Phys. Rev. **C56**, 1720 (1997)
9. J. Golak et al.: Nucl. Phys. **A707**, 365 (2002)
10. K. Dow et al.: Phys. Rev. Lett. **61**, 1706 (1988)
11. V.D. Efros, W. Leidemann, G. Orlandini: Phys. Rev. Lett. **78**, 432 (1997); Phys. Rev. **C58**, 582 (1998); Few-Body Syst. **26**, 251 (1999)
12. V.D. Efros, W. Leidemann, G. Orlandini: Phys. Rev. Lett. **78**, 4015 (1997); N. Barnea, V.D. Efros, W. Leidemann, G. Orlandini: Phys. Rev. **C63**, 057002 (2001)
13. V.D. Efros: Sov. J. Nucl. Phys. **41**, 949 (1985)
14. A. La Piana, W. Leidemann: Nucl. Phys. **A677**, 423 (2000)
15. K. Suzuki, S.Y. Lee: Prog. Theor. Phys. **64**, 2091 (1980)
16. P. Navrátil, B.R. Barrett: Phys. Rev. **C54**, 2986 (1996); Phys. Rev. **C59**, 1906 (1999)
17. H. Kamada et. al: Phys. Rev. **C64**, 044001 (2001)
18. S. Bacca, M. Marchisio, N. Barnea, W. Leidemann, G. Orlandini: Phys. Rev. Lett. **89**, 052502 (2002)
19. D.R. Thomson, M. LeMere, Y.C. Tang: Nucl. Phys. **A286**, 53 (1977); I. Reichstein, Y.C. Tang: Nucl. Phys. **A158**, 529 (1970)

Few-Body Systems Suppl. 14, 319–324 (2003)

Few-Body Systems

Variational Calculation of Three–Nucleon Electroweak Capture Reactions

L.E. Marcucci[1,2], M. Viviani[2,1], A. Kievsky[2,1], S. Rosati[1,2], R. Schiavilla[3,4]

[1] Physics Department, Pisa University, Via Buonarroti 2, I-56100 Pisa, Italy
[2] INFN, Sezione di Pisa, Via Buonarroti 2, I-56100 Pisa, Italia
[3] Jefferson Laboratory, Newport News, Virginia 23606, USA
[4] Physics Department, Old Dominion University, Norfolk, VA 23529, USA

Abstract. Recent advances in the study of the $p-d$ radiative and $\mu - {}^3$He weak capture processes are here presented and discussed.

1 Introduction

A number of electromagnetic (EM) and weak transitions in light nuclei have interesting astrophysical implications as well as important implications for an understanding of nuclear structure and dynamics. The theoretical description of these processes requires the knowledge of the initial and final nuclear states and the use of a realistic model for the EM and weak current operators. In particular, the trinucleon systems provide a unique "laboratory" due to the capability, achieved in the last few years, of obtaining very accurate bound and continuum nuclear wave functions [1, 2, 3]. Therefore, the study of EM and weak transitions in the three nucleon system does not suffer of uncertainties related to the computation of the wave functions and it is a direct test of the nuclear Hamiltonian H, from which the nuclear wave functions are obtained, and of the model used to describe the nuclear currents. Since the nuclear EM current is related to H through current conservation, it is clear that the two topics are inter-related. Furthermore, it is interesting to understand whether relativistic corrections as well as Δ-isobar and additional sub-nucleonic degrees of freedom play a role in these processes.

The model for the nuclear EM and weak current considered here has been tested in numerous few-nucleon processes and has been recently reviewed in refs. [4, 5, 6]. It includes one– and two–body operators. The one–body operators are obtained directly from the non-relativistic limit of the covariant single–nucleon vector and axial currents. In the study of the muon capture, the contribution coming from the induced pseudo–scalar term of the nucleon axial current has to be included (it gives a negligible contribution to β–decay processes). However, the experimental value of the corresponding form factor $G_{PS}(q_\sigma^2)$ is rather uncertain. Assuming pion–pole dominance, the partially conserved axial current (PCAC) hypothesis, and the Goldberger-Treiman relation, G_{PS} is predicted to be [7, 8, 9]

$$G_{PS}^{PCAC}(q_\sigma^2) = -\frac{2\,m_\mu\,m_N}{m_\pi^2 + q_\sigma^2}\,G_A(q_\sigma^2), \tag{1}$$

where q_σ is the four-momentum transferred to the nuclear system, m_N, m_μ and m_π indicates the nucleon, muon and pion mass, respectively, and G_A is the axial form factor. In our calculation, we have assumed

$$G_{PS}(q_\sigma^2) = R_{PS}\,G_{PS}^{PCAC}(q_\sigma^2), \tag{2}$$

where R_{PS} is a parameter which can be varied to study the sensitivity of our results to this form factor and to investigate to which extent $G_{PS}^{\rm expt} = G_{PS}^{\rm PCAC}$. As a starting point, we have assumed $R_{PS} = 1$.

The two–body EM current is separated in two terms. There is a "model-independent" part which is constructed consistently with the nucleon–nucleon interaction, in order to satisfy the current conservation relation [10]. The second part includes "model-dependent" contributions which come from the $\pi\rho\gamma$ and $\pi\omega\gamma$ processes and the Δ degrees of freedom. The latter contribution is included in the current and in the nuclear medium in an approximate way, by following the procedure described in ref. [11]. The two–body weak vector current is then obtained from the isovector part of the EM current, in accordance with the conserved-vector-current (CVC) hypothesis.

Two–body terms have been taken into account in both the axial charge and current operators. The two–body axial charge operator has been obtained consistently with the two–nucleon interaction model, following the methods of ref. [12]. The two–body axial current operators are derived from a meson-exchange model, including π- and ρ-exchanges and the $\rho\pi$-transition processes, as well as Δ-isobar excitation [5, 13]. The latter process gives the dominant contribution, and its strength is adjusted to reproduce the Gamow-Teller matrix element in tritium β-decay [5, 6, 13]. It has been shown that this procedure allows to significantly reduce the model dependence of the weak axial current contributions [5, 13].

The ^{3}He, ^{3}H bound and the $p-d$ continuum wave functions have been calculated by expanding on a basis of pair–correlated hyperspherical harmonic (PHH) functions [14]. Such a technique has been shown to be very accurate [1]. Note that with the PHH technique the inclusion of the Coulomb potential, clearly very important in the energy range considered here, does not present any difficulty.

We discuss in the following sections the $p-d$ radiative capture above deuteron breakup threshold (DBT) and the muon capture on ^{3}He in the tritium channel.

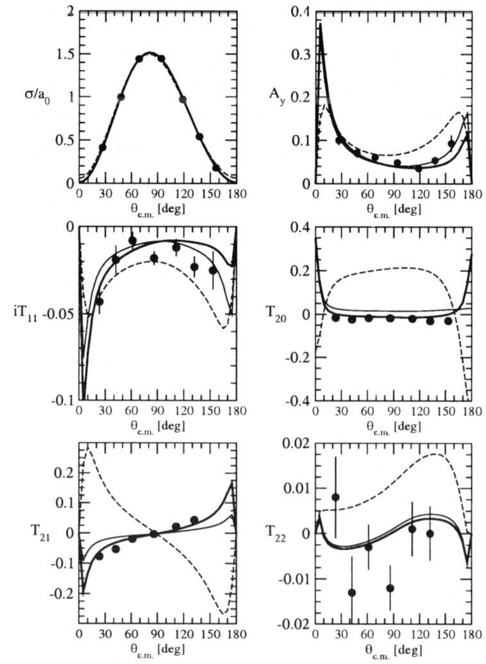

Figure 1. Differential cross section, proton vector analyzing power, and the four deuteron tensor analyzing powers for $p{-}d$ capture at $E_{\rm c.m.} = 3.33$ MeV, obtained with the AV18/UIX Hamiltonian model and one–body only (dashed lines) or both one– and two–body currents (thin solid lines). The experimental data are from ref. [18]. The results obtained in the LWA approximation for the spin-flip E_1 RME's are also shown (thick solid lines). In the first panel, $\mathrm{a}_0 = \int d\Omega\, \sigma(\Omega)/(4\pi)$.

2 $p - d$ Radiative Capture

Recently the PHH technique has been extended to compute $p - d$ scattering wave functions above the DBT [15] with comparable accuracy to those obtained below DBT. We can therefore compute $p{-}d$ capture observables at higher energies than previously published [16, 17]. We present here a preliminary study of the $p{-}d$ radiative capture at center-of-mass (c.m.) energy ($E_{\rm c.m.}$) of 3.33 MeV, for which high-quality experimental data, including differential cross sections, vector and tensor analyzing powers [18], exist. The theoretical calculation has been performed using the Argonne v_{18} (AV18) [19] two–nucleon and Urbana IX (UIX) [20] three–nucleon interactions. The results are shown in Fig. 1. By inspection of the figure, we note that the theoretical predictions for the differential cross section and the observables A_y and iT_{11} when one– and two–body contributions are included in the model for the current operator (thin solid line) are in good agreement with the experimental data. On the contrary, for the observables T_{20} and T_{21}, large discrepancies can be observed. A similar situation was found for the $E_{\text{c.m.}} = 2$ MeV observables [17]. The problem can be traced back to an overprediction of the spin-flip electric dipole E_1 reduced matrix elements (RME's). When the same RME's are computed in the long wavelength approximation (LWA) at leading order (thick solid lines in Fig. 1) also the observables T_{20} and T_{21} are rather well reproduced. Interestingly, an analysis of the next-to-leading order terms in the LWA performed in ref. [17] has shown that they give a sizeable contribution to the spin-flip E_1 RME's. Therefore, the use of the leading order only for the calculation of this small spin-flip transition matrix elements should be considered inadequate. The origin of the discrepancies observed in the T_{20} and T_{21} observables is currently under investigation.

3 Muon Capture

The muon weak capture on ^{3}He can occur through three different hadronic channels [21], but we have considered only the reaction $\mu^- + {}^3\text{He} \rightarrow {}^3\text{H} + \nu_\mu$. When the triton polarization is not detected, the differential capture rate is given by

$$\frac{d\Gamma}{d(\cos\theta)} = \frac{1}{2}\Gamma_0 \left[1 + A_v P_v \cos\theta + A_t P_t \left(\frac{3}{2}\cos^2\theta - \frac{1}{2}\right) + A_\Delta P_\Delta\right], \quad (3)$$

where Γ_0 is the total capture rate, A_v, A_t and A_Δ angular correlation parameters and θ is the angle between the muon polarization and the ^{3}H recoil direction. The coefficients P_v, P_t and P_Δ are linear combinations of the probabilities $P(f,\, f_z)$ of finding the μ^- ^{3}He system in the total-spin state $|f\, f_z\rangle$, and are defined in refs. [22, 23].

Table 1. Total capture rate Γ_0 (sec^{-1}) and angular correlation parameters A_v, A_t, and A_Δ, calculated using PHH wave functions corresponding to the AV18/UIX and AV14/TM Hamiltonian models, are compared with the experimental results. The theoretical uncertainties, shown in parenthesis, reflect the uncertainty in the determination of the $N\Delta$ transition axial coupling constant. The experimental values of Γ_0 and A_v have been taken from ref. [27] and [24], respectively. We assume $R_{PS} = 1$ in Eq. (2).

Observable	AV18/UIX	AV14/TM	Expt.
Γ_0	1484(8)	1486(8)	1496(4)
A_v	0.5350(14)	0.5336(14)	0.63(15)
A_t	–0.3650(9)	–0.3659(9)	
A_Δ	–0.1000(16)	–0.1005(17)	

The results for the total capture rate Γ_0 and angular correlation parameters A_v, A_t, and A_Δ are presented in Table 1. They are obtained using either the AV18/UIX interactions or the older Argonne v_{14} (AV14) two–nucleon interaction [25] in conjunction with the the Tucson-Melbourne (TM) three–nucleon interaction [26]. Note that both three–nucleon interactions have been adjusted to reproduce the triton binding energy.

The uncertainty (in parenthesis) in the predicted values is due to the uncertainty in the determination of the $N\Delta$ transition coupling constant, as discussed in the Introduction. The latter reflects the experimental error in the Gamow-Teller matrix element of tritium β-decay.

Inspection of Table 1 shows that the theoretical determination of the total capture rate Γ_0 is within 1 % of the recent experimental result [27]. Furthermore, the model dependence in the calculated observables is very weak: the AV18/UIX and AV14/TM results differ by less than 0.5 %. The agreement between theory and experiment and the weak model dependence mentioned above reflect, to a large extent, the fact that both the AV18/UIX and AV14/TM Hamiltonian models reproduce: *i*) the experimental binding energies as well as the charge and magnetic radii [28] of the trinucleons; *ii*) the Gamow-Teller matrix element in tritium β-decay. The value for the angular correlation parameter A_v listed in Table 1 is also in reasonable agreement with the corresponding experimental result, which however has a rather large error [24].

An important motivation of the present work has been to test the sensitivity of the muon capture observables to the induced pseudo-scalar form factor G_{PS} and, eventually, infer its value from the Γ_0 measurement. By repeating the calculation using different values of the parameter R_{PS}, defined in Eq. (2), we have found that the angular correlation parameters, in particular A_t and A_Δ, are more sensitive to changes in R_{PS} than the total capture rate, as first pointed out in ref. [22]. A precise measurement of these polarization observables could therefore be useful to ascertain the extent to which the induced pseudo-scalar form factors deviate from their PCAC values.

Finally, by enforcing perfect agreement between the experimental and theoretical values, taken with their uncertainties, for the total capture rate Γ_0, it is possible to obtain an estimate for the range of values allowed for R_{PS}, and we have found $R_{PS} = 0.94 \pm 0.06$. This 6 % uncertainty is smaller than that found in previous studies [23, 29, 30]. This substantial reduction in uncertainty can be traced back to the procedure used to constrain the (model-dependent) two–body axial currents discussed in the Introduction.

The work of R.S. is supported by the U.S. Department of Energy contract number DE-AC05-84ER40150 under which the Southeastern Universities Research Association (SURA) operates the Thomas Jefferson National Accelerator Facility.

References

1. A. Kievsky et al.: Phys. Rev. **C58**, 3085 (1998)
2. W. Glöckle et al.: Phys. Rep. **274**, 107 (1996)
3. A. Kievsky et al.: Nucl. Phys. **A607**, 402 (1996)
4. J. Carlson and R. Schiavilla: Rev. Mod. Phys. **70**, 743 (1998)

5. L.E. Marcucci et al.: Phys. Rev. Lett. **84**, 5959 (2000); Phys. Rev. **C63**, 015801 (2001)

6. L.E. Marcucci et al.: nucl-th/0112008, accepted in Phys. Rev. **C**

7. T.R. Hemmert et al.: Phys. Rev. **D51**, 158 (1995)

8. J.D. Walecka: In: *Muon Physics; Weak Interactions*, edited by V.W. Hughes and C.S. Wu, p. 114. New York: Academic Press 1975

9. J.D. Walecka: *Theoretical Nuclear and Subnuclear Physics* New York: Oxford University Press 1995

10. D.O. Riska: Phys. Scr. **31**, 471 (1985); A. Buchmann et al.: Nucl. Phys. **A443**, 726 (1985)

11. R. Schiavilla et al.: Phys. Rev. **C45**, 2628 (1992)

12. M. Kirchbach et al.: Nucl. Phys. **A542**, 616 (1992)

13. R. Schiavilla et al.: Phys. Rev. **C58**, 1263 (1998)

14. A. Kievsky et al.: Nucl. Phys. **A551**, 241 (1993); Nucl. Phys. **A577**, 511 (1994)

15. A. Kievsky et al.: Phys. Rev. **C64**, 024002 (2001)

16. M. Viviani et al.: Phys. Rev. **C54**, 534 (1996)

17. M. Viviani et al.: Phys. Rev. **C61**, 064001 (2000)

18. F. Goeckner et al.: Phys. Rev. **C45**, R2536 (1992)

19. R.B. Wiringa et al.: Phys. Rev. **C51**, 38 (1995)

20. B.S. Pudliner et al.: Phys. Rev. Lett. **74**, 4396 (1995)

21. D.F. Measday: Phys. Rep. **354**, 243 (2001)

22. W.-Y.P. Hwang, Phys. Rev. **C17**, 1799 (1978)

23. J.G. Congleton and H.W. Fearing: Nucl. Phys. **A552**, 534 (1992)

24. P.A. Souder et al.: Nucl. Instr. and Meth. in Phys. Res. **A402**, 311 (1998)

25. R.B. Wiringa et al.: Phys. Rev. **C29**, 1207 (1984)

26. S.A. Coon et al.: Nucl. Phys. **A317**, 242 (1979)

27. P. Ackerbauer et al.: Phys. Lett. **B417**, 224 (1998)

28. L.E. Marcucci et al.: Phys. Rev. **C58**, 3069 (1998)

29. J.G. Congleton and E. Truhlìk: Phys. Rev. **C53**, 956 (1995)

30. J. Govaerts and J.-L. Lucio-Martinez: Nucl. Phys. **A678**, 110 (2000)

Few-Body Systems Suppl. 14, 325–330 (2003)

Few-Body Systems

Extraction of the Static Magnetic Form-Factor and the Structure Function of the Neutron from Inclusive Scattering Data on Light Nuclei

A.S. Rinat

Weizmann Institute of Science, Department of Particle Physics, Rehovot 76100, Israel

Abstract. We show that quasi-elastic inclusive electron scattering data on light nuclei for medium Q^2 furnish information on $G_M^n(Q^2)$, whereas the deep-inelastic region for large Q^2, provides the Structure Function $F_2^n(x, Q^2)$. Common to the two extractions is the possibility to de-convolute medium effects, which is most accurately done for light targets. Results should be, and virtually are independent of the target.

1 Introduction

Most neutron observables can only indirectly be extracted from experiments on a nuclear medium, in which the n is embedded. We discuss below the neutron static magnetic form factor and its Structure Function (SF).

Consider the reduced cross section for inclusive scattering of unpolarized electron of energy E from non-oriented targets A over an angle θ

$$\frac{A^{-1}d^2\sigma_{eA}(E;\theta,\nu)/d\Omega\, d\nu}{\sigma_M(E;\theta,\nu)} = \left[\frac{2xM}{Q^2}F_2^A(x,Q^2) + \frac{2}{M}\tan^2(\theta/2)F_1^A(x,Q^2)\right]. \quad (1)$$

The two nuclear structure SF $F_k^A(x, Q^2)$ depend on $Q^2 = \boldsymbol{q}^2 - \nu^2$ ($\nu, \boldsymbol{q}$ are the energy-momentum transfer) and the Bjorken variable $x = Q^2/2M\nu, 0 \le x \le A$ (M is the nucleon mass). We invoke a relation between the SF of nuclei and of nucleons. For instance (for $Z = N$) [1]

$$F_k^A(x,Q^2) = \int_x^A \frac{dz}{z^{2-k}}[f^{PN,A}(z,Q^2)\frac{1}{2}\left[F_k^p\left(\frac{x}{z},Q^2\right) + F_k^n\left(\frac{x}{z},Q^2\right)\right], \quad (2)$$

which relates F^A, F^N to the SF $f^{PN,A}$ of a nucleus, composed of point-nucleons. A calculation of F_k^A thus requires data on F_k^p, an assumed form for F_k^n and in addition, a computed, unphysical $f^{PN,A}$.

We separate F_k^N in NE ($\gamma^* + N \to N$) and NI parts ($\gamma^* + N \to$ hadrons, partons), leading to the corresponding components $F_k^{A,NE}$ [2] ($\eta = Q^2/4M^2$)

$$F_1^{A,NE}(x) = \frac{f^{PN,A}(x)}{4} G_d^2 \left[(\alpha_p \mu_p)^2 + (\alpha_n \mu_n)^2 \right], \tag{3a}$$

$$F_2^{A,NE}(x) = \frac{x f^{PN,A}(x) G_d^2}{2(1+\eta)} \left[(\alpha_p \gamma)^2 + \left(\frac{\mu_n \eta}{1+5.6\eta} \right)^2 + \eta \left[(\alpha_p \mu_p)^2 + (\alpha_n \mu_n)^2 \right] \right], \tag{3b}$$

where reference to Q^2 has been dropped. Instead of the actual static electromagnetic form factors $G_{M,E}^N(Q^2)$, we use in Eq. (3) measured deviations from the standard dipole form [3, 4, 5].

$$\alpha_N \equiv G_M^N / \mu_N G_d \, ; \quad N = p, n \, , \tag{4a}$$

$$\gamma \equiv \frac{\mu_p G_E^p}{G_M^p} = \frac{G_E^p}{\alpha_p G_d} \, , \tag{4b}$$

$$\gamma \approx 1 + \theta(Q^2 - 0.3)[1 - 0.14(Q^2 - 0.3)] \, ; \quad Q^2 \lesssim 5.5 \, . \tag{4c}$$

For G_E^n we use the Galster parametrization [6]. Nuclear NI components completely dominate cross sections on the inelastic side $x \lesssim 1$ of the QEP, while for $x \gtrsim 1$ NE>NI. Those regions will be treated separately.

1.1 Quasi-Elastic Region $x \lesssim 1$: G_M^n

Consider first the x, Q^2 dependence of $F_k^{A,NE}(x, Q^2)$. The latter is primarily due to the form factors in Eqs. (3), which decrease with growing Q^2. The x-dependence resides in $f^{PN,A}(x, Q^2)$, which sharply decreases with growing $|1-x|$ away from the QEP at $x \approx 1$. $F_k^{A,NI}$ steadily increases with ν. From the above one concludes that $\ln[\sigma^{A,NE}/A]$ increases with ν (decreasing x for fixed Q^2). Generally for $A \geq 12$ there is a mere break in the slope in the QE region $|1-x| \ll 1$ for $A \geq 12$ [7] (Fig. 1a).

The unusual structure of the lightest nuclei, causes $f^{PN,A}(x, Q^2)$ to be narrow and sharply peaked. With no interference of NI, the above change in slope may for moderate Q^2 develop into a QE peak, as indeed observed for D [8] and ^{4}He [9] (Fig. 1b). For the same targets one can compute with great precision ground states [10] and non-diagonal target density matrices in the expression for $f^{PN,A}$ [11, 12].

Under the above conditions on kinematics and targets, for $x \leq 1$, $\sigma^{NI} \ll \sigma^{NE} \approx \sigma^{tot}$. With known $G_{E,M}^p$ and small G_E^n, this enables the extraction of G_M^n from NE. Tests for the above allocations are: i) Around $x \lesssim 1$, $\sigma^A/\sigma_M \propto f(x, Q^2)$, bell-shaped in $1-x$. ii) $G_M^n(Q^2)$ should be independent of the value of the individual x from which the one extracts G_M^n. iii) Idem for the chosen target.

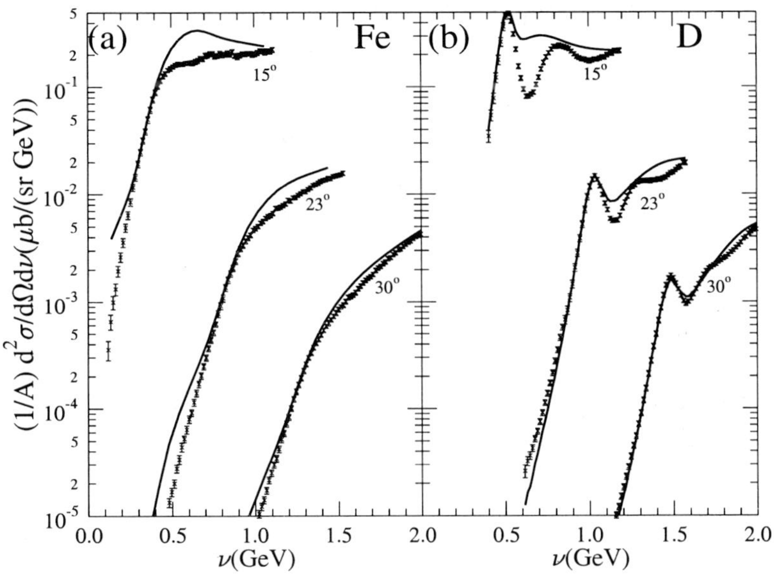

Figure 1. Partial data and predictions for inclusive cross sections ($E = 4.045$ GeV, $\theta = 15^\circ, 23^\circ, 30^\circ$) on Fe (a) and D (b).

We thus exploited suitable D, He data to extract G_M^n. Those satisfied the above criteria. The values followed the trend of previous measurements and added points previously measured values and adds points for intermittent Q^2. Hardly any target dependence has been detected (Fig. 2).

1.2 The Deep-Inelastic Region, $x \ll 1$: Extraction of $F_2^n(x, Q^2)$

That region is dominated by NI. We focus on $F_2^n(x, Q^2)$, commonly estimated from the 'primitive' ansatz $F_2^n \approx 2F_2^D - F_2^p$, which is only reliably for $x \lesssim 0.3$. Instead of a vehicle to compute F_k^A, we now consider Eq. (2) in the inverse sense: From data on σ^A, Eq. (1), known F_2^p and computed $f^{PN,A}$, we wish to extract $F_2^n(x, Q^2)$.

All modern data thus far [7, 8] appear to yield F_2^A in disjoint x, Q^2 regions, whereas the inversion of Eq. (2) requires data over a large x-range for the same Q^2. Only for $Q^2 \approx 3.5$ GeV^2 can one construct a data set for F_2^A for $x \lesssim 1.2$.

Virtually all previous methods addressed a D target (e.g. [13]). We have applied several inversion methods, in principle applicable to any target. All use a parametrization

$$\begin{aligned} F_2^n(x, Q^2) = F_2^n(x, Q^2; d_k) &= C(x, Q^2; d_k) F_2^p(x, Q^2) \,, \\ C(x, Q^2; d_k) &= \sum_{k \geq 0} d_k(Q^2)(1-x)^k \,, \end{aligned} \tag{5}$$

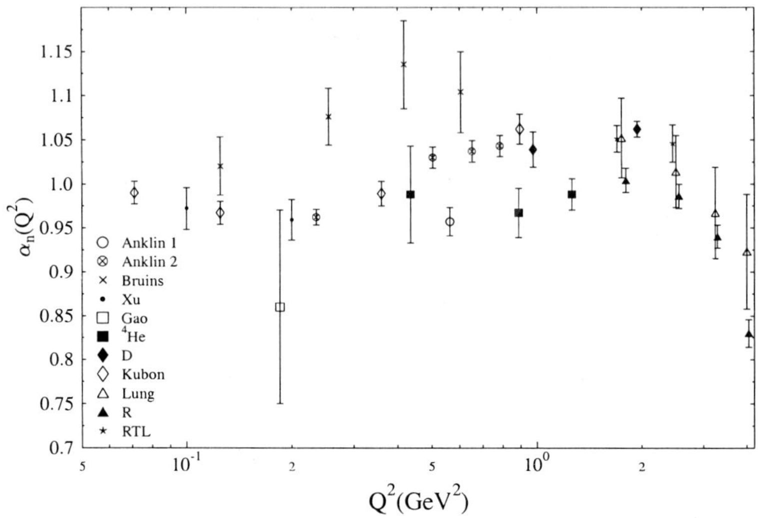

Figure 2. $\alpha_n = G_M^n/\mu_n G_d$ as function of Q^2. Shown are some previous representative results. Filled squares, diamonds, triangles and stars are our results.

with mildly constrained parameters. First we take $C(0) = 1$, to ensure a finite outcome for the Gottfried sum $S_G(Q^2) = \int_0^1 \frac{dx}{x}[F_2^p(x,Q^2) - F_2^n(x,Q^2)]$. Next we exploit the above 'primitive' ansatz for, say, $x = 0.2$. Taking $k_{max} = 2$ only one parameter is left, e.g. $d_0 = C(1)$. It moreover proved useful to parametrize F_2^p as follows

$$F_2^p(x, 3.5) = x^{-a^2} \sum_{m \geq 1} c_m(1-x)^m \, ; \quad x \geq 0.02 \, , \tag{6a}$$

$$F_2^p(x, 3.5) = 0.42 \, ; \quad x \leq 0.02 \, . \tag{6b}$$

In the region $0.02 \lesssim x \lesssim 0.9$, the above practically coincides with standard parametrizations [14]. Figure 3 shows our results for C, F_2^n for fixed $Q^2 = 3.5$ GeV^2 and given F_2^p. The band in C reflects results from several inversion methods and from different targets D,C,Fe. The value of C at the elastic point $x = 1$ has been the subject of several estimates with results, marked by small horizontal lines. All those, as well as our C, assumed smooth, i.e. resonance-averaged behaviour of F_2^N (cf. lower part of Fig. 3).

The above is an undesired feature of averaging: the lowest inelastic threshold of $F_2^N(x,Q^2)$, occurs at a mass $M + m_\pi$, or equivalently, at $x_{th}(Q^2) = [1 + 2Mm_\pi/Q^2]^{-1}$. In particular $x_{thr}(3.5) \approx 0.93$, which is marked in Fig. 3 by a vertical line. For $x_{th} < x < 1$, $F^N(x,Q^2)$ is strictly 0. In particular the mention prediction of C out to the elastic border, merely reflects the different approach to 0 of the p,n SF. As a consequence $C(x \to 1)$ is due to purely NE parts of

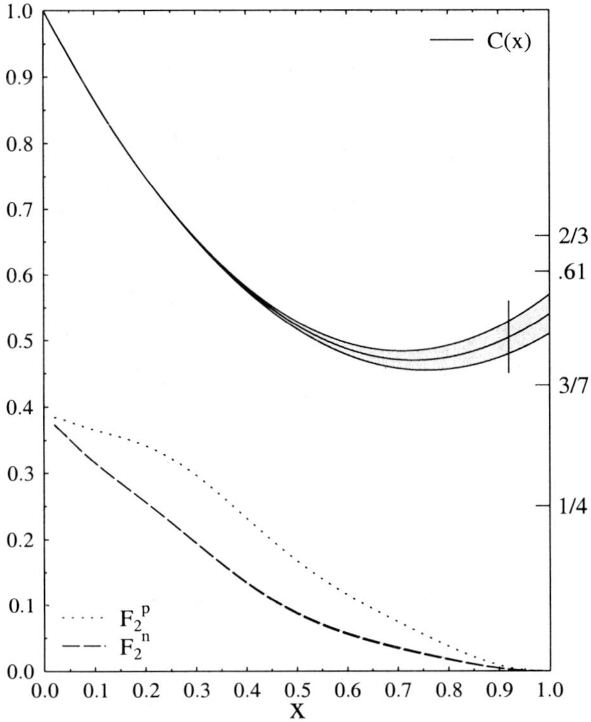

Figure 3. The ratio $C(x, 3.5) = F_2^n(x, 3.5)/F_2^p(x, 3.5)$ for $Q = 3.5$ GeV^2 from data on D, C, Fe. The drawn line corresponds to $C(1) = 0.54$ and the band represents the spread from averages over different targets and methods. The numbers on the right abscissa are standard quark model and QCD predictions for $C(1)$ with 0.61, the *NE* limit (7).

F_2^N, and equals (cf. Eq. (3b))

$$\lim_{x\to 1} C(x, Q^2) = \left[\frac{\mu_n \alpha_n(Q^2)}{\mu_p \alpha_p(Q^2)}\right]\left[1 + \frac{4M^2}{Q^2}\left(\frac{\gamma(Q^2)}{\mu_p}\right)^2\right]^{-1} . \tag{7}$$

From Eqs. (4), (7) one then *computes*

$$C(x = 1, 3.5) \approx 0.61, \tag{8}$$

surprisingly close to the *extracted* value as the ratio of the two F_2^N, which tend to 0 in a different way for $x \to 1$. More extensive reports can be found in refs. [15, 16].

Acknowledgement. Part of this work has been done in collaboration with M.F. Taragin and M. Viviani.

References

1. S.A. Gurvitz, A.S. Rinat: TR-PR-93-77/ WIS-93/97/Oct-PH; Prog. Nucl. Part. Phys. **34**, 245 (1995)
2. A.S. Rinat, M.F. Taragin: Phys. Rev. **C62**, 034602 (2000)
3. A.F. Sill et al.: Phys. Rev. **D48**, 29 (1993); L. Andivahis et al.: Phys. Rev. **D50**, 549 (1994)
4. E. Brash, A. Kozlov, Sh. Li, G.M. Huber: hep-ex/0111038v2
5. M. Jones et al.: Phys. Rev. Lett. **84**, 1398 (2000); Third Workshop on "Perspectives in Hadronic Physics", Trieste, Italy, 2001, IT, to be published; O. Gayou et al.: Phys. Rev. **C64**, 038202 (2001)
6. S. Galster et al.: Nucl. Phys. **B32**, 221 (1971)
7. J. Arrington et al.: Phys. Rev. Lett. **82**, 2056 (1999); PhD Thesis, CalTech, 1998
8. I. Niculescu et al.: Phys. Rev. Lett. **85**, 1182 (2000); J. Arrington: Private Communication
9. D.B. Day et al.: Phys. Rev. **C40**, 1849 (1993)
10. H. Kamado et al.: Phys. Rev. **C64**, 044001 (2001)
11. A.S. Rinat, M.F. Taragin: Phys. Rev. **C65**, 042201(R) (2002)
12. M. Viviani, A. Kievsky, A.S. Rinat: nucl-th/0111049, submitted to Phys. Rev. C
13. W. Melnitchouk, A.W. Thomas: Phys. Lett. **B377**, 11 (1996)
14. P. Amadrauz et al.: Phys. Lett **B295**, 159 (1992); M. Arneodo et al.: Phys. Lett. **B364**, 107 (1995)
15. A.S. Rinat, M.F. Taragin, M. Viviani: in preparation
16. A.S. Rinat, M.F. Taragin, nucl-th/0204071, submitted to Phys. Lett. B

Few-Body Systems Suppl. 14, 331–334 (2003)

Few-Body Systems

Search for Nucleon-Nucleon Correlations in the Photon-Asymmetry via the ^{4}He(γ,NN) Reaction

P. Grabmayr*, for the MAMI A2 Collaboration

Physikalisches Institut der Universität Tübingen
Auf der Morgenstelle 14, D-72076 Tübingen, Germany

Abstract. The two- and three-nucleon photodisintegration of the ^{4}He nucleus is studied in the energy range from 150 to 450 MeV in order to investigate the in-medium nucleon-nucleon interaction at short relative distances. Linearly polarised photons were employed for the study of nucleon-nucleon correlations because asymmetries are assumed to be less sensitive to the obscuring final state interactions.

1 Introduction

One of the current topics of interest in nuclear physics is centred around the strong force, in particular about its short range structure and the modifications in the nuclear medium. The free nucleon-nucleon interaction is extremely well described by several potentials extracted from fits to scattering data. However, even for the three-body system the experimental binding energies can not be reproduced; one is led to introduce three-body forces. In heavier systems where the strict microscopic descriptions can not be exercised the concept of nucleon-nucleon correlations is applied (e.g. see Refs. [1,2]). Correlations in nuclear wave functions are not observable directly, only by comparison of measured cross sections with model predictions the effects can be disentangled [3, 4]. Final state interactions (FSI) and meson-exchange currents (MEC) are the main background contributions. FSI is reduced for the lighter mass nuclei and there are predictions that asymmetries are less affected by FSI than cross sections. There is also an increased interest in modeling the photon-induced reactions for the few-body systems. These arguments lead to the investigation of beam asymmetries for the ^{4}He(γ,np) and ^{4}He(γ,pp) reactions. Note that ^{4}He has the highest central density of all nuclei which further should enhance the effects of nucleon-nucleon correlations.

* *E-mail address:* grabmayr@uni-tuebingen.de

2 Experiments

The experiments were performed at the electron accelerator MAMI at Mainz [5]. Linearly polarised photons were produced by coherent bremsstrahlung on a 100 μm thick diamond crystal which was positioned at the radiator position of the Glasgow tagging spectrometer [6]. Three angular settings of the diamond produced photon spectra which had an enhanced intensity of polarised photons in the range from 170 to 400 MeV. A maximum degree of polarisation of $\approx 50\%$ was reached. The neutrons and protons emitted from the liquid ^{4}He target [7] were detected by the scintillation detector systems PiP [8] and ToF [9]. The protons were detected by the hodoscope PiP covering a polar angular range of 70 to 110 degrees, while the energy of second nucleon, either proton or neutron, was determined via time-of-flight by ToF which spanned the polar angular range from 7 to 157 degrees; the azimuthal range was about 40 degrees. More technical details can be found in the discussion of previous experiments [4, 10–13].

3 Results

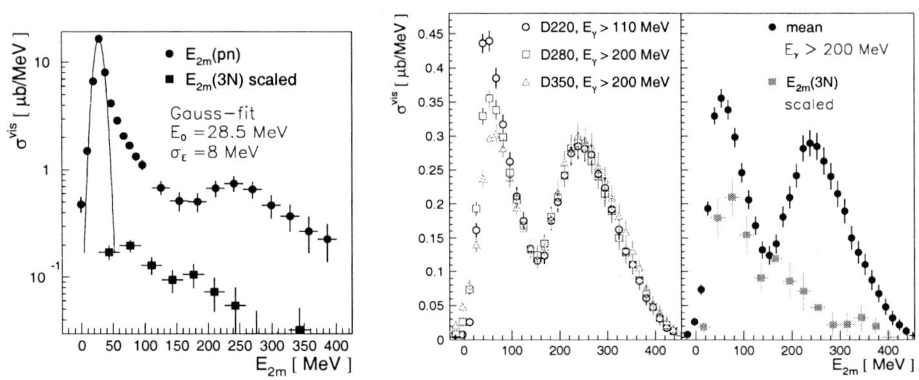

Figure 1. Cross section for the ^{4}He(γ,np), ^{4}He(γ,pp) and the (γ,npp) reactions as function of missing energy.

The missing energy spectra for both reactions are shown in Fig. 1. The much stronger yield for the *pn* channel as compared to *pp* is due to MEC. For *pn* a strong breakup peak at $\approx$28 MeV is observed (note the logarithmic scale) with marginal yield at high missing energies above E_m >200 MeV which is due to pion production. In contrast, the *pp* channel exhibits low yields everywhere, similar in strength to *np* at high E_m. In addition, we plot also the yield for 3 particles detected in the setup corrected for acceptance. It is obvious from the plots that the 3-particle contributions in *pp* is a much larger fractions as compared to the *np* channel. From this we argue that the peak in *np* at 28 MeV is due to two-body emission and that a valid range of integration extends to

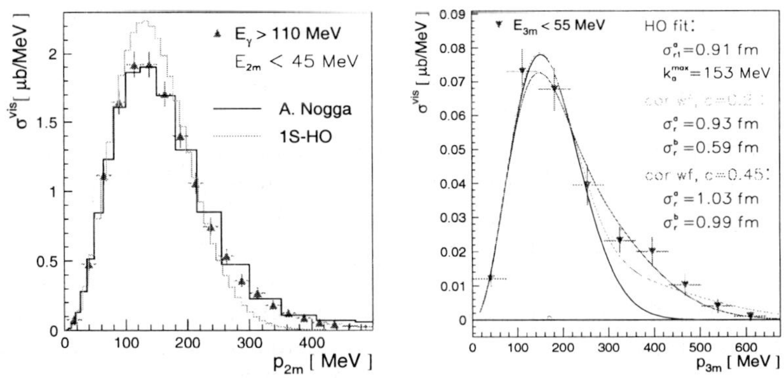

Figure 2. left: momentum distribution of the np-pair in comparison to model calculations; right: momentum distribution of the residual neutron from ^{4}He(γ,npp).

45 MeV.

In Fig. 2 we compare the missing momentum distribution for *np* and *npp* with model predictions. The simple harmonic oscillator model already provides a quite reasonable description of the data. The microscopic calculation by the Bochum group [14] reproduces the data exactly. The *npp* missing momentum is actually the momentum of the unobserved neutron. Several parameterisations have been tried.

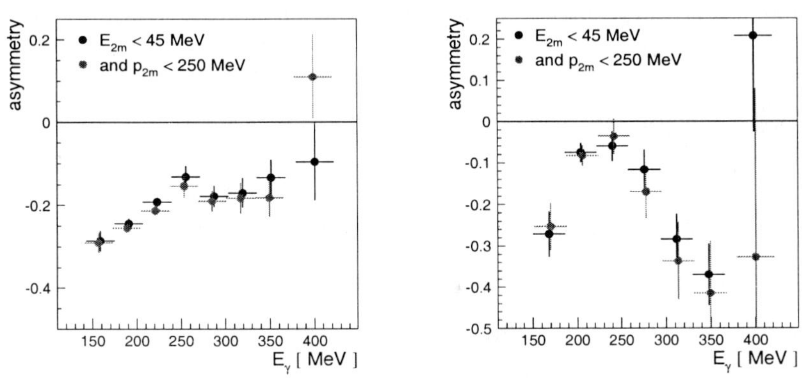

Figure 3. Photon asymmetries for the ^{4}He(γ,np) reaction (left) and for (γ,pp) leaving the residual at low excitation (E_m < 45 MeV).

Finally we present the asymmetries as function of photon energy. Except for the highest data point the additional restriction of having missing momentum $p_m < 250$ MeV does not change the data obtained for $E_m < 45$ MeV indicating a rather pure selection of 2N-events. The different behaviour of the *np* and the *pp* channel originates from the underlying reaction mechanism. There are indications from preliminary calculations in Gent [15] that the increase around

350 MeV in the pp channel is due to Δ contributions.

4 Conclusions

The study of nucleon emission on the light nucleus ^{4}He is quite illuminating because of its simple structure and high density. This nucleus plays a key role as it is positioned on the border between the few- and the many-body systems. The comparison with both type of calculations is in preparation.

Acknowledgement. I wish to express my gratitude to all the colleagues who have actively contributed to this work, in particular to F.A. Natter. This work is supported by grants from the Deutsche Forschungsgemeinschaft (GRK683 and SPP1034).

References

1. W.H. Dickhoff and H. Müther: Rep. Prog. Phys. **55**, 1947 (1992)
2. H.Müther and A. Polls: Prog. Part. Nucl. Phys. **45**, 243 (2000)
3. R.C. Carasco et al.: Nucl. Phys. **A570**, 701 (1994)
4. Th. Lamparter et al.: Z. Phys. **A355**, 1 (1996)
5. H. Herminghaus et al.: Nucl. Instr. Meth. **A187**, 103 (1981); T. Walcher, Prog. Part. Nucl. Phys. **24**, 189 (1990)
6. I. Anthony et al.: Nucl. Instr. Meth. **A301**, 230 (1991)
7. S. Oberkirsch et al.: Nucl. Instr. Meth. **A418**, 507 (1998)
8. I.J.D. MacGregor et al.: Nucl. Instr. Meth. **A382**, 479 (1996)
9. P. Grabmayr et al.: Nucl. Instr. Meth. **A402**, 85 (1998)
10. P. Grabmayr et al.: Phys. Lett. **B370**, 17 (1996)
11. I.J.D. MacGregor et al.: Phys. Rev. Lett. **80**, 245 (1998)
12. F.A. Natter: Prog. Part. Nucl. Phys. **44**, 461 (2000); A. Natter, PhD thesis, U. Tübingen 2002
13. P. Grabmayr: In: *Procs. Perspectives in Hadronic Physics*, Nucl. Phys. **A699**, 41c (2002)
14. A. Nogga: Private Communication; A. Nogga, H. Kamada, and W. Glöckle: Phys. Rev. Lett. **85**, 944 (2000)
15. W. van Nespen, J. Ryckebusch: Private Communication

Few-Body Systems Suppl. 14, 335–338 (2003)

Few-Body Systems

Search for Three-Body Effects in the $\gamma\,^3\mathrm{He} \rightarrow ppn$ Reaction Measured with the CLAS Detector

S. Niccolai*, for the CLAS Collaboration

Center for Nuclear Studies, The George Washington University
Washington, DC 20052, USA

Abstract. The three body photodisintegration of ^{3}He has been measured with the CLAS detector at Jefferson Lab, using tagged photons of energy up to 1.55 GeV. The acceptance of the spectrometer allowed for the first time to cover a wide momentum and angular range for the two outgoing protons. The goal of this experiment was to investigate the three-nucleon interaction in ^{3}He. According to calculations of Laget, by selecting certain kinematic regions the two- and three-body contributions can be distinguished. Preliminary results are discussed and compared with the theoretical predictions.

1 Introduction

The understanding of the strong interaction is a central issue in nuclear physics. In particular, the existence and nature of three-body forces has been a subject of intense debate in the last two decades. Although substantial progress has been made recently in the study of the three-nucleon system [1], and a wealth of low-energy data for cross sections and spin observables in *pd* and *nd* scattering and in 3N breakup is available (for example, see [2]), not much data exist at higher energies, except for measurements in restricted parts of the phase space [3].

In this paper, preliminary results for ^{3}He three-body photodisintegration cross section measurements, for $0.35 < E_\gamma < 1.55$ GeV, over an extended phase space domain, are presented [4]. The three-body photodisintegration of ^{3}He is an optimal tool for testing our theoretical knowledge of the three-nucleon interaction. The electromagnetic probe is well understood, and the ground-state wave function of ^{3}He can be calculated exactly with the Faddeev technique. At photon energies below the pion threshold, the $\gamma\,^3\mathrm{He} \rightarrow ppn$ reaction has been thoroughly studied. As the energy increases, on the other hand, the number

* *E-mail address:* silvia@jlab.org

of open channels becomes very large, and an exact and complete calculation based on realistic potentials becomes more and more difficult.

A different approach has been taken by Laget [5]. In his model, the amplitude for the $\gamma\,^3\mathrm{He} \to ppn$ reaction is expanded as a truncated sum of dominant diagrams, computed in momentum space. The diagrams include one-, two-, and three-nucleon absorption mechanisms. Final-state interactions are also taken into account. The contribution of the three-nucleon mechanisms is evaluated here by comparing our preliminary experimental cross sections with the predictions of Laget's model.

2 Experiment

The $\gamma\,^3\mathrm{He} \to ppn$ experiment was performed in Hall B at Jefferson Lab. A continuous electron beam from the CEBAF accelerator of energy 1.645 GeV produced a bremsstrahlung beam by hitting a thin radiator. The photon energy, determined by the Photon Tagger [6], ranged from 0.35 to 1.55 GeV. The tagging efficiency was measured once per day with a lead-glass detector, and its typical value was around 60%. The photons were incident on a cryogenic liquid ^{3}He target, with length 18 cm and density 0.067 g/cm^3.

Charged particles in the final state were detected in coincidence with a tagged photon by the CEBAF Large Acceptance Spectrometer (CLAS) [7]. Their charges, momenta, and angles were determined using the tracking information of the CLAS drift chambers; their velocities were extracted from time-of-flight measurements. This allowed us to identify the two protons in the final state by their reconstructed mass. The neutron was identified by the missing-mass technique (resolution $\sigma_m/m \simeq 2\%$). In this experiment, over 6 million ppn events were collected. The data cover a wide range in particle momentum and angle ($15° \leq \theta_p^{lab} \leq 125°$).

3 Results

In Fig. 1, the $\gamma^3\mathrm{He} \to ppn$ cross section is shown as a function of incident photon energy. The preliminary experimental results (triangles) are integrated over the CLAS acceptance and corrected for detection efficiency. We did not attempt to extract the total cross section, in order to avoid errors due to extrapolations of the measured data to regions not covered by the CLAS. The data sample presented here includes 30% of the data taken. The experimental results are compared with the predictions of Laget [5]. The theoretical cross section (open circles) was integrated over the experimental acceptance. The theory reproduces quite well the absolute value and energy dependence of the data, up to a photon energy of 800 MeV.

Figure 2 shows a more detailed analysis of the cross section for two selected photon energy bins, $E_\gamma = 0.4$ GeV (upper panel), and $E_\gamma = 0.7$ GeV (lower panel). The data are plotted as a function of reconstructed neutron momentum in the laboratory frame. The preliminary CLAS data (triangles) are compared

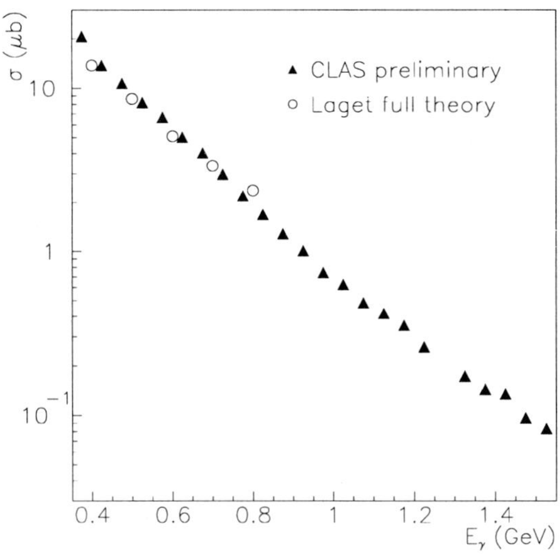

Figure 1. Preliminary results of the $\gamma\,^3$He $\rightarrow$ *ppn* cross section (triangles), integrated over the CLAS acceptance, as a function of incident photon energy. The open circles represent the theoretical predictions of Laget [5]. The statistical errors are smaller than the size of the symbols.

to the Laget full theory (solid line) and to the Laget calculations performed excluding the 3N mechanisms (dashed line). The full theory is generally in good agreement with the data, especially at low photon energies. At low momentum ($p_n \leq 200$ MeV), the 1N+2N mechanisms seem to contribute strongly to the cross section, while at higher neutron momenta the process is clearly dominated by 3N mechanisms. The relative importance of 3N mechanisms over the other processes seems greater at lower photon energies.

4 Conclusions

The $\gamma\,^3$He $\rightarrow$ *ppn* reaction was measured for the first time with the CLAS detector at photon energies between 0.35 and 1.55 GeV. The high statistics and wide kinematical coverage allow a thorough comparison with the available theoretical model. The preliminary results indicate the importance of three-nucleon mechanisms in this reaction. A more detailed analysis of the three-body breakup channel in selected kinematical regions is underway.

References

1. W. Glöckle et al.: Phys. Rep. **274**, 107 (1996)

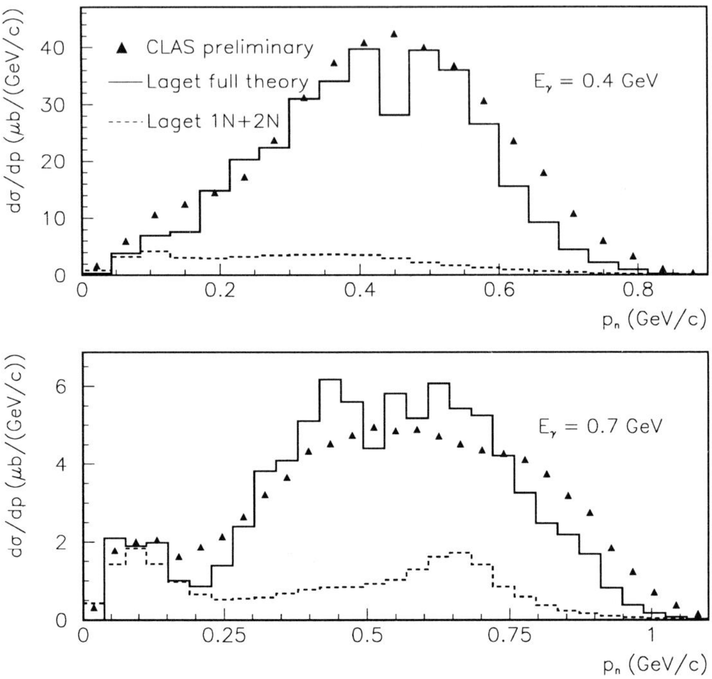

Figure 2. Neutron momentum distribution for the $\gamma\,^3\mathrm{He} \rightarrow ppn$ reaction, integrated over the CLAS geometry (triangles), for photon energy of 400 MeV (upper plot) and 700 MeV (lower plot). The solid line represents the full calculation of Laget [5], and the dashed line shows the model predictions when 3N mechanisms are excluded.

2. A.J. Sarty et al.: Phys. Rev. **C47**, 2 (1993)

3. G. Audit et al.: Nucl. Phys. **A614**, 461 (1997) 461; Phys. Lett. **B227**, 331 (1989); Phys. Lett. **B312**, 57 (1993); Phys. Rev. **C44**, R575 (1991)

4. B.L. Berman, P. Corvisiero, G. Audit (spokespersons): Jefferson Lab Experiment E93044

5. J.M. Laget: Nucl. Phys. **A446**, 489c (1985); Phys. Rev. **C35**, 832 (1987); J. Phys. **G62**, 1445 (1988); Nucl. Phys. **A497**, 391c (1989)

6. D.I. Sober et al.: Nucl. Instrum. Meth. **A440**, 263 (2000)

7. W. Brooks: Nucl. Phys. **A663**, 1077c (2000); B.A. Mecking et al.: to be submitted to Nucl. Instrum. Meth. (2002)

Few-Body Systems Suppl. 14, 339–342 (2003)

Few-Body Systems

Baryon Elastic and Transition Form Factors with a Poincaré Covariant Current Operator

E. Pace[1], G. Salmè[2], A. Molochkov[3,4]

[1] Dipartimento di Fisica, Università di Roma "Tor Vergata" and Istituto Nazionale di Fisica Nucleare, Sezione Tor Vergata, Via della Ricerca Scientifica 1, I-00133 Roma, Italy

[2] Istituto Nazionale di Fisica Nucleare, Sezione Roma I, P.le A. Moro 2, I-00185 Roma, Italy

[3] Bogoliubov Laboratory of Theoretical Physics, Joint Institute for Nuclear Research, 141980 Dubna, Russia

[4] RCNP, Osaka University, Osaka, 567-0047, Japan

Abstract. A Poincaré covariant electromagnetic current operator for systems of interacting particles has been proposed within the front-form Hamiltonian dynamics, to evaluate elastic and transition hadron form factors without ambiguities. Preliminary results for the nucleon elastic form factors in a constituent quark model are presented. The same current operator can be used to investigate the transition form factors from the nucleon to baryon resonances.

1 Introduction

Within the front-form Hamiltonian dynamics, an electromagnetic current operator for systems of interacting particles has been obtained in ref. [1] that : i) properly commutes with the representation operators of the Poincaré group; ii) satisfies hermiticity, parity and time reversal covariance, as well as current conservation. This operator is built up starting from the free current in the Breit frame where initial and final three-momenta of the system are directed along the spin quantization axis, z.

By using this covariant current, one gets rid of the well-known ambiguities, which in the front-form approach plague the extraction of elastic and transition form factors, when the free current is considered in the $q^+ = 0$ reference frame. Indeed, in our approach the number of independent matrix elements of the current is equal to the number of form factors [1], as it must be.

Our Poincaré covariant current operator for elastic and transition form factors is presented in Sec. 2. This current was already successfully applied to calculate the deuteron elastic form factors, by adopting front-form deuteron wave functions corresponding to a large set of realistic N-N interactions [1](b), [1](c),[2]. Preliminary results of our investigation of nucleon elastic form factors in a constituent quark (CQ) model are presented in Sec. 3.

2 Poincaré Covariant Current Operator

Let us define

$$\mathcal{J}^+(k\hat{e}_z; M_i, M_j) = \langle 0, K_i^{'+} | \Pi_i \; J^+_{free}(0) \; \Pi_j | 0, K_j^+ \rangle \,, \tag{1}$$

$$\mathcal{J}_\perp(k\hat{e}_z; M_i, M_j) = \langle 0, K_i^{'+} | \Pi_i \; \mathbf{J}_{\perp free}(0) \; \Pi_j | 0, K_j^+ \rangle \,, \tag{2}$$

$$\mathcal{J}^-(k\hat{e}_z; M_i, M_j) = \frac{M_j^2 \; K_i^{'+} - M_i^2 \; K_j^+}{2K_i^{'+} K_j^+ \; (K_i^{'+} - K_j^+)} \; \mathcal{J}^+(k\hat{e}_z; M_i, M_j) \,, \tag{3}$$

where Π_i is the projection operator onto the subspace $\mathcal{H}_i$ of the states with mass M_i and spin S_i, $|\mathbf{K}_\perp = 0, K_j^+\rangle$ is an eigenstate of the total momentum in the Breit frame where the three-momentum transfer $\mathbf{q}$ is directed along the spin quantization axis, z, $K_j^+ = [\sqrt{M_j^2 + k^2} - k]/\sqrt{2}$ and $K_i^{'+} = [\sqrt{M_i^2 + k^2} + k]/\sqrt{2}$ with $k = |\mathbf{q}|/2$. The operator $\mathcal{J}^\mu(k\hat{e}_z; M_i, M_j)$ is a matrix element between total momentum eigenstates, but an operator in the space of intrinsic states. Following ref. [1], it is easy to show that the current

$$\begin{aligned} j^\mu(k\hat{e}_z, M_i, M_j) \;&=\; \frac{1}{2} \left[\mathcal{J}^\mu(k\hat{e}_z, M_i, M_j) \; + \right. \\ &+ \left. L^\mu_\nu[R_x(-\pi)] \; e^{i\pi S_x} \; \mathcal{J}^\nu(k\hat{e}_z, M_j, M_i)^* \; e^{-i\pi S_x} \right] \end{aligned} \tag{4}$$

fulfills Poincaré, parity and time reversal covariance, hermiticity, current conservation, and charge normalization. The second term in Eq.(4), which implements hermiticity, implicitly introduces two-body terms in the current.

A distinct feature of our approach is that the form-factors of the interacting constituents in the system (i.e., the form-factors of CQ in the nucleon, in the present case) depend on the difference between the initial, p_j, and final, p'_j, momenta of the $j-th$ constituent, i.e. on $(p'_j - p_j)^2 \neq Q^2$, and cannot be factorized out [1](c).

3 Baryon Form Factors in a Constituent Quark Model

The nucleon form factors have been evaluated in a CQ model [3], which features: i) eigenstates of a mass operator which describes a large part of the baryon mass spectrum [4]; ii) phenomenological Dirac and Pauli form factors for the CQ. Using the Poincaré covariant current operator $j^\mu(k\hat{e}_z, M_i, M_j)$ defined in Sec. 2, the nucleon form-factors can be expressed by means of the matrix elements of the " + " and "x" components of the free current in the Breit reference frame where the three-momentum transfer is along the z axis:

$$G_E^\tau(Q^2) = \frac{1}{2} \; Tr[\mathcal{I}^+_{free}(\tau)] \quad , \quad G_M^\tau(Q^2) = -i\frac{M}{Q} \; Tr[\hat{\sigma}_y \; \mathcal{I}^x_{free}(\tau)] \,, \tag{5}$$

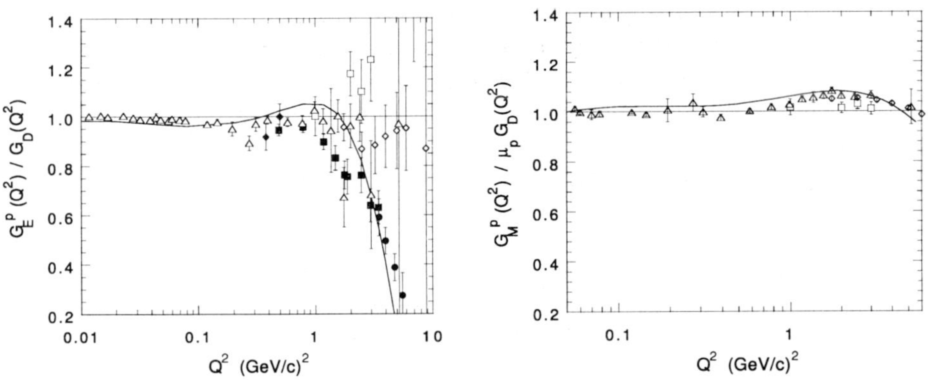

Figure 1. (a) The proton charge form-factor $G_E^p(Q^2)/G_D(Q^2)$ and (b) the proton magnetic form-factor $G_M^p(Q^2)/(\mu_p G_D(Q^2))$ vs Q^2. Triangles represent the experimental data of ref. [5], while full squares and full dots are obtained from the experimental ratio $G_E^p(Q^2)/G_M^p(Q^2)$ of Refs. [6](a) and (b) respectively, and the model of ref. [7] for $G_M^p(Q^2)$. The other experimental data are as in ref. [3].

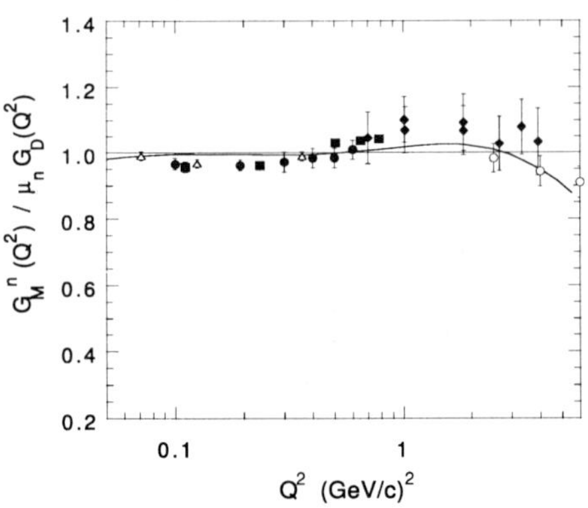

Figure 2. The neutron magnetic form-factor $G_M^n(Q^2)/(\mu_n G_D(Q^2))$ vs. Q^2. Full dots are from ref. [8], triangles from ref. [9], full squares from ref. [10], full diamonds from ref. [11] and open circles from ref. [12].

$$\mathcal{I}^r_{\sigma'\sigma\ free}(\tau) \equiv \langle \Psi^{\frac{1}{2}\tau}_{\frac{1}{2}\sigma'}, P'|\ J^r_{free}\ |\Psi^{\frac{1}{2}\tau}_{\frac{1}{2}\sigma}, P\rangle\ . \qquad (6)$$

In Eq. (6) $|\Psi^{\frac{1}{2}\tau}_{\frac{1}{2}\sigma}, P\rangle$ ($|\Psi^{\frac{1}{2}\tau}_{\frac{1}{2}\sigma'}, P'\rangle$) is the initial (final) nucleon state, $\tau = p, n$, and J^r_{free} is the $r-th$ component ($r = +, x$) of the sum of the CQ currents:

$$J^r_{free} = \sum_{j=1}^{3} I^r_j = \sum_{j=1}^{3} \left(e_j f_1^j[(p'_j - p_j)^2]\gamma^r\ +\ i\ \kappa_j f_2^j[(p'_j - p_j)^2]\frac{\sigma^{r\rho}q_\rho}{2m_j} \right)\ , \qquad (7)$$

where e_j (κ_j) is the charge (anomalous magnetic moment) of the j-th CQ quark, and $f^j_{1(2)}$ its Dirac (Pauli) form factor. To take into account in an effective way possible many-body current contributions, we allow for an effective value m^* of the mass of the quark interacting with the virtual photon.

Our aim is to determine m^* and the CQ form factors by a fit to the nucleon form-factors, and to use these CQ form-factors for a parameter-free evaluation of transition form-factors to baryon resonances. The CQ form-factors have been parametrized by the sum of a monopole and a dipole for the Dirac form-factors, and the sum of a dipole and an octupole for the Pauli form-factors. It results that $m^* = 329$ MeV and that the anomalous magnetic moments of the CQ, $\kappa_u = -0.372$ and $\kappa_d = 0.186$, are in agreement with isospin symmetry. Our preliminary results are shown in Figs. 1 and 2 (see also ref. [13]).

The same Poincaré covariant current, with the same constituent quark form factors and without any further free parameter, can be used to evaluate the transition form factors to the main baryonic resonances. However, the contributions of explicit two-body currents (e.g. the Z-graph current, see ref. [14]) should be investigated, before drawing definite conclusions.

References

1. (a) F. Lev, E. Pace, G. Salmè: Nucl. Phys. **A641**, 229 (1998); (b) Phys. Rev. Lett. **83**, 5250 (1999); (c) Phys. Rev. **C62**, 064004 (2000)
2. E. Pace and G. Salmè: Nucl. Phys. **A684**, 487c (2001); Nucl. Phys. **A689**, 441 (2001)
3. F. Cardarelli, E. Pace, G. Salmè, and S. Simula: Phys. Lett. **B357**, 267 (1995)
4. S. Capstick and N. Isgur: Phys. Rev. **D34**, 2809 (1986)
5. G. Höhler et al.: Nucl. Phys. **B114**, 505 (1970)
6. (a) M.K. Jones et al.: Phys. Rev. Lett. **84**, 1398 (2000); (b) O. Gayou et al.: Phys. Rev. Lett. **88**, 092301 (2002)
7. P.E. Bosted: Phys. Rev. **C51**, 409 (1995)
8. W. Xu et al.: Phys. Rev. Lett. **85**, 2900 (2000); F. Xiong et al.: Phys. Rev. Lett. **87**, 242501 (2002); W. Xu et al.: nucl-ex/0208007
9. G. Kubon et al.: Phys. Lett. **B524**, 26 (2002)
10. H. Anklin et al.: Phys. Lett. **B428**, 248 (1998)
11. W. Albrecht et al.: Phys. Lett. **B26**, 642 (1968)
12. S. Rock et al.: Phys. Rev. Lett. **49**, 1139 (1982)
13. E. Pace, G. Salmè, and A. Molochkov: Nucl. Phys. **A699**, 156c (2002)
14. J.P.B.C. de Melo et al.: Nucl. Phys. **A707**, 399 (2002)

Few-Body Systems Suppl. 14, 343–344 (2003)

Few-Body Systems

Benchmark Result on Total Three-Nucleon Photodisintegration

G. Orlandini[1], V.D. Efros[2], W. Leidemann[1], E.L. Tomusiak[3], J. Golak[4,5], R. Skibiński[5], W. Glöckle[4], H. Kamada[6], A. Nogga[7], H. Witała[5]

[1] Dipartimento di Fisica, Università di Trento and INFN, Italy

[2] Russian Research Centre "Kurchatov Institute", Russia

[3] Department of Physics and Astronomy, University of Victoria, Canada

[4] Institut für Theoretische Physik II, Ruhr-Universität Bochum, Germany

[5] Institute of Physics, Jagiellonian University, Cracow, Poland

[6] Department of Physics, Faculty of Engineering, Kyushu Institute of Technology, Japan

[7] Department of Physics, University of Arizona, Tucson, Arizona

We present the results of a benchmark calculation [1] of the three–nucleon photodisintegration cross section with modern realistic two– and three–nucleon forces, using both the Faddeev equations [2] and the Lorentz Integral Transform (LIT) method [3]. The comparison between the results obtained within the two methods gives an idea of the precision reached by three-body calculations involving continuum states and it may stimulate more accurate measurements.

The total unpolarized photodisintegration cross section is proportional to the transverse response function $R_T(E_\gamma)$

$$R_T(E_\gamma) = \int df |\langle \Psi_f | \mathbf{J_T} | \Psi_0 \rangle|^2 \delta(E_\gamma - E_f + E_0) \,, \tag{1}$$

where Ψ_0 and Ψ_f are the three–nucleon wave functions with energy E_0, and E_f, respectively. The way R_T is calculated in the Faddeev framework is described in ref. [2]. In the LIT approach the need to compute final state continuum wave functions is avoided. In fact the response function is obtained via evaluation and subsequent inversion of its LIT

$$L(\sigma_R, \sigma_I) = \int_{E_{th}}^{\infty} dE_\gamma \frac{R_T(E_\gamma)}{(E_\gamma - \sigma_R)^2 + \sigma_I^2} \,. \tag{2}$$

Because $L(\sigma_R, \sigma_I)$ is evaluated by solving an inhomogeneous differential equation with bound state boundary conditions it is evident that the Faddeev and

LIT methods are completely different. Therefore the comparison between the results obtained with the two methods is a very significant test of accuracy reached by these calculations.

In Fig. 1 the comparison between LIT and Faddeev results is presented in unretarded E1 approximation without (Fig. 1a) and with (Fig. 1b) the 3N force, thereby demonstrating the accuracy of present calculations.

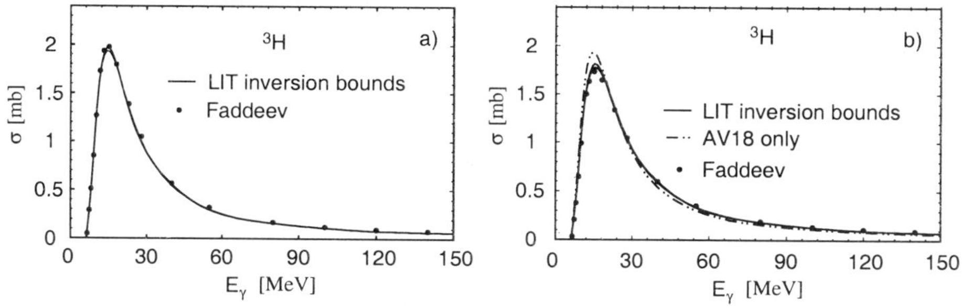

Figure 1. Faddeev (dots) and LIT (full curves) results in unretarded E1 approximation without (a) and with (b) 3N force (UIX, [4]). Dotted curve: with AV18 [4] only.

We have also investigated the contributions of magnetic and higher order electric multipoles, of retardation, meson exchange currents contributions as well as of Coulomb force. The results are in order here. The classical approximation for photonuclear cross sections, i.e. the unretarded E1 approximation, is extremely good below 50 MeV. It is also rather good at even higher energies (error at 100 MeV below 20 %), since one has a partial cancellation of the E1 retardation effects and contributions from additional multipoles. The Coulomb force effects on the cross section become negligible about 6-7 MeV above threshold but are important at lower energy.

The comparison with existing rather old data (see Fig. 7 in Ref. [1]) shows that the theoretical quality of the results is considerably higher than the experimental uncertainties. Thus new experimental activity in this field would appear to be timely.

References

1. J. Golak et al.: Nucl.Phys. **A707**, 365 (2002)
2. J. Golak et al.: Phys. Rev. **C52**, 1216 (1995); S. Ishikawa et al. Phys. Rev. **C57**, 39 (1998)
3. V.D. Efros, W. Leidemann, G. Orlandini: Phys. Lett. **B338**, 130 (1994)
4. R.B. Wiringa, V.G.J. Stoks, R. Schiavilla: Phys. Rev. **C51**, 38 (1995); B.S. Pudliner et al.: Phys. Rev. **C56**, 1720 (1997)

Few-Body Systems Suppl. 14, 345–346 (2003)

Few-Body Systems

The $^4He(\gamma,n)^3He$ Reaction with Full Final-State Interaction

S. Quaglioni[1], V.D. Efros[2], W. Leidemann[1], G. Orlandini[1]

[1] Dipartimento di Fisica, Università di Trento and INFN (Gruppo collegato di Trento), I-38050 Povo, Italy

[2] Russian Research Centre, "Kurchatov Institute", Kurchatov Square 1, 123182 Moscow, Russia

The total cross section of the process $^4He(\gamma,n)^3He$ is calculated. The four–nucleon problem is solved, and Final State Interaction (FSI) is taken rigorously into account both below and beyond the three-body breakup threshold. This is accomplished with the help of the Lorentz Integral Transform (LIT) method [1, 2]. The semi-realistic MTI-III potential [3] is taken as NN interaction. The excitation energy range up to 30 MeV is considered which includes the giant resonance peak, in particular. Like for the case of the total ^4He photoabsorbtion cross section in [4, 5] it is sufficient to take into account transitions induced by the unretarded dipole operator D [6]. The cross section is

$$\sigma_{(\gamma,n)} = (e^2/\hbar c)\omega_\gamma k\mu\pi^{-1}\int d\Omega_k \left|\langle \Psi^-(E_f)\,|D|\,\Psi_\alpha\rangle\right|^2, \quad E_f=\omega_\gamma+E_\alpha.$$

Here ω_γ is the incident photon energy, E_α and E_f are the energies of the initial and final state, Ψ_α and Ψ^- are the corresponding wave functions, and k and μ are the relative momentum and the reduced mass of the $(n,{}^3He)$ system.

In the LIT method [2] the transition matrix element is calculated from [1, 7]

$$\langle \Psi^-(E_f)\,|D|\,\Psi_\alpha\rangle = \left\langle \phi_1(E_f)\left|\hat{\mathcal{A}}D\right|\Psi_\alpha\right\rangle + \left\langle \phi_1(E_f)\left|V_1(E_f+i\varepsilon-H)^{-1}\hat{\mathcal{A}}D\right|\Psi_\alpha\right\rangle,$$

where H is the full nuclear Hamiltonian, ϕ_1 is the unperturbed wave function describing the nucleon 1 in the plane wave relative motion with respect to the nucleons 2, 3 and 4 bound to form ^3He, V_1 is the corresponding $n-{}^3He$ interaction part of H, and $\hat{\mathcal{A}}$ is the antisymmetrization operator. The first matrix element, $\left\langle \phi_1(E_f)\left|\hat{\mathcal{A}}D\right|\Psi_\alpha\right\rangle$, can be calculated directly. The second, FSI–dependent, matrix element can be expressed in terms of the function

$$F_{fi}(E) = \int d\gamma \langle\phi_1|V_1\hat{\mathcal{A}}\mid\Psi_\gamma\rangle\,\langle\Psi_\gamma\,|D|\,\Psi_\alpha\rangle\,\delta(E-E_\gamma),$$

where Ψ_γ is a complete set of states. One has

$$\left\langle \phi_1 (E_{\rm f}) \middle| V_1(E_{\rm f} + i\varepsilon - H)^{-1} \hat{A} D \middle| \Psi_\alpha \right\rangle = -i\pi F_{\rm fi}\,(E_{\rm f}) + \mathcal{P}\int_{E_{\rm th}}^{\infty} dE F_{\rm fi}(E)(E_{\rm f} - E)^{-1}.$$

The function $F_{\rm fi}$ is determined from its LIT, and the latter can be obtained as the overlap, $\left\langle \tilde{\Psi}_2 \mid \tilde{\Psi}_1 \right\rangle$, of the solutions $\tilde{\Psi}_1$ and $\tilde{\Psi}_2$ to the equations

$$\begin{aligned}(H - \sigma_{\rm R} + i\sigma_{\rm I}) \left|\tilde{\Psi}_1\right\rangle &= D \left|\Psi_\alpha\right\rangle \\ (H - \sigma_{\rm R} + i\sigma_{\rm I}) \left|\tilde{\Psi}_2\right\rangle &= \hat{A} V_1 \left|\phi_1\right\rangle .\end{aligned}$$

The important point is that both $\tilde{\Psi}_1$ and $\tilde{\Psi}_2$ are localized. Therefore bound–state type techniques can be applied to solve the above equations. As in [4] we make use of the correlated hyperspherical expansion. Our preliminary results are presented in Fig. 1 along with the results of the Born approximation. The comparison shows that FSI is of a great importance.

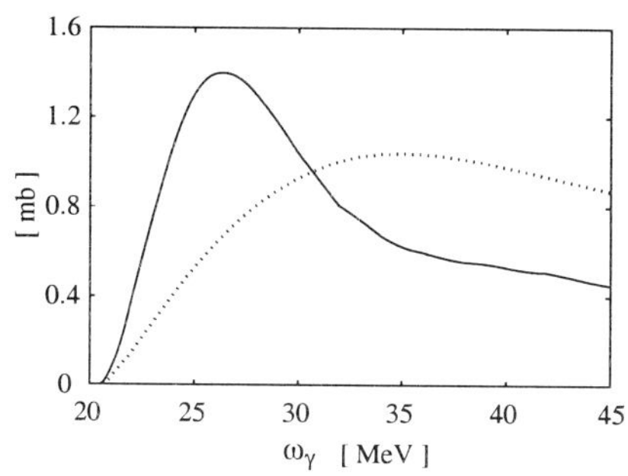

Figure 1. The calculated total $^4He(\gamma, n)^3He$ cross section. Solid curve: full results, dotted curve: Born approximation.

References

1. V.D. Efros: Sov. J. Nucl. Phys. **41**, 949 (1985); Phys. Atom. Nucl. **62**, 1833 (1999)
2. V.D. Efros, W. Leidemann, G. Orlandini: Phys. Lett. **B338**, 130 (1994)
3. R.A. Malfliet, J. Tjon: Nucl. Phys. **A677**, 161 (1969)
4. V.D. Efros, W. Leidemann, G. Orlandini: Phys. Rev. Lett. **78**, 4015 (1997)
5. N. Barnea, V.D. Efros, W. Leidemann, G. Orlandini: Phys. Rev. **C63**, 057002 (2001)
6. J. Golak et al.: Nucl. Phys. **A707**, 365 (2002)
7. A. La Piana, W. Leidemann: Nucl. Phys. **A677**, 423 (2000)

Few-Body Systems Suppl. 14, 347–348 (2003)

Few-Body Systems

Photodisintegration of the Deuteron between $E_\gamma = 2.4$ and 4.0 MeV

W. Tornow[1], N.G. Czakon[1], C.R. Howell[1], A. Hutcheson[1], J.H. Kelley[1], V.N. Litvinenko[2], S. Mikhailov[2], I.V. Pinayev[2], G.J. Weisel[3], H. Witała[4]

[1] Triangle Universities Nuclear Laboratory and Duke University, Durham, NC 27708

[2] Duke University Free-Electron Laser Laboratory, Duke University, Durham, NC 27708

[3] Penn State Altoona, Altoona, PA 16601

[4] Jagiellonian University, Cracow, Poland

Big-bang nucleosynthesis (BBN) is an observational cornerstone of the hot big-bang cosmology. According to [1] the $p(n,\gamma)d$ reaction is of special interest because the big-bang abundance of deuterium is most sensitive to the baryon density. However, in the most important energy range for BBN, i.e., from 25–200 keV in the c.m. system, experimental data are completely missing. In fact, published data exist only at thermal energies and at n-p c.m. energies above 275 keV. Therefore, theoretical models have been used [1] in the BBN energy range that are normalized to the high-precision thermal neutron capture cross-section data.

During a 100 hour run in May 2001 we measured the analyzing power for the time reversed reaction ^{2}H(γ,np) at 90° at seven energies between E_γ = 2.39 and 4.05 MeV corresponding to n-p c.m. energies of 165 keV to 1.82 MeV. The monoenergetic and linearly polarized γ rays were produced by Compton backscattering of relativistic electrons (300 to 375 MeV) from 670 nm FEL photons at the TUNL/DFELL High-Intensity γ-ray Source (HIγS). The collimated γ-ray beam of 2.6 cm diameter and average flux of about 5×10^5 γ/s was scattered off a 4 cm diameter and 6 cm long deuterated liquid scintillator (C_6D_{12}) contained in a thin-walled glass cylinder and viewed by a phototube (PMT). The axis of the scintillator-PMT arrangement coincided with the axis of the γ-ray beam. The energy spread $\Delta E/E$ of the γ-ray beam was about 3%.

The neutrons from the deuteron breakup were detected by four Bicron 501A liquid scintillator detectors, 2" in diameter and 2" in length, each viewed by a PMT. Two detectors were mounted at $\theta = 90°$ in the plane of the γ-ray polarization on opposite sides of the incident γ-ray beam, and the other two were mounted at $\theta = 90°$ in the plane perpendicular to the γ-ray polarization plane. The center-to-center distance between the deuterated scintillator and the neutron detectors was 17 cm. Neutron-gamma pulse-shape discrimination was applied. The neutron and proton energies varied from about 900 keV at $E_\gamma = 4.05$ MeV to 85 keV at $E_\gamma = 2.39$ MeV.

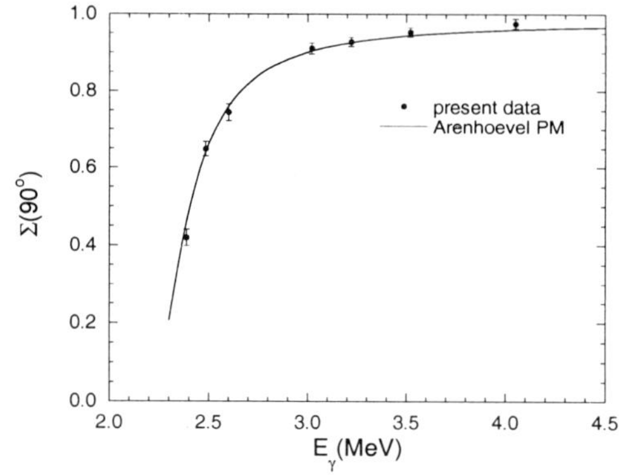

Figure 1. Excitation function of the photon analyzing power for the reaction $\gamma + d \rightarrow n + p$ at 90° in comparison to the theoretical prediction of Arenhövel based on the Bonn NN potential (in r-space).

To cancel instrumental asymmetries in our measurements, we either rotated the neutron detectors, which were mounted on a ring centered and positioned perpendicular to the γ-ray beam, through 90°, or we interchanged the detectors, i.e., the detectors in the horizontal plane were moved to the vertical plane and vice versa.

Extensive Monte-Carlo calculations have been performed to simulate the effects of our finite-geometry experimental setup on the data. Especially, scattering of the breakup neutron on its way out of the deuterated scintillator results in a considerably reduced asymmetry. Our data are compared in Fig. 1 to the theoretical prediction of Arenhövel [2]. Using the theoretical total $\gamma - d \rightarrow n + p$ cross section σ_T, the present data can be used to determine the M1 (s-wave) and E1 (p-wave) contributions to σ_T, which are of considerable theoretical interest, especially since the M1 piece can be used to obtain the cross-section difference $\sigma_p - \sigma_a$ that is needed to determine the Gerasimov-Drell-Hearn sum rule on the deuteron.

References

1. S. Burles et al.: Phys. Rev. Lett. **82**, 4176 (1999); astro-ph/9901157
2. H. Arenhövel: Private Communication (2001)

Few-Body Systems Suppl. 14, 349–354 (2003)

Few-Body Systems

Experimental Studies of the Gerasimov-Drell-Hearn Sum Rules

A. Deur*,†

University of Virginia

Abstract. The Gerasimov-Drell-Hearn sum rule, which was originally derived at $Q^2 = 0$, has been extended recently to finite Q^2. The measurement of its Q^2 behavior provides a powerful way to investigate the transition between the partonic and hadronic description of QCD. I recall first the theoretical grounds of GDH. Then new experimental results on both the original and extended sum rules are discussed. Here, I focus on the results obtained in Hall A at Jefferson Lab. I will conclude by presenting the experimental perspectives, both at low Q^2 and at large photon energies.

1 Introduction

The development of polarized beams and targets in the past decades has allowed further checks of QCD in its perturbative regime (pQCD) by providing additional constraints due to the spin degrees of freedom. At the same time, the nucleon spin structure in the Deep Inelastic Scattering (DIS) regime has been explored. The understanding of nucleon structure in this domain, where the relevant degrees of freedom are partons, is now relatively satisfactory. At lower energies, the strong interaction is described in term of hadronic degrees of freedom. The understanding of the transition between the high and low energy domains remains an outstanding problem for the strong interaction. Jefferson Lab (JLAB), with its energy ranging from ~1 to 6 GeV, is an ideal facility to study the transition region from the partonic to the hadronic degrees of freedom. In this context, I will explain why the extended GDH sum rule measured at the Q^2 accessible at JLAB is a particularly useful tool. Before talking about the extended GDH sum rule, I will recall briefly the original GDH sum rule.

*For the Jefferson Lab E94010 Collaboration
†*E-mail address:* deur@virginia.edu

2 The GDH Sum Rule

2.1 Basis

The original GDH sum rule was derived [1] at the real photon point, $Q^2 = 0$. For spin 1/2 targets

$$\int_{\nu_0}^{\infty} \sigma^{TT} \frac{d\nu}{\nu} = -\pi^2 \alpha \frac{\kappa^2}{M^2}, \tag{1}$$

where σ^{TT} is the transverse-transverse interference cross section, ν_0 the pion photoproduction threshold, ν the photon energy, κ the anomalous magnetic moment of the target and M its mass. In term of polarized photoproduction cross sections $\sigma^{TT} = \left(\sigma^{1/2} - \sigma^{3/2}\right)/2$ with 1/2 (3/2) meaning that the photon spin is anti-aligned (aligned) with the target spin.

The derivation of the sum rule follows three steps:

1. An unsubtracted dispersion relation links the real part of the forward Compton amplitude to the integral of the imaginary part, weighted by the photon energy. The dispersion relation assumes causality for the Compton amplitude.

2. The optical theorem expresses the imaginary part of the Compton amplitude in term of σ^{TT}. The optical theorem relies on unitarity.

3. A Low Energy Theorem expands in ν the real part of the Compton amplitude. The Low Energy Theorem assumes relativity and gauge invariance.

We can see that the derivation rests on solid theoretical grounds. The only assumption that might be open to questions is in step 1: the validity of the non-subtraction hypothesis. However, experimental data as well as theoretical bounds suggest that this hypothesis is correct [2].

2.2 Experimental Status

As shown, the GDH sum rule is a fundamental quantity. It has been used for analysis and for theoretical predictions. It is thus important to check its validity. Another reason for experimental checks of the sum rule is that the first indirect estimates [3] of the GDH integral (the left-hand side of Eq. (1)), using single-pion photoproduction, have shown that the sum rule is violated. However, it is not clear whether the violation is real or comes from the limitation of using unpolarized and singly polarized pion data to perform this indirect estimate.

The check of the sum rule is being done at MAMI ($0.14 < \nu < 0.80$ GeV) and ELSA ($0.7 < \nu < 3.0$ GeV). In both cases, doubly polarized pion photoproduction is detected in a 4π detector. The MAMI result on the proton is published [4] and the complement at ELSA is being analyzed. Preliminary results can be found in [5]. The MAMI data together with the ELSA preliminary analysis up to $\nu = 1.8$ GeV show an over-saturation of the sum rule. One reason could be that in spite of the $1/\nu$ weighting, the large energy part of the integral is significant. Clearly, data at higher photon energy have to be taken. This is the goal of E159 at SLAC ($5 < \nu < 40\,\mathrm{GeV}$) [6].

With this experimental program, the validity of the original sum rule on the proton should be clarified in the next few years. As far as the neutron is concerned, no data are available yet.

3 The Extended GDH Sum Rule

Anselmino et al. [7] suggested that the GDH integral extended to finite Q^2 will help in understanding the transition from perturbative to non-perturbative QCD. Also pointed out was the connection between the generalized GDH integral and the Bjorken integral. The generalization consists of going from the photoproduction to the electroproduction cross sections. However, several extensions are possible, depending on the choices of the virtual photon flux and how the spin structure function g_2 is included [8]. The definitions are equivalent at the photon point and in the Bjorken domain but differ at intermediate Q^2.

Among the different GDH extensions, the one of Ji and Osborn [9] stands out because it generalizes not only the integral side but the full sum rule. Hence it keeps the checking power of a sum rule that was lost with the former definitions. In addition Ji and Osborn showed that the Bjorken and the GDH sum rules are different aspects of their generalized GDH sum rule. It is written

$$\int_{\nu_0}^{\infty} G_{1(2)} \frac{d\nu}{\nu} = \overline{S_{1(2)}}\,, \tag{2}$$

where $\overline{S_{1(2)}}$ are the forward Compton amplitudes without the elastic contribution. $\overline{S_{1(2)}}$ are calculable using chiral perturbation theory at low Q^2, higher twist expansion at larger Q^2 and eventually at any Q^2 with lattice QCD. Hence, the extended GDH sum rule is calculable on the full QCD spectrum, in particular in the transition region, which makes it an unique tool to understand the transition between the quark/gluon and nucleon/meson degrees of freedom of the strong interaction. SLAC [10] and HERMES [11] have provided data for the generalized GDH integral, mainly in the DIS region. Several experiments at Jefferson Lab have been or will be performed to measure the extended integral at low and intermediate Q^2. The measurement of the extended GDH integral on the neutron at intermediate Q^2 has been done in Hall A (experiment E94-010 [12]). Two other measurements of GDH(Q^2) in a similar Q^2 range as E94-010 have been performed using the 4π detector CLAS in Hall B and frozen spin targets (hydrogen and deuterium). Preliminary results can be found in [18].

4 The Hall A E94-010 Experiment

4.1 Experimental Apparatus

We measured the inclusive $^3\overrightarrow{He}(\mathrm{e},\mathrm{e}')\mathrm{X}$ reaction in Jefferson Lab Hall A. We used the two nearly identical High Resolution Spectrometers (HRS) independently. The comparison of the data from both arms reduces and helps to estimate our systematic errors. Each HRS contains a set of wire chambers for tracking and momentum measurement ($\Delta p/p \simeq 10^{-4}$), two planes of scintillators for triggering and a set of preshower, shower and Cerenkov detectors for particle identification (π/e rejection inefficiency at the few 10^{-4} level).

We used up to 15 μA of CW beam polarized in average at 71%. The polarization is known at the 4% level (relative) from a Møller polarimeter. The beam current was measured at the percent level by a set of cavities sandwiching an Unser monitor. Finally, the beam energy was measured both by accurate tracking of the H(e,e′p) reaction or by the measurement of the bending of the beam in the magnetic arc leading the beam to Hall A. Both methods reach an accuracy of $2.\times10^{-4}$.

Due to the lack of free neutron targets, we used a polarized ^{3}He target as an effective polarized neutron target. It consists of a 40 cm glass cylinder of gaseous ^{3}He at $\sim$12 atm in running conditions. The target length and high density allow a reasonable counting rate in spite of the gaseous nature of the target. The luminosity reached was approximately 10^{36} cm^{-2}s^{-1}. The target polarization averaged 35% and is known at better than 4% relative. It was measured by NMR calibrated using two independent techniques: the known proton thermal polarization of a water filled cell and EPR polarimetry.

4.2 From ^{3}He to Neutron

About 90% of the ^{3}He wavefunction is in a S-wave state. Here, the two proton spins are anti-parallel due to the Pauli principle and their contribution to the nuclear spin cancels. This is not the case for the remaining 10% that consists of D and S' states. This leads to a small contribution of the proton spins which can be accounted for by using the effective polarizations of the protons and neutron within the ^{3}He nucleus. Fermi motion and binding are taken into account with a convolution model [13]. This method is reliable for the DIS. It can also be used in the resonance domain but only for integrated quantities such as sum rules [14].

It should be noted that the sum rule for the ^{3}He nucleus can also be computed and its comparison to data will improve our understanding of the ^{3}He nuclear structure.

4.3 The Experiment

The Q^2 coverage of E94-010 ranges from 0.1 GeV^2 to 1.0 GeV^2. The integration goes up to W=2 GeV. The kinematics domain covered is mainly the resonance region with the beginning of the DIS. The quasi-elastic was covered as well for nuclear studies.

The four inclusive cross sections $\sigma^{\uparrow\uparrow}(W,Q^2)$, $\sigma^{\uparrow\downarrow}(W,Q^2)$, $\sigma^{\rightarrow\uparrow}(W,Q^2)$ and $\sigma^{\rightarrow\downarrow}(W,Q^2)$ were measured (the first (second) arrow gives the beam (target) helicity direction). The detection angle was 15.5° and data were taken at the beam energies of 0.86 GeV, 1.72 GeV, 2.59 GeV, 3.39 GeV, 4.24 GeV and 5.07

GeV. A linear combination of these cross sections provides σ^{TT}. The extended GDH integral, defined as $\int d\nu(1-x)\sigma^{TT}/\nu$, is shown in Fig. 1.

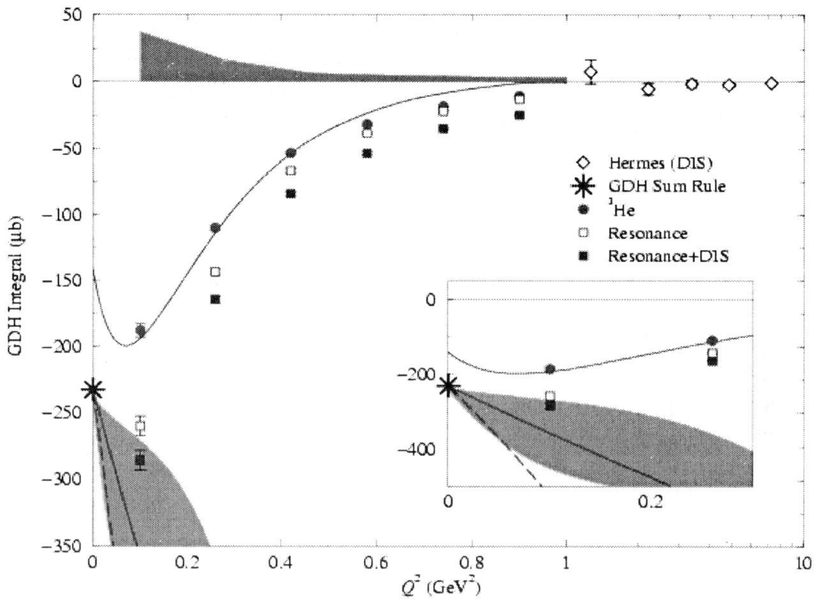

Figure 1. The GDH integral on ^{3}He (full circles) without the quasi-elastic contribution. The neutron (empty squares) is extracted using [14]. The contribution from W>2 is estimated with [15]. Added to our data, we obtain the full squares. The band above the horizontal axis gives the systematic errors. The thin line is the MAID model [8] (to be compared with the empty squares). Note the log scale for the HERMES data. The star is the original GDH sum rule ($Q^2 = 0$) and the straight thick lines are two Chiral Perturbation Theory predictions [16,17].

5 GDH at Small Q^2

The goal of E94-010 was to provide a benchmark measurement of the Q^2 evolution of GDH within the transition domain. The data also constrain the models describing the neutron structure. However, it is also interesting to gather data at lower Q^2. It will allow further check of the Chiral Perturbation calculations in a region where the validity of the chiral symmetry approximation is well established. In addition, low Q^2 data will map the turn-over at small Q^2. This will permit the extrapolation to $Q^2 = 0$ to check the validity of the original GDH sum rule on the neutron. The measurement of the turn-over will also provide a strong constraint to the models. Such an experiment [19] is scheduled to run soon in Hall A. New septum magnets will allow forward angle detection (6° and 9°), reaching a Q^2 as low as 0.02 GeV^2.

6 Summary and Conclusion

Studies of the original GDH sum rule are under way, in particular at ELSA, MAMI, JLAB and SLAC. The case of the original sum rule should be settled within the next few years. As far as the extended sum rule is concerned, high-precision data have been gathered in the transition region from the partonic to hadronic degrees of freedom of the strong interaction. These data will improve our understanding of such a transition. At lower Q^2, benchmark measurements for Chiral Perturbation calculations will be available soon. This set of data at low and intermediate Q^2 will be a reference for the lattice QCD calculations.

References

1. S. Gerasimov: Yad. Fiz. **2**, 598 (1965); S. D. Drell and A. C. Hearn: Phys. Rev. Lett. **16**, 908 (1966)
2. S. Bass: Z. Phys. **A355**, 77 (1996)
3. I. Karliner: Phys. Rev. **D7**, 9 (1973); Workman and Arndt. Phys. Rev. **D45**, 1789 (1992)
4. J. Ahrens et al.: Phys. Rev. Lett. **87**, 022003 (2001)
5. http://wwwa2.kph.uni-mainz.de/publications/
6. P. E. Bosted and D. Crabb (spokespersons): SLAC proposal E159
7. M. Anselmino, B.L. Ioffe and E. Leader, Sov. J. Nucl. Phys. **49**, 136 (1989)
8. D. Drechsel et al.: Phys. Rev. **D63**, 114010 (2001)
9. X. Ji and J. Osborn: J. Phys **G27**, 127 (2001)
10. K. Abe et al. (E143 collaboration): Phys. Rev. Lett. **78**, 815 (1997)
11. K. Ackerstaff et al. (HERMES collaboration): Phys. Lett. **B444**, 531 (1998)
12. M. Amarian et al.: Submitted to Phys. Rev. Lett.; nucl-ex/0205020
13. C. Ciofi degli Atti et al.: Phys. Rev. **C48**, R968 (1993)
14. C. Ciofi degli Atti and S. Scopetta: Phys. Let. **B404**, 223 (1997)
15. N. Bianchi and E. Thomas: Phys. Lett. **B450**, 439 (1999)
16. V. Bernard et al.: hep-ph/0203167
17. X. Ji et al.: Phys. Lett. **B472**, 1 (2000)
18. http://www.jlab.org/baryons2002/papers/devita.pdf
19. J.-P. Chen, A. Deur, and F. Garibaldi (spokespersons): JLab Proposal E97110

Few-Body Systems Suppl. 14, 355–358 (2003)

Few-Body Systems

Electron Scattering from Polarized ^{3}He: Results of Full Faddeev Calculations

J. Golak[1,2,*], W. Glöckle[1], H. Kamada[3], H. Witała[2], R. Skibiński[2], A. Nogga[4]

[1] Institut für Theoretische Physik II, Ruhr Universität Bochum, D-44780 Bochum, Germany

[2] M. Smoluchowski Institute of Physics, Jagiellonian University, PL-30059 Kraków, Poland

[3] Department of Physics, Faculty of Engineering, Kyushu Institute of Technology, 1-1 Sensuicho, Tobata, Kitakyushu 804-8550, Japan

[4] Department of Physics, University of Arizona, Tucson, Arizona 85721, USA

Abstract. We study within a Faddeev framework two processes, in which electrons (polarized or not) are scattered inelastically from polarized ^{3}He. For the helicity asymmetry A in the $^3\vec{\mathrm{He}}(\vec{e}, e')$ process results of first calculations with a three-nucleon force are compared to recent data. Next, some predictions for the analyzing power A_0 in the process $^3\vec{\mathrm{He}}(e, e'N)$ are shown. Finally we consider the spin dependent momentum distributions in polarized ^{3}He from the reaction $^3\vec{\mathrm{He}}(e, e'\vec{p})$.

1 Introduction

Three-nucleon (3N) systems are an excellent test ground for our understanding of nuclear forces. Exact nonrelativistic calculations are available both for the ground states of ^{3}He and ^{3}H and the 3N continuum. In the Faddeev framework a few modern high-precision nucleon-nucleon (NN) realistic potentials have been employed [1]. Also calculations combining different NN and 3N force models have been recently performed [2, 3]. Although further investigations in the pure 3N system are necessary to establish the final form of the 3N Hamiltonian, it is mandatory to use already now the existing formalism and study electromagnetic processes with three nucleons.

* *E-mail address:* ufgolak@cyf-kr.edu.pl

2 Results for the $^3\vec{\text{He}}(\vec{e}, e')$ Reaction

Beside the deuteron, polarized ^{3}He appears to be a useful target to obtain information on the electromagnetic neutron form factors. For this purpose the spin-dependent helicity asymmetry A is measured, since, for two different target spin orientations, it is very sensitive either to G_M^n or to G_E^n. In [4] the magnetic neutron form factor G_M^n has been extracted from the process $^3\vec{\text{He}}(\boldsymbol{e}, e')$ at q^2= 0.1 and 0.2 $(\text{GeV/c})^2$. The analysis of the data relied on precise solutions of the 3N Faddeev equations for ^{3}He and the 3N continuum, thereby using modern nuclear forces and consistent mesonic exchange currents (MEC). The resulting values for G_M^n agreed perfectly with results extracted from the cross section ratio $d(e, e'n)/d(e, e'p)$ [5].

In this contribution we demonstrate that our results are not significantly changed by inclusion of a 3NF. This effect is demonstrated in Fig. 1, where the preliminary calculations with the Urbana IX 3NF [6] (solid line) are compared to calculations used before to extract G_M^n (dashed line).

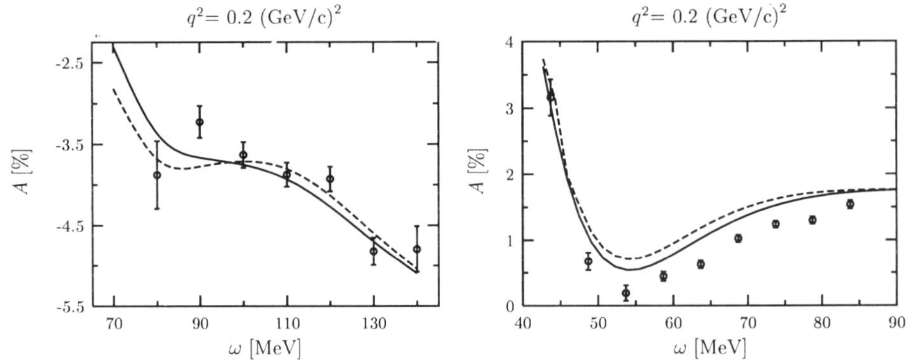

Figure 1. The data for the spin-dependent asymmetry A from [4] (left) and from [9] (right) in comparison to the results of full Faddeev calculations based on the AV18 NN force [7] alone (dashed curves) and with inclusion of the Urbana IX 3N force [6] (solid curves). The π- and ρ-like MEC's are included according to the Riska prescription [8].

In [9] the first precision measurement of the spin-dependent asymmetry in the threshold region of $^3\vec{\text{He}}(\boldsymbol{e}, e')$ at q^2= 0.1 and 0.2 $(\text{GeV/c})^2$ has been analyzed. We have found rather good agreement between the data and our nonrelativistic Faddeev calculations for q^2= 0.1 $(\text{GeV/c})^2$ but also a clear discrepancy at q^2= 0.2 $(\text{GeV/c})^2$ has been observed. As we show also in Fig. 1 that discrepancy can be partly removed by inclusion of a 3N force.

3 Predictions for the Analyzing Power A_0 in the $^3\vec{\mathrm{He}}(e,e'N)$ Process

In this section we would like to comment on another interesting quantity: the analyzing power A_0, for which unpolarized electrons are scattered off ^{3}He with a polarization axis normal to the electron scattering plane. In the final state both the outgoing electron and a nucleon N are detected. This observable, which is strictly equal to 0 in a plane wave impulse approximation and thus is ideal to study rescattering effects, has been recently measured at NIKHEF [10].

In this measurement carried out at $0.10 < q^2 < 0.23$ $(\mathrm{GeV/c})^2$ large observed values of A_0 for the neutron channel (≈ 0.4 for $q^2 = 0.2$ $(\mathrm{GeV/c})^2$) demonstrate the importance of rescattering effects in this kinematical regime. Furthermore, theoretical results of our Faddeev calculations systematically underpredict the observed A_0. Inclusion of dominant (π- and ρ-like) meson-exchange parts in the electromagnetic current operator modifies the calculations of A_0 by about 0.03, which is insufficient to remove the discrepancy. Thus we are left with an interesting situation, which cannot at the moment be explained. Relativistic effects, for example, are not supposed to remove the disagreement between the data and theory in this kinematical region.

4 Momentum Distributions of $\vec{p}\vec{d}$ Clusters in Polarized ^{3}He

Within our formalism we would also like to access the spin-dependent momentum distribution of $\vec{p}\vec{d}$ clusters in polarized ^{3}He. We have to restrict ourselves to a nonrelativistic regime, where the 3N c.m. energy of the final state is below the pion production threshold. In that kinematical regime we have calculated in Ref. [11] the longitudinal and transverse response functions W_L and W_T for different final deuteron momenta and at various three momentum transfer values. All the spins and momenta are chosen parallel or anti-parallel to the photon momentum. While in PWIA W_L and W_T up to known factors directly yield the sought spin-dependent momentum distribution, FSI and MEC's preclude in most cases the direct access to that distribution. As shown in Fig. 2, it is only at small pd relative momenta $p \leq 200$ MeV/c that the investigated momentum distribution can be accessed within our constrained kinematics. This purely theoretical outcome definitely requires an experimental verification.

5 Conclusions

Within our Faddeev framework the reaction mechanism for various processes and features of the 3N current can be studied. As examples we analyze recent experimental data on the spin dependent helicity asymmetry A for the process $^3\vec{\mathrm{He}}(\vec{e},e')$ and on the analyzing power A_0 for the reaction $^3\vec{\mathrm{He}}(e,e'N)$. We find in all considered cases important effects of the final state interactions. First results of our calculations indicate that inclusion of a 3NF might be necessary in some kinematical regions. Discrepancies observed for some observables indicate

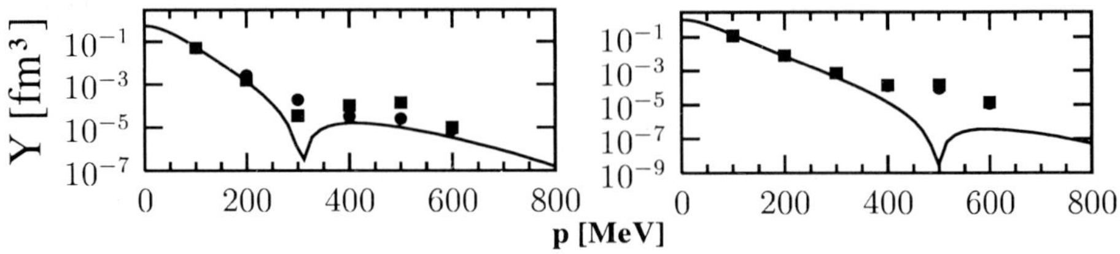

Figure 2. The spin-dependent momentum distributions Y (solid curve) as a function of the relative proton-deuteron momentum p together with the values extracted from W_L (squares) and from W_T (circles), as described in ref. [11]. In the left panel the proton and ^{3}He spins are parallel, in the right - anti-parallel.

that the present dynamical picture might be not complete. New experiments starting from low energy and momentum transfers and investigating different reaction channels and observables would be of great importance.

Acknowledgement. This work was supported by the Deutsche Forschungsgemeinschaft (J.G.), the Polish Committee for Scientific Research under Grants No. 2P03B02818 and 2P03B05622, and by NFS grant # PHY0070858. R.S. is a holder of scholarship from the Foundation for Polish Science and acknowledges its financial support. The numerical calculations have been performed on the Cray T90 of the NIC in Jülich, Germany, and on the PVP cluster of the NERSC Computing Facility, USA.

References

1. W. Glöckle et al.: Phys. Rep. **274**, 107 (1996)
2. H. Witała et al.: Phys. Rev. **C63**, 024007 (2001)
3. J. Kuroś-Żołnierczuk et al.: Phys. Rev. **C66**, 024003 (2002); Phys. Rev. **C66**, 024004 (2002)
4. W. Xu et al.: Phys. Rev. Lett. **85**, 2900 (2000)
5. H. Anklin et al.: Phys. Lett **B428**, 248 (1998)
6. B. S. Pudliner et al.: Phys. Rev. **C56**, 1720 (1997)
7. R. B. Wiringa et al.: Phys. Rev. **C51**, 38 (1995)
8. D. O. Riska: Phys. Scr. **31**, 107 (1985); **31**, 471 (1985)
9. F. Xiong et al.: Phys. Rev. Lett. **87**, 242501 (2001)
10. H. J. Bulten for the 94-05 experiment at NIKHEF: Private Communication
11. J. Golak et al.: Phys. Rev. **C66**, 024008 (2002)

Few-Body Systems Suppl. 14, 359–360 (2003)

Few-Body Systems

Measurement of Spin Dependent Total Cross Sections in Neutron-Deuteron Scattering

Z. Doležal[1,*], N.S. Borisov[2], G.M. Gurevich[1,3], M.I. Ivanov[1,2], P. Kodyš[1], E.S. Kuzmin[2], A.A. Lukhanin[1,4], I.L. Pisarev[2], Yu.A. Plis[2], A.I. Tsvetkov[1], Yu.A. Usov[2], I. Wilhelm[1]

[1] Charles University, Prague, Institute of Particle and Nuclear Physics, V Holešovičkách 2, CZ-180 00 Prague 8, The Czech Republic

[2] Dzhelepov Laboratory of Nuclear Problems, JINR, Dubna, Russia

[3] Institute for Nuclear Research, Russian Acad. of Science, Moscow, Russia

[4] Kharkov Institute of Physics and Technology, Kharkov, Ukraine

Since the beginning of nuclear physics, a central issue has been understanding the nucleon-nucleon (NN) force. Due to the progress in the development of new theoretical models as well as due to new experimental data our understanding of the NN system increased significantly during last 15 years. However, when these models were used to predict properties of three-nucleon (3N) system, they failed to provide a similarly exact picture.

Recently new successful attempts to this problem appeared. The most promising approach has been presented by the Bochum-Krakow group in their report (see [1]). Here the theoretical predictions are compared to the experimental database and an overall agreement is found. However, the database contains mostly unpolarized cross sections, because there are only very few polarization data and of insufficient quality.

The only two-spin $n-d$ experiment has been performed at TUNL [2]. The group measured spin-dependent total cross section difference $\Delta\sigma_L$ at 4 energies (5, 7, 9 and 12 MeV) and compared them to the predictions mentioned above. These comparisons yield discrepancies on the level of 50% of the measured amplitudes.

It is evident that for further development of 3N system description new experimental data are absolutely necessary. Our experimental group is well equipped for measurements of the same observable at slightly higher energy of 16 MeV. Predictions for spin-dependent cross sections $\Delta\sigma_L$ and $\Delta\sigma_T$ are 70

* *E-mail address:* Zdenek.Dolezal@mff.cuni.cz

mb and 250 mb, respectively (see [3]). These values are measurable using our setup, considerably upgraded during last years.

The cross section differences $\Delta\sigma_L$ and $\Delta\sigma_T$ will be measured using classic transmission method. The polarized neutron beam is produced as a secondary beam via the ^{3}H(d,n)^{4}He reaction. Deuteron beam from the van de Graaff electrostatic accelerator strikes a Ti-T target. The produced neutron beam is partially polarized. To achieve a monoenergetic collimated beam, the associated particle method is used with the help of time-of-flight neutron spectroscopy. Neutron beam incident on the polarized target will be monitored by two thin scintillator detectors. Behind the target two scintillator detectors detect transmitted neutrons. All detectors are equipped with the $n - \gamma$ pulse-shape discrimination circuitry.

The whole data acquisition system is based on CAMAC modules and controlled by computer via a ROOT-based computer code. A detailed description of the neutron production and detection and data acquisition can be found in [4] and [5].

The beam polarization at 16 MeV is $P_b = (13.5 \pm 1.4)\%$ and intensity is up to 10^4 s^{-1}. The neutron spin can be rotated from the original transverse direction to the longitudinal one with a special permanent magnet. This, together with the longitudinal polarizability of the target allows measurements of both $\Delta\sigma_L$ and $\Delta\sigma_T$.

The polarized target consists of a refrigerator based on dilution of ^{3}He in ^{4}He and a super-conducting magnetic system for frozen spin polarization of a sample (see [6]). During last years it has been modified to enable the deuteron polarization. Measurements in 2001-2002 showed the degree of polarization of 40%. The possibility of polarizing ^{6}LiH and ^{6}LiD has also been shown and its use for existing device is under preparation.

Acknowledgement. Part of the work has been funded by the Grant Agency of the Czech Republic under the project Nr. 202/00/0899.

References

1. W. Glöckle et al.: Phys. Rep. **274**, 107 (1996)
2. R.D. Foster et al.: Annual Report TUNL XL 2000-01, 24 (2000)
3. H. Witała et al.: Phys. Lett. **B447**, 216 (1999)
4. I. Wilhelm, M. Parsa, Z. Dolezal: Nucl. Instrum. and Meth. in Phys. Res. **A317**, 553 (1992)
5. J. Broz et al.: Z. Phys. **A354**, 401 (1996)
6. N.S. Borisov et al.: Nucl. Instrum. and Meth. in Phys. Res. **A345**, 421 (1994)

Few-Body Systems Suppl. 14, 361–366 (2003)

Few-Body Systems

Clusters and Condensates in Fermi Systems

M. Beyer*

Fachbereich Physik, Universität Rostock, 18051 Rostock, Germany

Abstract. Superconductivity, superfluidity, condensation, cluster formation, etc. are phenomena that might occur in many-particle systems. These are due to residual interactions between the particles. To explain these phenomena consistently in a microscopic approach, at some point, one needs to solve few-body equations that are modified because of the Pauli principle and the interactions of the many particles around.

1 Introduction

Correlations in many-particle systems are responsible for a number of interesting and exciting properties of the system. In an infinite, equilibrated system the presentation of these properties (or phases of matter) are usually organized in a phase diagram. This is a plot of the dominant state of matter separated by "critical lines" depending on the two basic variational parameters of quantum statistics relevant for a grand canonical ensemble, i.e. the temperature T and the chemical potential μ. Instead of the chemical potential μ that is difficult to access experimentally one might by use of proper thermodynamic relations chose the particle density $n(\mu, T)$. Note, however, that the relation between μ and $n(\mu, T)$ is not simple and depends on the specific system considered.

The phase diagram becomes particularly rich when the chemical potential μ is in the order of MeV to GeV. This is at rather large particle densities, where nuclei are close together. This is the domain where Quantum Chromodynamics (QCD) dominates the dynamics of the system. Since nucleons are made of quarks it is obvious that the relevant degrees of freedom to describe the system might change from nucleons (nuclear matter) to quarks (quark matter) at some point. In addition gluons might appear as additional degrees of freedom that are otherwise absent at lower densities. Because of the free (color) charge this phase is called a (quark gluon) plasma. Indeed, lattice QCD and model calculations suggest a variety of phenomena in the QCD dominated phase of matter. These are superfluidity, critical points, first and second order phase transitions, color

* *E-mail address:* michael.beyer@physik.uni-rostock.de

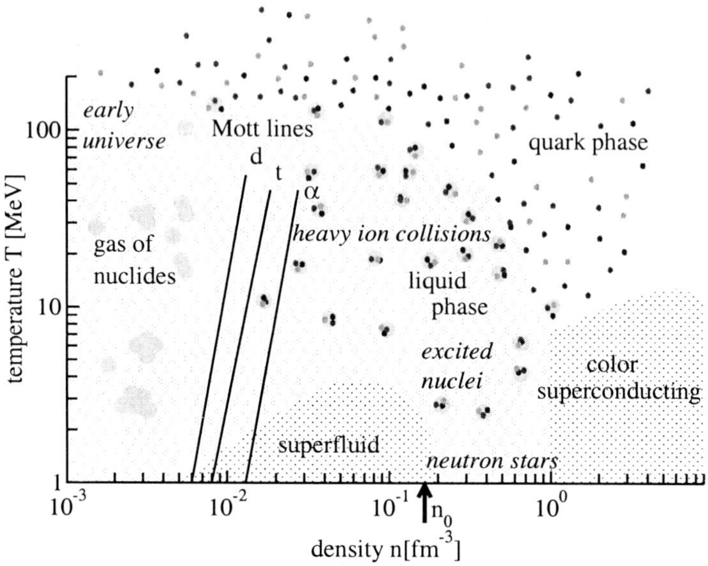

Figure 1. Schematic view of the phase diagram of nuclear matter. The phase diagram is empirical accessible by heavy ion collisions, excited nuclei, observation of neutron stars and the early universe as indicated in the diagram. New plans at GSI aim at exploring the color superconducting phase as well.

super conductivity, plasma phase, among others. A sketch of the phase diagram is given in Fig. 1. To explain all these different states of matter one needs a treatment that goes beyond the simple quasiparticle picture.

To tackle the complicated many-particle system the Green function approach is a good starting point [1]. Equations for Green functions are hierarchally coupled and hence an approximate solution of the problem is unavoidable. If we are interested in two-body correlations only, the three-body Green functions are approximated. Quite a few phenomena are already accessible at this level. However, there are good reasons to go beyond two-particle correlations, e.g.:

- Particle production even in a dense environment such as deuteron formation in a heavy ion reaction, need a third particle to conserve energy-momentum [2].
- To study the properties of α-particles [3] or determine the critical temperature of a possible α-particle condensate [4, 5, 6] needs an in-medium four-body equation.
- Recent results in the Hubbard model indicate, that three-particle contributions may lead to a different (lower) critical temperature compared to the simple Thouless criterion [7]. Question of this type have not been addressed for nuclear matter.

- The chiral phase transition is often discussed along with a confinement-deconfinement transition based on investigating mesons (quark-antiquark states). Does this transition happen for nucleons (three-quark states) at the same density/temperature?

Some of these issues related to the nuclear matter phase are addressed in the present paper. The ones related to quark matter are given by Mattiello et al [8].

2 In-Medium Few-Body Equations

To arrive at suitable few-body equations that describe clusters in medium and provide the critical temperatures and densities where condensation occurs we utilize the Dyson equation approach, see e.g. ref. [9]. This way it is possible to decouple the Green function hierarchy and write Green functions in the n-body space, where $n = 2, 3, 4, \ldots$ is the number of particles in the considered cluster.

In Hartree-Fock approximation the solution of the one-particle problem leads to the following quasi-particle energy

$$\varepsilon_1 = \frac{k_1^2}{2m_1} + \sum_2 V_2(12, \widetilde{12}) f_2 \simeq \frac{k_1^2}{2m_1^{\rm eff}} + \Sigma^{\rm HF}(0). \tag{1}$$

The last equation introduces the effective mass that is a valid concept for the rather low densities considered here and $\mu^{\rm eff} \equiv \mu - \Sigma^{\rm HF}(0)$. The Fermi function $f_i \equiv f(\varepsilon_i)$ for the i-th particle is given by

$$f(\varepsilon_i) = \frac{1}{e^{\beta(\varepsilon_i - \mu)} + 1}. \tag{2}$$

The resolvent G_0 for n noninteracting quasiparticles is

$$G_0(z) = (z - H_0)^{-1} N \equiv R_0(z) N, \qquad H_0 = \sum_{i=1}^{n} \varepsilon_i\,, \tag{3}$$

where G_0, H_0, and N are formally matrices in n particle space. The Matsubara frequency z_λ has been analytically continued into the complex plane, $z_\lambda \to z$ [1]. The Pauli-blocking for n-particles is

$$N = \bar{f}_1 \bar{f}_2 \ldots \bar{f}_n \pm f_1 f_2 \ldots f_n, \qquad \bar{f} = 1 - f\,, \tag{4}$$

where the upper sign is for Fermi-type and the lower for Bose type clusters. The full resolvent $G(z)$ is given by

$$G(z) = (z - H_0 - V)^{-1} N, \qquad V \equiv \sum_{\text{pairs } \alpha} N_2^\alpha V_2^\alpha. \tag{5}$$

Note that $V^\dagger \neq V$. For the two-body case as well as for a two-body subsystem embedded in the n-body cluster the standard definition of the t matrix leads to the Feynman-Galitskii equation for finite temperature and densities [1],

$$T_2^\alpha(z) = V_2^\alpha + V_2^\alpha N_2^\alpha R_0(z) T_2^\alpha(z). \tag{6}$$

Introducing the Alt Grassberger Sandhas (AGS) transition operator $U_{\alpha\beta}(z)$ the effective inhomogeneous in-medium AGS equation reads

$$U_{\alpha\beta}(z) = (1 - \delta_{\alpha\beta}) R_0^{-1}(z) + \sum_{\gamma \neq \alpha} N_2^\gamma T_2^\gamma(z) R_0(z) U_{\gamma\beta}(z). \tag{7}$$

The homogeneous in-medium AGS equation uses the form factors defined by

$$|F_\beta\rangle \equiv \sum_\gamma \bar{\delta}_{\beta\gamma} N_2^\gamma V_2^\gamma |\psi_{B_3}\rangle \tag{8}$$

to calculate the bound state ψ_{B_3}

$$|F_\alpha\rangle \quad = \quad \sum_\beta \bar{\delta}_{\alpha\beta} N_2^\beta T_2^\beta(B_3) R_0(B_3) |F_\beta\rangle. \tag{9}$$

Finally, the four-body bound state is described by

$$|\mathcal{F}_\beta^\sigma\rangle = \sum_{\tau\gamma} \bar{\delta}_{\sigma\tau} U_{\beta\gamma}^\tau(B_4) R_0(B_4) N_2^\gamma T_2^\gamma(B_4) R_0(B_4) |\mathcal{F}_\gamma^\tau\rangle, \tag{10}$$

where $\alpha \subset \sigma, \gamma \subset \tau$ and σ, τ denote the four-body partitions. The two-body input is given in (6) and the three-body input by (7). The numerical solution requires some mild approximations that are however well understood in the context of the isolated few-body problem. The above equations, although written in a way familiar to the isolated few-body case, systematically includes all the dominant medium effects. These are the self energy corrections (hence changes in binding energies) and the Pauli blocking factors that changes the effective potential acting between the quasi particles.

3 Discussion of Results

Pieces of nuclear matter can be explored by heavy ion collisions. If the energy is high enough a transition to quark-gluon matter is expected. Presently we focus on intermediate to low scattering energies. We consider the collision of ^{129}Xe and ^{119}Sn at 50 MeV/A by the INDRA collaboration [10]. The microscopic simulation of the heavy ion collision is done by solving a generalized Boltzmann equation (Boltzmann Uehling Uhlenbeck (BUU) simulation) [2]. The reaction rates appearing in the collision integrals are *a priori* medium dependent. The input into the collision integrals is taken from a solution of the AGS equations (7). For details on the specific interaction model see ref. [11]. We investigate the influence of medium dependent rates in the BUU simulation of the heavy ion collision as compared to the use of isolated rates. Figure 2 shows that the net effect (gain-minus-loss) of deuteron production becomes larger for by using in-medium rates (solid) compared to the isolated rates (dashed). The change is significant, however, a comparison with experimental data is difficult since deuterons may also be evaporating from larger clusters that has not been taken into account in the present calculation so far. The ratio of protons to deuterons

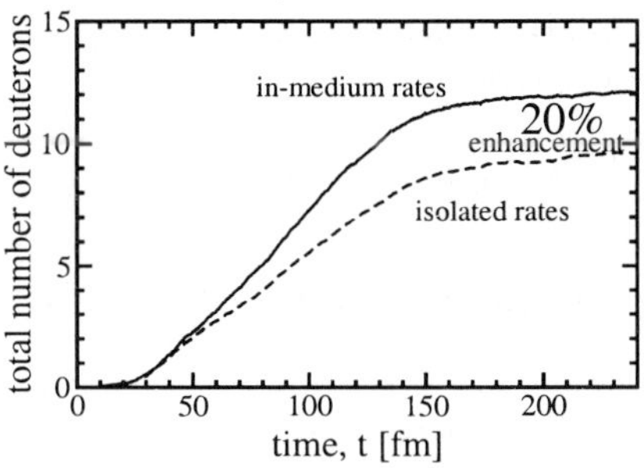

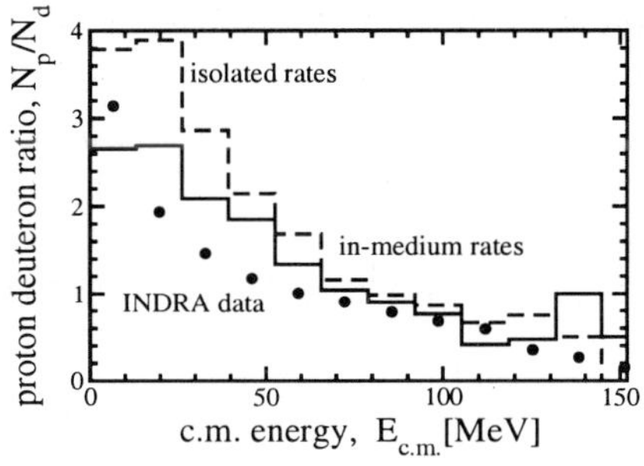

Figure 2. Time evolution of deuteron formation during the collision of ^{129}Xe+^{119}Sn at 50 MeV/A. Using in-medium rates lead to a 20% enhancement.

Figure 3. Ratio of proton to deuteron numbers as a function of c.m. energy. The experimental data are from the INDRA collaboration. In-medium results are closer to experiments.

may be better suited for a comparison to experiments that is shown in Figure 3. The use of in-medium rates (solid) lead to a shape closer to the experimental data (dots) than the use of isolated rates (dashed).

Previously, we have shown that for infinite nuclear matter the use of in-medium rates leads to significant changes, i.e. faster time scales as has been shown in detail within linear response theory in refs. [12, 13].

As does the nucleon, see eq. (1), the cluster changes its mass. The binding energy changes as well until the Pauli blocking is too strong for bound states to exist (Mott effect). This effect depends on the momentum of the cluster. This important effect for modeling of heavy ion collisions is shown in Figure 4.

In Figure 5 part of the phase diagram of nuclear matter is shown. The condition for the onset of superfluidity for α-particles is $B(T_c, \mu_c, P = 0) = 4\mu_c$,

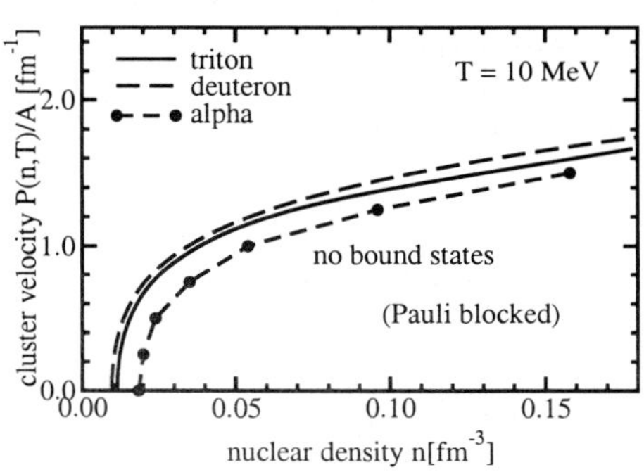

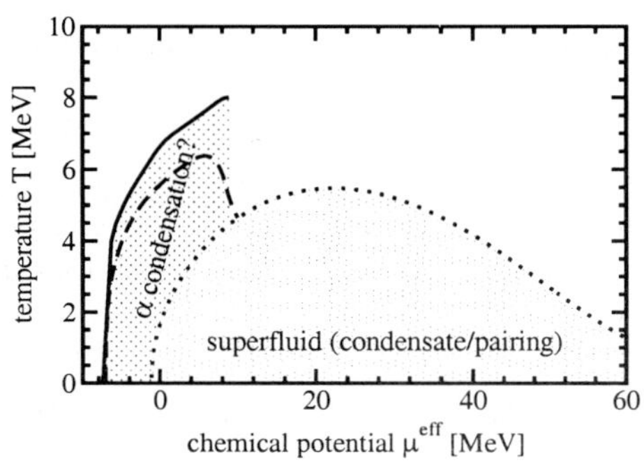

Figure 4. Due to the Mott effect clusters exist only above the respective curves. Below no bound states exist because of Pauli blocking.

Figure 5. Part of the phase diagram with lines of critical temperatures. Solid line four-body AGS type calculation. Others earlier calculation [4]

where B is the binding energy. The critical temperature found by solving the homogeneous AGS equation for $\mu < 0$ confirms the onset of α condensation even at higher values (solid line) as given earlier (dashed, from [4]) which was based on a variational calculation using the 2+2 component of the α particle. For $\mu > 0$ the condition $E = 4\mu$ for the phase transition can also be fulfilled. However, the significance for a possible quartetting needs further investigation.

To conclude, strongly correlated Fermi systems such as nuclear/quark matter provide an excellent field for few-body techniques. Experiments with light fragments (up to the α-particle) in the final state are particularly suited to test the in-medium few-body equations.

Acknowledgement. Work supported by the Deutsche Forschungsgemeinschaft.

References

1. L.P. Kadanoff, G. Baym: *Quantum statistical mechanics.* New York: Benjamin 1962; A.L. Fetter, J.D. Walecka: *Quantum Theory of Many-Particle Systems*, New York: McGraw-Hill 1971
2. C. Kuhrts, M. Beyer, P. Danielewicz and G. Röpke: Phys. Rev. **C63**, 034605 (2001) and references therein
3. M. Beyer, S.A. Sofianos, C. Kuhrts, G. Röpke and P. Schuck: Phys. Lett. **B488**, 247 (2000)
4. G. Röpke, A. Schnell, P. Schuck, P. Nozieres: Phys. Rev. Lett. **80**, 3177 (1998)
5. A. Tohsaki, H. Horiuchi, P. Schuck and G. Röpke: Phys. Rev. Lett. **87**, 192501 (2001)
6. Y. Suzuki and M. Takahashi: Phys. Rev. **C65**, 064318 (2002)
7. M. Himmerich and M. Letz: Phys. Rev. **B64**, 144519 (2001)
8. S. Mattiello, M. Beyer, T. Frederico, H.J. Weber: These Proceedings
9. J. Dukelsky, G. Röpke, and P. Schuck: Nucl. Phys. **A628**, 17 (1998)
10. D. Gorio et al. (INDRA Collaboration): Eur. Phys. J. **A7**, 245 (2000)
11. M. Beyer, G. Röpke and A. Sedrakian: Phys. Lett. **B376**, 7 (1996)
12. M. Beyer and G. Röpke: Phys. Rev. **C56**, 2636 (1997)
13. C. Kuhrts, M. Beyer and G. Röpke: Nucl. Phys. **A668**, 137 (2000)

Few-Body Systems Suppl. 14, 367–372 (2003)

Few-Body Systems

Measuring NN Correlations with ^{3}He(e,e′pp)n

L.B. Weinstein*, R. Niyazov, for the CLAS Collaboration

Physics Dept., Old Dominion University, Norfolk, VA 23529, USA

Abstract. Measuring two nucleon knockout reactions from light nuclei can test our knowledge of nuclear bound state and continuum wave functions as well as our models of one- and two-nucleon currents. We have used the Jefferson Lab CLAS, a 4π spectrometer, to measure kinematically complete $3N$ breakup of ^{3}He with 2.2 and 4.4 GeV electrons. When one nucleon carries away most of the kinetic energy, there is a large peak at 180° in the opening angle of the other two nucleons, indicating the presence of NN correlations. These correlations, unlike those measured in other reactions, appear to be unaffected by two nucleon currents such as MEC or IC.

1 Introduction

Previous measurements of proton knockout from nuclei, such as O$(e,e'p)$ [1], have measured the momentum distribution for valence knockout and the momentum and energy distribution for more deeply bound nucleons. Valence knockout is well described by single nucleon knockout calculations that include the effects of final state interactions. However, the fact that a) we only see about 70% of the 'expected' number of nucleons and b) there is a large cross section at large excitation energies of the residual nucleus indicates that multi-nucleon processes play a significant role. While there have been many $(e,e'p)$ measurements from nuclei, there have been very few $(e,e'pp)$ measurements.

There are two general sources of two-nucleon knockout from nuclei. One source is the interaction of the virtual photon with two nucleons, typically through meson exchange currents or isobar configurations. This is referred to as a two-body current. The second one is the interaction of the virtual photon with one nucleon of a 'correlated pair'. (Note that the distinction between correlations [in the wave function] and currents [in the operator] is blurred by the fact that unitary transformations can transform one into the other.)

A correlated pair is typically a pair of nucleons that have large relative momentum (because they are at short range) and small total momentum. Thus,

* *E-mail address:* weinstei@physics.odu.edu

one signature of correlations is finding two nucleons with large relative momentum and small total momentum in the initial state. Unfortunately, the effects of NN correlations are frequently obscured by the effects of two body currents [2]. In order to disentangle these competing effects, a series of comprehensive measurements are needed.

The seven experiments took data simultaneously in Spring 1999, measuring approximately 500 million $A(e,e'X)$events, using the Jefferson Lab Large Acceptance Spectrometer (CLAS), a 4π magnetic spectrometer, with 1.1, 2.2 and 4.4 GeV polarized electrons incident on targets from ^{3}He to ^{56}Fe.

2 Measuring ^{3}He(e,e′pp)n

We studied ^{3}He(e,e′pp)n by measuring electron induced two proton knockout reactions from ^{3}He using the CLAS detector and cutting on the missing mass. Figures 1a and b show the electron acceptance and undetected neutron missing mass resolution for $E_{beam} = 2.2$ GeV. The threshold of the CLAS is approximately 250 MeV/c for protons.

Note that all data shown here are **preliminary**.

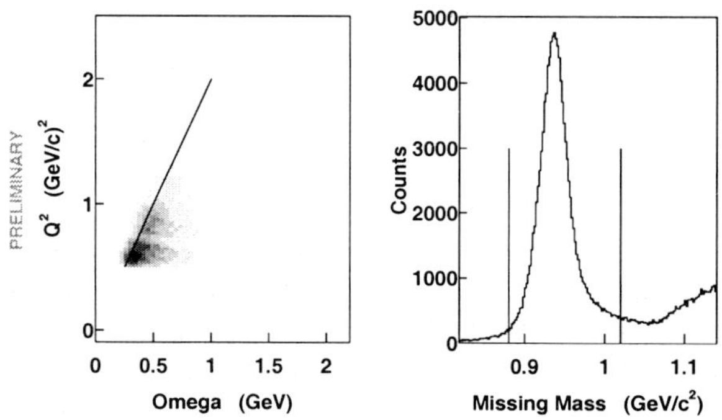

Figure 1. a) Q^2 vs ω for ^{3}He(e,e′pp)n at $E_{beam} = 2.2$ GeV. Note the huge kinematic acceptance. b) Missing mass for ^{3}He(e,e′pp). We cut at the indicated lines to select (e,e′pp)n events.

Because this is the first time that ^{3}He(e,e′pp)n has been measured using an almost 4π detector, our data analysis philosophy is to follow and understand the dominant features of the data.

In order to understand the energy sharing in the reaction, we plotted the kinetic energy divided by the energy transfer of the first proton (T_{p1}/ω) versus that of the second proton (T_{p2}/ω) for each event (a lab-frame Dalitz plot). See Fig. 2. The threshold for proton detection in CLAS is $p_p \geq 250$ MeV/c and we set the same threshold for neutrons $p_n \geq 250$ MeV/c.

We see three peaks at the three corners of the plot, corresponding to events where two 'fast' nucleons each have less than 20% of the energy transfer and the third 'leading' nucleon has the remainder. We call the two nucleons 'fast' because $p \gg p_{fermi}$. These peaks are much more pronounced at $E_{beam} = 4.4$ GeV (not shown). We cut on these peaks where the two fast nucleons each have less than 20% of the energy transfer.

Then we looked at the opening angle of the two fast nucleons. Figure 3-a

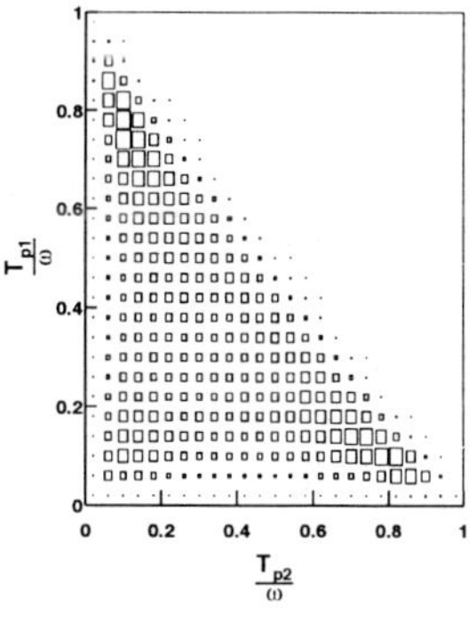

Figure 2. Nucleon kinetic energy distribution for 2.2 GeV ^{3}He(e,e′pp)n. The kinetic energy of proton 1 divided by ω (T_{p1}/ω) is plotted versus the same for proton 2. (Note $p_p, p_n \geq 250$ MeV/c.) Note the peaks in the corners.

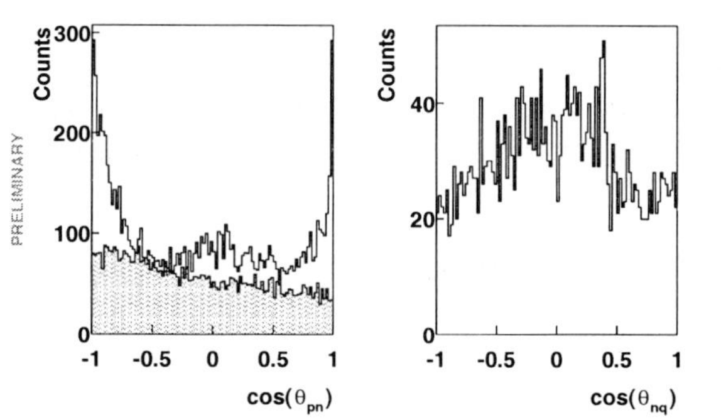

Figure 3. a) Opening angle of the fast *pn* pairs for events in the upper left and lower right corners of Fig. 2-b. The histogram shows the data, the filled histogram shows the fire ball phase space simulation (with arbitrary normalization). b) The angle between the neutron in the fast *pn* pair and $\boldsymbol{q}$ where $p_\perp < 300$ MeV/c.

shows the pair opening angle for fast *pn* pairs with a leading proton. Note the large peak at 180 degrees ($\cos\theta_{NN} \approx -1$). The distribution for fast *pp* pairs with a leading proton is identical. The peak is not due to the cuts, since we do not see it in a fire ball phase space simulation assuming three body absorption of the virtual photon and phase space decay. It is also not due to the CLAS acceptance since we see it both for leading protons (which we detect) and leading neutrons (which we infer from missing mass). This back-to-back peak is a strong indication of correlated NN pairs.

3 Studying Correlated Pairs

Now consider these presumably correlated pairs. Since we believe that we have observed events where the leading nucleon absorbed the virtual photon and the two fast nucleons are emitted back to back, we cut on the perpendicular momentum of the leading nucleon to deemphasize rescattering ($p_\perp < 300$ MeV/c). This cut selects the back-to-back events very cleanly. Unfortunately, there are only 3400 fast *pn* and 1100 fast *pp* events remaining in the entire 2.2 GeV data set (and ten times fewer at 4.4 GeV).

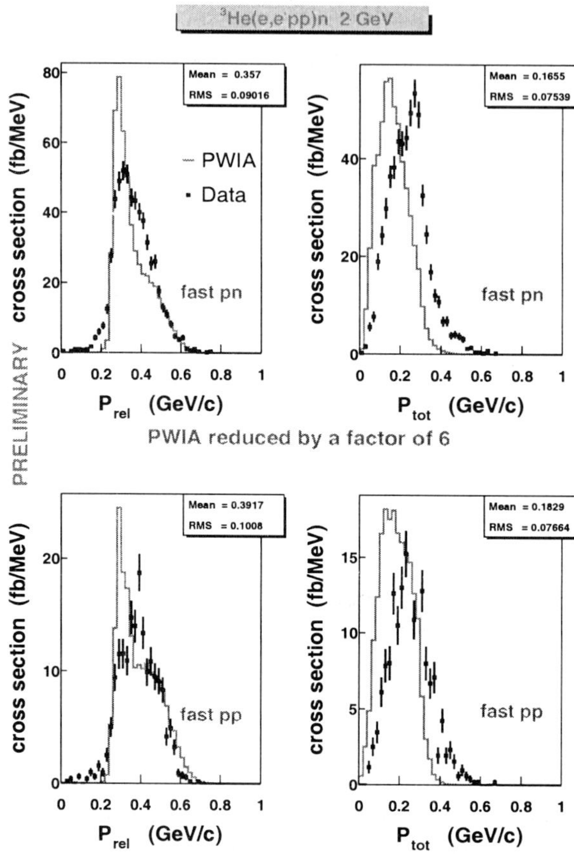

Figure 4. Comparison of data and PWIA cross sections for $^3\mathrm{He}(e, e'pp)n$ events at $E_{beam} = 2.2$ GeV. a) leading proton and a fast pn pair: pair relative momentum p_{rel}; b) leading proton and a fast pn pair: total momentum p_{tot}; c) leading neutron and a fast pp pair: p_{rel}; d) leading neutron and a fast pp pair: p_{tot}. PWIA cross sections are decreased by a factor of six.

If the fast back-to-back NN pairs are really uninvolved in the photon absorption, then they should be distributed isotropically (the angular distribution of the neutrons with respect to $\boldsymbol{q}$ is shown in Fig. 3-b). Further evidence that the fast NN pair is a spectator comes from the average momentum of the pair along $\boldsymbol{q}$. This is about 0.07 GeV/c for $E_{beam} = 2.2$ GeV and about 0.1 GeV/c for $E_{beam} = 4.4$ GeV, much less than the average momentum transfers of $Q^2 = 0.7$ and 1.4 $(\mathrm{GeV/c})^2$ respectively.

The fast NN pair relative ($p_{rel} = \frac{1}{2}|\boldsymbol{p}_1 - \boldsymbol{p}_2|$) and total ($p_{total} = |\boldsymbol{p}_1 + \boldsymbol{p}_2|$) momentum distributions are shown in Fig. 4 and for fast pn pairs and fast pp pairs at $E_{beam} = 2.2$ GeV. Note how similar the pp and pn distributions are. The 4.4 GeV distributions (not shown) are similar.

Thus, because when we select a quasifree leading nucleon, the fast NN pairs are back to back, isotropic and have small average momentum along $\boldsymbol{q}$, we conclude that the fast NN pair is a spectator. Because we measure similar p_{total} and p_{rel} distributions for pp and pn pairs and $0.5 < Q^2 < 1$ ($E_{beam} = 2.2$ GeV) and $1 < Q^2 < 2$ $(\mathrm{GeV/c})^2$ ($E_{beam} = 4.4$ GeV), we conclude that we have measured bound state NN correlations.

We appear to have measured NN correlations in ^{3}He by striking the **third** nucleon and detecting the correlated pair. This is similar to other proposed correlation searches where you strike one nucleon of a correlated pair and detect the other nucleon leaving the nucleus. However, these other searches suffer from the weakness that their proposed signal can also be due to two body currents (e.g. photon absorption on an exchanged meson).

4 Comparison to Theory

Calculations by W. Glöckle [3] at lower energy strengthen this conclusion. He calculated the ^{3}He(e,e′pp)n cross section where the leading nucleon has momentum $\boldsymbol{p}_N = \boldsymbol{q}$ and the other two nucleons have total momentum $p_{total} = 0$ for various values of the momentum transfer, $400 \leq |\boldsymbol{q}| \leq 600$ MeV/c, and relative momentum. He found that MEC did not contribute, rescattering of the leading nucleon did not contribute, and the continuum state interaction of the outgoing NN pair decreased the cross section by a factor of approximately 10 relative to the PWIA result. Thus, he found that this reaction is a very clean way to measure the overlap integral between the NN continuum state and the same two nucleons in the bound state.

C. Ciofi degli Atti and L. Kaptari also found that the continuum interaction of the outgoing pair significantly decreased the cross section [4].

We compared our results to three other calculations, 1) a Plane Wave Impulse Approximation (PWIA) calculation by M. Sargsian [5] using Glöckle's bound state wave function with no final state interactions, 2) a calculation by J.-M. Laget [6] using a Faddeev wave function from P. Sauer and including one-, two-, and three- body mechanisms as well as rescattering terms, and 3) a home-made model of pion production on the struck proton followed by pion absorption on the remaining pn pair. We averaged all of the models over the CLAS acceptances and cuts using a monte carlo.

The pion production and rescattering model convoluted pion production cross sections [7], pion absorption on deuterium cross sections [8], and proton momentum distributions from ^{3}He(e,e′p) measurements [9]. This model predicted too large an average energy transfer (since it is Δ dominated) and too large a relative momentum. In addition, the ratio of the number of fast pn pairs to fast pp pairs was much lower than the data (1 instead of 3). Thus, while this mechanism might be very important for other three nucleon knockout experiments, it does not explain this data.

Preliminary calculations from Laget describe the kinetic energy, p_{rel} and p_{total} distributions very well for pn pairs and not quite as well for pp pairs. They indicate that one body knockout plus pair distortion is dominant for the fast pn pair but that three-body mechanisms are also significant for the pp pair. However, the virtual photon distribution for the data is much closer to the calculated one-body distribution than to the three-body distribution.

The PWIA calculation of Sargsian has Q^2 vs ω, NN pair opening angle, p_{rel}, and p_{total} distributions that are consistent with the data (see Fig. 4).

It is a factor of 6 larger than the data which is consistent with the expected effects of the NN continuum state interaction calculated by Glöckle. However, it predicts 5 pn pairs for each pp pair versus 3 in the data and it predicts a ratio of 4 for $\sigma(E_{beam} = 2.2)/\sigma(E_{beam} = 4.4)$ versus 11 for the data.

5 Summary

We have studied the ^{3}He(e,e′pp)n reaction, selecting events where one nucleon has most of the kinetic energy and has less than 300 MeV/c of momentum perpendicular to $\boldsymbol{q}$. When we do this, we see isotropic, back-to-back, fast NN pairs with small average momentum along $\boldsymbol{q}$. We have measured the total and relative momentum distributions of these pairs and found that they do not depend significantly on isospin (pp vs pn pairs) or on momentum transfer.

PWIA calculations reproduce many features of the data. Calculations by Glöckle at lower energy indicate that the cross section depends primarily on the overlap integral between the continuum state and bound state of the NN pair. Neither meson exchange currents nor the final state rescattering of the leading nucleon appear to contribute to the cross section. Preliminary calculations by Laget confirm these results, at least for fast pn pairs.

Thus, by measuring ^{3}He(e,e′pp)n, we appear to have directly measured NN correlations without any significant contamination from other processes by striking the **third** nucleon and detecting the spectator correlated pair.

Acknowledgement. I thank Claudio Ciofi degli Atti, Walter Glöckle, Jean-Marc Laget, and Misak Sargsian for their calculations and physics insight. This work was supported by a grant from the US Department of Energy.

References

1. J. Gao et al.: Phys. Rev. Lett. **84**, 3265 (2000)
2. S. Janssen et al.: Nucl. Phys. **A672**, 285 (2000)
3. W. Glöckle, In: Proceedings of the "*5th Workshop on Electromagnetically Induced Two-Hadron Emission*", Lund, Sweden, 2001
4. C. Ciofi degli Atti, L.P. Kaptari: Phys. Rev. **C66**, 044004 (2002)
5. M. Sargsian: Private Communication
6. G. Audit et al.: Nucl. Phys. **A614**, 461 (1997);
   J.M. Laget: J. Phys. G **14**, 1445 (1988)
7. D. Drechsel et al.: Nucl. Phys. **A645**, 145 (1999)
8. C.H. Oh et al.: Phys. Rev. **C56**, 635 (1997)
9. E. Jans et al.: Nucl. Phys. **A475**, 687 (1987)

Few-Body Systems Suppl. 14, 373–376 (2003)

Few-Body Systems

Two-Body Correlations and the Structure of Bose-Einstein Condensates

O. Sørensen, D.V. Fedorov, A.S. Jensen

Department of Physics and Astronomy, University of Aarhus, DK-8000 Aarhus C, Denmark

Abstract. We investigate systems of identical bosons with the focus on two-body correlations. We use a hyperspherical adiabatic method and apply a Faddeev type of decomposition of the wave function. We discuss the qualitative structure of a condensate in the regime of large scattering length a_s.

1 Short Introduction to the Method

We describe N identical, trapped bosons of masses m located at positions $\boldsymbol{r}_i$, $i = 1, \ldots, N$, by first removing dependence on the center of mass coordinates $\boldsymbol{R}_{\text{c.m.}} = \sum_i \boldsymbol{r}_i/N$ and then parameterizing remaining relative degrees of freedom by hyperspherical coordinates. The essential coordinate is the hyperradius ρ which is related to the particle coordinates by

$$\rho^2 = \frac{1}{N}\sum_{i<j} r_{ij}^2 = \sum_i r_i^2 - NR_{\text{c.m.}}^2 \,, \qquad r_{ij} = |\boldsymbol{r}_j - \boldsymbol{r}_i| \,. \tag{1}$$

The hyperradius ρ is a measure of the size of the system. The details of the method can be found in [1]. In short, we make a hyperspherical adiabatic expansion of the wave function. We assume a Faddeev-like decomposition of the angular degrees of freedom, where only s-wave dependence is maintained. The regime of short-range interactions makes our model especially efficient. The essential equation is the familiar-looking hyperradial equation:

$$\left(-\frac{\hbar^2}{2m}\frac{d^2}{d\rho^2} + U(\rho)\right) f(\rho) = E f(\rho) \,. \tag{2}$$

Here $f(\rho)$ is a radial wave function that contains information about average properties of the system; e.g. the rms distance between particles is given by $\bar{r^2} = 2\langle f|\rho^2|f\rangle/(N-1)$. The radial potential $U(\rho)$ is a sum of three terms:

$$U(\rho) = \frac{\hbar^2}{2m}\left(\frac{\lambda(\rho)}{\rho^2} + \frac{\rho^2}{b_{\text{t}}^4} + \frac{(3N-6)(3N-4)}{4\rho^2}\right) , \tag{3}$$

where $b_t = \sqrt{\hbar/(m\omega)}$ is the length scale of an external harmonic trap of frequency $\nu = \omega/(2\pi)$ holding the particles together. The first term λ/ρ^2 is an adiabatic potential due to the two-body interactions $V(r_{ij})$. The third term is a kinetic energy centrifugal barrier.

2 Qualitative Results

Previous investigations have shown the qualitative behaviour of the adiabatic potential for large hyperradius. In the absence of bound two-body states (negative two-body s-wave scattering length a_s) the angular eigenvalue approaches zero like a_s/ρ as the hyperradius diverges (very dilute gas) [2, 3]. If the two-body attraction holds a two-body bound state ($a_s > 0$) the angular eigenvalue diverges proportional to ρ^2 [2]. On the threshold for a two-body bound state ($|a_s| \to \infty$) λ approaches a constant value [2].

In the opposite limit ($\rho \to 0$) λ vanishes proportional to ρ^2, so λ/ρ^2 can be neglected in comparison with the ρ^{-2} centrifugal term in the radial potential.

With these qualitative results as asymptotic values we model the adiabatic potential in the case of unbound two-body systems ($a_s < 0$).

$$\lambda(\rho) \to \begin{cases} mV(0)N^2\rho^2/\hbar^2 & \text{when } \rho \ll b \,, \\ -5N^2 & \text{when } b \ll \rho \ll N^{3/2}|a_s| \,, \quad [2] \,, \\ 3/2\sqrt{3/\pi}N^{7/2}a_s/\rho & \text{when } N^{3/2}|a_s| \ll \rho \,, \quad [3] \,, \end{cases} \tag{4}$$

where b is the range of the two-body interactions. For the radial potential we choose $b_t/b = 1442$ and only vary N and a_s.

Figure 1 shows the radial potential as a function of the hyperradius with the model of equation (4). In (a)-(d) $N = 6000$ and only the scattering length a_s varies. In (a) $a_s = 0$ and there is effectively no interaction ($\lambda = 0$), which gives a simple radial potential with only one extremum. In (b) $a_s = -0.05b$ and the attraction between the bosons is sufficiently strong to create a second minimum in the radial potential at smaller hyperradius. A barrier, due to the centrifugal term in the radial potential, separates the minima at small and large hyperradii. For increased attraction in (c) the barrier has just vanished, and only the minimum at small hyperradius with the minimum value at negative energy exists. Fixing the value of $a_s = -0.35b$ and decreasing the number of particles from $N = 6000$, the sequence is (d)-(f). In (e), $N = 3000$, and a barrier is about to occur, and in (f), $N = 500$, an intermediate barrier between a minimum at small and a minimum at large hyperradius is again present.

The combination $N|a_s|/b_t$ has been shown theoretically [3] and experimentally [4] to be limited to values smaller than approximately 0.5 for stable condensates. This value is in (c) 0.75, just when the barrier vanishes. However, the critical point for collapse should be reached before (smaller $|a_s|$), since macroscopic tunneling through a barrier may occur. Other adiabatic channels, apart from the lowest, may also play an important role in the dynamics.

In recent experiments with a system of Rubidium-85 atoms performed by Donley *et al.* [5] it was possible to tune the effective scattering length to almost

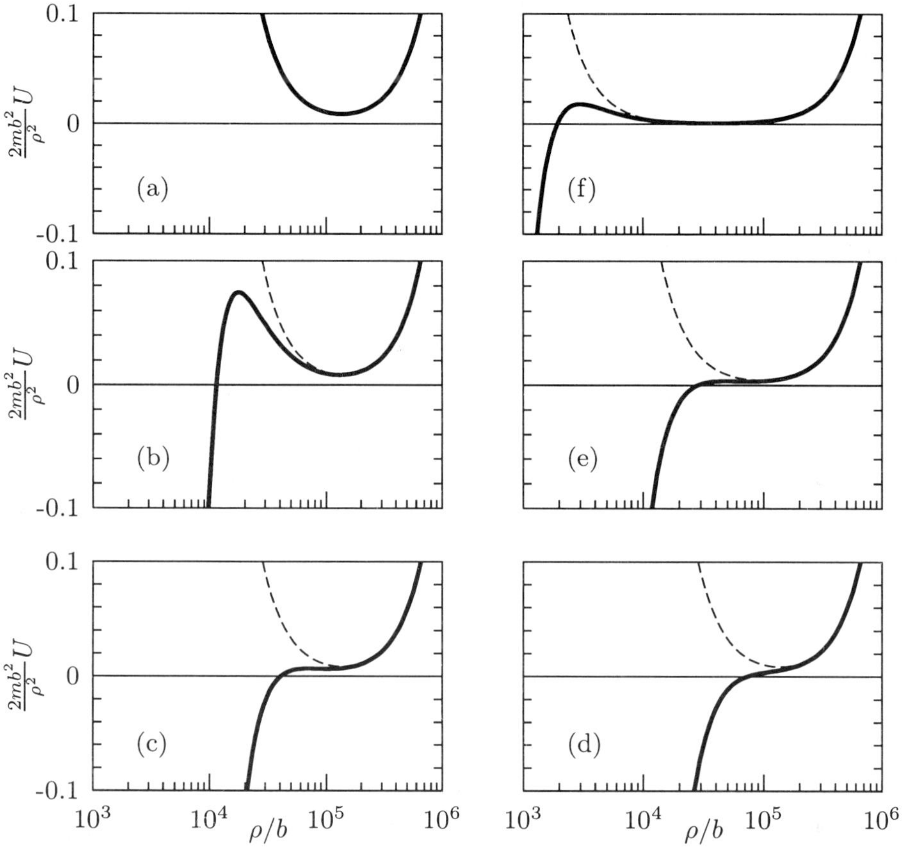

Figure 1. Radial potentials with $b_t/b = 1442$ and (a) $N = 6000$, $a_s \sim 0$; (b) $N = 6000$, $a_s/b = -0.05$; (c) $N = 6000$, $a_s/b = -0.18$; (d) $N = 6000$, $a_s/b = -0.35$; (e) $N = 3000$, $a_s/b = -0.35$; (f) $N = 500$, $a_s/b = -0.35$. The dashed lines are with $a_s = 0$. The divergence $U(\rho) \to +\infty$ when $\rho \to 0$ is not shown.

any value by the use of a Feshbach resonance. A condensate of N_0 atoms was first created with effectively zero interactions. Figure 1(a) shows the radial potential in this case. The condensate will form similar to the ground state in this potential. Then the scattering length was tuned to a large and negative value (sudden change from (a) to (d)) for some time τ_{evolve} before switching back. This initiates a contraction of the initial condensate. After a period of time smaller values of ρ will therefore be available.

The two-body recombination rate grows rapidly with the density [6]. Since smaller hyperradius means larger density, the contracted condensate has a much larger recombination rate than the initial condensate. This leads to rapid loss of particle pairs from the condensate. If, however, after changing (a) to (d)

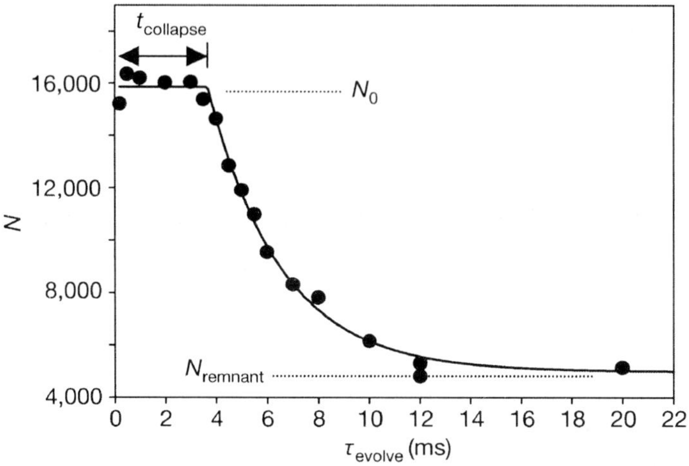

Figure 2. The number of remaining atoms in a condensate after removal of the barrier during the time τ_{evolve}; reproduced from Donley *et al.* [5].

one immediately changes back to (a), nothing happens, since the condensate has had no time to contract. Therefore the attraction must last for at least some time, t_{collapse}, before loss of particles sets in. This effect has been verified experimentally [5] and is shown in figure 2. The remnant number of particles can be explained by the sequence in figure 1(d)-(f). At lower particle numbers a barrier emerges. This prevents the remaining wave packet at larger hyperradii from further contraction, and recombination processes with loss of particles therefore stops, leaving N_{remnant} atoms in the condensate.

In conclusion, we described N bosons by a hyperspherical adiabatic method, where the hyperradius is the essential coordinate. We discussed a collapse sequence of a Bose-Einstein condensate using qualitative results of the model.

References

1. O. Sørensen, D.V. Fedorov, and A.S. Jensen: Phys. Rev. **A66**, 032507 (2002)
2. O. Sørensen, D.V. Fedorov, and A.S. Jensen: Phys. Rev. Lett. **89**, 73002 (2002)
3. J.L. Bohn, B.D. Esry, and C.H. Greene: Phys. Rev. **A58**, 584 (1998)
4. J.L. Roberts et al.: Phys. Rev. Lett. **86**, 4211 (2001)
5. E.A. Donley et al.: Nature (London) **412**, 295 (2001)
6. E. Nielsen and J.H. Macek: Phys. Rev. Lett. **83**, 1566 (1999)

Few-Body Systems Suppl. 14, 377–378 (2003)

Few-Body Systems

Study of Nuclear Matter Distributions of Halo Nuclei ^{6}He and ^{8}He with Elastic Proton Scattering

F. Aksouh[1], A. Bleile[1], O.V. Bochkarev[2], L.V. Chulkov[2], D. Cortina-Gil[1,*], A.V. Dobrovolsky[1,3], P. Egelhof[1], H. Geissel[1], M. Hellström[1], N.B. Isaev[3], O.A. Kiselev[1,3,†], B.G. Komkov[3], M. Mátos[1], F.V. Moroz[3], G. Münzenberg[1], M. Mutterer[4], V.A. Mylnikov[3], S.R. Neumaier[1], V.N. Pribora[2], D.M. Seliverstov[3], L.O. Sergueev[3], A. Shrivastava[1,**], K. Sümmerer[1], H. Weick[1], M. Winkler[1], V.I. Yatsoura[3]

[1] Gesellschaft für Schwerionenforschung mbH, D-64291 Darmstadt, Germany

[2] Russian Research Centre *Kurchatov Institute*, RU-123182 Moscow, Russia

[3] Petersburg Nuclear Physics Institute, RU-188350 St. Petersburg, Russia

[4] Institut für Kernphysik, Technische Universität, D-64289 Darmstadt, Germany

1 Introduction

The study of neutron-rich light nuclei near the drip line has attracted much attention as they exhibit a particular nuclear structure, namely an extended distribution of the valence neutrons surrounding a compact core. Recently the elastic proton scattering from ^{6,8}He and ^{8,9,11}Li at 700 MeV/u using an inverse kinematics was successfully measured at GSI [1,2]. A high-pressure hydrogen-filled ionization chamber was used as the target and a proton detector. The previous experiments were performed in the small range $0.002 \leq |t| \leq 0.05$ $(\mathrm{GeV/c})^2$ of the four-momentum transfer squared t and have yielded valuable information on the nuclear sizes and radial structure of nuclear matter density distributions with the aid of the Glauber multiple scattering theory. Recently, a novel experimental approach has been accomplished with the aim to deduce the differential p ^{6,8}He cross sections in the t-range $0.05 \leq |t| \leq 0.25$ $(\mathrm{GeV/c})^2$ close to the expected first diffraction minimum.

* *Present address:* Depto. de Fisica de Particulas, Universidade de Santiago de Compostela, E-15706 Santiago de Compostela, Spain

† Corresponding author, *E-mail address:* O.Kisselev@gsi.de

** *Present address:* Nuclear Physics Division, Bhabha Atomic Research Centre, Trombay, Mumbai IN-400085, India

2 Experimental Setup and Results

The experimental setup allowed to track and identify the projectile nuclei in coincidence with the recoil protons. The major difference with respect to the previous experiments was that instead of the active gaseous target, a 600 mg/cm^2 liquid hydrogen target was used and a position sensitive scintillator wall measured the recoil proton energies via time-of-flight [3]. The experimental arrangement allowed for a very low-background data taking. At present, only

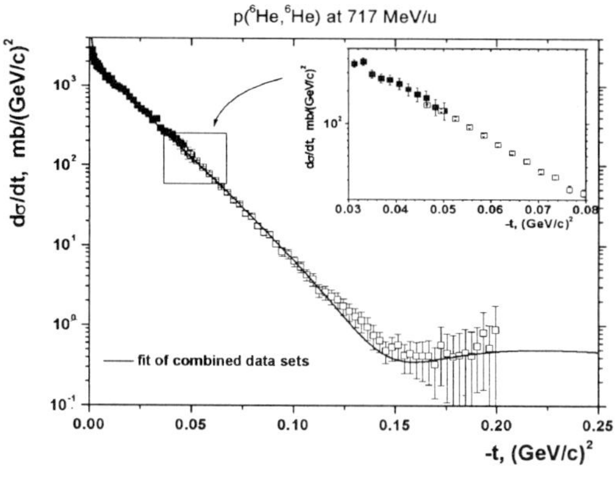

Figure 1. Experimental differential cross section of p ^{6}He elastic scattering versus momentum transfer squared t. Full symbols are the results of the low momentum transfer experiment [1], empty symbols - actual measurement. The insert shows the area of overlap.

preliminary results for the p^{6,8}He differential cross sections at the larger t-range have been evaluated. An example of the measured cross section is shown on Fig. 1 together with the fit of the combined data sets at low and high momentum transfer squared by a Glauber model using a double gaussian parametrisation for modeling the nuclear matter distribution in core and halo. According to theoretical estimations [4], the recently measured data together with the data from the previous experiments should allow to deduce the size and radial shape of the core in ^{6,8}He with high precision.

References

1. G.D. Alkhazov et al.: Phys. Rev. Lett. **78**, 2313 (1997); S.R. Neumaier et al.: Nucl. Phys. **A** (2002) (in print)
2. P. Egelhof et al.: Europ. Phys. J. A **15**, 27 (2002)
3. F. Aksouh et al.: GSI Scientific Report 2000, ISSN 0174-0814, 193 (2001)
4. L.V. Chulkov et al.: Nucl. Phys. A **578**, 291 (1995)

Few-Body Systems Suppl. 14, 379–380 (2003)

Few-Body Systems

Three Quark Clusters in Hot and Dense Matter

S. Mattiello[1,*], M. Beyer[1], T. Frederico[2], H.J. Weber[3]

[1] Fachbereich Physik, Universität Rostock, D-18051 Rostock, Germany

[2] Dep. de Física, ITA, CTA, 12.228-900 São José dos Campos, Brazil

[3] Dept. of Physics, University of Virginia, Charlottesville, VA 22904, U.S.A.

The phase diagram of quantum chromodynamics (QCD) shows a rich structure as lattice simulations and model calculations indicate [1]. One very interesting aspect is the transition from quarks to nucleons as relevant degrees of freedom (Mott transition [2, 3]). This is related to the transition between the hadronic matter and the quark-gluon plasma phase. Another particularly interesting possibility is quark pairing that leads to the color superconductivity analogous to Cooper pairing. An aspect that has hardly been investigated in this context is the appearance of three-quark correlations that, in the vicinity of the phase transition, should play an important role and might as well influence quark pairing. Here we address the issue of three-quark clusters in a medium of finite temperatures and densities. This necessarily involves relativity as well as effects of the medium. To implement relativity we use the light front approach and utilize a Green function formalism. We consider first the isolated case and then medium effects. Using a zero-range interaction, the two-body propagator is $t = (i\lambda^{-1} - B(M_2))^{-1}$, where $B(M_2)$ denotes the loop momentum integral [4]. The logarithmic divergence appearing in the integral can be eliminated by introducing an invariant cut-off Λ. The requirement is that Λ is smaller than the intermediate mass of the virtual two-quark subsystem, i.e. $M_{20}^2 < \Lambda^2$ and hence $t \to t_\Lambda$. We introduce a similar regularisation for the three-body equation ($M_{30}^2 < \Lambda^2$) that leads to an equation for the vertex function Γ_Λ that depends parametrically on the cut-off Λ,

$$\Gamma_\Lambda(M_3; y, \boldsymbol{q}_\perp) = \frac{i}{(2\pi)^3}\, t_\Lambda(M_2) \int_{x_{\min}}^{1-y} \frac{dx}{x(1-y-x)}$$

$$\int d^2k_\perp \frac{\theta\left(M_{03}^2 - \Lambda^2\right)}{M_3^2 - M_{03}^2} \Gamma_\Lambda(M_3; x, \boldsymbol{k}_\perp).$$

* *E-mail address:* stefano.mattiello@physik.uni-rostock.de

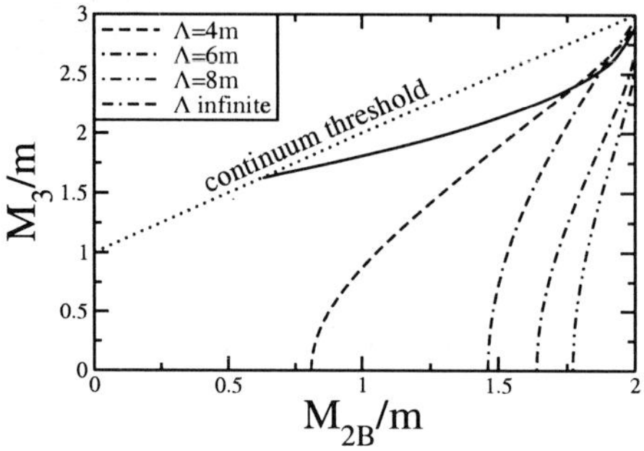

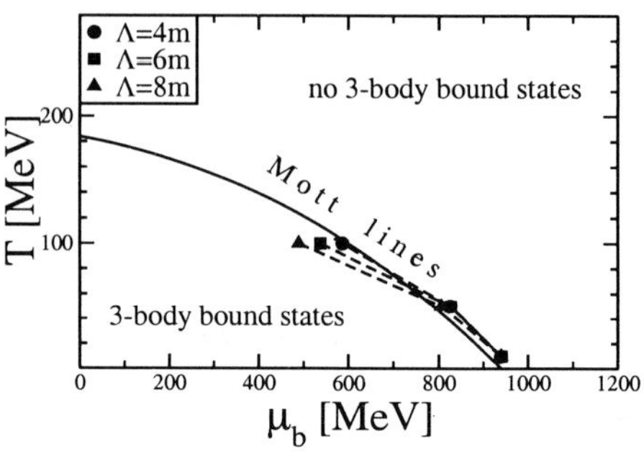

Figure 1. M_3 as function of the mass of the "diquark" M_{2B}.

Figure 2. Phase-diagram of QCD. See text for explanation.

In Fig. 1 we show the mass of the three-quark cluster as function of the diquark mass for different cut-offs in units of the quark mass. Note that $M_3 \to 0$ for finite diquark masses which is the relativistic analog of the Thomas collapse [5]. The results using the previous regularisation [4] are shown as solid line.

To investigate medium effects we fit the parameters (Λ, λ) to reproduce the proton mass. The three-body equation systematically includes the dominant medium effects due to Pauli blocking and self energy corrections, $m \to m(T, \mu)$. For a given temperature and a given choice of the fit parameters, the three-quark system becomes more weakly bound, as the baryonic chemical potential μ_b increases. The continuum is reached for binding energy $B_3 = 0$ (Mott transition). The Mott lines, i.e. the values of T and μ_b where this transition occurs are given in Fig. 2. We show our previous result (solid line) along with the ones using the invariant cut-off. From other models investigating $q\bar{q}$ states the confinement-deconfinement phase transition has been estimated as well. We find a similar qualitative behavior for the three-body Mott lines. A closer comparison indicates differences at larger densities that are currently investigated. The next aim is to investigate the phase transition between the color superconducting phase and the quark matter considering the influence of three-quark-correlations.

References

1. K. Rajagopal, F. Wilczek: In: *Handbook of QCD*, M. Shifman ed., World Scientific 2001
2. S. Mattiello et al.: Few Body Syst. **31**, 159 (2002)
3. M. Beyer, S. Mattiello, T. Frederico, H.J. Weber: Phys. Lett. **B521**, 33 (2001)
4. T. Frederico: Phys. Lett. **B282**, 409 (1992)
5. L.H. Thomas: Phys. Rev. **47**, 903 (1935)

Few-Body Systems Suppl. 14, 381–386 (2003)

Few-Body Systems

The Calculation of Form Factors in Light-Front Dynamics

B.L.G. Bakker

Department of Physics and Astronomy, Faculty of Exact Sciences, Vrije Universiteit, De Boelelaan 1081, NL 1081 Amsterdam, The Netherlands

Abstract. Form factors are extracted form current matrix elements. In a light-front calculation based on a simple field-theoretical model, we show that care must be exercised in the choice of the current components and the reference frame, lest the extracted form factors are unreliable.

1 Introduction

In nuclear physics the calculation of form factors of bound systems seems relatively straightforward. If the appropriate current is known, one can proceed with taking the matrix elements of that current with respect to the relevant bound-state wave functions. However, it is well known that in practice this simple recipe is not so easy to follow, as in most interesting cases neither the current not the wave functions are known exactly. Consequently, not only may the calculated form factors be numerically inaccurate, but it may also happen that the current matrix elements violate some basic symmetry, e.g., current conservation or relativistic covariance.

In situations where a nonrelativistic treatment is believed to be adequate, methods to improve the calculation of form factors have been developed that can be shown explicitly to conserve the most important symmetries: current conservation and rotational covariance. If such methods are applied, a consistent set of form factors can be extracted from calculated matrix elements.

In the case where Poincaré invariance really matters, the situation is more complicated. In a quantum-mechanical framework it has been known since Dirac's classic paper [1] that besides the energy operator, other generators of the Poincaré group must contain interaction. Moreover, if a current operator is constructed in a specific reference frame, it may change its form in another frame, since in any form of dynamics either some rotations or boosts or both contain interactions.

A relevant case is the quark model, in particular as applied to light hadrons, where it is clear that the quarks must be described with relativistic dynamics. In order to explore the possible pitfalls one may encounter, we studied systems of a constituent quark and antiquark, that interact through a contact interaction [2]. The covariant vertex function is given in this case by a simple spin structure times a constant. In this simple field theoretical model only one-body currents occur and the calculation of the current matrix elements and form factors are relatively straightforward.

In order to make contact with a quantum-mechanical picture, we compared our fully covariant calculation with one in light-front dynamics (LFD). This form of dynamics is singled out for several reasons. An important one is that a set of so-called light-front boosts exists that is independent of any interaction and allows one to connect a system with any momentum in a purely kinematical way with the same system at rest. This property of LFD enables one to express the wave functions of bound systems in relative coordinates, which means that they express intrinsic properties of those systems, independent of the chosen reference frame. Equally important is the fact that in LFD the vacuum is relatively simple, because massive particles cannot be created from the vacuum, which makes an expansion in Fock sectors well grounded.

LFD has, however, one disadvantage: *manifest* rotational invariance is broken. This does not imply that one could not use LFD to correctly calculate any property that relies on rotational symmetry, e.g., the form factors of particles with spin, but it means that only at the end of a LFD calculation one can–and must–check that rotational invariance is maintained.

In this paper we present our calculations of the electromagnetic properties of pseudoscalar and vector mesons, built out of spin-1/2 quarks. We use the simple zero-range interaction referred to above for the strong interaction and the simple Dirac structure $\bar{u}\gamma^{\mu}u$ for the quark current. We show in particular that higher Fock sectors may be relevant, depending on what component of the current is used to calculate the form factors. In our simple model the two-body Fock sector can be represented by a so-called valence vertex Ψ, while the other Fock sectors are found to lead to a nonvalence vertex Ψ'. The covariant amplitudes are found to be the sums of the nonvalence and valence LFD amplitudes. In the present paper we give a few examples of the relative numerical importance of these two contributions. More examples and many details may be found in our previously published work [2, 3, 4, 5].

2 Model Calculations

The interaction Lagrangians for pseudoscalar and vector mesons we took are

$$\mathcal{L}_{\text{int}}^{\text{PS}} = \bar{\psi}\gamma_5\psi\phi, \quad \mathcal{L}_{\text{int}}^{\text{V}} = \bar{\psi}\gamma_\mu\psi\phi^\mu. \tag{1}$$

The covariant amplitude is given in general by

$$G^\mu = iN_c g^2 \Lambda^4 \int \frac{d^4k}{(2\pi)^4} \frac{S^\mu}{D_\Lambda(k-p)D(k-p)D(k)D(k-p')D_\Lambda(k-p')}. \tag{2}$$

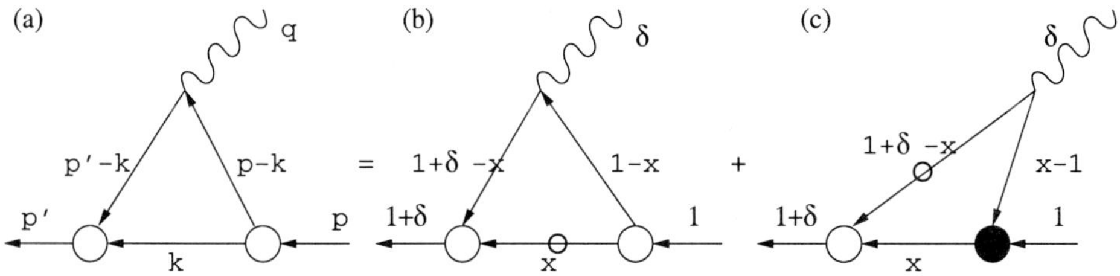

Figure 1. Covariant triangle diagram (a) split into valence (b) and nonvalence (c) contributions.

S^μ is the trace over the spin factors in the fermion propagators. The D's are the denominators from the fermion propagators and the smearing of the EM vertex. The smearing is done by multiplying the EM vertex γ^μ with the monopole factors $D_\Lambda(k-p)\ D_\Lambda(k-p')$ to make the matrix elements finite. They are given by

$$D(k) = k^2 - m^2 + i\epsilon, \quad D_\Lambda(k-p) = (k-p)^2 - \Lambda^2 + i\epsilon\,. \tag{3}$$

In LFD a covariant amplitude is split into light-front time-ordered amplitudes by integrating over the momentum component k^- (we use the notation $P^\pm = (P^0 \pm P^3)/\sqrt{2}$). This is pictorially given by Fig. 1. The LF valence diagram is defined in $0 < k^+ < p^+$ and the LF nonvalence diagram is defined in $p^+ < k^+ < p'^+$. We define $x = k^+/p^+$, $\delta = q^+/p^+ = p'^+/p^+ - 1$. The LF valence vertex Ψ is given by open circles in parts (b) and (c), the nonvalence vertex is given by a black circle.

The fermion propagator has a LF propagating part and an instantaneous part

$$\not{P} + m_q = (\not{P}_{\rm on} + m_q) + \gamma^+(P^- - P^-_{\rm on})\,. \tag{4}$$

The corresponding split of the trace is

$$S^\mu = (S^\mu)_{\rm on} + (S^\mu)_{\rm inst}\,. \tag{5}$$

If the integration range of the nonvalence region, $p^+ < k^+ < p^+ + q^+$, shrinks to zero in the $q^+ \to 0$ limit, the nonvalence contribution may not vanish, but give a finite contribution,

$$\lim_{q^+\to 0} \int_{p^+}^{p^+ + q^+} dk^+ (\cdots) \equiv \lim_{\delta \to 0} \int_1^{1+\delta} dx (\cdots) \neq 0. \tag{6}$$

It is called the *zero-mode* contribution. Such a contribution can occur only if the integrand contains a $1/q^+$ singularity.

2.1 Reference Frame

In our calculations we used three different kinematics. We give here results for the *Drell-Yan-West* (DYW) frame only. Other cases can be found in the cited

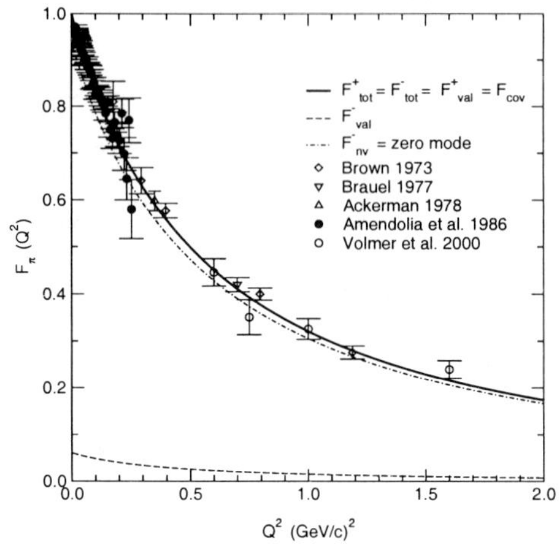

Figure 2. Pion form factor in the DYW frame. Data from [6, 7, 8, 9, 10]

literature. The momenta are in the DYW frame given by

$$\begin{aligned} p &= (p^+, 0, 0, m^2/(2p^+)) , \\ p' &= (p^+, Q\cos\phi, Q\sin\phi, (Q^2+m^2)/(2p^+)) , \end{aligned}$$

where $q^2 = -Q^2$. (Notation: $p^\mu = (p^+, p_x, p_y, p^-)$.)

3 Results

3.1 Pseudoscalar Current

If the symmetries hermiticity, covariance, and time-reversal are applied one finds that the current can be expressed in terms of a single real, invariant form factor

$$\langle p', h'|J^\mu|p, h\rangle = ie_m(p'^\mu + p^\mu)F(q^2) . \tag{7}$$

As an example we give the pion form factor calculated in the Drell-Yan-West frame. The model parameters are $m_q = m_{\bar{q}} = 0.25$, $\Lambda_q = \Lambda_{\bar{q}} = 0.90$ in units GeV/c^2. Fig. 2 shows that if we extract the form factor from the plus-component of the current, the nonvalence part vanishes. Had we used the minus component J^-, then the nonvalence part would contribute and even dominate, which in the DYW frame means that there is a substantial zero mode contribution to J^-.

3.2 Vector Meson Current and Angular Conditions

Applying the same symmetries as before we find that the current of a spin-1 meson can be expressed in terms of three real, invariant form factors

$$\begin{aligned} G^{\mu}_{h'h} &= \langle p', h'|J^{\mu}|p, h\rangle = -\,\epsilon^{*}_{h'} \cdot \epsilon_h \,(p+p')^{\mu}\; F_1(Q^2) \\ &\quad + (\epsilon^{\mu}_{h}\; q\cdot \epsilon^{*}_{h'} - \epsilon^{*\mu}_{h'}\; q\cdot\epsilon_h)\; F_2(Q^2) \\ &\quad + \frac{(\epsilon^{*}_{h'}\cdot q)(\epsilon_h \cdot q)}{2m^2}(p+p')^{\mu}\; F_3(Q^2). \end{aligned} \tag{8}$$

The physical form factors G_C (charge monopole), G_M (magnetic dipole), and G_Q (electric quadrupole), can be defined as linear combinations of the invariant ones

$$\begin{aligned} G_C &= \left(1+\frac{2}{3}\eta\right) F_1 + \frac{2}{3}\eta\; F_2 + \frac{2}{3}\eta(1+\eta)\; F_3\,, \\ G_M &= -F_2\,, \quad G_Q = F_1 + F_2 + (1+\eta)\; F_3\,, \\ \eta &= \frac{Q^2}{4m^2}\,, \quad Q^2 = -(p'-p)^2\,. \end{aligned} \tag{9}$$

We work here only with the plus component of the current. Then we may choose to extract the form factors from G^{+}_{++}, G^{+}_{-+}, and either G^{+}_{0+} (scheme *b*), G^{+}_{+0} (scheme *c*), or G^{+}_{00} (scheme *d*). In all these schemes F_1 and F_3 are the same, but F_2 depends on the scheme, so we use the notation F_2^b etc. There are apparently two independent consistency conditions, $F_2^b = F_2^c$ and $F_2^b = F_2^d$. They are the so-called *angular conditions.* In the literature sometimes other schemes are used, [11, 12, 13], but they are linearly related to our schemes.

For illustrative reasons we give in Fig. 3 the results for the different schemes applied in the DYW reference frame. Only the valence contribution is retained. Other examples can be found in the cited literature. It is peculiar for the DYW reference frame that schemes *b* and *c* give the full covariant results with vanishing nonvalence contributions. This means that the angular condition $F_2^b = F_2^c$ is fulfilled. Scheme *d* needs a zero mode contribution, so $F_2^b = F_2^d$ is not fulfilled by the valence contributions to the current. In fact, G_M^d and G_Q^d diverge for $Q^2 \to 0$. In other reference frames both angular conditions may be violated.

4 Conclusions

Rotational invariance is restored if both valence parts and nonvalence parts are included. The minus component J^- will contain zero-mode contributions, but the plus component J^+ may be free of zero modes. For vector mesons the schemes *b* or *c* in connection with the DYW reference frame must be used to avoid zero modes completely. This is equivalent to the Grach-Kondratyuk scheme. Large errors, violations of covariance, may occur if any of these conditions is not met.

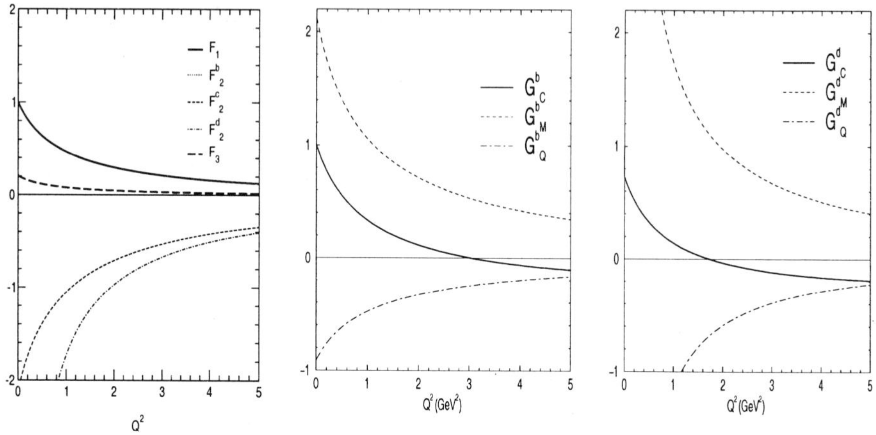

Figure 3. Invariant form factors F (left) and physical form factors G^b (middle) and G^d (right).

References

1. P.A.M. Dirac: Rev. Mod. Phys. **21**, 392 (1949)
2. B.L.G. Bakker, C.-R. Ji: Phys. Rev. **D62**, 074014 (2000)
3. B.L.G. Bakker, H.-M. Choi, C.-R. Ji: Phys. Rev. **D63**, 074014 (2001)
4. B.L.G. Bakker, C.-R. Ji: Phys. Rev. **D65**, 073002 (2002)
5. B.L.G. Bakker, H.-M. Choi, C.-R. Ji: Phys.Rev. **D65**, 116001 (2002)
6. C.N. Brown, et al.: Phys. Rev. **D8**, 92 (1973)
7. P. Brauel et al.: Phys. Lett. **B69**, 253 (1977)
8. H. Ackermann et al.: Nucl. Phys. **B137**, 294 (1978)
9. S.R. Amendolia et al.: Nucl. Phys. **B277**, 168 (1986)
10. J. Volmer et al.: Phys. Rev. Lett. **86**, 1713 (2001)
11. I.L. Grach, L.A. Kondratyuk: Sov. J. Nucl. Phys. **39**, 198 (1984)
12. P.L. Chung, F. Coester, B.D. Keister, W.N. Polyzou: Phys. Rev. **C37**, 2000 (1988)
13. S.J. Brodsky, J.R. Hiller: Phys. Rev. **D46**, 2141 (1992)

Few-Body Systems Suppl. 14, 387–392 (2003)

Few-Body Systems

Structure and Applications of Point Form Relativistic Quantum Mechanics

W.H. Klink

Department of Physics and Astronomy, University of Iowa, Iowa City, Iowa, USA

Abstract. The framework of point form relativistic quantum mechanics is used to construct mass and current operators for hadronic systems with finite degree of freedom. For the point form all of the interactions are in the four-momentum operator and, since Lorentz transformations are kinematic, the theory is manifestly covariant.

In the Bakamjian-Thomas version of the point form the four-momentum operator is written as a product of the four-velocity operator and mass operator, where the mass operator is the sum of free and interacting mass operators. Interacting mass operators can be constructed from vertices, matrix elements of local field operators evaluated at the space-time point zero, where the states are eigenstates of the four-velocity. Applications include the study of the spectra and widths of vector mesons, viewed as bound states of quark-antiquark pairs.

Besides mass operators, current operators are needed to compute form factors. Form factors are matrix elements of current operators on mass operator eigenstates and are often calculated with one-body current operators (in the point form this is called the point form spectator approximation); but in a properly relativistic theory there must also be many-body current operators. Minimal currents needed to satisfy current conservation in the presence of hadronic interactions (called dynamically determined currents) are shown to be easily calculated in the point form.

1 Point Form Relativistic Quantum Mechanics

The goal of all forms of relativistic dynamics is to formulate a relativistic quantum mechanics for finite degree of freedom systems by starting with the commutation relations of the Poincaré algebra, which assures the correct relativistic invariance of the theory. The generators of the Poincaré algebra are modified by putting interactions in the mass operator in such a way that the Poincaré commutation relations are still satisfied. There is a systematic procedure for doing this in all forms of dynamics, called the Bakamjian-Thomas procedure [1], but in the point form it is particularly simple.

In the point form all interactions are in the four-momentum operator and all the Lorentz generators are interaction independent. It is then more convenient to deal with global Lorentz transformations, rather than the Lorentz generators. Since the Lorentz generators contain no interactions, the point form is manifestly Lorentz covariant. If U_Λ is a unitary operator representing the Lorentz transformation Λ, then the four-momentum operator must satisfy the following point form equations:

$$\begin{aligned} [P_\mu, P_\nu] &= 0\,, &(1)\\ U_\Lambda P_\mu U_\Lambda^{-1} &= (\Lambda^{-1})_\mu^\nu P_\nu\,. &(2)\end{aligned}$$

These equations are simply one way of writing the Poincaré commutation relations in which the relations of the four-momentum operators are emphasized. The first equation states that energy and momentum can be simultaneously measured, even though all the operators contain interactions, while the second equation states that the four-momentum operator transforms as a four-vector under Lorentz transformations. The mass operator is defined to be $M := \sqrt{P \cdot P}$ and must have a nonnegative spectrum.

One of the main reasons for developing the point form is to formulate a covariant many-body relativistic quantum mechanics that is between an infinite degree of freedom field theory and nonrelativistic quantum mechanics. It is thus important to introduce the many-body aspect at the beginning, which is most simply done by introducing creation and annihilation operators. Let $a^\dagger(p,\sigma)$ be the operator that creates the state $|p,\sigma>$ from the vacuum, where p is the four-momentum with $p \cdot p = m^2$ and σ is a spin projection ranging between j and $-j$. If $a(p,\sigma)$ is its adjoint, these operators must satisfy the following relations:

$$\begin{aligned} [a(p,\sigma), a^\dagger(p',\sigma')]_\pm &= 2E\delta^3(p-p')\,\delta_{\sigma,\sigma'} &(3)\\ U_a a^\dagger(p,\sigma)U_a^{-1} &= e^{ip\cdot a} a^\dagger(p,\sigma), &(4)\\ P_\mu(fr) &= \sum \int \frac{d^3p}{2E} p_\mu a^\dagger(p,\sigma) a(p,\sigma), &(5)\\ U_\Lambda a^\dagger(p,\sigma)U_\Lambda^{-1} &= \sum a^\dagger(\Lambda p,\sigma')D^j_{\sigma',\sigma}(R_W). &(6)\end{aligned}$$

Here $P_\mu(fr)$ is the free four-momentum operator and satisfies the point form equations. It plays a role analogous to the free Hamiltonian in nonrelativistic quantum mechanics. U_a in Eq. (4) is the unitary operator representing the four-translation a and R_W is a Wigner rotation [2]. To introduce interactions, the four-momentum operator is written as $P_\mu = MV_\mu$, with the mass operator the sum of free and interacting mass operators, $M = M_{fr} + M_I$, and V_μ the free four-velocity operator. With such a four-momentum operator, the point form equations become a mass eigenvalue equation:

$$M\Psi = m\Psi, \qquad (7)$$

which gives the bound and scattering wave functions.

One of the ways of constructing an interacting mass operator is from a vertex, acting on a truncated Fock space, which is the model hadronic Hilbert space. As an example, if the creation and annihilation operators are those for (bare) pions and nucleons, the truncated Fock space could be chosen as the direct sum of two nucleon spaces with zero to n pions, appropriate for the scattering of two nucleons as well as the production of pi mesons.

The creation and annihilation operators of Eq. (3) can also be used to form local fields; the general procedure for this construction is given in ref. [3] and is reviewed in detail in [4]. Moreover ref. [3] shows how to take products of arbitrary spin fields and couple them together to form Lorentz scalar densities, which, following this reference will be denoted by $\mathcal{H}(x)$. Examples of such scalar densities of products of local fields are the familiar $\bar{\Psi}\gamma_5\Psi\phi$ for pion-nucleon vertices and $\bar{\Psi}\gamma_\mu\Psi A^\mu$ for electron-photon vertices.

The fact that $\mathcal{H}(x)$ is a Lorentz scalar density means that $U_\Lambda \mathcal{H}(x) U_\Lambda^{-1} = \mathcal{H}(\Lambda x)$. Therefore at the space-time point $x = 0$, $U_\Lambda \mathcal{H}(0) U_\Lambda^{-1} = \mathcal{H}(0)$, so that $\mathcal{H}(0)$ is a Lorentz scalar. Taking matrix elements of $\mathcal{H}(0)$ on the truncated Fock space, $< p_1^{'}, ...p_{n'}^{'} |\mathcal{H}(0)| p_1 ... p_n > |_{v'=v}$, where the initial and final four-velocities are the same, generates the kernel of an interacting mass operator M_I on the truncated Fock space, because it commutes with the four-velocity operator (since it is diagonal in the four-velocity, by construction) and commutes with Lorentz transformations. In fact, as shown in ref. [5], M_I actually is independent of the four-velocity. Also, since M_I arises from products of fields, it always has only off-diagonal matrix elements in the truncated Hilbert space. An application of interacting mass operators generated from vertices comes from vector mesons considered as bound states of quark-antiquark pairs. The truncated Hilbert space is the direct sum of quark-antiquark pairs plus quark-antiquark-pseudoscalar meson Hilbert spaces, and the vertex is of the form $\mathcal{H}(x) = \bar{\Psi}(x)\gamma_5\Psi(x)\phi(x)$, which couples the two Hilbert spaces. Computations giving masses are given in ref. [6] and decay widths in ref. [7].

Along with interacting mass operators generated from vertices as sketched above, there are a number of other ways in which interacting mass operators can be constructed. Reference [8] lists some of the publications in which interacting mass operators are constructed in the context of point form dynamics.

Besides the mass operator, the other operator needed to compute form factors is the current operator. Current operators must satisfy general properties such as Poincaré covariance and current conservation. In the point form the current operator at the space-time point 0 plays a special role in that it determines the Poincaré covariance and conservation properties at an arbitrary space-time point. In fact, if $J_\mu(0)$ satisfies

$$U_\Lambda J_\mu(0) U_\Lambda^{-1} = (\Lambda_\mu^\nu)^{-1} J_\nu(0) , \tag{8}$$

$$[P^\mu, J_\mu(0)] = 0 , \tag{9}$$

then $J_\mu(x) := e^{iP\cdot x} J_\mu(0) e^{-iP\cdot x}$ is Poincaré covariant and conserved. Thus it suffices to construct $J_\mu(0)$ in order that the current operator at an arbitrary space-time point have the correct properties. One of the main advantages of the point form is that current operators can be consistently defined over different few-body relativistic systems. The same two-body current operator used for deuteron calculations can be used for other few-body systems.

Form factors are current operator matrix elements. If the states are chosen to be eigenstates of the four-momentum operator, then the covariance properties of the states and current operators make it possible to greatly simplify the structure of the form factors. As shown in ref. [9] current operators are irreducible tensor operators of the Poincaré group, so that a generalized Wigner-Eckart theorem can be used to decompose current matrix elements into Clebsch-Gordan coefficients times reduced matrix elements, which are the invariant form factors.

The goal then is to construct conserved electromagnetic current operators on a hadronic Hilbert space as the sum of a one-body current operator (defined in [4]), and a many-body current operator,

$$J_\mu(0) : \; = \; J^1_\mu(0) + J^{many-body}_\mu(0), \tag{10}$$

itself the sum of dynamically determined and model dependent pieces. In this note only the dynamically determined currents are considered and they will be built from free current operators. Free current operators are conserved under the free hadronic four-momentum operator, but not in the presence of hadronic interactions. That is, if the (hadronic) four-momentum operator is the sum of free and interacting four-momentum operators, $P^\mu(h) = P^\mu(fr) + P^\mu(I)$, (satisfying Eqs. (1,2)), current conservation, Eq. (9), will not hold unless a term (denoted by "DD" for dynamically determined) is added to the one-body current operator:

$$[P^\mu(fr) + P^\mu(I), J^1_\mu(0) + J^{DD}_\mu(0)] \; = \; 0. \tag{11}$$

If the interacting four-momentum operator, $P^\mu(I)$ does not commute with the one-body current operator, $J^1_\mu(0)$, there must be additional (dynamically determined) currents present in order to satisfy current conservation, exactly as in the non-relativistic case [10]. Making use of the fact that the one-body current operator is conserved with respect to the free four-momentum operator, $[P^\mu(fr), J^1_\mu(0)] = 0$, and taking eigenfunctions of the hadronic four-momentum operator to construct current operator matrix elements, Eq. (11) becomes

$$Q^\mu < p^{'}|J^{DD}_\mu(0)|p > \; = \; - < p^{'}|[P^\mu(I), J^1_\mu(0)]|p >, \tag{12}$$

where Q^μ is the four-vector momentum transfer, $Q^\mu = (p^{'} - p)^\mu$.

A general solution to Eq. (12) is given by

$$< p^{'}|J^{DD}_\mu(0)|p > \; = \; -\frac{(Q_\mu + Q^\perp_\mu)}{Q^2} < p^{'}|[P^\nu(I), J^1_\nu(0)]|p > \tag{13}$$

$$= \; -\frac{(Q_\mu + Q^\perp_\mu)}{Q^2} Q^\nu < p^{'}|J^1_\nu(0)|p >; \tag{14}$$

here $Q^{\perp}$ is a four-vector perpendicular to $Q, Q \cdot Q^{\perp} = Q^{\mu}Q^{\perp}_{\mu} = 0$, and is determined by the requirement that there should be no pole at $Q^2 = 0$ [11]. As can be seen from Eq. (13), if the interacting four-momentum operator is zero, the dynamically determined current operator also is zero. Then the total current operator (without model dependent currents, see [11]) has the form

$$J_\mu(0) = J^1_\mu(0) + J^{DD}_\mu(0); \tag{15}$$

$$< p^{'}|J_\mu(0)|p> = < p^{'}|J^1_\mu(0)|p> - \frac{Q_\mu + Q^{\perp}_\mu}{Q^2} Q^\nu < p^{'}|J^1_\nu(0)|p>, \tag{16}$$

and from Eq. (16) it follows that $J_\mu(0)$ is conserved, $Q^\mu < p^{'}|J_\mu(0)|p>= 0$.

At present no form factor calculations have included the effects of dynamically determined currents. Rather the existing form factor calculations make use of the so-called point form spectator approximation, discussed in ref. [9], where the current operator for a many-body system is approximated by a one-body current for the struck constituent, while the momentum of the unstruck constituents remains unchanged. It can be shown that while such a current operator is not conserved in the presence of hadronic interactions, the results for electric and magnetic form factors are not affected by the lack of current conservation.

To date there have been four types of form factor calculations carried out involving two or three particle constituents. Two of these give good results as compared with experiment (see the pion form factor calculation by T. Allen and W.H. Klink, as well as the three quark calculations by S. Boffi et al. in ref. [8]), while two give discrepant results (see the deuteron form factor calculation by T. Allen, W.H. Klink, and W.N. Polyzou in ref. [8], as well the the calculations of form factors using different forms of dynamics by B. Desplanques et al., also in ref. [8], and the contribution by L. Theußl [12] in these proceedings). It appears likely that an important parameter distinguishing these results is the matter radius of the bound state wave function; if it is small, meaning significant relativistic momenta in the wave function, the point form spectator approximation gives good results. A number of calculations are presently under way that are designed to give a better insight into the nature of these and other aspects of point form relativistic quantum mechanics.

References

1. B. Bakamjian, L.H. Thomas: Phys. Rev. **92**, 1300 (1953); see also B.D. Keister, W.N. Polyzou: In: Advances in Nuclear Physics **20**, (Edited by J.W. Negele and E.W. Vogt), p. 225. New York: Plenum 1991
2. W.H. Klink: Phys. Rev. **C58**, 3617 (1998)
3. S. Weinberg: Phys. Rev. **133**, B1318 (1964)
4. W.H. Klink: "Current Operators for Arbitrary Spin Particles", Nucl. Phys. A. (to be published); nucl-th/0209064

5. W.H. Klink: "Constructing Point Form Mass Operators from Vertex Interactions", Nucl. Phys. **A** (to be published); nucl-th/0012031

6. A. Krassnigg, W. Schweiger, W.H. Klink: "A Relativistic Point-Form Approach to Quark-Antiquark Systems", these proceedings

7. A Krassnigg, W. Schweiger, W.H. Klink: "Hadronic Vector-Meson Decay Widths from a Relativistic Coupled-Channel Approach", these proceedings

8. M. Rogers, W.H. Klink: Phys. Rev. **C58**, 3605 (1998) (separable mass operator); T. Allen, W.H. Klink: Phys. Rev. **C58**, 3670 (1998) (harmonic oscillator ground state wave function); F. Coester, D.O. Riska: Few Body Systems **25** 29 (1998) (phenomenological mass operator to fit spectrum); T. Allen, W.H. Klink, W.N. Polyzou: Phys. Rev. **C63**, 034002 (2001) (nonrelativistic deuteron potential converted to mass operator); S. Boffi, L. Glozman, W.H. Klink, W. Plessas, M. Radici, R. Wagenbrunn: Phys. Lett. **B511**, 33 (2001); Phys. Lett. **B516**, 319 (2001); Eur. Phys. J. **A14**, 17 (2002) (semi-relativistic Hamiltonian with linear confinement and hyperfine spin-flavor interaction converted to mass operator); B. Desplanques, L. Theußl, S. Noguera: Eur. Phys. J. **A13**, 461 (2002); Phys. Rev. **C65**, 038202 (2002) (Wick-Cutkosky field theory motivated mass operator)

9. W.H. Klink: Phys. Rev. **C58**, 3587 (1998)

10. See for example, D.O. Riska: Phys. Rep. **181**, 207 (1989)

11. W.H. Klink: "Point Form Electrodynamics and the Construction of Current Operators", nucl-th/0012033 (submitted)

12. L. Theußl: These Proceedings

Few-Body Systems Suppl. 14, 393–398 (2003)

Few-Body Systems

Comparison of Different Boost Transformations for the Calculation of Form Factors in Relativistic Quantum Mechanics

L. Theußl[1], A. Amghar[2], B. Desplanques[3], S. Noguera[1]

[1] Departamento de Fisica Teorica, Universidad de Valencia, E-46100 Burjassot (Valencia), Spain

[2] Faculté des Sciences, Université de Boumerdes, 35000 Boumerdes, Algeria

[3] Institut des Sciences Nucléaires (UMR CNRS/IN2P3–UJF), F-38026 Grenoble Cedex, France

Abstract. The effect of different boost expressions, pertinent to the instant, front and point forms of relativistic quantum mechanics, is considered for the calculation of the ground-state form factor of a two-body system in simple scalar models. Results with a Galilean boost as well as an explicitly covariant calculation based on the Bethe-Salpeter approach are given for comparison. It is found that the present so-called point-form calculations of form factors strongly deviate from all the other ones. This suggests that the formalism which underlies them requires further elaboration. A proposition in this sense is made.

1 Introduction

Among the forms of relativistic quantum mechanics originally proposed by Dirac [1], the point form is probably the least known and also the one least used. This may be due to the non-linear constraints on coordinates when quantization is performed on a hyperboloïd. On the other hand, the dynamics is particularly appealing in this form, because the interaction is solely contained in the momentum operators, i.e., only four generators of the Poincaré group are dynamic, while, in particular, rotations and boosts are kinematic. Interest in the point form has been revived recently by the work of Klink [2] and its applications to the form factors of the pion [3], the deuteron [4] and the nucleon [5]. In the deuteron case, there is no significant improvement with respect to a non-relativistic calculation. In the nucleon case, the agreement with experimental data is at first sight spectacular, especially at the lowest Q^2 range considered by the authors.

However, analysing the calculations in detail, some questions arise. The well-known vector-meson dominance mechanism, which explains to a large part the nucleon form factors at low Q^2 via the coupling of the photon to the nucleon by ρ and ω exchanges, is not accounted for at all, and nothing indicates that this might be achieved in a hidden way by incorporated relativistic effects. Furthermore, no intrinsic quark form factors (which could account for ρ and ω exchanges implicitly) are employed, while they are necessary for the construction of the quark-quark interaction.

From these works, one may deduce an expression for the mean squared radius that scales like $1/M^2$, where M is the total mass of the bound system. This would have the surprising consequence that the root mean squared radius increases with decreasing M, and, in particular, diverges in the limit $M \to 0$. This is contrary to physical intuition, where a smaller total mass, usually brought about by increasing the interaction, leads to a more compact system. Note that the limit $M \to 0$ is not a completely academic one since it applies to the pion. A simple scaling argument would lead to a squared radius of the pion of $3/m_\pi^2 \simeq 6\,\text{fm}^2$ (!).

The success of the calculation in ref. [5] is attributed by the authors to relativistic boost effects. Curiously, taking into account the Lorentz-contraction effect by a simple replacement of the argument of the form factor, one gets an effect that increases the form factor for a given Q^2. This is the opposite of what was found in ref. [5].

In a series of recent papers, motivated by the above mentioned observations, the reliability of the employed point-form implementation has been tested on some simple scalar systems, which are academic, but offer the advantage of eliminating uncertainties, like spin effects and intrinsic form factors of the constituents, while allowing partly analytic results. These works include a two-body system composed of scalar particles exchanging a zero-mass boson [6] and a system corresponding to a zero-range interaction [7]. A systematic comparison with predictions of the other forms of dynamics (instant and front form), as well as with a non-relativistic and an explicitly covariant calculation based on the Bethe-Salpeter approach has been presented in ref. [8], where also the sensitivity of the results on the mass operator was investigated.

We shall only outline in this contribution the procedure that was followed in ref. [8], concentrating on the construction of the mass operator, where another questionable feature of recent point form applications arises. An example of numerical results for form factors will confirm the unusual results obtained in the more realistic calculations of refs. [4, 5].

2 Mass Operator

The starting point of every calculation is a dynamical equation (a mass operator), from which a wave function may be determined that is subsequently used for the calculation of observables. The important constraint to be fulfilled by such an equation is its covariance, i.e., the resulting spectrum should be invariant with respect to Lorentz boosts. We choose an equation with a quadratic form of the energy- and the phase-space factors, which facilitates the calculations:

$$\left(E_P^2 - (e_{p_1} + e_{p_2})^2\right) \Phi(\boldsymbol{p}_1, \boldsymbol{p}_2) = \iint \frac{d\boldsymbol{p}_1'}{(2\pi)^3} \frac{d\boldsymbol{p}_2'}{(2\pi)^3}$$

$$\frac{\sqrt{2\,(e_{p_1} + e_{p_2})}}{\sqrt{2\,e_{p_1}}\,\sqrt{2\,e_{p_2}}} V_{int}(\boldsymbol{p}_1, \boldsymbol{p}_2, \boldsymbol{p}_1', \boldsymbol{p}_2') \frac{\sqrt{2\,(e_{p_1'} + e_{p_2'})}}{\sqrt{2\,e_{p_1'}}\,\sqrt{2\,e_{p_2'}}} \Phi(\boldsymbol{p}_1', \boldsymbol{p}_2'), \quad (1)$$

where $E_P = \sqrt{M^2 + \boldsymbol{P}^2}$, $e_p = \sqrt{m^2 + \boldsymbol{p}^2}$. The quantities M and $\boldsymbol{P}$ represent the total mass and the total momentum of the system under consideration.

One can easily determine the constraints that ensure the invariance of the mass spectrum of this equation. The same constraints also provide a direct way to relate a wave function calculated in a moving frame (with a finite momentum $\boldsymbol{P}$), to the rest frame wave function. Provided $V_{int}(\boldsymbol{p}_1, \boldsymbol{p}_2, \boldsymbol{p}_1', \boldsymbol{p}_2')$ is appropriately chosen, the above equation can always be transformed into the center of mass by a simple change of variables:

$$(M^2 - 4\,e_k^2)\,\phi_0(\boldsymbol{k}) = \int \frac{d\boldsymbol{k}'}{(2\pi)^3} \frac{1}{\sqrt{e_k}} V_{int}(\boldsymbol{k}, \boldsymbol{k}') \frac{1}{\sqrt{e_{k'}}} \phi_0(\boldsymbol{k}'). \quad (2)$$

This equation is of the form $M^2 = M_0^2 + V$, i.e., M^2 gives an invariant mass operator whose solutions can be used in any form of relativistic quantum mechanics. It is just the relation between the set of particle momenta, $\boldsymbol{p}_1$, $\boldsymbol{p}_2$, to the set of 'internal' momenta, $\boldsymbol{k}$, $\boldsymbol{P}$, that will be different from one form to the other. In ref. [8] this transformation has been explicitly demonstrated for the instant form of dynamics. It is given by an expression that is equivalent to the one obtained from the Bakamjian-Thomas construction [9]:

$$\begin{aligned} \boldsymbol{p}_1 &= \boldsymbol{k} - \boldsymbol{P}\,\frac{\boldsymbol{k}\cdot\boldsymbol{P}}{P^2} + \boldsymbol{P}\,\frac{\boldsymbol{k}\cdot\boldsymbol{P}}{P^2}\,\frac{\sqrt{4\,e_k^2 + P^2}}{2\,e_k} + e_k\,\frac{\boldsymbol{P}}{2\,e_k}, \\ e_{p_1} &= e_k\,\frac{\sqrt{4\,e_k^2 + P^2}}{2\,e_k} + \boldsymbol{k}\cdot\frac{\boldsymbol{P}}{2\,e_k}, \end{aligned} \quad (3)$$

together with similar expressions for particle 2, where the change, $\boldsymbol{k} \to -\boldsymbol{k}$ has to be made. In the instant form, the momenta of the constituent particles are related to the total momentum $\boldsymbol{P}$ by the equality $\boldsymbol{p}_1 + \boldsymbol{p}_2 = \boldsymbol{p}_1' + \boldsymbol{p}_2' = \boldsymbol{P}$, consistently with the property that the momentum has a kinematical character in this form. When the dynamics is described on a surface different from the instant-form one, $t = \tau$, other relations between momenta are obtained. This is easiest to see by integrating plane waves over the hyper-surface under consideration. For a hyper-plane, $\lambda \cdot x = \tau$ with $\lambda^2 = 1$, for instance, one obtains:

$$\int d^4x\,\delta(\lambda\cdot x)\,e^{i(p_1+p_2-P)\cdot x} =$$

$$(2\pi)^3\frac{1}{\lambda^0}\delta\left(\boldsymbol{p}_1 + \boldsymbol{p}_2 - \boldsymbol{P} - \frac{\boldsymbol{\lambda}}{\lambda^0}(e_1 + e_2 - E_P)\right), \quad (4)$$

while for a hyperboloïd ($x^2 = \tau$), one gets for the special case $\tau = 0$:

$$\int d^4x\, \delta(x \cdot x)\, e^{i(p_1+p_2-P)\cdot x} = \frac{4\pi}{(\boldsymbol{p}_1 + \boldsymbol{p}_2 - \boldsymbol{P})^2 - (e_1 + e_2 - E_P)^2 + i\epsilon}. \quad (5)$$

Equation (4) would apply to an instant form of dynamics when $\boldsymbol{\lambda} = 0$, and we recover correctly $\boldsymbol{p}_1 + \boldsymbol{p}_2 = \boldsymbol{P}$ in this case. However, in front form (Eq. (4) with $\boldsymbol{\lambda}/\lambda^0$ equal to a unit vector $\boldsymbol{n}$) and in point form, Eq. (5), the relations are more complicated. In particular, in neither of these forms we may set the sum of the particle momenta ($\boldsymbol{p}_1 + \boldsymbol{p}_2$) equal to zero in the rest frame of the system ($\boldsymbol{P} = 0$). This contrasts with the prescription used in recent point form applications, where a relation of the form

$$\boldsymbol{p}_1 + \boldsymbol{p}_2 = \frac{2e_k}{M}\boldsymbol{P} \quad (6)$$

is applied, i.e., $\boldsymbol{p}_1 + \boldsymbol{p}_2 = 0$ in the center of mass, which can never be obtained from a relation of the form of Eq. (5). The relation between particle and 'internal' momenta obtained in that way rather resembles the one in instant form given by Eq. (3), with factors $2e_k$ replaced by the total mass M. This suggests, like Eq. (6), that we might expect troubles in the limit $M \to 0$ in this case.

3 Results for Elastic Form Factors

Expressions for elastic charge form factors have been given in ref. [8] in the different forms of relativistic quantum mechanics and for different choices of the two-body interaction model. We reproduce in Table 1 results obtained for a zero-range interaction (corresponding to an infinite-mass exchange boson). This model has the advantage that some calculations may be carried out analytically [7], which is useful for checking certain properties like the exact logarithmic dependence of the high Q^2 behavior of the form factors. The other extreme of a zero mass exchange boson has been discussed in ref. [6].

Inspecting the results, two features are noticed immediately:

- The form factors in point form depart significantly from all the others, especially for high Q^2 and in the limit of small M.
- The root mean squared radius of the bound state (proportional to the slope of the form factor at the origin) diverges in the limit $M \to 0$.

These features are qualitatively the same as the ones found in recent works, where the so-called point form formalism was applied to the calculation of the nucleon form factor [5]. In particular, because of the wrong power law behavior at high Q^2, the form factors in point form miss the Born amplitude, contrary to all the other approaches. This is a severe shortcoming and one has to wonder about the reasons for these peculiarities.

One possible explanation would be that two-body currents play a much more dominant role in the point form than in all the other approaches. This

Table 1. Elastic vector- and scalar form factors, $F_1(Q^2)$ and $F_0(Q^2)$, for a system bound by the exchange of an infinite-mass boson (zero-range interaction) for two values of the total mass M. The wave function used in the instant form (I.F.), front form (F.F.) and point form (P.F.) cases is issued from Eq. (2). B.S. gives the results for a Bethe-Salpeter calculation, while Gal. corresponds to a calculation employing a Galilean boost. Asymptotic behaviors for $F_1(Q^2)$ are Q^{-2} $(\log Q)^2$, Q^{-2} $(\log Q)^2$, Q^{-4}, Q^{-2} $(\log Q)^2$, and Q^{-1} for I.F., F.F., P.F., B.S. and Gal., respectively.

Q^2/m^2		0.01	0.1	1.0	10.0	100.0
$M = 1.6\,m$						
F_1	I.F.	0.999	0.990	0.917	0.594	0.208
F_0	I.F.	1.325	1.309	1.176	0.658	0.187
F_1	F.F.	0.999	0.989	0.908	0.566	0.191
F_0	F.F.	1.325	1.309	1.176	0.659	0.187
F_1	P.F.	0.999	0.986	0.871	0.353	0.207-01
F_0	P.F.	0.998	0.976	0.800	0.236	0.108-01
F_1	B.S.	0.999	0.989	0.908	0.566	0.191
F_0	B.S.	1.325	1.309	1.176	0.659	0.187
$F_1 = F_0$	Gal.	0.999	0.994	0.947	0.699	0.320
$M = 0.1\,m$						
F_1	I.F.	0.999	0.996	0.963	0.759	0.343
F_0	I.F.	1.498	1.487	1.389	0.920	0.320
F_1	F.F.	0.999	0.995	0.954	0.723	0.315
F_0	F.F.	1.498	1.487	1.389	0.920	0.320
F_1	P.F.	0.839	0.222	0.82-02	0.141-03	0.20-05
F_0	P.F.	0.699	0.130	0.42-02	0.071-03	0.10-05
F_1	B.S.	0.999	0.995	0.954	0.723	0.315
F_0	B.S.	1.498	1.487	1.389	0.920	0.320
$F_1 = F_0$	Gal.	0.999	0.998	0.980	0.846	0.475

possibility is currently investigated [10]. In view of the unitary equivalence of all relativistic quantum mechanics approaches, and because of the huge effect, one might wonder whether this is the most efficient way to proceed.

An alternative to this approach was sketched in ref. [7]. It is based on the observation that, in practice, recent applications of the point-form [4, 5, 6] rely on employing wave functions issued from a mass operator whose solutions can also be identified with instant-form ones in the center of mass system. As noticed by Sokolov [11], this point-form approach is not identical to the one proposed by Dirac, where quantization is performed on a hyperboloïd, $x \cdot x = \tau$.

When the system at rest described on the hyper-plane, $\lambda_0 \cdot x = \tau$, is kinematically boosted to get initial and final states with four-momenta, P_i^μ and P_f^μ, these ones appear as described (quantized) on different surfaces, $\lambda_i \cdot x = \tau$ and $\lambda_f \cdot x = \tau$, where $\lambda^\mu_{i,f} \propto P^\mu_{i,f}$ with $\lambda^2_{i,f} = 1$. This feature results from the identification of point- and instant-form wave functions in the center of mass. It does not correspond to the usual description of a process which, generally, relies on the same definition of the surface at all steps.

What is understood as point form in recent works is usually defined without any reference to surfaces at all. This is not really in the spirit of Dirac, but it is true that one does not need surfaces in order to construct a set of generators that satisfies the Poincaré algebra. However, the problem pointed out here concerns less the construction of the generators of the Poincaré algebra in terms of the total momentum $\boldsymbol{P}$ and the internal variable $\boldsymbol{k}$, but rather the relation of this set of variables to the physical ones. A point form approach in the Dirac sense, with quantization performed on a hyperboloïd, would at least resolve the issues of relations between variables that were raised in this paper.

Acknowledgement. This work has been supported by the EC-IHP Network ESOP, under contract HPRN-CT-2000-00130 as well as the MCYT (Spain) under contract BFM2001-3563-C02-01.

References

1. P.A. Dirac: Rev. Mod. Phys. **21**, 392 (1949)
2. W.H. Klink: Phys. Rev. **C58**, 3587 (1998)
3. T.W. Allen, W.H. Klink: Phys. Rev. **C58**, 3670 (1998)
4. T.W. Allen, W.H. Klink, W.N. Polyzou: Phys. Rev. **C63**, 034002 (2001)
5. R.F. Wagenbrunn, S. Boffi, W. Klink, W. Plessas, M. Radici: Phys. Lett. **B511**, 33 (2001)
6. B. Desplanques, L. Theußl: Eur. Phys. J. **A13**, 461 (2002)
7. B. Desplanques, L. Theußl, S. Noguera: Phys. Rev. **C65**, 038202 (2002)
8. A. Amghar, B. Desplanques, L. Theußl: arXiv:nucl-th/0202046, to appear in Nucl. Phys. **A**
9. B. Bakamjian, L.H. Thomas: Phys. Rev. **92**, 1300 (1953)
10. L. Theußl, B. Desplanques: in preparation
11. S.N. Sokolov: Theor. Math. Phys. **62**, 140 (1985)

Few-Body Systems Suppl. 14, 399–402 (2003)

Few-Body Systems

Locality in Relativistic Quantum Mechanics

F. Coester

Physics Division, Argonne National Laboratory, Argonne IL 60439, USA

Abstract. Manifolds of essentially localized states can be defined within the framework of relativistic quantum mechanics without reference to infinite systems.

The fundamental requirements of relativistic quantum theory can be stated as follows:

1. *Implementation of relativistic space-time symmetry.* This requirement is realized by unitary representations of the inhomogeneous Lorentz group (Poincaré group) including space reflections, and anti-unitary representations of time reversal [1].

2. *Einstein causality: Events localized in relatively space-like space-time regions must be causally independent.* This requires an appropriate definition of states localized in finite space-time regions [2].

Implementation of the locality requirements has been discussed extensively in the literature in the context of the algebraic formulation of quantum field theories [2, 3]. The basic ingredients are operator algebras, nets of localized sub-algebras with the covariance, union, intersection and inclusion properties of the associated space-time regions. "Essentially localized" state vectors obtain by application of local operators to the vacuum state. Essential features of this framework are infinitely many degrees of freedom and inequivalent Hilbert-space representations of the states depending on the dynamics. Finite systems necessarily appear as subsystems of infinite systems [2].

In relativistic quantum mechanics [4] the emphasis has been on the implementation of the Poincaré symmetry by unitary representations of the Poincaré group on the Hilbert space of states which is a finite tensor product of irreducible representations (single-particle states), or a finite direct sum of such spaces. The symmetries include anti-unitary representations of time-reversal and unitary representations of space reflections and charge conjugation. The mass operator $\mathcal{M}$ and the spin j^2 are functions of the ten Poincaré generators $\mathcal{G} := \{P^\mu, J^{\mu\nu}\}$. Eight other functions of the generators, $\{F_i, \Delta F_i\}$, which commute with the mass and spin operators can be defined such that the ten Poincaré operators can be expressed as functions of $\mathcal{M}, j^2, \{F_i, \Delta F_i\}$. Realistic interactions are introduced by modifications of only the mass operator. While the spectra of these mass operators extend to infinity, any particular model constructed in this manner is meant to be applicable only for a limited range of the mass spectrum. The choice of the operators $\{F_i, \Delta F_i\}$, e.g. $\{j_x, j_z, \boldsymbol{P}, \boldsymbol{X}\}$, determines the form of kinematics, that is a subset of generators, if any, which are independent of the mass operator.

Einstein causality is realized asymptotically by the properties of cluster separability [5, 6]. For all partitions into subsystems $\{a_1, \ldots, a_n\}$ and large spacelike separation of the subsystems the Poincaré generators $\mathcal{G}$ must become equivalent to the sum of the subsystem generators, $\mathcal{G} \to \sum_i \mathcal{G}_{a_i}$. Cluster separability has been realized as a strong operator limit of the unitary operators $U(\Lambda, d)$,

$$\|\Big(\mathcal{G} - \sum_i \mathcal{G}_{a_i}\Big) \otimes_i U_{a_i}(\mathbf{1}, d_i)\Psi\| \to 0 \,, \tag{1}$$

for $(d_i - d_j)^2 \to \infty$. The rate of convergence depends on the states Ψ as well as the properties of the generators. In practice this property would be useless unless the limit is approached sufficiently rapidly for products of states relatively localized in finite space-time regions.

Recent mathematical developments [3, 7, 8, 9] indicate how manifolds of localized states can be defined strictly within the frame-work of relativistic quantum mechanics without reference to infinite systems. This construction involves the domains of unbounded operators defined as functions of the Poincaré generators.

States $\Psi(\mathcal{O})$ localized in finite convex space-time regions should satisfy the following properties.

A. Linear Manifold. States localized in $\mathcal{O}$ should form a linear manifold $\mathcal{K}(\mathcal{O})$. For any $\Psi_1 \in \mathcal{K}(\mathcal{O})$ and $\Psi_2 \in \mathcal{K}(\mathcal{O})$ any linear combination $C_1\Psi_1 + C_2\Psi_2 \in \mathcal{K}(\mathcal{O})$.

B. Covariance. For any Poincarŕe transformation $g = \{\Lambda, d\}$ and any $\Psi(\mathcal{O}) \in \mathcal{K}(\mathcal{O})$ transformed states $U(g)\Psi(\mathcal{O})$ should be localized in the region $g\mathcal{O}$, $U(g)\Psi(\mathcal{O}) \in \mathcal{K}(g\mathcal{O})$.

C. Union, Intersection, Inclusion. $\mathcal{K}(\mathcal{O}_1) \cup \mathcal{K}(\mathcal{O}_2) = \mathcal{K}(\mathcal{O}_1 \cup \mathcal{O}_2)$, $\mathcal{K}(\mathcal{O}_1) \cap \mathcal{K}(\mathcal{O}_2) = \mathcal{K}(\mathcal{O}_1 \cap \mathcal{O}_2)$, $\mathcal{K}(\mathcal{O}_1) \subset \mathcal{K}(\mathcal{O}_2)$ if $\mathcal{O}_1 \subset \mathcal{O}_2$.

D. Causal independence If all points in $\mathcal{O}'$ are relatively space-like to all points in $\mathcal{O}$ the the intersection of $\mathcal{K}(\mathcal{O})$ and $\mathcal{K}(\mathcal{O}')$ is empty, $\mathcal{K}(O) \cap \mathcal{K}(\mathcal{O}') = \{0\}$.

The basic geometrical ingredients are wedge regions and their intersections. The relevant definitions and properties of wedges can be briefly summarized as follows. The standard wedge W_1 is defined by the set of points $x_1 > |x^0|$ and $\{x_2, x_3\}$ arbitrary. The wedge reflection R_{W_1}, $R_{W_1} x = \{-x^0, -x_1, x_2, x_3\}$, produces the "causal complement" W_1' of the wedge W_1, that is every point in W_1' is relatively space-like to every point in W_1. The anti-unitary operator J_{W_1} representing the wedge reflection is given by the CPT operator times a rotation by π about the 1-axis. It follows that $J_{W_1}^2 = \mathbf{1}$.

Positive translations $T(d_1)x = \{x^0, x_1 + d_1, x_2, x_3\}$ produce a wedge that is a subset of W_1. The wedge W_1 is invariant under Lorentz boosts Λ_{W_1}, $\Lambda_{W_1}(\chi)x = \{x^0 \cosh\chi + x_1 \sinh\chi, x^0 \sinh\chi + x_1 \cosh\chi, x_2, x_3\}$, for all values of χ. The Lorentz boosts Λ_{W_1} commute with the wedge reflection R_{W_1}.

Every other wedge W is a Poincaré transformation of W_1. Convex, "causally complete" space-time regions $\mathcal{O}$ obtain as intersections of sets $\mathcal{W}$ of wedges,

$$\mathcal{O} = \bigcap_{\mathcal{W} \supset \mathcal{O}} \mathcal{W} \,. \tag{2}$$

A region is called "causally complete" if it includes all points that are relatively space-like to its causal complement.

The relevance of wedges for the implementation of localization requirements relies on the wedge separation property:

- For any two space-like separated, convex, causally complete regions $\mathcal{O}_1$ and $\mathcal{O}_2$ there exists a wedge W such that $\mathcal{O}_1 \subset W$ and $\mathcal{O}_2 \subset W'$.

The anti-unitary wedge reflection operator J_{W_1} commutes with the Lorentz transformations $U(\Lambda_{W_1}) = e^{iK_1\chi}$ that leave the wedge W_1 invariant.

The construction of manifolds of localized states depends on the properties of the Tomita operators

$$S_W := J_W \Delta_W^{\frac{1}{2}} = U^\dagger(\Lambda, d) S_{W_1} U(\Lambda, d) \,, \qquad W = \{\Lambda, d\} W_1 \,, \tag{3}$$

where $\Delta_{W_1} := \exp(2\pi K_1)$ is a self-adjoint operator with a range equal to its domain [7]. Essential properties of the Tomita operator are summarized as follows.

1. The Tomita operator S_W is a densely defined, anti-linear operator on $\mathcal{H}$ with the range $R(S_W)$ equal to the domain $D(S_W)$.

2. The anti-unitary wedge reflection operator J_{W_1} anti-commutes with the generator K_1. Thus $J_W S_W = S_W^{-1} J_W$

3. The operator S_W^2 acts as the identity on its domain, $S_W^2 \subset \mathbf{1}$.

For all $X \in D(S_W)$

$$\forall\, \Psi := X + S_W X \quad S_W \Psi = \Psi \qquad\quad \forall \Phi := X - S_W X \quad S_W \Phi = -\Phi. \tag{4}$$

The manifold $\mathcal{H}_R(W)$ of vectors Ψ satisfying $S_W \Psi = \Psi$ is a real linear manifold. For $\Psi_1 \in \mathcal{H}_R(W)$ and $\Psi_2 \in \mathcal{H}_R(W)$ and real coefficients c_1, c_2 the linear combination $c_1\Psi_1 + c_2\Psi_2 \in \mathcal{H}_R(W)$.

$$\mathcal{H}_R(W) + i\mathcal{H}_R(W) = D(S_W) \,, \qquad \mathcal{H}_R(W) \cap i\mathcal{H}_R(W) = \{0\} \,. \tag{5}$$

If $X \in \mathcal{H}_R(W) \cap i\mathcal{H}_R(W)$ then $X = S_W X$ and $X = -S_W X$. Hence $X = 0$.

Relevant properties of the manifold $\mathcal{H}_R(W)$ are summarized as follows:

1. $\mathcal{H}_R(W)$ is Poincaré covariant

$$U(\Lambda, d)\mathcal{H}_R(W) = \mathcal{H}_R(\{\Lambda, d\}W) , \qquad J_W\mathcal{H}_R(W) = \mathcal{H}_R(W') , \tag{6}$$

2. $\mathcal{H}_R(W)$ satisfies the inclusion property,

$$W_1 \subset W_2 \qquad \text{implies} \qquad \mathcal{H}_R(W_1) \subset \mathcal{H}_R(W_2) . \tag{7}$$

3. $\mathcal{H}_R(W)$ satisfies"wedge duality" $\mathcal{H}_R(W) \cap \mathcal{H}'_R(W) = \{0\}$.

From the definition $\Psi' := J_W\Psi = \Delta^{1/2}\Psi$ for some Ψ satisfying $\Psi = S_W\Psi$ it follows that $S_W\Psi' = J_W\Delta_W^{1/2}J_W\Psi = \Delta^{-1/2}\Psi = \Delta^{-1}\Psi' \neq \pm\Psi'$.

For regions $\mathcal{O}$ that are intersections of wedges the $\mathcal{H}_R(\mathcal{O})$ is defined by $\mathcal{H}_R(W_1 \cap W_2) := \mathcal{H}_R(W_1) \cap \mathcal{H}_R(W_2)$. Since causally complete convex regions are intersections of wedges, the map $W \to \mathcal{H}_R(W)$ extends to causally complete convex regions $\mathcal{O}$ by

$$\mathcal{H}_R(\mathcal{O}) = \bigcap_{W \supset \mathcal{O}} \mathcal{H}_R(W) . \tag{8}$$

The properties summarized above depend only on the properties of the Poincaré representations. They for hold for massive single-particle states and of composite systems. For interacting representations the Tomita operator has the same cluster properties as the Poincaré representations, providing for relative localization of subsystems. This suggests that the manifold of states $\Psi(\mathcal{O}) \in \mathcal{H}_R(\mathcal{O}) + i\mathcal{H}_R(\mathcal{O})$ may provide localization of particle states in the framework of relativistic quantum mechanics without reference to infinite systems.

Acknowledgement. I am indebted to Bert Schroer for valuable discussions. Supported in part by the U.S. Department of Energy, Nuclear Physics Division, contract W-31-109-ENG-38.

References

1. E. P. Wigner, Ann. Math. **40**, 141 (1939)
2. R. Haag, "Local Quantum Physics", Springer Verlag (1993)
3. H.J. Borchers J. Math. Phys. **41**, 3604 (2000)
4. B.D. Keister and W. N. Polyzou, Adv. in Nuc. Phys. **20**, 225 (1991)
5. F. Coester and W.N. Polyzou, Phys. Rev. **D26** 1348 (1982)
6. W.N. Polyzou, J. Math. Phys. to be published.
7. R. Brunetti, D. Guido and R. Longo, "Modular Localization and Wigner Particles"; math-ph/0203021v2
8. J. Mund, Annales Henri Poincaré **2** 907 (2001)
9. L. Fasarella and B. Schroer: J. Phys. A: Math. Gen. **35** (2002)

Few-Body Systems Suppl. 14, 403–406 (2003)

Few-Body Systems

A Relativistic Point-Form Approach to Quark-Antiquark Systems

A. Krassnigg[1], W. Schweiger[1], W.H. Klink[2]

[1] Institute for Theoretical Physics, University of Graz, Universitätsplatz 5, A-8010 Graz, Austria

[2] Department for Physics and Astronomy, The University of Iowa, Iowa City, IA 52242, USA

Abstract. We develop a formalism for the treatment of a two-particle system interacting via one-boson exchange in the point form of relativistic quantum mechanics. A Poincaré invariant coupled-channel treatment leads to a generalized eigenvalue equation for the mass operator of the system. The formalism is applied to vector mesons in a constituent-quark picture including confinement for the quark and antiquark degrees of freedom. A dynamical equation which yields a mass spectrum is solved by the use of a resonance condition. The results are used to examine relativistic effects as compared to earlier calculations and they are also compared to experimental data.

1 Introduction

In the recent development of Constituent Quark Models (CQMs) special relativity has taken on a prominent role. Results for particle spectra as well as hadronic and electromagnetic baryon properties have shown considerable improvement whenever relativistic effects have been incorporated into the formalism. One can perform Poincaré invariant calculations for observables such as electromagnetic form factors or hadronic baryon decay widths [1]. The next important step is to also treat the hyperfine interaction by going beyond the commonly used instantaneous approximation.

In the present contribution we present a relativistic coupled-channel treatment for a two-particle system of one (constituent) quark and one (constituent) antiquark which form vector mesons. The particular model employed for the hyperfine interaction is the chiral CQM of ref. [2]. This model has been used with great success for baryons. For vector mesons [3], the results are not as good. In the work presented here a fully relativistic point-form calculation with the exchange-meson channel explicitly included is presented and compared to a semirelativistic approach, in which the meson-exchange is treated in an instantaneous approximation.

For more details on the point form formalism we refer to refs. [4]. States labeled by the overall four-velocity [5] are used as a basis for the quantum-mechanical treatment of few-particle systems in point-form. The setup of the coupled two-channel problem, the structure of the dynamical equation, and the implementation of confinement are discussed in Sec. 2. Section 3 contains some comments on the numerics and the specific solution method employed in the calculations. The results of the calculations are vector meson masses and some branching ratios for their hadronic decay widths. The numbers for the masses are presented in Sec. 4, while the results for the widths are given in ref. [6].

2 Coupled-Channel Setup and the Eigenvalue Equation

The dynamical equation to be solved is the eigenvalue equation for the mass operator $M|\Psi\rangle = m|\Psi\rangle$. In a coupled-channel approach the mass operator has a matrix structure and the eigenvalue equation reads

$$M|\Psi\rangle = \begin{pmatrix} D_2^0 & K^\dagger \\ K & D_{2+1}^0 \end{pmatrix} \begin{pmatrix} |\Psi_2\rangle \\ |\Psi_{2+1}\rangle \end{pmatrix} = m \begin{pmatrix} |\Psi_2\rangle \\ |\Psi_{2+1}\rangle \end{pmatrix} , \tag{1}$$

describing a system with an 2-particle plus 2+1-particle Hilbert space, coupled by a transition (vertex) operator K. D_2^0 denotes the free mass operator for 2 particles and is in our case replaced by M_2^c, which provides for the confinement of the quark-antiquark system. For convenience we have chosen a harmonic oscillator confining potential in the two-particle diagonal part of the square of the coupled-channel mass operator, which produces a pseudolinear confinement interaction in M. D_{2+1}^0 is replaced by $D_{2+1}^c = \sqrt{(M_2^c)^2 + \kappa^2} + \sqrt{m_\Pi^2 + \kappa^2}$, where κ is the relative momentum between the quark-antiquark cluster and the free pseudoscalar meson Π. The harmonic-oscillator eigenfunctions are used later as a basis for expanding the quark-antiquark wave functions. As a result, one can discretize the dynamical equation and obtain a set of coupled algebraic equations that can be solved using standard techniques. Our hyperfine interaction arises from the well-known pseudoscalar interaction Hamiltonian density $\bar{\psi}\gamma_5 \boldsymbol{\lambda}_F \psi \boldsymbol{\phi}$ in the way described in refs. [4]. The resulting mass operator is of Bakamjian-Thomas type [7] and acts on a Hilbert space which is the direct sum of a quark-antiquark and a quark-antiquark-pseudoscalar-meson Hilbert space.

Writing out the matrix eigenvalue equation for the mass operator (1) results in two coupled equations for $|\Psi_2\rangle$ and $|\Psi_{2+1}\rangle$, from which the latter can be eliminated to yield

$$M_2^c|\Psi_2\rangle + K^\dagger(m - D_{2+1}^c)^{-1}K|\Psi_2\rangle = m|\Psi_2\rangle . \tag{2}$$

Because of the appearance of the eigenvalue m also in the denominator of the interaction term, Eq. (2) is actually a generalized eigenvalue equation. It is subsequently transformed into an integral equation by the use of velocity states.

3 Comments on the Numerics and Solution Strategy

After the manipulations described above one is left with a set of algebraic equations, where the coefficients of the variables are nine-dimensional integrals. These are solved using standard Monte-Carlo techniques. Numerical errors are kept smaller than one MeV, so that a presentation of numbers rounded to MeV is reasonable. The computer time used for the solution of the generalized eigenvalue problem for one particular flavor state of a vector meson in the current configuration is approximately one day on a standard PC.

As a solution strategy for the generalized eigenvalue equation we adopt a resonance-condition approach used, e.g., in ref. [8]. There one presets several mass-eigenvalues for the system and inserts them, one at a time, into the interaction term. For each of those values the equation can then be solved like an ordinary eigenvalue equation and yields a spectrum, which will be complex if the preset eigenvalue lies above the threshold for the emission of the exchange particle. Plotting the spectrum as a function of the preset eigenvalue m_{pre} and matching the real parts of the eigenvalues as functions of m_{pre} with m_{pre} one determines the positions of the actual eigenvalues of the generalized problem. The widths of the resonances are then determined via the imaginary parts of the eigenvalue functions at the resonance positions found earlier.

4 Results and Discussion

Most of the parameters (constituent-quark and pseudoscalar-meson masses, octet coupling strength $g_{\mathrm{PS}}^2 = 8.42$, and ratio of singlet/octet coupling $(g_0/g_8)^2 = 1.34$) in our calculation are taken from ref. [2]. We adjust the confinement parameters $a = 312\,\mathrm{MeV}$ and $V_0 = -1.045\,\mathrm{GeV}^2$, whose meaning can be best seen from the eigenvalues of the corresponding harmonic oscillator mass operator. The eigenvalues are $M_{nl} = \sqrt{8\,a^2\left(2n + l + \frac{3}{2}\right) + V_0 + 4\bar{m}^2}$, where $\bar{m}^2$ is the mean value of m^2 for the confined quark and antiquark and V_0 is an overall shift constant. The oscillator parameters are fixed to reproduce position and splitting of the ϱ ground and first excited states.

The results presented in this paper have been obtained using oscillator states up to $n = 5$ which gives fully converged results. Nonzero orbital angular momentum states have not been included in this calculation, and in this regard our results are preliminary. Nevertheless the spectrum for the vector mesons is already quite reasonable, especially concerning the ω ground state, which comes out much better than when the instantaneous approximation is used for the hyperfine interaction. The numbers (in MeV) for particle masses are given in Fig. 1.

5 Conclusions and Outlook

Our calculations have shown that a fully relativistic coupled-channel treatment of an interacting two-particle system is possible within the point-form

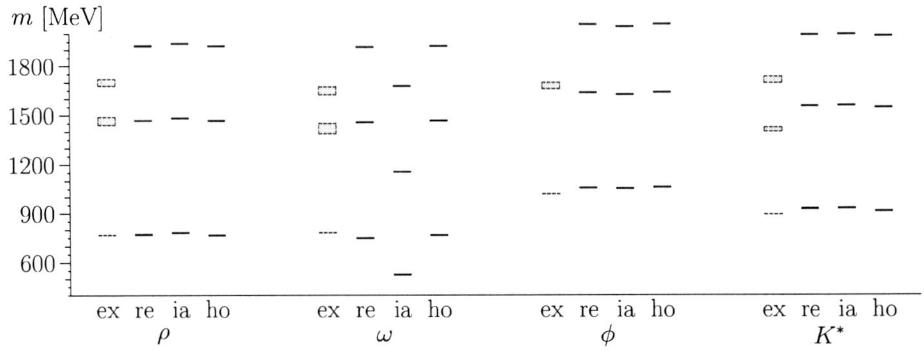

Figure 1. Results for vector-meson masses of ground, first, and second excited states: Experimental data, point form, instantan. approximation, harmonic oscillator only.

approach. The results for vector-meson spectra are considerably better than those using an instantaneous approximation for the hyperfine interaction. Also as compared with the instantaneous approximation, the convergence of the full calculation with increasing numbers of basis states is clearly superior. We therefore conclude that relativistic effects are essential in the treatment the hyperfine interaction. Inclusion of states with higher orbital momentum are expected to further improve the results. The present formalism should also be useful in treating other few-body decaying systems, such as positronium.

Acknowledgement. This work was conducted under FWF-project Nr. P14794 and supported in part by the Dekanat der Naturwissenschaftlichen Fakultät der Universität Graz.

References

1. R.F. Wagenbrunn et al.; T. Melde et al.: both in this volume
2. L.Ya. Glozman et al.: Phys. Rev. **D58**, 094030 (1998)
3. T. Thonhauser: Master's thesis. Univ. Graz 1998 (unpublished)
4. W.H. Klink: This volume; W.H. Klink: nucl-th/0012031, to be published in Nucl. Phys. **A**
5. W.H. Klink: Phys. Rev. **C58**, 3617 (1998)
6. W. Schweiger et al.: This Volume
7. B. Bakamjian and L.H. Thomas: Phys. Rev. **92**, 1300 (1953); B.D. Keister and W.N. Polyzou: In: Adv. Nucl. Phys. Vol. 20, p. 225. New York: Plenum Press 1991
8. W.H. Blask: Ph.D. Thesis. Univ. Bonn 1990 (unpublished)

Few-Body Systems Suppl. 14, 407–410 (2003)

Few-Body Systems

Poincaré Invariance, Cluster Properties, and Particle Production

W. Polyzou*,†

Department of Physics and Astronomy, The University of Iowa, Iowa City, IA 52242, USA

Abstract. A method is presented for constructing a class of Poincaré invariant quantum mechanical models of systems of a finite number of degrees of freedom that satisfy cluster separability, the spectral condition, but do not conserve particle number. The class of models includes the relativistic Lee model [1] and relativistic isobar models.

Any reasonable quantum theory of strongly interacting particles should be Poincaré invariant, and should satisfy cluster properties and the spectral condition. These conditions are physical requirements. In addition, the theory should be able to model reactions that change particle number. These conditions require the existence of a unitary representation of the Poincaré group $\hat{U}(\Lambda, a)$ [2] on the model Hilbert space $\mathcal{H}$ that satisfies a cluster separability condition and the spectral condition.

While these requirements are dictated by physics, it is difficult to find well-defined mathematical models for realistic systems that are consistent with all of these properties. Relativistic quantum theory of N-particles [1, 3] is a mathematically well-defined theory where these conditions can be satisfied. A mathematically well-defined theory is required to formulate high-precision ab-initio calculations, which are important for strongly interacting systems.

This talk explains how to extend the relativistic quantum theory of N particles [1, 3] to treat a class of models with particle production. This class of models has a bounded number of degrees of freedom and includes the relativistic Lee model [1] and relativistic isobar models.

The key observation is that mass is not superselected in relativistic quantum theory. To contrast this with non-relativistic quantum theory, consider a

* *E-mail address:* polyzou@uiowa.edu

†This work Supported by the U.S. Department of Energy, Nuclear Physics Division, contract DE-FG02-86ER40286.

reaction involving a fixed number of constituent particles, where a bound target is broken up into a number of subsystems. In this reaction, the mass and number of particles in the initial and final states is different. This reaction cannot be consistently modeled with a non-relativistic quantum theory because it violates Galilean invariance if the Galilean and inertial masses are identified.

The difference between the above reaction and a general production reaction is the presence of a conserved number of constituent particles, which limits the number of degrees of freedom. Motivated by this observation, this talk considers production reactions that have a set of conserved quantum numbers which are associated with particles called "atomic partons". Additional bare particles with *composite* atomic-parton quantum numbers are introduced. The model Hilbert space is a direct sum of tensor products of bare and atomic particle subspaces with the same atomic-parton numbers. The dynamics is restricted to conserve atomic-parton number.

For example, a model where the atomic partons are a nucleon and two pions could have a bare Δ, with the quantum numbers of a nucleon and a pion, and a bare ρ, with the quantum numbers of two pions. Introducing bare particles does not mean that these particles exist as stable particles. Stable particles correspond to point eigenstates of the mass operator. This talk investigates the modifications of the fixed N-construction [1, 3] which are needed to treat this type of particle production.

In the above example the conserved atomic-parton numbers are $\mathcal{N} := \{N_N, N_\pi\}$. The Hilbert space, $\mathcal{H}_\mathcal{N}$, is a direct sum of tensor products of irreducible representation spaces of the Poincaré group, where the particles in each tensor product have atomic parton number $\mathcal{N}$

$$\mathcal{H}_\mathcal{N} = \oplus_i(\otimes\mathcal{H}_{a_i}); \text{ i.e. } \quad \mathcal{H}_{\{1,1\}} = (\mathcal{H}_N \otimes \mathcal{H}_\pi) \oplus \mathcal{H}_\Delta .$$

There is a natural partial ordering on the quantum numbers $\mathcal{N}$ given by $\mathcal{N} \leq \mathcal{N}'$ if $N_i \leq N_i'$ for each type, i, of atomic parton. The space $\mathcal{H}_\mathcal{N}$ has a natural non-interacting representation of the Poincaré group, $\hat{U}_0(\Lambda, a)$

For each partition, b, of the atomic partons into disjoint equivalence classes, b_i, there is a factorization of the model Hilbert space, $\mathcal{H}_\mathcal{N}$, into a direct sum of a tensor product, $\mathcal{H}_b$, of n_b proper subsystem Hilbert spaces, $\mathcal{H}_{b_i}$, with $\mathcal{N}_i < \mathcal{N}$, and a residual space, $\mathcal{H}^b$, corresponding to bare particles with quantum numbers in more than one equivalence class:

$$\mathcal{H}_\mathcal{N} = \mathcal{H}_b \oplus \mathcal{H}^b \qquad \mathcal{H}_b := \otimes_{b_i \in b}\mathcal{H}_{b_i} .$$

The difference between the construction outlined in this talk and the construction for fixed N in [1, 3] involves the modifications that are necessary to treat the residual space $\mathcal{H}^b$. The asymptotic dynamics and translation operators associated with partition b are needed to formulate cluster properties. On $\mathcal{H} = \mathcal{H}_b \oplus \mathcal{H}^b$ these operators can be expressed in terms of subsystem representations $\hat{U}_{b_i}(\Lambda, a)$ as:

$$\hat{U}_b(\Lambda, a) := \begin{pmatrix} \otimes \hat{U}_{b_i}(I, y_i) & 0 \\ 0 & \hat{I}^b \end{pmatrix} \qquad \hat{T}_b(y_1, \cdots, y_k) := \begin{pmatrix} \otimes_i \hat{U}_{b_i}(I, y_i) & 0 \\ 0 & \hat{I}^b \end{pmatrix}. \tag{1}$$

The projection on $\mathcal{H}_b$,

$$\hat{\Pi}_b := \begin{pmatrix} \hat{I} & 0 \\ 0 & 0 \end{pmatrix},$$

is used to define a modified formulation of the cluster property

$$\lim_{\min(y_i - y_j)^2 \to \infty} \| [\hat{U}(\Lambda, x) - \hat{U}_b(\Lambda, x)] \hat{T}_b(y_1, \cdots, y_k) \hat{\Pi}_b |\psi\rangle \| = 0. \tag{2}$$

This condition requires more than a range condition on the interaction; it also requires that part of the operator that survives asymptotically is a tensor product of proper-subsystem representations [3].

In this formulation $\mathcal{N}$-body interactions, $\hat{V}_{\mathcal{N}}$, satisfy

$$\lim_{\min(y_i - y_j)^2 \to \infty} \| \hat{V}_{\mathcal{N}} \hat{T}_b(y_1, \cdots y_k) \hat{\Pi}_b |\psi\rangle \| = 0 \quad \forall b.$$

The dynamics of all proper subsystems determine the dynamics of the full system up to an $\mathcal{N}$-body interaction.

The construction of a two atomic parton dynamics uses *Wigner's forms of the dynamics* [4, 3]; Clebsch-Gordan coefficients of the Poincaré group are used to construct an irreducible basis for the non-interacting system. Interactions are added to the non-interacting mass that are block-diagonal in the non-interacting spin, $\hat{j}_0$, and commute with and are independent of the commuting functions, $\hat{F}_0$, of the Poincaré generators that label vectors in irreducible subspaces. It follows that simultaneous eigenstates of the mass, $\hat{M}$, $\hat{j}_0$, and $\hat{F}_0$ define an irreducible basis for an interacting representation of the Poincaré group. In this representation the mass eigenvalue and non-interacting spin are the new Casimir operators.

The many-body construction is recursive. The representations $\hat{U}_b(\Lambda, a)$ are constructed from tensor products of proper subsystem, $(\mathcal{N}_i < \mathcal{N})$, representations. If $\hat{U}_b(\Lambda, a)$ is decomposed into irreducible representations, the spin ,$\hat{j}$, and the vector operators, $\hat{F}$, depend on the interaction. If the order of constructing irreducible representations and adding interactions is reversed, the result is a mass operator, $\bar{M}_b$, that commutes with $\hat{j}_0$, and $\hat{F}_0$, and is independent of $\hat{F}_0$. Simultaneous eigenstates of $\bar{M}_b$, $\hat{j}_0$, and $\hat{F}_0$ transform irreducibly and define a representation, $\bar{U}_b(\Lambda, a)$, of the Poincaré group. The representations $\bar{U}_b(\Lambda, a)$ and $\hat{U}_b(\Lambda, a)$ are related by a unitary scattering equivalence $\hat{B}_b$ [1, 4] of the form:

$$\hat{B}_b := \begin{pmatrix} \hat{A}_b & 0 \\ 0 & \hat{I}^b \end{pmatrix} \qquad \bar{U}_b(\Lambda, a) = \hat{B}_b \hat{U}_b(\Lambda, a) \hat{B}_b^\dagger \tag{3}$$

which satisfies

$$\lim_{t\to\pm\infty} \|[(\hat{B}_b - \hat{I})\hat{\Pi}_0\hat{U}_0(I,t)|\psi\rangle\| = 0,$$

where $\hat{\Pi}_0$ is the projector on the $\mathcal{N}$-atomic parton subspace of $\mathcal{H}$. For systems of more than three atomic partons, the freedom to choose the operators $\hat{B}_b$ must be utilized to ensure that the $\hat{B}_b$ are also consistent with cluster properties. The techniques used the N-particle case[1, 4] can be applied to the modified operators (1) and (3).

All interactions appear in the $\bar{U}_b(\Lambda, a)$ as b ranges over all partitions with at least two equivalence classes, except the $\mathcal{N}$-body interactions. The mass operators $\bar{M}_b$ for the different $\bar{U}_b(\Lambda, a)$ can be combined, using an Ursell cluster expansion, to define

$$\bar{M} := \sum_b (-)^{n_b}(n_b - 1)!\hat{B}_b\hat{M}_b\hat{B}_b^\dagger,$$

which commutes with the non-interacting $\mathcal{N}$-body spin $\hat{j}_0$ and the non-interacting $\hat{F}_0$. Simultaneous eigenstates of $\bar{M}$, $\hat{j}_0$, $\hat{F}_0$ define a basis that transforms irreducibly and thus a representation $\bar{U}(\Lambda, a)$ of the $\mathcal{N}$-body dynamics.

The representation $\bar{U}(\Lambda, a)$ fails to satisfy the cluster property (2) because of the transformations $\hat{B}_b$ appearing in Eq. (3). This can be remedied by defining

$$\hat{B} = \frac{1 - i\hat{\beta}}{1 + i\hat{\beta}} \qquad \hat{\beta} = \sum_b (-1)^{n_b}(n_b - 1)!\hat{\beta}_b$$

$$\hat{\beta}_b := i\frac{\hat{B}_b - \hat{I}}{\hat{B}_b + \hat{I}}.$$

When the Cayley transforms, $\hat{\beta}_b$, are bounded the operator $\hat{B}$ is a scattering equivalence and the representation

$$\hat{U}(\Lambda, a) := \hat{B}^\dagger\bar{U}(\Lambda, a)\hat{B}$$

satisfies cluster properties (2) for suitable short-ranged [3] interactions.

This construction shows how the method for constructing relativistic quantum theory of a fixed number of particles can be extended to treat a class reactions with particle production. Conservation of atomic parton number allowed the separation of the difficulties associated with particle production from the difficulties associated with having an infinite number of degrees of freedom.

References

1. F. Coester and W.N. Polyzou: Phys. Rev. **D26**, 1348 (1982)
2. E.P. Wigner: Ann. Math. **40**, 141 (1939)
3. W.N. Polyzou: J. Math. Phys. (in press); nucl-th/0201013
4. W. Polyzou: Ann. Phys. **193**, 367 (1989)

Few-Body Systems Suppl. 14, 411–414 (2003)

Few-Body Systems

Nucleon Electromagnetic and Axial Form Factors in Point-Form Relativistic Quantum Mechanics

R.F. Wagenbrunn[1], S. Boffi[2], L.Ya. Glozman[1], W.H. Klink[3], W. Plessas[1], M. Radici[2]

[1] Institut für Theoretische Physik, Universität Graz, A-8010 Graz, Austria

[2] Dipartimento di Fisica Nucleare e Teorica, Università di Pavia, and Istituto Nazionale di Fisica Nucleare, Sez. di Pavia, Via Bassi 6, I-27100 Pavia, Italy

[3] Department of Physics and Astronomy, University of Iowa, Iowa City, IA 52242, USA

Abstract. Results for the proton and neutron electric and magnetic form factors as well as the nucleon axial form factor are presented for constituent quark models, based on either one-gluon-exchange and Goldstone-boson-exchange dynamics. The calculations are performed in a covariant framework using the point-form approach to relativistic quantum mechanics. The only input to the calculations is the nucleon wave function of the corresponding constituent quark model. A comparison is given to results of the instanton-induced constituent quark model treated with the Bethe-Salpeter equation.

It has become fairly obvious that nucleon electromagnetic as well as axial form factors need a relativistic treatment. In order to satisfy all requirements of Lorentz covariance one can either proceed along relativistic field theory or relativistic (Poincaré-invariant) quantum mechanics. The latter approach appears promising whenever one can deal with a fixed number of particles (or a finite number of degrees of freedom). Usually one follows one of the three forms of dynamics, where the generators of the Poincaré group are minimally affected by interactions. Dirac defined them as instant, front, and point forms [1].

The point form is very convenient for practical calculations since only the components of the four-momentum operator $\hat{P}^{\mu}$ contain interactions. Therefore all Lorentz transformations remain purely kinematic and the theory is manifestly covariant (for more details see ref. [2]). In order to construct the interacting four-momentum operators one follows the Bakamjian-Thomas construction [3], where the interaction is introduced into the mass operator $\hat{M}$ by adding to the free mass operator $\hat{M}_{\rm fr}$ the interacting term $\hat{M}_{int}$. Through multiplication with the free four-velocity operator $\hat{V}^{\mu}_{\rm fr}$, which is not affected by interactions, one obtains

$$\hat{P}^{\mu} = \hat{M}\hat{V}^{\mu}_{\mathrm{fr}} = (\hat{M}_{\mathrm{fr}} + \hat{M}_{int})\hat{V}^{\mu}_{\mathrm{fr}}. \tag{1}$$

Poincaré invariance implies that $\hat{M}$ commutes with $\hat{V}^{\mu}_{\mathrm{fr}}$ and is a scalar under Lorentz transformations. Therefore eigenstates of the four-momentum operator are simultaneous eigenstates of both the mass and the velocity operators. As a consequence the motion of the system as a whole and the internal motion are separated. The latter is described by a wave function containing only the internal degrees of freedom. It can be found by solving the eigenvalue problem for the mass operator

$$\hat{M}\Psi = M\Psi. \tag{2}$$

The electromagnetic and axial form factors of the nucleons are obtained by sandwiching the electromagnetic and axial current operators between eigenstates $|P, \Sigma\rangle$, where P and Σ are the eigenvalues of the total momentum and the z-component of the total angular momentum (for details see refs. [4]). The calculation boils down to matrix elements of the current operators between free three-particle states $|p_1, p_2, p_3; \sigma_1, \sigma_2, \sigma_3\rangle$, where p_i are the individual quark four-momenta and σ_i their spin projections. At the present stage we cannot yet deal with the full current operators but have to truncate them to one-body operators by the so-called point-form spectator approximation (PFSA) [2]

$$\begin{aligned} &\langle p'_1, p'_2, p'_3; \sigma'_1, \sigma'_2, \sigma'_3 | \hat{J}^{\mu}(0) | p_1, p_2, p_3; \sigma_1, \sigma_2, \sigma_3\rangle = \\ &2E_2\delta(\boldsymbol{p}'_2 - \boldsymbol{p}_2)\delta_{\sigma'_2\sigma_2} 2E_3\delta(\boldsymbol{p}'_3 - \boldsymbol{p}_3)\delta_{\sigma'_3\sigma_3} \langle p'_1, \sigma'_1 | \hat{j}^{\mu}(0) | p_1, \sigma_1\rangle , \end{aligned} \tag{3}$$

and similarly for the axial current $\hat{\boldsymbol{A}}^{\mu}(0)$ with $\hat{j}^{\mu}(0)$ replaced by $\hat{\boldsymbol{a}}^{\mu}(0)$. By the small letters we indicate free one-body currents of the constituent quarks. The axial current is denoted as a vector in isospin space. With regard to the PFSA it is important to notice that the impulse $\tilde{q} = p'_1 - p_1$ delivered to the single constituent quark is different from the impulse $q = P' - P$ delivered to the nucleon as a whole. The momentum transfer $\tilde{q}$ can be uniquely determined from q and the two spectator conditions $\boldsymbol{p}'_2 = \boldsymbol{p}_2$ and $\boldsymbol{p}'_3 = \boldsymbol{p}_3$. For the one-body current matrix elements in the above equation we employ the usual expressions for electromagnetic and axial currents of pointlike spin-$\frac{1}{2}$ particles (see refs. [4]).

The results for all electromagnetic and axial form factors from the Goldstone-boson-exchange (GBE) CQM of ref. [5] have already been published in refs. [4]. There it was found that relativistic effects are most important. The direct predictions of the GBE CQM in PFSA come remarkably close to the experimental data in all instances. This observation has recently been confirmed also with regard to the electric radii and magnetic moments not only of the nucleons but all (measured) octet and decuplet ground states [6].

Here we present a comparison of nucleon form factor results from different CQMs and different relativistic approaches. First, we compare the PFSA predictions of the GBE CQM with analogous results from a CQM whose hyperfine interaction is based on one-gluon exchange (OGE), namely a relativized version

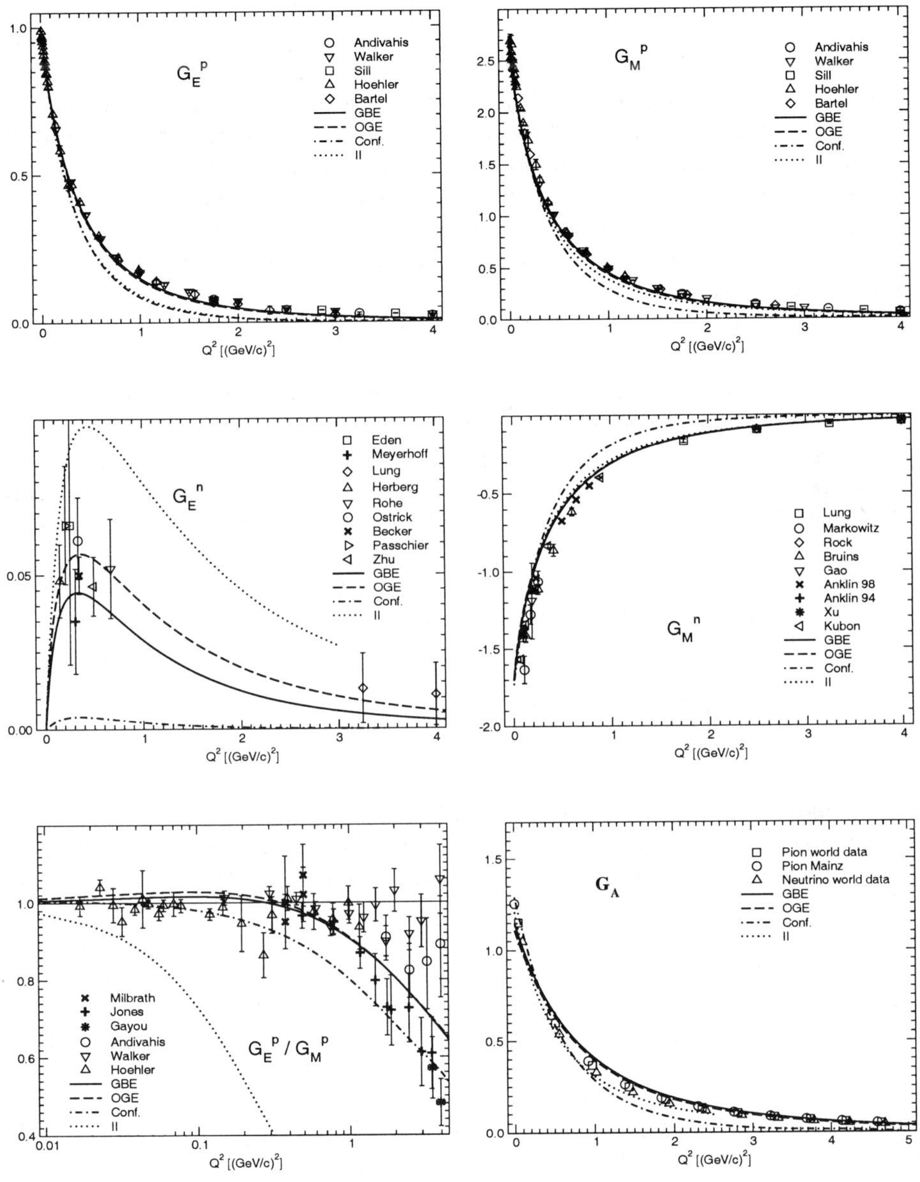

Figure 1. Predictions of different CQMs for the nucleon electromagnetic and axial form factors. The solid and dashed lines represent our PFSA results for the GBE and OGE CQMs, respectively; the dash-dotted line refers to the case with confinement only. The dotted lines show the results of the II CQM within the Bethe-Salpeter approach after ref. [9].

of the Bhaduri-Cohler-Nogami CQM [7] as parametrized in ref. [8]. Then we also provide a comparison to the results of an instanton-induced (II) CQM as obtained by the Bonn group within a Bethe-Salpeter (BS) formalism [9].

As is immediately evident from the results collected in Fig. 1, the overall behaviour of the relativistic predictions appears quite reasonable in all cases. This confirms the previous findings that the inclusion of relativity is most important for the nucleon form factors. For their gross properties dynamical effects are of lesser relevance. Even a three-quark wave function that relies solely on confinement produces the right features, except for the neutron. In this case a realistic wave function is required, with the mixed-symmetry spatial components taken into account. For the proton form factors and also the nucleon axial form factor there is a striking similarity of the results obtained in PFSA and in the BS approach, where the latter also uses a single-particle approximation for the current operators. Only for the subtle details of the neutron electric form factor and the ratio of the proton electric to magnetic form factors the predictions of the II CQM fall short compared to experimental data. A comparison of the CQM results for electric radii and magnetic moments is given in ref. [6], yielding a picture congruent with the one found here.

Acknowledgement. This work was supported by the Austrian Science Fund (Project P14806).

References

1. P.A.M. Dirac: Rev. Mod. Phys. **21**, 392 (1949)
2. W.H. Klink: Phys. Rev. **C58**, 3587 (1998)
3. B. Bakamjian and L.H. Thomas: Phys. Rev. **92**, 1300 (1953)
4. R.F. Wagenbrunn, S. Boffi, W. Klink, W. Plessas, and M. Radici: Phys. Lett. **B511**, 33 (2001); L.Ya. Glozman, M. Radici, R.F. Wagenbrunn, S. Boffi, W. Klink, and W. Plessas: Phys. Lett. **B516**, 183 (2001); S. Boffi, L.Ya. Glozman, W. Klink, W. Plessas, M. Radici, and R.F. Wagenbrunn: Eur. Phys. J. **A14**, 17 (2002)
5. L.Ya. Glozman, W. Plessas, K. Varga, and R.F. Wagenbrunn: Phys. Rev. **D58**, 094030 (1998)
6. K. Berger: Diploma Thesis, Univ. Graz (2002); K. Berger, W. Plessas, and R.F. Wagenbrunn: Prog. Part. Nucl. Phys., to appear; see also K. Berger, W. Plessas, and R.F. Wagenbrunn: These Proceedings
7. R.K. Bhaduri, L.E. Cohler, and Y. Nogami, Nuovo Cim. A **65**, 376 (1981)
8. L. Theußl, R.F. Wagenbrunn, B. Desplanques, and W. Plessas: Eur. Phys. J. **A12**, 91 (2001)
9. D. Merten, U.Löring, K. Kretzschmar, B. Metsch, and H.R. Petry: Eur. Phys. J. **A14**, 477 (2002)

Few-Body Systems Suppl. 14, 415–416 (2003)

Few-Body Systems

Heuristic Model of Two-Quark Relativistic System with Scalar–Vector Interaction

A. Duviryak

Institute for Condensed Matter Physics, Lviv, UA-79011, Ukraine

I. Todorov has observed that the Klein-Gordon equation

$$\left\{\mathbf{p}^2 + [m + V_{\rm s}(r)]^2 - [E - V_{\rm v}(r)]^2\right\}\Psi(\mathbf{r}) = 0 \tag{1}$$

(here $\mathbf{p} = -\mathrm{i}\,\boldsymbol{\nabla}$) can be transformed, by means of the substitution

$$E \to E_M = \frac{M^2 - m_1^2 - m_2^2}{2M}, \qquad m \to m_M = \frac{m_1 m_2}{M}, \tag{2}$$

into the quasipotential equation describing the scalar $V_{\rm s}(r)$ and vector $V_{\rm v}(r)$ interactions of two spinless particles of the rest masses m_a $(a = 1,2)$ [1]. In the case of Coulomb potentials this equation reproduces the QFT spectrum of the total mass M of the system up to α^4–terms of coupling constant expansion.

Here we construct the relativistic potential model of mesons by the choice

$$V_{\rm s}(r) = ar, \qquad V_{\rm v}(r) = -\alpha/r \tag{3}$$

(with $a > 0$ and $\alpha > 0$). It leads in the nonrelativistic limit to the funnel potential model which is appropriate to the description of heavy quarkonia. In the relativistic domain one obtains the asymptotically linear Regge trajectories:

$$M^2 = 8a\left[\ell + 2n_r + \tfrac{3}{2} - \sqrt{2}\alpha\right] + 2\left(m_1^2 + m_2^2 + \sqrt{2}m_1 m_2\right) + O(\ell^{-1}), \tag{4}$$

where $n_r = 0, 1, \ldots$ and $\ell = 0, 1, \ldots$ are the radial and angular quantum numbers, and $\ell \gtrsim n_r$. This structure and $(\ell + 2n_r)$–degeneracy of the spectrum (4) are characteristic of actual light meson spectra, up to fine splitting.

One can account spin effects in Eqs. (1)-(3) by $V \to \tilde{V} = V + W$, where

$$\begin{aligned} W_{\rm s} &= -\frac{1}{4}\mathbf{L}\cdot\left(\frac{\boldsymbol{\sigma}_1}{m_1^2} + \frac{\boldsymbol{\sigma}_2}{m_2^2}\right)\frac{V_{\rm s}'(r)}{r}, \\ W_{\rm v} &= \frac{1}{4}\mathbf{L}\cdot\left(\left[\frac{1}{m_1^2} + \frac{2}{m_1 m_2}\right]\boldsymbol{\sigma}_1 + \left[\frac{1}{m_2^2} + \frac{2}{m_1 m_2}\right]\boldsymbol{\sigma}_2\right)\frac{V_{\rm v}'(r)}{r} \\ &\quad + \frac{1}{12 m_1 m_2}\left(\frac{1}{r}V_{\rm v}'(r) - V_{\rm v}''(r)\right)T + \frac{1}{6 m_1 m_2}\boldsymbol{\sigma}_1\cdot\boldsymbol{\sigma}_2\,\Delta V_{\rm v}(r) \end{aligned} \tag{5}$$

are spin-dependent parts of the Breit-Fermi Hamiltonian [2], $\mathbf{L} = -\mathrm{i}\,\mathbf{r} \times \boldsymbol{\nabla}$ is the orbital momentum operator, $\boldsymbol{\sigma}_a$ is the spin operator acting on the ath particle spin variable, and $T = 3(\boldsymbol{\sigma}_1 \cdot \mathbf{r})(\boldsymbol{\sigma}_2 \cdot \mathbf{r})/r^2 - \boldsymbol{\sigma}_1 \cdot \boldsymbol{\sigma}_2$. In the case of heavy mesons which are weakly relativistic systems the spin operators can be treated perturbatively, as in refs. [2, 3]. In the case of light mesons which are considered as strongly relativistic systems two difficulties occur. First, the spin corrections (5) appear to be of the same order as the potentials (3). But actual spin effects in light meson spectra are small. Second, the operators (5) are divergent if m_1 or/and $m_2 \to 0$. Consequently, divergent terms appear in the mass spectrum. The possible way to avoid these two problems is the following modification of operators (5): the rest masses of quarks involved in these operators should be replaced by the "constituent" masses:

$$m_a \to M_a = \sqrt{m_a^2 + E_M^2 - m_M^2} = \frac{M^2 + m_a^2 - m_{\bar{a}}^2}{2M}, \qquad \begin{array}{c} a = 1, 2, \\ \bar{a} = 3 - a \end{array} \tag{6}$$

possessing the properties 1) $M_1 + M_2 = M$, 2) $M_a \approx M/2$ if $M \gg m_{1(2)}$, and 3) $M_a \to m_a$ in the nonrelativistic limit. Using the modified operators $\overline{W}$ in the equation (1)-(3) (by means of the substitution $V \to \overline{V} = V + \overline{W}$) we come to the asymptotically linear Regge trajectories (4) with the same slope parameter $8a$, but each trajectory splits into three ones by parallel shift

$$\Delta M^2 = 2\eta a \quad \text{for} \quad \ell = j + \eta, \quad \text{where} \quad \eta = 0, \pm 1. \tag{7}$$

This mass splitting is unambiguously fixed by the typical hadron scale $8a$. It does not depend on the total angular momentum j. Qualitatively this result correlates with actual light meson spectra as well as with theoretical results following from the string models [4]. Numerically, the mass splitting (7) (i.e., 25% of $8a$) exceeds actual experimental data (about 5÷6%; see [4]). A better fit can be achieved by introducing some adjustable parameter in addition to the minimal set a, α and m_a ($a = 1, 2$) used in the present model.

References

1. I.T. Todorov: Phys. Rev. **D3** 2351, (1971); V.A. Rizov, H. Sazdjian, I.T. Todorov: Ann. Phys. (NY) **165**, 59 (1985)

2. M.G. Olsson, K.J. Miller: Phys. Rev. **D28**, 674 (1983); R.W. Childers: Phys. Rev. **D36**, 606, 3813 (1987); W. Lucha, F.F. Schöberl: *Die Starke Wechselwirkung. Eine Einführung in nichtrelativistische Potentialmodelle.* Mannheim: Bibliographisches Institut & F. A. Brockhaus 1989

3. M. Hirano, K. Iwata, K. Kato, T. Murota: Prog. Theor. Phys. **67**, 1251 (1982); D.D. Brayshaw: Phys. Rev. **D36**, 1465 (1987); S.N. Grudtsin: *Potential Models and Heavy Quarkonia Spectroscopy.* Preprint IHEP 88-38. Serpukhov 1988

4. E.B. Berdnikov, G.P. Pron'ko: Sov. J. Nucl. Phys. **54**, 763 (1991); E.B. Berdnikov, G.G. Nanobashvili, G.P. Pron'ko: ibid **55**, 203 (1992)

Few-Body Systems Suppl. 14, 417–418 (2003)

Three-Boson Relativistic Bound States with Zero-Range Two-Body Interaction

V.A. Karmanov[1,*], J. Carbonell[2,†]

[1] Lebedev Physical Institute, Leninsky Prospekt 53, 119991 Moscow, Russia

[2] Institut des Sciences Nucléaires, 53 Avenue des Martyrs, 38026 Grenoble, France

Zero-range two-body interaction, applied to the nonrelativistic three-body system, results in the Thomas collapse [1]. The latter means that the three-body binding energy tends to $-\infty$ when the interaction radius tends to zero. Relativistic three-body calculations with zero-range interaction, performed in a minimal relativistic model [2] and in the framework of the Light-Front Dynamics [3], have shown that, due to relativistic repulsion, the three-body binding energy remains finite and the Thomas collapse is consequently avoided.

We have considered [4] in the field-theoretical framework, the problem of three equal mass (m) bosons interacting via zero-range forces. Few points are worth to be precise.

(*i*) In quantum field theory the number of particles is not conserved. The three-body relativistic system is, therefore, an idealization. Our aim is to study whether this idealized relativistic system still exists for zero range interaction.

(*ii*) The vacuum remains bare (no vacuum fluctuations) only in the Light-Front Dynamics. That is why we work in this approach.

(*iii*) In field theory all elementary interactions are point-like and, in this sense, are zero-range. We consider the Hamiltonian $H_{int} = \lambda\varphi^4(x)$ restricted to two-body system. It generates a two-body kernel which is simply the constant λ. This is the precise meaning of zero-range two-body interaction used in our study.

A relativistic Faddeev equation is derived, which provides a parameter-free relation between the two-body bound state mass M_2 and the three-body one M_3. It coincides with equation (11) from [3] except for the integration limits.

In Fig. 1(a) our calculations (solid line) of the three-body ground state mass M_3 as a function of the two-body one M_2 together with the dissociation

* *E-mail addresses:* karmanov@sci.lebedev.ru, karmanov@isn.in2p3.fr
† *E-mail address:* carbonel@isn.in2p3.fr

threshold $M_3 = M_2 + m$ are displayed (in m units). In the zero two-body binding limit $B_2 = 2m - M_2 \to 0$ the three-boson system has a binding energy $B_3 \approx 0.012\, m$. Dash line is the result obtained using the integration limits

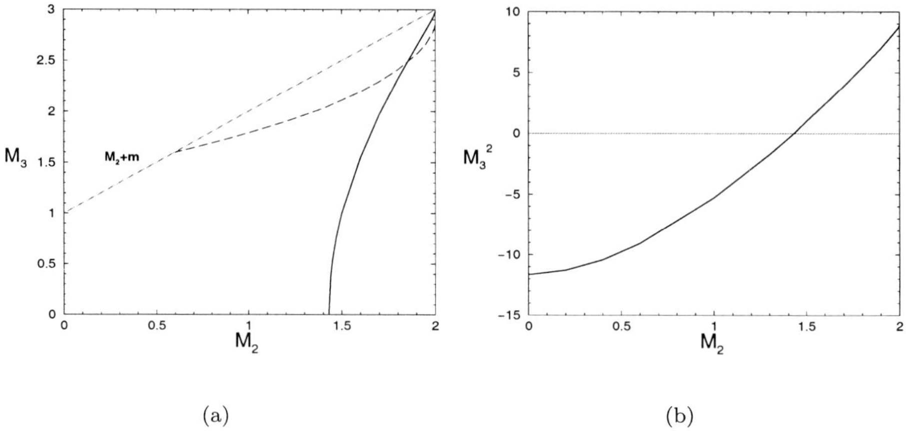

Figure 1. (a) Three-body bound state mass M_3 v.s. the two-body one M_2. Solid line corresponds to present work and dash line – to calculations using the integration limits of [3]. (b) Three-body bound state mass squared M_3^2 v.s. M_2.

of [3]. In both cases, like in [2], the repulsive relativistic effects produce a natural cutoff, leading to a finite spectrum and – in the Thomas sense – an absence of collapse. However, the behaviours of the solid and dashed curves qualitatively differ from each other. When M_2 decreases, we found that the three-body mass M_3 (solid line) decreases very quickly and vanishes at the two-body mass value $M_2 = M_c \approx 1.43\, m$.

Results for M_3^2 are given in Fig. 1(b). For $M_2 \leq M_c$, M_3^2 becomes negative and there are no physical solutions with real M_3 mass. However M_3^2 remains finite in all the two-body mass range $M_2 \in [0, 2]$ with $M_3^2 \approx -(3.41\, m)^2$ in the limit $M_2 \to 0$.

We conclude that in the system of three relativistic bosons with zero-range interaction, a relativistic counterpart of Thomas collapse takes place for strong enough two-body forces such that $M_2 < M_c$. If in nonrelativistic treatment the system ceases to exist when its binding energy tends to $-\infty$, in the relativistic approach the system does not exist when its mass squared becomes negative.

References

1. L.H. Thomas: Phys. Rev. **47**, 903 (1935)
2. J.V. Lindesay and H.P. Noyes: Preprint SLAC-PUB-2932 (1986)
3. T. Frederico: Phys. Lett. **B282**, 409 (1992)
4. J. Carbonell, V.A. Karmanov: nucl-th/0207073 (submitted for publication)

Few-Body Systems Suppl. 14, 419–420 (2003)

Few-Body Systems

Hadronic Vector-Meson Decay Widths from a Relativistic Coupled-Channel Approach

A. Krassnigg[1], W. Schweiger[1], W.H. Klink[2]

[1] Institut für Theoretische Physik, Universität Graz, Universitätsplatz 5, A-8010 Graz, Austria

[2] Department of Physics and Astronomy, University of Iowa, Iowa City, Iowa 52242, USA

When calculating meson and baryon spectra within constituent quark models, not only ground states but also excited states are usually treated as stable quark-antiquark and three-quark configurations. Resonance widths are then often obtained in first-order perturbation theory by employing a particular model for the decay vertex [1]. Moreover, the constituent dynamics is frequently treated in a non-relativistic or, at most, semirelativistic way.

The relativistic coupled channel approach which is outlined in ref. [2] partly overcomes these problems. This formalism not only accounts for relativistic particle kinematics, but it also covers recoil effects in particle exchange and it gives non-perturbative predictions for decay widths. Results for vector-meson masses, where the hyperfine interaction between quark and antiquark arises from the chiral constituent-quark model [3] (i.e. pseudoscalar meson exchange), are quoted in ref. [2]. In the present contribution we will concentrate on the decay widths of vector mesons and we will try to shed some light on how these are influenced by non-perturbative and relativistic effects.

In ref. [2] the exchange-particle dynamics is explicitly taken into account by considering a vector meson as a superposition of a $|\mathrm{q\bar{q}}\rangle$ and a $|\mathrm{q\bar{q}M}_\Pi\rangle$ (q...quark, $\mathrm{\bar{q}}$...antiquark, M_Π...pseudoscalar meson) state rather than a pure $|\mathrm{q\bar{q}}\rangle$ state. A vertex operator K which can be derived from the usual field theoretical coupling of pseudoscalar fields to spin 1/2 fields [4] provides then the transition between the $|\mathrm{q\bar{q}}\rangle$ and $|\mathrm{q\bar{q}M}_\Pi\rangle$ states. If the two-channel problem for the mass operator is reduced to a one-channel problem for the $\mathrm{q\bar{q}}$ channel, one ends up with a non-linear eigenvalue problem, since the resulting optical potential also depends on the mass eigenvalue m_i

$$V_{\mathrm{opt}}(m_i) = K^\dagger (m_i - D^{\mathrm{c}}_{2+1} + i\epsilon)^{-1} K\,. \tag{1}$$

Furthermore the optical potential and correspondingly the mass eigenvalue m_i

Table 1. Experimental decay widths (in MeV) versus model predictions.

decay channel	Γ_{exp}	Γ_{full}	Γ_{pert}	Γ_{nr}
$\rho(1450) \to \omega\pi,\ \rho\eta$	<6+12	≈ 0.3	≈ 0.2	0
$\rho(1700) \to \omega\pi,\ \rho\eta$	seen	≈ 0	≈ 0.1	0
$\omega(1420) \to \rho\pi$	174±60	≈ 0.9	≈ 0.2	0
$\omega(1600) \to \rho\pi$	seen	≈ 1.1	≈ 0.6	0

may become complex for an excited state. This means that an excited vector meson state has a finite lifetime (the decay width is $\Gamma_{\text{full}} = 2Im(m_i)$) and it decays into a lower lying state by emission of a pseudoscalar meson. If one does not want to solve the full (non-linear) eigenvalue problem as in ref. [2] to determine decay widths, one could also think of treating them perturbatively. One way is to consider the whole optical potential as a perturbation to the confinement part. In this case the decay width is given by (nl label the eigenstates of the confinement potential)

$$\Gamma_{\text{pert}}(M^{\text{c}}_{nl}) = 2|\langle nl|\text{Im}(V_{\text{opt}}(M^{\text{c}}_{nl}))|nl\rangle| \,. \tag{2}$$

A more common approximation is to take the non-relativistic limit of the vertex operator K and of the particle kinematics and to calculate the decay width via

$$\Gamma^{n'l',nl}_{\text{nr}} \propto |\langle n'l'|K_{\text{nr}}|nl\rangle|^2 \,. \tag{3}$$

Results for the full calculation and for these two approximations are given in Table 1. Both relativistic and non-perturbative effects are seen to be sizable. Apart from the $\omega(1450)$ the predicted (non-perturbative) decay widths have at least the right order of magnitude. For the $\omega(1450)$ a strong final-state interaction, which is not taken into account in our model, is likely to play a role. Our simple model, however, could be improved by adding those contributions to the optical potential in which the emitted meson is again absorbed by the emitting (anti)quark. This would also increase the number of possible decay modes.

Acknowledgement. This work was supported by the Austrian Science Fund (FWF-project Nr. P14794·TPH).

References

1. A. Le Yaouanc et al.: *Hadron Transitions in the Quark Model.* New York: Gordon and Breach 1988
2. A. Krassnigg et al.: these proceedings
3. W. Plessas: these proceedings
4. W.H. Klink: these proceedings

List of participants

Name, affiliation	E-mail address
Yoshinori Akaishi, Tsukuba	yoshinori.akaishi@kek.jp
Taylan Akdogan, MIT	akdogan@mit.edu
Erwin O. Alt, U. Mainz	erwin.alt@uni-mainz.de
Miron Amusia, Jerusalem	amusia@vms.huji.ac.il
Bryon Anderson, Kent State U.	anderson@hpacq.kent.edu
Hartmuth Arenhövel, U. Mainz	arenhoevel@kph.uni-mainz.de
Edward Armour, Nottingham U.	edward.armour@nottingham.ac.uk
Sonia Bacca, U. Trento	bacca@science.unitn.it
Bernard L.G. Bakker, V.U. Amsterdam	blgbkkr@nat.vu.nl
Thomas Barford, U. Manchester	barford@a35.ph.man.ac.uk
Nir Barnea, Jerusalem	nir@wiener.fiz.huji.ac.il
Raul Barrachina, Bariloche	barra@cab.cnea.gov.ar
Katharina Berger, U. Graz	katharina.berger@uni-graz.at
Michael Beyer, U. Rostock	michael.beyer@physik.uni-rostock.de
Raymond F. Bishop, U. Manchester	r.f.bishop@umist.ac.uk
Tancredi Botto, MIT	tancredi@mitlns.mit.edu
Saša Ceci, Zagreb	ceci@rudjer.irb.hr
Fritz Coester, Argonne	coester@anl.gov
Jurij Darewych, Toronto	darewych@yorku.ca
Arnoldas Deltuva, U. Hannover	deltuva@itp.uni-hannover.de
Bertrand Desplanques, Grenoble	bertrand.desplanques@isn.in2p3.fr
Alexandre Deur, U. Virginia	deur@virginia.edu
Wytse van Dijk, Ontario	vandijk@physics.mcmaster.ca
Pal Doleschall, Budapest	doles@rmki.kfki.hu
Zdenek Doležal, Prague	zdenek.dolezal@mff.cuni.cz
Askold Duviryak, Lviv	duviryak@ph.icmp.lviv.ua
Dmitri Fedorov, U. Aarhus	fedorov@ifa.au.dk
Antonio Fonseca, U. Lisboa	fonseca@alf4.cii.fc.ul.pt
Avraham Gal, Jerusalem	avragal@vms.huji.ac.il
Igor Gašparič, Zagreb	igorg@hogar.irb.hr
Shalev Gilad, MIT	sgilad@lns.mit.edu
Jacek Golak, Bochum/Krakow	golakj@tp2.ruhr-uni-bochum.de
Bojan Golli, U. Ljubljana, IJS	bojan.golli@ijs.si
Pedro Gonzalez, U. Valencia	gonzalez@gtae2.ist.utl.pt
Peter Grabmayr, U. Tuebingen	grabmayr@uni-tuebingen.de
Rafael Guardiola, U. Valencia	rafael.guardiola@uv.es
Amal Hemmdan, Ruhr-U. Bochum	amel.hemdan@tp2.ruhr-uni-bochum.de
Douglas W. Higinbotham, JLab	doug@jlab.org
Emiko Hiyama, Tsukuba	emiko.hiyama@kek.jp
Jiři Horaček, Prague	jiri.horacek@mff.cuni.cz
Yordanka Ilieva, George Washington U.	jordanka@jlab.org

Damijan Janc, IJS	damijan.janc@ijs.si
Nasser Kalantar, Groningen	nasser@kvi.nl
Masayasu Kamimura, Kyushu U.	kami2scp@mbox.nc.kyushu-u.ac.jp
Vladimir Karmanov, Moscow	karmanov@isn.in2p3.fr
Murat Kaskulov, U. Tuebingen	kaskulov@pit.physik.uni-tuebingen.de
Alejandro Kievsky, Pisa	alejandro.kievsky@pi.infn.it
Mladen Kiš, Groningen	mkis@kvi.nl
Oleg Kisselev, Darmstadt	o.kisselev@gsi.de
Stanislaw Kistryn, Cracow	skistryn@if.uj.edu.pl
Dubravko Klabučar, U. Zagreb	klabucar@phy.hr
William Klink, U. Iowa	william-klink@uiowa.edu
Andreas Krassnigg, U. Graz	andreas.krassnigg@uni-graz.at
Rajmund Krivec, IJS	rajmund.krivec@ijs.si
Vladimir Kukulin, Moscow	kukulin@anna19.npi.msu.su
Timo Lähde, Helsinki	talahde@pcu.helsinki.fi
Rimantas Lazauskas, Grenoble	lazauskas@isn.in2p3.fr
Winfried Leidemann, U. Trento	leideman@science.unitn.it
Leslie Lekala, Pretoria	lekala@science.unisa.ac.za
Sergey Levin, St Petersburg State U.	levin@physto.se
Alexander Machavariani, Dubna/Tbilisi	alexm@cv.jinr.ru
Masoud Mahjour-Shafiei, Groningen	shafiei@kvi.nl
Victor Mandelzweig, Jerusalem	victor@phys.huji.ac.il
Mario A. Marchisio, U. Trento	marchisi@science.unitn.it
Laura E. Marcucci, U. Pisa	marcucci@df.unipi.it
Stefano Mattiello, U. Rostock	stefano@darss.mpg.uni-rostock.de
Alexander Matveenko, Dubna	amatv@thsun1.jinr.ru
Thomas Melde, U. Graz	thomas.melde@uni-graz.at
Harald Merkel, U. Mainz	merkel@kph.uni-mainz.de
Marco Mirazita, Frascati	mirazita@lnf.infn.it
Hidekatsu Nemura, Tsukuba	hidekatsu.nemura@kek.jp
Silvia Niccolai, George Washington U.	silvia@jlab.org
Igor Obukhovsky, Moscow State U.	obukh@nucl-th.sinp.msu.ru
Giuseppina Orlandini, U. Trento	orlandin@science.unitn.it
Shinsho Oryu, Tokyo U. of Science	oryu@ph.noda.sut.ac.jp
Eulogio Oset, U. Valencia	oset@ific.uv.es
Emanuele Pace, U. Rome	pace@roma2.infn.it
Zoltan Papp, Debrecen	zpapp@sue.natsci.csulb.edu
Thomas Pattard, Dresden	tpattard@mpipks-dresden.mpg.de
Willibald Plessas, U. Graz	plessas@uni-graz.at
Wayne Polyzou, U. Iowa	polyzou@uiowa.edu
Milan Potokar, IJS	milan.potokar@ijs.si
Sofia Quaglioni, U. Trento	quaglion@alpha.science.unitn.it
Jean-Marc Richard, Grenoble	jean-marc.richard@isn.in2p3.fr
Avraham Rinat, Rehovot	fnrinat@weizmann.weizmann.ac.il
Dan-Olof Riska, U. Helsinki	dan-olof.riska.@Helsinki.fi
Sergio Rosati, Pisa	rosati@df.unipi.it

Mitja Rosina, U. Ljubljana	mitja.rosina@ijs.si
Olga Rubtsova, Moscow State U.	rubtsova@nucl-th.sinp.msu.ru
Werner Sandhas, U. Bonn	sandhas@physik.uni-bonn.de
Kenji Sasaki, Tsukuba	kenjis@post.kek.jp
Peter U. Sauer, U. Hannover	sauer@itp.uni-hannover.de
Michael Schwamb, U. Mainz	schwamb@kph.uni-mainz.de
Wolfgang Schweiger, U. Graz	wolfgang.schweiger@uni-graz.at
Claude Semay, Mons	claude.semay@umh.ac.be
Bianka Sengl, U. Graz	bianka.sengl@uni-graz.at
Bernard Silvestre-Brac, Grenoble	silvestre@isn.in2p3.fr
Simon Širca, U. Ljubljana, IJS	simon.sirca@ijs.si
Ole Sørensen, U. Aarhus	oles@ifa.au.dk
Florica Stancu, U. Liège	fstancu@ulg.ac.be
Juris P. Svenne, U. Manitoba	svenne@physics.umanitoba.ca
Lukas Theußl, U. Valencia	lukas.theussl@uv.es
Werner Tornow, Duke U.	tornow@tunl.duke.edu
Kalman Varga, Oak Ridge NL	varga@surfdec.ssd.ornl.gov
Michele Viviani, Pisa	michele.viviani@pi.infn.it
Branislav Vlahovic, NCCU/JLab	branko@jlab.org
Robert F. Wagenbrunn, U. Graz	robert.wagenbrunn@uni-graz.at
Hans Weda, U. Utrecht	weda@phys.uu.nl
Lawrence Weinstein, Old Dominion U.	weinstei@physics.odu.edu
Henryk Witała, Cracow	witala@if.uj.edu.pl
Pavel Zarubin, Dubna	zarubin@lhe.jinr.ru
Zilu Zhou, MIT	zzhou@loon.mit.edu
Matjaž Žitnik, IJS	matjaz.zitnik@ijs.si

Author index

Springer-Verlag and the Environment

We at Springer-Verlag firmly believe that an international science publisher has a special obligation to the environment, and our corporate policies consistently reflect this conviction.

We also expect our business partners – printers, paper mills, packaging manufacturers, etc. – to commit themselves to using environmentally friendly materials and production processes.

The paper in this book is made from no-chlorine pulp and is acid free, in conformance with international standards for paper permanency.